# Applied Analysis by the Hilbert Space Method

## An Introduction with Applications to the Wave, Heat, and Schrödinger Equations

**Samuel S. Holland, Jr.**

*University of Massachusetts*
*Amherst, Massachusetts*

**Dover Publications, Inc.**
Mineola, New York

*Bibliographical Note*

This Dover edition, first published in 2007, is a corrected republication of the work originally published in the "Monographs and Textbooks in Pure and Applied Mathematics" series by Marcel Dekker, Inc., New York, in 1990.

*Library of Congress Cataloging-in-Publication Data*

Holland, Samuel S., 1928–
    Applied analysis by the hilbert space method : an introduction with applications to the wave, heat, and Schrödinger equations / Samuel S. Holland, Jr.
        p. cm.
    Originally published: New York : M. Dekker, 1990, in series: Monographs and textbooks in pure and applied mathematics.
    Includes index.
    ISBN-13: 978-0-486-45801-4
    ISBN-10: 0-486-45801-6
    1. Differential equations. 2. Hilbert space. 3. Differential operators.
I. Title.

QA372.H76 2007
515'.35—dc22
                                                            2006103551

Manufactured in the United States by LSC Communications
45801604    2019
www.doverpublications.com

*To Chester Brown and John Byrne,*
*teachers of Physics and Algebra at Methuen High School,*
*Methuen, Massachusetts, 1945*

# Preface

This book, designed as a text for a one-year undergraduate applied mathematics course, features Hermitian differential operators on $L^2$ spaces and the eigenvalues and eigenfunctions of these operators. The operator theory, which is presented here in a format accessible to an undergraduate student, allows for a conceptually unified modern exposition of most of the commonly covered topics, such as: ordinary linear differential equations; Fourier series; Legendre, Hermite, and Laguerre polynomials; Bessel functions; spherical harmonics; the wave, heat, and Schrödinger equations; and the Fourier transform. In addition, the $L^2$ spaces and Hermitian operators have independent value as important concepts of modern mathematics, concepts of general application. No presently available undergraduate textbook appears to offer a comparable treatment of these important concepts.

The book can also serve as a reference for scientists who want an elementary exposition of the basic theory of unbounded Hermitian operators on Hilbert space.

The prerequisites for the use of this text are three semesters of calculus, one semester of linear algebra, and basic physics. No prior training in differential equations is required, as first and second order linear differential equations are covered in the first two chapters. Because the book includes the usual material on ordinary differential equations, students need not take a separate course in that subject, so may take this course earlier in their undergraduate careers.

A substantial number of the exercises involve numerical calculations. All the computations required in these exercises are within range of a good hand-held programmable scientific electronic calculator. In practice, many students will know a scientific programming language, like BASIC, "C," FORTRAN, or PASCAL, and will have access to a computer. These students will have little trouble writing the simple programs required for these exercises. But the text does not require knowledge of a programming language nor does it require access to a computer. All the numerical exercises can be done on a good, programmable hand-held calculator. To show the feasibility of this, I have included several sample programs for the Hewlett-Packard HP-15C.

I should like to thank reviewers Peter J. Gingo of the University of Akron, G. V. Ramanathan of the University of Illinois at Chicago, and Eugene R. Speer of Rutgers for their insightful reviews and very helpful comments.

Special thanks are due the many students who patiently suffered through the various early experimental versions of this course. Whatever success this book may achieve in making the hitherto recondite topic of unbounded operators on Hilbert space accessible to today's undergraduate may in large measure be attributable to them. They were the arbiters who guided the evolution of the presentation to its current form. There are far too many to name individually, but I do want to single out Christopher A. Davis and Dae Song Im, two students who made a special contribution in the Spring of 1984.

Also, my sincere thanks to Ms. Maria Allegra, Associate Acquisitions Editor at Marcel Dekker, Inc.

This book has evolved from lecture notes prepared over the years for a one-year course entitled "Applied Analysis," given at the University of Massachusetts. More often than not, the notes were produced lecture-by-lecture and were passed out to the student hot off the presses. Working under such frantic deadlines, Mrs. Marguerite Bombardier cheerfully, swiftly, and accurately typed each section, most several times over, did the figures and tables, and corrected my errors, all the while maintaining an unfailing good humor. She also flawlessly typed the final manuscript. Without Peg, there would not have been any book. To her, my deepest gratitude.

*Samuel S. Holland, Jr.*

# A Note on Method

One of the most troublesome features inherent in the theory of differential operators on $L^2$ spaces is the fact that these operators cannot be everywhere defined. Providing an accessible explanation of how one defines the domain of an unbounded operator is the principal challenge facing any expositor of this theory. The generally used procedure is to first define the differential operator on a convenient dense domain, verify its Hermitian character there, then survey all possible selfadjoint extensions (if any). This complicated process is impossible to explain to an undergraduate. Moreover, it also appears as an almost insurmountable hurdle for a physicist or engineer who wishes to get quickly to the applications. In this text I have used a more direct method. For each of the classical differential expressions (Bessel, Hermite, Laguerre, Laplace, Legendre), I begin with the maximal operator associated with that expression. The concept of the maximal operator is explained well in J. Weidmann, *Linear Operators in Hilbert Spaces* (Grad. Texts in Math. 68, Springer-Verlag, New York, 1980, Section 8.4). When this maximal operator is Hermitian, it is already selfadjoint (Weidmann's Theorem 8.22). Among the classical cases, this occurs only for Hermite's operator and for the Laplacian on $L^2(-\infty, \infty)$. (The proof for Hermite's operator is in the Appendix to my Chapter 4; that for the Laplacian in Theorem 5.4 of Chapter 8.) For the other cases, when the maximal operator is not Hermitian, I narrow the domain by the imposition of boundary conditions to produce a Hermitian operator. This procedure seems to be accessible and intuitively appealing to my intended undergraduate audience. It also leads quickly to the applications. For the case of Legendre's operator, the connection of this approach and the classical process referred to above has been worked out in detail by N. I. Akhiezer and I. M. Glazman in their book *Theory of Linear Operators in Hilbert Space* (Frederick Ungar, New York, 1963, vol II, Appendix II, §9).

# Contents

# Contents

# Applied Analysis by the Hilbert Space Method

## An Introduction with Applications to the Wave, Heat, and Schrödinger Equations

# 1

## First Order Linear Differential Equations

### 1.1  THE EQUATION $a(x)y' + b(x)y = h(x)$

Consider the following problem: Find a function $y$ that satisfies the equation

$$\frac{dy}{dx} = e^x$$

That is, find a function $y$ whose derivative is $e^x$. We may multiply both sides of the equation by $dx$, the differential of the independent variable $x$, then integrate both sides to get

$$\int \frac{dy}{dx}\, dx = \int e^x\, dx$$

or

$$y = e^x + C$$

where $C$ is an arbitrary constant. Thus we find to our *differential equation* $y' = e^x$ a one-parameter family of solutions: $y = e^x$, $y = e^x - 2$, $y = e^x + 3.6$, $y = e^x + 12.501$, etc. If, for some fixed value of the independent variable, say $x_0$, we ask that our solution $y$ take on a specified value $y_0$, then the resulting so-called *initial value problem* has a unique solution:

$$y_0 = e^{x_0} + C \quad \text{so} \quad C = y_0 - e^{x_0}$$

and $y = e^x + y_0 - e^{x_0}$ is the unique solution to our differential equation that satisfies $y(x_0) = y_0$.

Let us consider another example:

$$x\frac{dy}{dx} + y = e^x$$

We note that $(d/dx)(xy) = x(dy/dx) + y$ so that our differential equation can be written

$$\frac{d}{dx}(xy) = e^x$$

and we may integrate as in the first example to get the solution

$$xy = e^x + C \quad \text{or} \quad y = \frac{e^x + C}{x}$$

where $C$ is an arbitrary constant. And, as above, we can solve the initial value problem, except if $x_0 = 0$.

The solution to our differential equation that includes an arbitrary constant $C$ we refer to as the *general solution*. In the example just completed, our general solution is singular at $x = 0$, which is traceable to the fact that in our differential equation the coefficient $x$ of $dy/dx$ vanishes at $x = 0$. Note, however, that for the particular value $C = -1$ of the arbitrary constant $C$, we have

$$y = \frac{e^x - 1}{x} = \frac{(1 + x + x^2/2! + \cdots) - 1}{x} = \frac{x + x^2/2! + \cdots}{x}$$

$$= 1 + \frac{x}{2!} + \frac{x^2}{3!} + \cdots = \sum_{n=0}^{\infty} \frac{x^n}{(n + 1)!}$$

which is well behaved for all $x$. So, while we can solve the initial value problem by prescribing arbitrarily the value of our solution at any point $x_0 \neq 0$, at $x_0 = 0$ there is precisely one nonsingular solution $y$ and that one necessarily satisfies $y(0) = 1$.

Let us look at another example,

$$x\frac{dy}{dx} - y = x^2 e^x$$

Here we observe that

$$\frac{d}{dx}\left(\frac{1}{x}y\right) = \frac{1}{x}y' - \frac{1}{x^2}y$$

so that

$$x^2 \frac{d}{dx}\left(\frac{1}{x}y\right) = xy' - y$$

which is the left side of our equation. Thus the equation becomes

$$x^2 \frac{d}{dx}\left(\frac{1}{x}y\right) = x^2 e^x$$

Upon canceling and integrating, we find

$$\frac{1}{x}y = e^x + C \qquad \text{or} \qquad y = xe^x + Cx$$

which is well behaved everywhere, even though the coefficient of $y'$ vanishes at $x = 0$. Thus, while the vanishing of the leading coefficient may produce a singularity, it does not guarantee one.

Our solution to the general equation of the form

$$a(x)y' + b(x)y = h(x) \tag{1.1.1}$$

where $a$, $b$, and $h$ are given functions of the independent variable $x$, and $y$ is the unknown function we seek, is modeled on the previous examples. First we divide by $a(x)$ to get $y' + (b/a)y = h/a$; then we attempt to make the left side a perfect differential by choosing a function $f$ so that

$$\frac{1}{f}(fy)' = y' + \left(\frac{b}{a}\right)y$$

As $(1/f)(fy)' = y' + (f'/f)y$, the sought-for $f$ must satisfy

$$\frac{f'}{f} = \frac{b}{a}$$

Recalling the important formula from calculus

$$(\ln f)' = \frac{f'}{f}$$

we get $(\ln f)' = b/a$, so $\ln f = \int (b/a)\, dx$, or $f(x) = \exp(\int (b/a)\, dx)$. Thus we may always write down an explicit formula for such an $f$ and so recast our equation in the form

$$\frac{1}{f}(fy)' = \frac{h}{a} \qquad \text{or} \qquad (fy)' = \frac{fh}{a}$$

then integrate to get $fy = \int (fh/a)\, dx + C$ or

$$y = \frac{1}{f(x)} \int \frac{f(x)h(x)}{a(x)} dx + \frac{C}{f(x)} \quad \text{where } f(x) = \exp\left(\int \frac{b(x)}{a(x)} dx\right)$$

as the explicit general solution to our equation (1.1.1). Noting the presence of $a(x)$ in the denominators, we see again that there may be difficulties extending the solution through regions where $a(x) = 0$, and for this reason we shall usually restrict attention to regions where $a(x) \neq 0$.

**Exercises**

1. Solve the following differential equations:

   (a) $\dfrac{dy}{dx} = x$   (b) $\dfrac{dy}{dx} = \dfrac{1}{x}$   (c) $\dfrac{dy}{dx} = \dfrac{\ln x}{x}$

2. Solve the following initial value problems:

   (a) $x\dfrac{dy}{dx} + y = x, \quad y(1) = 2$

   (b) $x\dfrac{dy}{dx} - y = 1, \quad y(1) = 0$

3. Find the function $f$:

   (a) $\dfrac{1}{f}(fy)' = y' + \dfrac{1}{x}y$

   (b) $\dfrac{1}{f}(fy)' = y' - (\sin x)y$

4. Find the general solution:

   (a) $y' + 2xy = e^{-x^2}$
   (b) $(3x^2 + 1)y' - 2xy = 6x$

## 1.2 FIRST ORDER LINEAR DIFFERENTIAL EXPRESSIONS; THE KERNEL

We may view the left side of our first order linear equation $ay' + by = h$ as an operation performed on the function $y$. For example, if our equation is $3xy' - y = h$, we can view the left side of the equation as the value taken by an expression $l(\cdot) \equiv 3x(\cdot)' - (\cdot)$ acting on the function $y$: $l(y) \equiv 3x(y)' - (y)$. The sign "$\equiv$" means "defined by." For each particular function $y$, the first

**Table 1.1**

| $y$ | $l(y) \equiv 3xy' - y$ |
|-----|------------------------|
| 1   | $-1$                   |
| $x$ | $2x$                   |
| $e^x$ | $(3x - 1)e^x$        |

order differential expression $l(y) \equiv 3xy' - y$ produces a new function $3xy' - y$. Some examples are contained in Table 1.1.

For example, we get the third entry this way: if $y = e^x$, then $l(y) = 3x(e^x)' - e^x = 3xe^x - e^x = (3x - 1)e^x$.

From this point of view, our differential equation $3xy' - y = h$ can be rephrased as the following problem: given a function $h$, find a function $y$ so that $l(y) = h$.

We say that our differential expression $l$ is *linear* if it satisfies these two conditions:

Additive: $l(y_1 + y_2) = l(y_1) + l(y_2)$ for any two functions $y_1$ and $y_2$.
Homogeneous: $l(Cy) = Cl(y)$ for any constant $C$ and any function $y$.

Our first order linear differential expression $l(y) = a(x)y' + b(x)y$ does satisfy these conditions.

$$l(y_1 + y_2) = a(y_1 + y_2)' + b(y_1 + y_2) = a(y_1' + y_2') + b(y_1 + y_2)$$

$$= ay_1' + by_1 + ay_2' + by_2 = l(y_1) + l(y_2)$$

and

$$l(Cy) = a(Cy)' + b(Cy) = aCy' + bCy = Cl(y)$$

These calculations are just an application of two rules: (1) the derivative of the sum is the sum of the derivatives and (2) the derivative of a constant times a function is the constant times the derivative.

The property of linearity severely restricts the possibilities for our expression. For example, $l(y) \equiv y^2$ fails both tests for linearity. A linear differential expression can contain no term $(y')^2$, no term $y^2$, no term $y'y$, etc.

In linear algebra we have encountered linear transformations and their matrices. There is a close analogy between the problem of solving systems of linear equations and the problem of solving linear differential equations. Given a vector $m$ and a matrix $A$ (the matrix of coefficients), linear algebra concerns itself with the problem of finding a vector $x$ such that $Ax = m$. In differential equations, given a function $h$ and a linear differential expression $l$,

**Table 1.2**

| Linear algebra | Differential equations |
|---|---|
| Vectors | Functions |
| Matrix $A$ | Linear differential expression $l$ |
| Given $m$, find $x$ such that $Ax = m$; | Given $h$, find $y$ such that $l(y) = h$; |
| i.e., solve | i.e., solve |
| $\begin{bmatrix} a_{11} & a_{12} \\ a_{21} & a_{22} \end{bmatrix}\begin{bmatrix} x_1 \\ x_2 \end{bmatrix} = \begin{bmatrix} m_1 \\ m_2 \end{bmatrix}$ for $x = \begin{bmatrix} x_1 \\ x_2 \end{bmatrix}$ | $a(x)y' + b(x)y = h$ for $y$ |

we wish to find a function $y$ such that $l(y) = h$. Table 1.2 shows the connection.

Note this implication of linearity: any linear expression will take the zero function to the zero function; in symbols, $l(0) = 0$. Because $l(0) = l(0 + 0) = l(0) + l(0)$, subtracting $l(0)$ from both sides, we get $l(0) = 0$.

An equation $l(y) = 0$ is called a *homogeneous* equation (right side 0). As we have just noted, it always has the solution $y = 0$. But there are always nonzero solutions. $l(y) = 0$ means $ay' + by = 0$, and to solve this equation we can just appeal to the general theory of Section 1.1 or, better, note that the equation can be simply transformed to $y'/y = -b/a$ or $(\ln y)' = -b/a$, so $\ln y = -\int (b/a)\,dx + A$ where $A$ is an arbitrary constant of integration. Then $y = Cf$, where $f(x) = \exp(-\int (b/a)\,dx)$ and $C = e^A$. Note that the $f$ we have used here is $1/f$ for the $f$ in Section 1.1.

The set of solutions of the equation $l(y) = 0$ is called the *kernel* of the expression $l$ and denoted ker($l$). The kernel of a linear expression is always a subspace of functions. That means that if two functions $y_1$ and $y_2$ belong to the kernel, then so does their sum $y_1 + y_2$, and if $y$ belongs to the kernel so does $Cy$ for any constant $C$. This is the case because the expression $l$ is linear. If $l(y_1) = 0$ and $l(y_2) = 0$, then, by additivity, $l(y_1 + y_2) = l(y_1) + l(y_2) = 0 + 0 = 0$. Similarly, $l(Cy) = Cl(y) = C \cdot 0 = 0$. Thus the kernel of any linear expression is always a subspace.

We have established more specific information about the kernel of a first order linear differential expression $l(y) \equiv ay' + by$. We have shown that its kernel is one-dimensional. This means that there is one nonzero function, namely $f(x) = \exp(-\int (b/a)\,dx)$, such that any function $y$ in the kernel of $l$ is a scalar multiple of $f$, $y = Cf$ for some constant $C$. We say that the kernel is *spanned* by $f$. We summarize the result of our study thus far:

*The first order linear differential expression $l(y) \equiv ay' + by$ has a one-dimensional kernel spanned by the function $f(x) = \exp(-\int (b/a)\,dx)$.*

Note that the function $a(x)$ occurs in the denominator, so there may be trouble at points $x$ where $a(x) = 0$.

Of what use is the kernel? It provides us with a very helpful way to find the general solution of the equation $l(y) = h$.

**Theorem 1.1** *Suppose $l(y) \equiv ay' + by$ is a first order linear differential expression, and the kernel of $l$ is spanned by the function $f$. Suppose further that $y_p$ is some solution to $l(y) = h$. Then every solution $y$ of $l(y) = h$ has the form $y = Cf + y_p$ for a suitable constant $C$.*

*Proof.* Suppose that $y$ is a solution of $l(y) = h$. According to the hypothesis of our theorem, $y_p$ is also a solution. Then, by linearity of $l$, we have

$$l(y - y_p) = l(y) - l(y_p) = h - h = 0$$

Hence $y - y_p$ belongs to the kernel of $l$. As the kernel of $l$ is spanned by $f$, we have $y - y_p = Cf$ for some constant $C$. Thus $y = Cf + y_p$.

The solution $Cf + y_p$ is the *general solution* to $l(y) = h$ and thus is the same in principle as the boxed formula in Section 1.1. However, this new method is much better for two reasons:

First, it gives us a useful overall perspective on how the general solution of $l(y) = h$ is made up. The general solution $y$ is the sum of two functions. One involves the kernel of the expression $l$. This function $f$ depends only on $l$ and not on the right side, $h$. The second term in the sum is *any* particular solution of $l(y) = h$.

Second, the particular solution $y_p$ can often be calculated by special quick methods. For example, suppose we wish to solve $3xy' - y = \ln x + 1$. The kernel of the expression $l(y) \equiv 3xy' - y$ is spanned by

$$f(x) = \exp\left(-\int \frac{-1}{3x}\,dx\right) = \exp\left(\frac{1}{3}\ln x\right) = x^{1/3}$$

So the general solution will have the form $y = Cx^{1/3} + y_p$. To find a particular solution $y_p$, we might try guessing a solution of the form $y_p = A\ln x + B$ where $A$ and $B$ are constants. Substituting our guess into $l$ and solving for the constants $A$ and $B$, we find $A = -1$, $B = -4$. So the general solution is $y = Cx^{1/3} - \ln x - 4$.

Now if we want to solve $3xy' - y = x$, the kernel remains the same because we have the same expression on the left. So our general solution has the form $y = Cx^{1/3} + y_p$ and we can find $y_p$ by the same sort of trick (Exercise 5).

## Exercises

1. Check the entries in the following tables, and fill in the blank entries:

| $y$ | $l(y) \equiv e^x y' + y$ |
|---|---|
| $x$ | $e^x + x$ |
| $x^n$ | $x^{n-1}(ne^x + x)$ |
| $e^{-x}$ | |
| $\sin x$ | |

| $y$ | $l(y) \equiv (y')^2 - 2\sin(y')$ |
|---|---|
| $x$ | $1 - 2\sin 1$ |
| $x^n$ | $n^2 x^{2(n-1)} - 2\sin(nx^{n-1})$ |
| $e^x$ | |
| $\cos x$ | |

2. For each of the following expressions, determine whether $l$ is additive [i.e., $l(y_1 + y_2) = l(y_1) + l(y_2)$ for any two functions $y_1$ and $y_2$] and homogeneous [i.e., $l(Cy) = Cl(y)$ for any constant $C$ and any function $y$]. Determine which $l$ are linear.

   (a) $l(y) \equiv y' + 1$ [Soln.: $l(y_1 + y_2) = (y_1 + y_2)' + 1 = y_1' + y_2' + 1$. But $l(y_1) + l(y_2) = (y_1' + 1) + (y_2' + 1) = y_1' + y_2' + 2$. Hence $l(y_1 + y_2) \neq l(y_1) + l(y_2)$, so $l$ is not additive. $l(Cy) = (Cy)' + 1 = Cy' + 1$. But $Cl(y) = C(y' + 1) = Cy' + C$. If $C \neq 1$, then $l(Cy) \neq Cl(y)$. Since for $l$ to be homogeneous, we must have $l(Cy) = Cl(y)$ for *every* constant $C$, $l$ is not homogeneous either. Thus $l$ is not linear.]

   (b) $l(y) \equiv y'$.

   (c) $l(y) \equiv A$ (constant) (treat the cases $A = 0$ and $A \neq 0$ separately).

   (d) $l(y) \equiv (y')^2$.

   (e) $l(y) \equiv x^2 y' + 2xy$.

   (f) $l(y) = yy'$.

   (g) $l(y) \equiv \dfrac{\cos x}{\sin x} y' + x^2 e^x y$.

3. The first order linear expression $l(y) \equiv a(x)y' + b(x)y$ satisfies

$$l(1) = -2x, \qquad l(e^{-x^2}) = 0$$

   Find $a(x)$.

4. Find the kernel:

   (a) $l(y) \equiv y'$

   (b) $l(y) \equiv (3x^2 + 1)y' - 2xy$

   (c) $l(y) \equiv 3xy' - y$

   (d) $l(y) \equiv (x^2 + 1)y' - (1 - x)^2 y$

5. Knowing the kernel of $l(y) \equiv 3xy' - y$ (see Exercise 4c or last part of this section), find the general solution to $3xy' - y = x$ by finding a particular solution of the form $y_p = Ax + B$, with $A$ and $B$ suitable constants.

## 1.3 FINDING A PARTICULAR SOLUTION BY VARIATION OF PARAMETERS

In Section 1.2 we brought out the point that the general solution to the linear first order differential equation $ay' + by = h$ is the sum of two parts. One part consists of a general function from the kernel of the expression $l(y) \equiv ay' + by$. This part depends only on $l$ and not on the right side, $h$. The second part consists of *any* solution $y_p$ to $l(y) = h$. This part is called a *particular solution*.

In this section, we study a method for finding particular solutions that is called *variation of parameters* or, sometimes, *variation of constants*. The name of the method is reasonably descriptive of the idea behind it. If $f$ is a solution to $l(y) = 0$, then $Cf$, where $C$ is any constant, is also a solution to $l(y) = 0$. The method of variation of parameters is to replace the constant $C$ by a function, say $g(x)$, and attempt to find $g(x)$ so that $y = g(x)f(x)$ solves $l(y) = h$. Thus, since we replace the constant $C$ by a function $g(x)$, we are "varying the constant."

In detail, the method of variation of parameters goes like this. Let $l(y) \equiv ay' + by$ be a first order linear differential expression with kernel spanned by the function $f$. Set $y = g(x)f(x)$. We would like to find $g$ so that $y$ is a particular solution to $l(y) = h$, with $h$ some function given in advance. Then

$$l(y) = ay' + by = a(gf)' + b(gf)$$

$$= a(g'f + gf') + bgf \qquad \text{(product rule for differentiation)}$$

$$= ag'f + g(af' + bf)$$

But $af' + bf = l(f)$ is zero because $f$ spans the kernel of $l$. Hence $l(y) = ag'f = h$, so

$$g' = \frac{h}{af} \qquad \text{or} \qquad g = \int \frac{h}{af}\,dx$$

Thus we have a particular solution $y_p = gf = f(x)\int (h/af)\,dx$, where $f$ spans the kernel of $l$.

For example, to solve $xy' + 2y = x$, we begin by finding the kernel of $l(y) \equiv xy' + 2y$. This one-dimensional space is spanned by $f(x) = 1/x^2$. Then

$$g(x) = \int \frac{h}{af}\,dx = \int \frac{x}{x(1/x^2)}\,dx = \int x^2\,dx = \frac{1}{3}x^3$$

Hence a particular solution to $xy' + 2y = x$ is

$$y_p = gf = \left(\frac{1}{3}x^3\right)\left(\frac{1}{x^2}\right) = \frac{x}{3}$$

and the general solution is

$$y = \frac{x}{3} + \frac{C}{x^2}, \qquad C \text{ any constant}$$

**Exercises**

1. If $l(y) \equiv (3x^2 + 1)y' - 2xy$, show that ker($l$) consists of all constant multiples of the function $(3x^2 + 1)^{1/3}$ (Exercise 4b of Section 1.2). Using the method of variation of parameters, show that the general solution to

$$(3x^2 + 1)y' - 2xy = h(x)$$

is

$$y = c(3x^2 + 1)^{1/3} + (3x^2 + 1)^{1/3} \int (3x^2 + 1)^{-4/3} h(x)\, dx$$

Solve this explicitly for $h(x) = 6x$ (Exercise 4b of Section 1.1).

2. If $l(y) \equiv (x^2 + 1)y' - (1 - x)^2 y$, show that ker($l$) consists of all constant multiples of the function

$$\frac{e^x}{1 + x^2} \qquad \text{(Exercise 4d of Section 1.2)}$$

Using the method of variation of parameters, show that the general solution to

$$(x^2 + 1)y' - (1 - x)^2 y = h(x)$$

is

$$y = \frac{ce^x}{1 + x^2} + \frac{e^x}{1 + x^2} \int e^{-x} h(x)\, dx$$

Solve this explicitly for $h(x) = xe^{-x}$ and, for this case, find the unique solution that satisfies $y(0) = 1$.

## 1.4  POWER SERIES REVIEW

In calculus we studied power series expansions of the basic functions. For example,

$$e^x = 1 + x + \frac{x^2}{2!} + \frac{x^3}{3!} + \cdots + \frac{x^n}{n!} + \cdots = \sum_{k=0}^{\infty} \frac{x^k}{k!}$$

$$\sin x = x - \frac{x^3}{3!} + \frac{x^5}{5!} - \cdots + (-1)^n \frac{x^{2n+1}}{(2n+1)!} + \cdots = \sum_{k=0}^{\infty} (-1)^k \frac{x^{2k+1}}{(2k+1)!}$$

$$\cos x = 1 - \frac{x^2}{2!} + \frac{x^4}{4!} - \cdots + (-1)^n \frac{x^{2n}}{(2n)!} + \cdots = \sum_{k=0}^{\infty} (-1)^k \frac{x^{2k}}{(2k)!}$$

and

$$\frac{1}{1-x} = 1 + x + x^2 + \cdots + x^n + \cdots = \sum_{k=0}^{\infty} x^k, \qquad |x| < 1$$

The first three series converge for all values of $x$, the last one for $|x| < 1$ as indicated. But, before discussing convergence, let us remind ourselves of the procedures for doing arithmetic with power series and for differentiating and integrating them. The procedures that we shall discuss are usually referred to as "formal," as they do not address questions of convergence.

As for arithmetic, power series are added, subtracted, multiplied, and divided just like polynomials. For example, suppose we wish to find the first three nonzero terms for the power series expansion of $e^x \sin x$.

$$e^x \sin x = \left(1 + x + \frac{x^2}{2} + \cdots\right)\left(x - \frac{x^3}{6} + \cdots\right) = x + x^2 + \left(\frac{1}{2} - \frac{1}{6}\right)x^3 + \cdots$$

$$= x + x^2 + \frac{1}{3}x^3 + \cdots$$

Finding the first three nonzero terms of $e^x/\cos x$ involves long division as we do it for polynomials.

$$
\begin{array}{r}
1+x+x^2+\cdots \\
1 - \dfrac{x^2}{2} + \dfrac{x^4}{24} - \cdots \overline{\big)\, 1+x+\dfrac{x^2}{2}+\dfrac{x^3}{6}+\dfrac{x^4}{24}+\cdots} \\
1 \qquad -\dfrac{x^2}{2} \qquad +\dfrac{x^4}{24} - \cdots \\
\overline{\qquad x+x^2+\dfrac{x^3}{6}+0+\cdots} \\
x \qquad -\dfrac{x^3}{2} \qquad +\dfrac{x^5}{24}+\cdots \\
\overline{\qquad x^2+\dfrac{2x^3}{3}+\cdots} \\
x^2 \qquad -\dfrac{x^4}{2}+\cdots \\
\hline
\end{array}
$$

Thus $e^x/\cos x = 1 + x + x^2 + \cdots$.

We may often compute the power series of a composite function $f(g(x))$ by substituting $g$ for the variable in the power series of $f$. For example,

$$e^{-x^2} = 1 + (-x^2) + \frac{(-x^2)^2}{2!} + \frac{(-x^2)^3}{3!} + \cdots$$

$$= 1 - x^2 + \frac{x^4}{2!} - \frac{x^6}{3!} + \cdots$$

As for differentiation and integration of power series, the formal procedure is to differentiate and integrate them term by term, just like polynomials. For example,

$$\frac{d}{dx}\left(\frac{1}{1-x}\right) = \frac{d}{dx}(1 + x + x^2 + x^3 + \cdots + x^n + \cdots)$$

$$= \frac{d}{dx}(1) + \frac{d}{dx}(x) + \frac{d}{dx}(x^2) + \frac{d}{dx}(x^3) + \cdots + \frac{d}{dx}(x^n) + \cdots$$

$$= 0 + 1 + 2x + 3x^2 + \cdots + nx^{n-1} + \cdots$$

$$= \sum_{n=1}^{\infty} nx^{n-1}$$

Note that also $(d/dx)(1 - x)^{-1} = (-1)(1 - x)^{-2}(-1) = (1 - x)^{-2}$, so that we have

$$\left(\frac{1}{1-x}\right)^2 = \sum_{n=1}^{\infty} nx^{n-1}$$

a result that we could have gotten by multiplying the series for $1/(1 - x)$ by itself. But this way is easier.

When you integrate a power series term by term, do a definite integration starting at $x = 0$. (Start at $x = a$ for series in powers of $x - a$.) This avoids the awkward arbitrary constant. For example, starting with

$$\cos t = \sum_{k=0}^{\infty} (-1)^k \frac{t^{2k}}{(2k)!}$$

we get

$$\int_0^x \cos t \, dt = \int_0^x \left[\sum_{k=0}^{\infty} (-1)^k \frac{t^{2k}}{(2k)!}\right] dt$$

$$= \sum_{k=0}^{\infty} \frac{(-1)^k}{(2k)!} \int_0^x t^{2k} \, dt = \sum_{k=0}^{\infty} \frac{(-1)^k}{(2k)!} \frac{t^{2k+1}}{2k+1}\Big|_0^x$$

$$= \sum_{k=0}^{\infty} \frac{(-1)^k}{(2k+1)!} x^{2k+1}$$

The last series we recognize as sin $x$. Hence power series procedures have proved for us the formula

$$\int_0^x \cos t \, dt = \sin x$$

Note that in the last calculation we worked with the "sigma" notation, rather than writing out the series term by term. Practice both ways. A power series expansion of a function $f(x)$,

$$f(x) = a_0 + a_1 x + a_2 x^2 + \cdots + a_n x^n + \cdots = \sum_{k=0}^{\infty} a_k x^k$$

is said to *converge* to $f(x)$ for a particular value of $x$, say $x = x_0$, if the difference between the value of the function at $x = x_0$, namely $f(x_0)$, and the sum of the first $N$ terms of the series,

$$a_0 + a_1(x_0) + a_2(x_0)^2 + a_3(x_0)^3 + \cdots + a_N(x_0)^N$$

can be made as small as we want by taking $N$ large enough. This is called pointwise convergence, because we are working at a particular point, $x = x_0$.

The facts of convergence are these. Given any power series, there is associated with that power series a real number $R$, $0 \leqslant R \leqslant \infty$, such that the series converges for all $x$ such that $|x| < R$ and does not converge when $|x| > R$. If we regard $x$ as a real variable, then the region of convergence $|x| < R$ is either an open interval $-R < x < R$, the entire real line $\mathbf{R}$, or just the point $x = 0$. If we regard $x$ as a complex variable, in which case, following tradition, we write $z$ instead of $x$, then the region of convergence $|z| < R$ is either an open disk centered at the origin, the entire complex plane, or just the point $z = 0$. Hence, at any point in the disk (or interval) of convergence we may compute the value of the function represented by the power series to arbitrary accuracy by computing the value of a sufficiently long partial sum of the series. In principle, anyway.

The series for $e^x$, sin $x$, and cos $x$ converge for all values of $x$ (use the ratio test), so for these series we have $R = \infty$. The series for $1/(1 - x)$ has $R = 1$. A function which is expandable in a power series with positive radius of convergence, $R > 0$, is called *analytic at* $x = 0$. Thus $e^x$, sin $x$, cos $x$, and $1/(1 - x)$ are all analytic at $x = 0$. Instead of expanding in powers of $x$, we may expand in powers of $x - a$. Then the region of convergence is an interval or disk centered at $x = a$ instead of $x = 0$. If this region of convergence does not reduce to a point, we say the function it represents is *analytic at* $x = a$. Almost all the functions that arise in practice are analytic except at isolated points. The functions $e^x$, sin $x$, and cos $x$ are analytic everywhere; such functions are called *entire*. The function $1/(1 - x)$ is analytic everywhere except at $x = 1$.

As to the computation of a function from a partial sum of its power series expansion, there is a difference between "in principle" and "in practice." In practice, we would like to get sufficient accuracy by taking a reasonable number of terms of the series, say six or fewer. In the general power series expansion,

$$f(x) = a_0 + a_1 x + a_2 x^2 + \cdots + a_n x^n + \cdots = \sum_{k=0}^{\infty} a_k x^k \qquad (1.4.1)$$

the $n$th coefficient $a_n$ is determined by the $n$th derivative of $f(x)$ evaluated at $x = 0$,

$$a_n = \frac{f^{(n)}(0)}{n!}, \qquad n = 0, 1, 2, \ldots \qquad (1.4.2)$$

(If one forgets formula (1.4.2), there is an easy way to remember it. In series (1.4.1) put $x = 0$ to get $f(0) = a_0$. Then differentiate series (1.4.1) term by term, put $x = 0$ again to get $f'(0) = a_1$, etc.) Since all the information that determines the coefficients $a_n$ comes from the values of $f(x)$ and its derivatives at $x = 0$, one would expect series (1.4.1) to better represent the function $f(x)$ the closer $x$ is to 0. In fact, if any power series (1.4.1) converges at $x_0$, then it will converge more rapidly at values of $x$ that satisfy $|x| < |x_0|$. So we expect that if we are to get acceptable accuracy in approximating a function by a partial sum of its power series expansion, we must stick to reasonably small values of $x$. Let's see how these matters work out in practice.

How well does the six-term partial sum ($n = 0$ to $n = 5$) of the series for $e^x$ represent the function? We want to compare $e^x$ and

$$1 + x + \frac{x^2}{2!} + \frac{x^3}{3!} + \frac{x^4}{4!} + \frac{x^5}{5!}$$

On the HP-15C, as on any scientific electronic calculator, there is an $e^x$ button which gives the value of $e^x$, accurate to eight or more significant figures, almost instantaneously. As for the polynomial, the HP-15C Owner's Handbook (p. 79) suggests that Horner's method is swiftest.

$$1 + x + \frac{x^2}{2!} + \frac{x^3}{3!} + \frac{x^4}{4!} + \frac{x^5}{5!} = 1 + x\left(1 + \frac{x}{2!} + \frac{x^2}{3!} + \frac{x^3}{4!} + \frac{x^4}{5!}\right)$$

$$= 1 + x\left(1 + x\left(\frac{1}{2!} + \frac{x}{3!} + \frac{x^2}{4!} + \frac{x^3}{5!}\right)\right)$$

$$= 1 + x\left(1 + x\left(\frac{1}{2} + x\left(\frac{1}{6} + \frac{x}{4!} + \frac{x^2}{5!}\right)\right)\right)$$

$$= 1 + x\left(1 + x\left(.5 + x\left(\frac{1}{6} + x\left(\frac{1}{4!} + \frac{x}{5!}\right)\right)\right)\right)$$

The following programs, written for the HP-15C, compute this polynomial. The first program utilizes the stack (p. 32 of the Owner's Handbook); the second uses the conventional method of storing the variable in a memory register. The stack is supposed to be faster, but both programs take the same amount of time, about 3 seconds per value.

| | |
|---|---|
| g P/R | g P/R |
| f LBL A | f LBL B |
| ENTER, ENTER, ENTER | STO 0 |
| 5 fx! ÷ | 5 fx! ÷ |
| 4 fx! 1/x + | 5 fx! 1/x + |
| × 6 1/x + | RCL × 0 |
| × .5 + | 6 1/x + |
| × 1 + | RCL × 0 |
| × 1 + | .5 + |
| g RTN | RCL × 0 |
| g P/R | 1 + |
| | RCL × 0 |
| | 1 + |
| | g RTN |
| | g P/R |

One can also program the HP-15C, using the "$y^x$" key, to sum the power series as it is written, without using Horner's method. But this program takes about 50% more time to run.

To run the programs, enter the value of $x$; then push either fA or fB. Table 1.3 compares the values of $e^x$ with the $n = 5$ partial sum of its power series for $0 \leqslant x \leqslant 1$.

Table 1.3

| $x$ | $e^x$ | $\sum_{k=0}^{5} x^k/k!$ |
|---|---|---|
| 0 | 1 | 1 |
| 0.1 | 1.105170918 | 1.105170917 |
| 0.2 | 1.221402758 | 1.221402667 |
| 0.3 | 1.349858808 | 1.349857750 |
| 0.4 | 1.491824698 | 1.491818667 |
| 0.5 | 1.648721271 | 1.648697917 |
| 0.6 | 1.822118800 | 1.822048000 |
| 0.7 | 2.013752707 | 2.013571417 |
| 0.8 | 2.225540928 | 2.225130666 |
| 0.9 | 2.459603111 | 2.458758250 |
| 1.0 | 2.718281828 | 2.716666667 |

Note that at $x = 0.1$ the six-term partial sum of the power series agrees with
the value of $e^x$ to one digit in the ninth decimal place! Even at $x = 1$, the error
is 0.06%. The percent error is computed as follows:

$$\text{error} = \frac{|\text{actual value} - \text{approximate value}|}{\text{actual value}} \times 100.$$

To see how the partial sum compares with the function over a wider range,
it is best to use a graphical comparison. In Figure 1.1 we have plotted on
semilog paper the function $e^x$ and the six-term partial sum of its power series

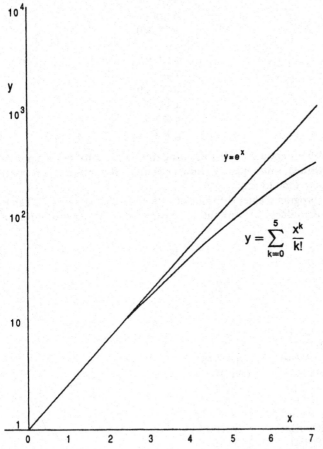

**Figure 1.1**   Comparison of $e^x$ and the six-term partial sum of its power series.

over the range $0 \leqslant x \leqslant 7$. On semilog paper, the function $y = e^x$ plots as a straight line, which makes this the graph paper of choice for plots that involve exponentials. Figure 1.1 shows clearly how the six-term partial sum becomes a poorer and poorer approximation the larger the $x$. At $x = 7$, the partial sum is less than one-third of the actual value. Of course, the more terms we take, the better we will do.

**Exercises**

1. (a) By subtracting 1 from the series for $e^x$ show that

$$e^x - 1 = \sum_{n=1}^{\infty} \frac{x^n}{n!}$$

   (b) Using the result in (a), find the series for $(e^x - 1)/x$.
2. Find the first four nonzero coefficients.

   (a) $\left(1 + x - \dfrac{x^2}{2} + x^3 + \cdots\right)\left(2 - x - x^2 - \dfrac{1}{3}x^3 + \cdots\right) =$

   (b) $\dfrac{1 + x + x^2/2 + x^3/6 + \cdots}{1 + 2x} =$

3. (a) By multiplying the series for $\sin x$ by that for $\cos x$, find the first four nonzero terms in the power series expansion of $f(x) = \sin x \cos x$.
   (b) By substituting $t = 2x$ in the power series expansion for $\sin t$, determine the general (sigma notation) power series for $\frac{1}{2}\sin 2x$. How does this answer relate to your answer in part (a)?
4. By dividing the series for $\sin x$ by that for $\cos x$, determine the first three nonzero terms in the power series expansion for $\tan x$.
5. By differentiating the series for $\sin x$, prove the formula $(d/dx)(\sin x) = \cos x$.
6. By integrating the series

$$\frac{1}{1 - t} = 1 + t + t^2 + \cdots + t^n + \cdots, \qquad |t| < 1$$

term by term from $t = 0$ to $t = x$, prove the formula

$$\ln\left(\frac{1}{1 - x}\right) = x + \frac{x^2}{2} + \cdots + \frac{x^n}{n} + \cdots, \qquad |x| < 1$$

7. Approximate $e^{-x^2}$ in two ways:
   (a) By computing the sum of the first six nonzero terms of its power series expansion,

$$1 - x^2 + \frac{x^4}{2!} - \frac{x^6}{3!} + \frac{x^8}{4!} - \frac{x^{10}}{5!}$$

(b) By computing the sum of the first six nonzero terms of the power series expansion of $e^{x^2}$,

$$1 + x^2 + \frac{x^4}{2!} + \frac{x^6}{3!} + \frac{x^8}{4!} + \frac{x^{10}}{5!}$$

and then inverting the result.

Compare, in one table, the values of $e^{-x^2}$ and the approximations in (a) and (b) over the range $0 \leqslant x \leqslant 3$ in increments of 0.2.

8. Approximate $\sin x$ by the five-term partial sum of its power series

$$x - \frac{x^3}{3!} + \frac{x^5}{5!} - \frac{x^7}{7!} + \frac{x^9}{9!}$$

Compare the values in a table, $0 \leqslant x \leqslant \pi$, in increments of $0.1\pi$, and compare them graphically over the range $0 \leqslant x \leqslant 1.7\pi$. (Remember to put your calculator in radian mode. Use ordinary (linear) 10 mm to cm graph paper.)

## 1.5 THE INITIAL VALUE PROBLEM FOR A FIRST ORDER LINEAR DIFFERENTIAL EQUATION

Let us assume that the coefficient functions $a(x)$ and $b(x)$ and the right side $h(x)$ of our equation $a(x)y' + b(x)y = h(x)$ are expandable in power series:

$$a(x) = a_0 + a_1 x + a_2 x^2 + a_3 x^3 + \cdots$$
$$b(x) = b_0 + b_1 x + b_2 x^2 + b_3 x^3 + \cdots$$
$$h(x) = h_0 + h_1 x + h_2 x^2 + h_3 x^3 + \cdots$$

We seek to determine coefficients $\alpha_i$ in a formal power series expansion for our unknown function

$$y = \alpha_0 + \alpha_1 x + \alpha_2 x^2 + \alpha_3 x^3 + \cdots + \alpha_n x^n + \cdots$$

so that our equation is satisfied. We differentiate $y$ term by term,

$$y' = \alpha_1 + 2\alpha_2 x + 3\alpha_3 x^2 + \cdots + n\alpha_n x^{n-1} + \cdots$$

multiply this series by the series for $a(x)$ and collect terms,

$$a(x)y' = a_0\alpha_1 + (2a_0\alpha_2 + a_1\alpha_1)x + (3a_0\alpha_3 + 2a_1\alpha_2 + a_2\alpha_1)x^2 + \cdots$$

do the same for $b(x)y$, and then add term by term. Our equation will be satisfied if coefficients of the power series just computed equal those for the expansion of $h(x)$. Equating these coefficients, we get an infinite system of equations:

$$a_0\alpha_1 = h_0 - b_0\alpha_0$$

$$2a_0\alpha_2 = h_1 - (a_1 + b_0)\alpha_1 - b_1\alpha_0$$

$$3a_0\alpha_3 = h_2 - (2a_1 + b_0)\alpha_2 - (a_2 + b_1)\alpha_1 - b_2\alpha_0$$

$$4a_0\alpha_4 = h_3 - (3a_1 + b_0)\alpha_3 - (2a_2 + b_1)\alpha_2 - (a_3 + b_2)\alpha_1 - b_3\alpha_0$$

$$\vdots$$

$$na_0\alpha_n = h_{n-1} - ((n-1)a_1 + b_0)\alpha_{n-1} - ((n-2)a_2 + b_1)\alpha_{n-2} - \cdots$$

$$- (a_{n-1} + b_{n-2})\alpha_1 - b_{n-1}\alpha_0$$

$$\vdots$$

$$(1.5.1)$$

Using expansions in powers of $x$ means we are expanding our functions around $x_0 = 0$. (The general case is handled by expanding in powers of $x - x_0$.) We have $a(0) = a_0$, and we shall assume $a(0) \neq 0$, as we have already noted that difficulties arise at a point $x_0$ where the leading coefficient $a(x)$ equals 0. Then, once the first coefficient $\alpha_0 = y(0)$ is specified, we can solve equations (1.5.1) for all other coefficients $\alpha_i$ of our unknown function $y(x)$:

$$\alpha_1 = \frac{1}{a_0}(h_0 - b_0\alpha_0)$$

$$\alpha_2 = \frac{1}{2a_0}(h_1 - (a_1 + b_0)\alpha_1 - b_1\alpha_0)$$

etc.

In this book we shall be concerned mainly with coefficient functions that are analytic plus possibly some terms of the form $1/(z - z_0)^n$. We shall stress the formal aspects of the computations and be less concerned with computations of radii of convergence. With that proviso, we may state the result we have just proved as follows.

**Theorem 1.2.** *Suppose the functions $a(x)$, $b(x)$, $h(x)$ are analytic at $x = 0$, and $a(0) \neq 0$. Then, given any number $y_0$, there is one, and only one, function $y(x)$ analytic at $x = 0$, satisfying $ay' + by = h$ and $y(0) = y_0$.*

The coefficients of the power series expansion of $y(x)$ are gotten by solving equations (1.5.1) by the method explained in the paragraph following those equations. This method is called the *method of undetermined coefficients* because initially the function $y(x)$ is expanded as a power series with undetermined or unknown coefficients, and these coefficients are then determined recursively so that $y$ satisfies $ay' + by = h$.

If we expand in powers of $(x - x_0)$, then by the same method we may prove

**Theorem 1.3.** *Suppose the functions* $a(x)$, $b(x)$, $h(x)$ *are analytic at* $x = x_0$, *and* $a(x_0) \neq 0$. *Then, given any number* $y_0$, *there is one, and only one, function* $y(x)$ *analytic at* $x = x_0$, *satisfying* $ay' + by = h$ *and* $y(x_0) = y_0$.

When you apply the method of undetermined coefficients in practice, work from first principles; don't use the general formulas derived above. For example, suppose we want to find the first four nonzero coefficients of the power series expansion of the solution $y = y(x)$ to $(3x^2 + 1)y' - 2xy = x$ satisfying $y(0) = 3/2$. Start with the formal power series for $y(x)$ with undetermined coefficients:

$$y = \alpha_0 + \alpha_1 x + \alpha_2 x^2 + \alpha_3 x^3 + \alpha_4 x^4 + \cdots$$

Compute the first few derivatives and do the indicated multiplications to get

$$3x^2 y' = 0 + 0x + 3\alpha_1 x^2 + 6\alpha_2 x^3 + 9\alpha_3 x^4 + 12\alpha_4 x^5 + \cdots$$

$$y' = \alpha_1 + 2\alpha_2 x + 3\alpha_3 x^2 + 4\alpha_4 x^3 + 5\alpha_5 x^4 + 6\alpha_6 x^5 + \cdots$$

$$-2xy = 0 - 2\alpha_0 x - 2\alpha_1 x^2 - 2\alpha_2 x^3 - 2\alpha_3 x^4 - 2\alpha_4 x^5 - \cdots$$

Then add:

$$(3x^2 + 1)y' - 2xy = \alpha_1 + 2(\alpha_2 - \alpha_0)x + (\alpha_1 + 3\alpha_3)x^2 + 4(\alpha_2 + \alpha_4)x^3$$
$$+ (6\alpha_3 + 5\alpha_5)x^4 + (10\alpha_4 + 6\alpha_6)x^5 + \cdots$$

This power series must equal that for $h(x) = x = 0 + 1x + 0x^2 + 0x^3 + \cdots$. Equate coefficients:

$$\alpha_1 = 0$$
$$2(\alpha_2 - \alpha_0) = 1$$
$$\alpha_1 + 3\alpha_3 = 0$$
$$4(\alpha_2 + \alpha_4) = 0$$
$$6\alpha_3 + 5\alpha_5 = 0$$
$$10\alpha_4 + 6\alpha_6 = 0$$

The initial condition $y(0) = 3/2$ gives $\alpha_0 = 3/2$. The first equation says $\alpha_1 = 0$, the second furnishes $\alpha_2 = 2$, the third, $\alpha_3 = 0$, etc. Our answer is

$$y = \frac{3}{2} + 2x^2 - 2x^4 + \frac{10}{3}x^6 + \cdots$$

Follow this pattern in doing the first three exercises below. Remember to include the ellipsis dots (three dots) when you write down the first few terms of a series, unless you know that all subsequent terms are zero.

**Exercises**

1. By the method of undetermined coefficients, find the first four nonzero coefficients of the power series expansion of the solution to

$$(x^2 + 1)y' - (1 - x)^2 y = xe^{-x}, \qquad y(0) = 1$$

2. (Continuation) In Exercise 2 of Section 1.3 we found the solution to the initial value problem in Exercise 1, namely

$$y = \frac{1}{4(1 + x^2)}(5e^x - (2x + 1)e^{-x})$$

   By expanding this function in a power series, check your answer to Exercise 1.

3. By the method of undetermined coefficients, find the first four nonzero terms in the power series solution to

$$y' - 2xy = 1, \qquad y(0) = 0$$

4. Suppose that in equations (1.5.1) of this section we have $a_0 = 0$ and $na_1 + b_0 \neq 0$ for $n = 0, 1, 2, \ldots$. Show that there is then a unique solution for $y(x)$. Compare with an example from Section 1.1.

# 2

# Second Order Linear Differential Equations

## 2.1 BASIC CONCEPTS OF LINEAR ALGEBRA FOR FUNCTION SPACES

### 2.1.1 Equality of Functions

We use equality of functions commonly in two different ways. When we write

$$x^3 + 2x^2 - x - 1 = (x - 1)^3 + 5(x - 1)^2 + 6(x - 1) + 1 \qquad (2.1.1)$$

we mean that the left side is identically equal to the right side, which means that if we expand the powers of $x - 1$ on the right, then combine and rearrange terms in descending powers of the indeterminate $x$, we get the polynomial on the left side. It follows that if we substitute for $x$ any real or complex number whatever, the left side and right side of (2.1.1) will give equal results. So if we consider the left side and the right side of equation (2.1.1) as separate functions

$$f(x) = x^3 + 2x^2 - x - 1, \qquad g(x) = (x - 1)^3 + 5(x - 1)^2 + 6(x - 1) + 1$$

then the equality expressed in equation (2.1.1), $f(x) = g(x)$, means that $f$ and $g$ are equal *as functions*: for *any* real (or complex) number $x$, the function value $f(x)$ equals the function value $g(x)$. Since $f$ and $g$ are equal as functions, they have the same derivative, $f' = g'$, and the same integral, $\int_0^x f(t)\, dt = \int_0^x g(t)\, dt$.

The other common use of the sign "$=$" is represented by the example

$$x^2 - 4x + 3 = 0 \tag{2.1.2}$$

This does not express equality of functions, but rather here we are asking for the "roots" of this equation, namely those real or complex numbers for which it is true. Equation (2.1.2) holds for only two values of $x$, namely $x = 1$ and $x = 3$. As the functions on the left side, $f(x) = x^2 - 4x + 3$, and the right side, $g(x) = 0$, are different, if we differentiate or integrate equation (2.1.2) we will get an incorrect result.

We will be using equality of functions in the first sense. When we write $f = g$ for real or complex functions $f$ and $g$ we mean that $f(x) = g(x)$ for every substituted value of $x$ in some real interval $a \leqslant x \leqslant b$ where $a < b$, or for all $z$ in some disk $|z - z_0| < \rho$, $\rho > 0$, in the complex plane. We may have $a = -\infty$, $b = +\infty$. Also $\rho = \infty$.

### 2.1.2 Function Spaces

By the term *vector space of functions* we mean a set of functions that has the same formal properties as the vector space of $n$-tuples that we learned about in linear algebra: namely closure under sum and closure under scalar multiplication. Thus a set $V$ of functions is a vector space of functions if $V$ contains, along with functions $f$ and $g$, their sum $f + g$ and $cf$ for any constant $c$. A subset $W$ of $V$ is a *subspace* (of functions) if $W$ is a vector space in its own right.

For example, consider the set $C[0, 1]$ of real-valued continuous functions on the interval $0 \leqslant x \leqslant 1$. By "continuous" we mean that the function has an unbroken graph. If we wanted to be more precise, we could use the "$\varepsilon-\delta$" definition of continuity. Either way, we easily convince ourselves that if $f$ and $g$ are continuous, so are $f + g$ and $cf$ for any constant $c$. Thus $C[0, 1]$ is a vector space of functions.

Another example: let $V$ be the set of complex-valued functions that have convergent power series in $|z| < 1$. As the power series for $f + g$ is obtained by adding term by term the power series for $f$ and $g$, $V$ contains $f + g$ if it contains both $f$ and $g$. Similarly, it contains $cf$ along with $f$ for any complex constant $c$. So $V$ is a vector space of complex-valued functions. The set $W$ of functions in $V$ that satisfy also $f(0) = 0$ is a subspace of $V$.

### 2.1.3 Linear Independence and Linear Dependence

The concepts of linear independence and linear dependence apply to vector spaces of functions. These concepts are so important that we shall make some formal definitions.

**Definition 2.1.** A set of functions $f_1, f_2, \ldots, f_n$ is *linearly dependent* if it is possible to express one of the functions, say $f_i$, as a linear combination of the remaining ones, i.e., if

$$f_i = a_1 f_1 + \cdots + a_{i-1} f_{i-1} + a_{i+1} f_{i+1} + \cdots + a_n f_n$$

for some constants $a_j$.

There is another equivalent way to put this condition: the functions $f_1, f_2, \ldots, f_n$ are linearly dependent provided that there exist constants $c_1, c_2, \ldots, c_n$, *not all zero*, such that $c_1 f_1 + c_2 f_2 + \cdots + c_n f_n = 0$. Exercise 4 asks you to prove the equivalence of these conditions. Note that *some* of the $c_i$ may be zero, just as long as they are not *all* zero.

The opposite of linear dependence is linear independence.

**Definition 2.2.** A set of functions $f_1, f_2, \ldots, f_n$ is *linearly independent* if the only constants $c_i$ for which we have

$$c_1 f_1 + c_2 f_2 + \cdots + c_n f_n = 0$$

are $c_1 = c_2 = \cdots = c_n = 0$.

For example, the functions $\sin x$ and $\cos x$ are linearly independent. To prove this, suppose we had an equation

$$c_1 \sin x + c_2 \cos x = 0 \tag{2.1.3}$$

Remember we are using the equal sign here as identical equality of functions as discussed at the beginning of this section. In this case we are assuming that (2.1.3) holds for all real $x$, $-\infty < x < \infty$. (We could just as well have taken $0 \leqslant x \leqslant 2\pi$. Why?) Put $x = \pi/2$ in equation (2.1.3). As $\sin(\pi/2) = 1$, $\cos(\pi/2) = 0$, we get $c_1 \cdot 1 + c_2 \cdot 0 = 0$ so $c_1 = 0$. Then we put $x = 0$ to get $c_2 = 0$. Thus the only constants for which (2.1.3) holds are $c_1 = c_2 = 0$, so $\sin x$ and $\cos x$ are, by Definition 2.2, linearly independent.

On the other hand, the functions $f_1(x) = \sin 2x$ and $f_2(x) = \sin x \cos x$ are linearly dependent because

(1) $\sin 2x + (-2) \sin x \cos x = 0$

These concepts, linear dependence and linear independence, described in Definitions 2.1 and 2.2 are taken over verbatim from linear algebra. There are two more concepts from linear algebra that will be of vital importance for us here.

The first is the concept of a spanning set. Suppose $V$ is a vector space of functions. The functions $f_1, f_2, \ldots, f_n$ within $V$ are said to *span* $V$ if every function $f$ in $V$ can be expressed as a linear combination of $f_1, f_2, \ldots, f_n$.

That is, given any $f$ in $V$, there exist constants $c_1, c_2, \ldots, c_n$ such that

$$f = c_1 f_1 + c_2 f_2 + \cdots + c_n f_n$$

The second concept is that of a basis. A set of functions $f_1, f_2, \ldots, f_n$ in a vector space of functions $V$ is said to be a *basis* for $V$ if

(i) $f_1, f_2, \ldots, f_n$ span $V$ and
(ii) they are linearly independent.

Recall the following key facts from linear algebra. (For convenience we shall henceforth refer to a vector space of functions as simply a "function space.")

**Theorem 2.3.** *Any two bases of a function space have the same number of elements.*

This number is called the *dimension* of the space.

**Theorem 2.4.** *Let $V$ be a function space of dimension $n$. Then*
(i) *Any $n$ linearly independent functions in $V$ span $V$, thus constitute a basis.*
(ii) *Any $n$ functions in $V$ that span $V$ are linearly independent, thus constitute a basis.*
(iii) *Given any $k$ linearly independent functions in $V$, $k < n$, these may be extended to a basis.*

Our principal application of these results is to the kernel of a second order linear differential expression, which we shall soon prove to be a subspace of dimension 2. Thus, if we find two linearly independent functions in the kernel, then any function in the kernel is a linear combination of these two. Also, if we have one function in the kernel, then we know that there exists a second function in the kernel, independent of the first, and the two together span the kernel.

### 2.1.4 The Wronskian

We turn finally to an important criterion for linear independence in function spaces based on the so-called Wronskian determinant. Suppose we have an equation

$$c_1 f_1(x) + c_2 f_2(x) = 0 \qquad (2.1.4)$$

for constants $c_1, c_2$ and functions $f_1, f_2$. Equation (2.1.4) says that the function on the left is identically zero. It thus will have a zero derivative, so we get

$$c_1 f_1'(x) + c_2 f_2'(x) = 0 \qquad (2.1.5)$$

We write (2.1.4) and (2.1.5) together as a system of two equations in two unknowns $c_1$ and $c_2$.

$$c_1 f_1(x) + c_2 f_2(x) = 0$$
$$c_1 f_1'(x) + c_2 f_2'(x) = 0 \qquad (2.1.6)$$

The values $c_1 = c_2 = 0$ solve the homogenous system (2.1.6). If (2.1.6) is to have a nonzero solution $[c_1, c_2]$, then the determinant

$$\begin{vmatrix} f_1(x) & f_2(x) \\ f_1'(x) & f_2'(x) \end{vmatrix} \qquad (2.1.7)$$

must be zero. This determinant is called the *Wronskian* of the functions $f_1$ and $f_2$. We denote it $W(f_1, f_2)$ or sometimes $W(f_1, f_2)(x)$, to stress that it is a function of $x$. Our discussion shows that if $W(f_1, f_2)(x) \neq 0$, even at one point, then the system (2.1.6), thus also the equation (2.1.4), can have only the solution $c_1 = c_2 = 0$. Thus we have the following *criterion for linear independence*:

**Theorem 2.5.** *If the Wronskian $W(f_1, f_2)(x)$ is $\neq 0$, even for one value of $x$, then $f_1$ and $f_2$ are linearly independent.*

For example, if $f_1(x) = \sin x$, $f_2(x) = \cos x$, then

$$W(f_1, f_2) = \begin{vmatrix} \sin x & \cos x \\ \cos x & -\sin x \end{vmatrix} = -\sin^2 x - \cos^2 x = -1$$

so $\sin x$ and $\cos x$ are linearly independent, as we concluded earlier.

The Wronskian determinant of $n$ functions $f_1, f_2, \ldots, f_n$ is the determinant of an $n \times n$ matrix:

$$W = \begin{vmatrix} f_1 & f_2 & \cdots & f_n \\ f_1' & f_2' & \cdots & f_n' \\ f_1'' & f_2'' & \cdots & f_n'' \\ \vdots & & & \\ f_1^{(n-1)} & f_2^{(n-1)} & \cdots & f_n^{(n-1)} \end{vmatrix}$$

All the formal apparatus and concepts that we have introduced in this lecture are valid for complex functions and complex scalars. We have the following generalization of Theorem 2.5.

**Theorem 2.6.** *Let $f_1, f_2, \ldots, f_n$ be real or complex functions. If their Wronskian is $\neq 0$, even at one point, then they are linearly independent (over the real or complex numbers, respectively).*

When dealing with complex functions, certain useful facts should be borne in mind. The complex exponential $e^z$ satisfies the two important formulas

$$e^{z_1 + z_2} = e^{z_1} e^{z_2} \quad \text{and} \quad e^{iz} = \cos z + i \sin z \quad \text{(Euler)}$$

In particular, if $z = y$ is real, then $e^{iy} = \cos y + i \sin y$ so $|e^{iy}|^2 = \cos^2 y + \sin^2 y = 1$. Thus, if $z = x + iy$, $e^z = e^{x+iy} = e^x e^{iy}$, so $|e^z| = e^x$. As the real exponential $e^x$ is never zero, it follows that the complex exponential, $e^z$, which is entire (analytic in the entire complex plane), is also never zero. For this material, one may consult the book *Complex Variables* by G. Polya and G. Latta.

### Exercises

1. Which of the following represent function equality, and which ask for roots?
   (a) $x^2 - 2x + 2 = (x - 1)^2 + 1$
   (b) $2x^2 - 2x - 1 = (x - 2)^2 - 6$
   (c) $(x^2 + y^2)(z^2 + w^2) = (xw + yz)^2 + (yw - xz)^2$ (4 variables)
2. Determine the constants $a$, $b$, $c$, $d$ so that the following is a function equality:

$$x^3 - 3x^2 - x + 6 = a(x - 2)^3 + b(x - 2)^2 + c(x - 2) + d$$

3. Let $V$ be the vector space of functions on $0 \leqslant x \leqslant 1$. For each of the following subsets $D$ of $V$ determine (1) which are closed under addition and (2) which are closed under scalar multiplication. Which are subspaces of $V$?
   (a) $D = \{f : f(\frac{1}{2}) = 1\}$. [Soln.: If $f(\frac{1}{2}) = 1$ and $g(\frac{1}{2}) = 1$, then if $h = f + g$, $h(\frac{1}{2}) = f(\frac{1}{2}) + g(\frac{1}{2}) = 1 + 1 = 2$. Hence $h(\frac{1}{2}) \neq 1$, so $h = f + g$ does not belong to $D$. Thus $D$ is not closed under addition. Neither is $D$ closed under scalar multiplication. Not a subspace.]
   (b) $D = \{f : f(\frac{1}{2}) = 0\}$.
   (c) $D = \{f : f(0) = f(1) = 0\}$
   (d) $D = \{f : f'(0) = f'(1) = 0\}$
   (e) $D = \{f : |f(x)| \leqslant 1, 0 \leqslant x \leqslant 1\}$
   (f) $D = \{f : f(x) \geqslant 0, 0 \leqslant x \leqslant 1\}$
   (g) $D = \{f : f(0) = f(1), f'(0) = f'(1)\}$
4. Prove the assertion after Definition 2.1: if the functions $f_1, f_2, \ldots, f_n$ are linearly dependent then there exist constants $c_1, c_2, \ldots, c_n$, not all zero, such that

$$c_1 f_1 + c_2 f_2 + \cdots + c_n f_n = 0$$

And conversely, if the latter condition holds then $f_1, f_2, \ldots, f_n$ are linearly dependent in the sense of Definition 2.1.
5. In each case, prove that the set of functions $f_1, f_2, \ldots, f_n$ is linearly dependent by finding (real or complex) constants $c_1, c_2, \ldots, c_n$, not all zero,

such that $c_1f_1 + c_2f_2 + \cdots + c_nf_n = 0$.

(a) $f_1(x) = e^x$, $f_2(x) = e^{-x}$, $f_3(x) = \sinh x$

(b) $f_1(x) = \sin^2 x$, $f_2(x) = \cos^2 x$, $f_3(x) = 1$

(c) $f_1(x) = x$, $f_2(x) = x^2$, $f_3(x) = 0$

(d) $f_1(x) = \ln(x^2)$, $f_2(x) = \ln x$, $f_3(x) = e^x$

(e) $f_1(z) = e^{iz}$, $f_2(z) = \cos z$, $f_3(z) = \sin z$

(f) $f_1(z) = \arctan z$, $f_2(z) = \ln\dfrac{1 + iz}{1 - iz}$

6. (a) Show that the functions $1, x, x^2, \ldots, x^n$ are linearly independent.

    (b) Show that the space of polynomials of degree $\leqslant n$ has dimension $n + 1$, and find a basis for it.

7. If $\alpha$ and $\beta$ are different complex constants, prove that $e^{\alpha z}$ and $e^{\beta z}$ are linearly independent.

8. Let $g_1 = f_1 - f_2$, $g_2 = f_1 + f_2$.

    (a) Prove that if $f_1$ and $f_2$ are linearly independent, so are $g_1$ and $g_2$.

    (b) Prove that $g_1$ and $g_2$ span the same subspace as $f_1$ and $f_2$.

9. Are $x$ and $|x|$ equal as real functions? Deal separately with the intervals $0 \leqslant x < \infty$ and $-\infty < x < \infty$.

## 2.2 THE INITIAL VALUE PROBLEM FOR A SECOND ORDER LINEAR HOMOGENEOUS DIFFERENTIAL EQUATION

We take up now the study of second order linear differential expressions which have the form

$$l(y) \equiv a(x)y'' + b(x)y' + c(x)y \tag{2.2.1}$$

where $'$ denotes differentiation with respect to the independent variable $x$. The essence of the meaning of *linearity* of $l$ lies in the formal relations we have considered in Chapter 1:

Additive: $l(y_1 + y_2) = l(y_1) + l(y_2)$ for any two functions $y_1$ and $y_2$.

Homogenous: $l(cy) = cl(y)$ for any constant $c$ and any function $y$.

Our second order expression $l$ given in (2.2.1) does satisfy these rules. As in the first order case, to verify that $l$ is additive and homogenous (=linear) is simply a matter of applying the rules "the derivative of the sum is the sum of the derivatives" and "the derivative of a constant times a function is the constant times the derivative." We shall leave the detailed verification for the exercises. All the expressions of classical mathematical physics are linear, and for that reason we focus our attention on that class.

In this section we shall be seeking solutions $y$ of the homogeneous equation $l(y) = 0$ that have a preassigned value $y(x_0)$ and preassigned derivative $y'(x_0)$ at a specified point $x_0$. This is the so-called *initial value problem*. For a first order equation, a solution is uniquely determined by the value $y(x_0)$. In the second order case, two pieces of data, the value $y(x_0)$ and the derivative $y'(x_0)$, serve to determine uniquely a solution $y(x)$.

**Theorem 2.7.** *Suppose $a(x)$, $b(x)$, $c(x)$ are analytic at $x = x_0$ and $a(x_0) \neq 0$. Let $y_0$, $y_1$ be any two fixed numbers. Then the equation*

$$a(x)y'' + b(x)y' + c(x)y = 0 \tag{2.2.2}$$

*has one, and only one, solution $y(x)$ that satisfies*

$$y(x_0) = y_0, \qquad y'(x_0) = y_1$$

*This solution is analytic in the largest open disk, centered at $x_0$, in which both $b(x)/a(x)$ and $c(x)/a(x)$ are analytic.*

*Proof.* The proof is by the method of undetermined coefficients (Section 1.5), which also provides a practical procedure for determining the first few coefficients of the power series expansion of the solution. For simplicity in the proof we shall take $x_0 = 0$. In the general case, just expand in powers of $x - x_0$.

The power series for the coefficients

$$a(x) = \sum_{n=0}^{\infty} a_n x^n, \qquad b(x) = \sum_{n=0}^{\infty} b_n x^n, \qquad c(x) = \sum_{n=0}^{\infty} c_n x^n$$

are assumed known, and $a(0) = a_0 \neq 0$. We assume an expansion $y = \sum_{n=0}^{\infty} y_n x^n$ with $y_0 = y(0)$ and $y_1 = y'(0)$ given. Then

$$y' = \sum_{n=0}^{\infty} n y_n x^{n-1} = \sum_{n=1}^{\infty} n y_n x^{n-1} = \sum_{n=0}^{\infty} (n+1) y_{n+1} x^n$$

and

$$y'' = \sum_{n=0}^{\infty} n(n+1) y_{n+1} x^{n-1} = \sum_{n=1}^{\infty} n(n+1) y_{n+1} x^{n-1} = \sum_{n=0}^{\infty} (n+1)(n+2) y_{n+2} x^n$$

Our equation (2.2.2) becomes

$$\left( \sum_{n=0}^{\infty} a_n x^n \right) \left( \sum_{n=0}^{\infty} (n+1)(n+2) y_{n+2} x^n \right) + \left( \sum_{n=0}^{\infty} b_n x^n \right) \left( \sum_{n=0}^{\infty} (n+1) y_{n+1} x^n \right)$$

$$+ \left( \sum_{n=0}^{\infty} c_n x^n \right) \left( \sum_{n=0}^{\infty} y_n x^n \right) = 0 \tag{2.2.3}$$

Write out the first few terms:

$$(a_0 + a_1x + a_2x^2 + \cdots)(1 \cdot 2y_2 + 2 \cdot 3y_3x + 3 \cdot 4y_4x^2 + \cdots)$$
$$+ (b_0 + b_1x + b_2x^2 + \cdots)(y_1 + 2y_2x + 3y_3x^2 + \cdots)$$
$$+ (c_0 + c_1x + c_2x^2 + \cdots)(y_0 + y_1x + y_2x^2 + \cdots) = 0$$

then collect terms:

$$(2a_0y_2 + b_0y_1 + c_0y_0) + (2 \cdot 3a_0y_3 + 2(a_1 + b_0)y_2 + (b_1 + c_0)y_1 + c_1y_0)x$$
$$+ (3 \cdot 4a_0y_4 + (2 \cdot 3a_1 + 3b_0)y_3 + (2(a_2 + b_1) + c_0)y_2$$
$$+ (b_2 + c_1)y_1 + c_2y_0)x^2 + \cdots = 0$$

If this series is to solve our equation, as we desire, then the coefficients of each power of $x$ must all be zero. Setting the coefficients equal to zero leads to the following system of equations (remember $a_0 \neq 0$):

$$y_2 = -\frac{b_0y_1 + c_0y_0}{2a_0}$$

$$y_3 = -\frac{2(a_1 + b_0)y_2 + (b_1 + c_0)y_1 + c_1y_0}{2 \cdot 3a_0} \qquad (2.2.4)$$

$$y_4 = -\frac{3(2a_1 + b_0)y_3 + (2(a_2 + b_1) + c_0)y_2 + (b_2 + c_1)y_1 + c_2y_0}{3 \cdot 4a_0}$$

$$\vdots$$

Since $y_0$ and $y_1$ are given, the first equation in the system (2.2.4) determines $y_2$. Then, with $y_2$ determined, the second equation determines $y_3$, the next $y_4$, etc. (see Exercises 4, 5, and 6). Thus we have proved our result, at least formally, that $ay'' + by' + cy = 0$ has a solution, and that solution is uniquely determined by the two pieces of data $y(x_0)$ and $y'(x_0)$.

We will not prove here the last assertion in the theorem, namely that with the coefficients determined by equations (2.2.4), the series $\Sigma_{n=0}^{\infty} y_n x^n$ con-verges in the common disk of convergence of the power series expansions of $b(x)/a(x)$ and $c(x)/a(x)$. A proof is given in Appendix 2.A. Also, one might want to consult Problem 7.12, p. 262, in Polya-Latta. Nice discussions are also contained in Chapter VI, §3, of L. Ahlfor's book *Complex Analysis* and in Chapter XII of L. Ince, *Ordinary Differential Equations*. Note that while the theorem gives a smallest disk of convergence of the series $y(x) = \Sigma y_n x^n$, the function $y(x)$ itself may be analytic in a much larger domain.

When you apply the method of undetermined coefficients in practice, work from first principles; don't use equations (2.2.4). For example, suppose we want to find the first few terms of the power series of the solution to

$$(1 - x^2)y'' - 2xy' + 6y = 0$$

that satisfies $y(0) = 1$, $y'(0) = 0$. Let

$$y = 1 + y_2x^2 + y_3x^3 + y_4x^4 + y_5x^5 + \cdots$$

where we have included the data $y_0 = 1$, $y_1 = 0$. Then

$$y' = 2y_2x + 3y_3x^2 + 4y_4x^3 + 5y_5x^4 + \cdots$$

and

$$y'' = 2y_2 + 6y_3x + 12y_4x^2 + 20y_5x^3 + \cdots$$

So

$$
\begin{aligned}
(1 - x^2)y'' - 2xy' + 6y = {}& 2y_2 + 6y_3x + 12y_4x^2 + 20y_5x^3 + \cdots \\
& - 2y_2x^2 - 6y_3x^3 - 12y_4x^4 + \cdots \\
& - 4y_2x^2 - 6y_3x^3 - \cdots \\
& + 6 + 6y_2x^2 + 6y_3x^3 + \cdots \\
= {}& (2y_2 + 6) + (6y_3)x + 12y_4x^2 + \cdots = 0
\end{aligned}
$$

Thus $y_2 = -3$, $y_3 = y_4 = 0$, so $y = 1 - 3x^2$ through terms of degree 4. But if we substitute $y = 1 - 3x^2$ in the original equation, we see that it *is* a solution. As it also satisfies $y(0) = 1$, $y'(0) = 0$ we see further, by uniqueness, that it is *the* solution for these initial conditions, so all coefficients of $x^n$, $n \geqslant 3$, are zero. So the unique solution to the initial value problem, $(1 - x^2)y'' - 2xy' + 6y = 0$, $y(0) = 1$, $y'(0) = 0$, is the simple polynomial $y = 1 - 3x^2$.

Our main result in this section is Theorem 2.7 which asserts the following principle: *The initial value problem for a second order linear homogenous differential equation is uniquely solvable.* While we have used the method of undetermined coefficients to derive this principle, it is the principle itself rather than its method of derivation that will be our main focus for the next few sections. As to computing solutions, it is best to get the solutions in terms of the familiar elementary functions where possible. When this is not possible, we will return to power series computations using the method of undetermined coefficients.

The principle cited above has this important corollary:

**Corollary 2.8.** *Under the same hypotheses as in Theorem 2.7, the only solution to the initial value problem*

$$a(x)y'' + b(x)y' + c(x)y = 0, \qquad y(x_0) = y'(x_0) = 0$$

*is the identically zero solution $y = 0$.*

Obviously, $y = 0$ is a solution to this initial value problem and, by the uniqueness part of Theorem 2.7, is the only solution.

## Exercises

1. Check the entries in the following tables, and fill in the blank entries:

(a)

| $y$ | $l(y) \equiv y'' + y$ |
|---|---|
| $x$ | $x$ |
| $x^n$ | $n(n-1)x^{n-2} + x^n$ |
| $\sin x$ | |
| $\cos x$ | |

(b)

| $y$ | $l(y) \equiv e^x y'' + y' + y$ |
|---|---|
| $x$ | $1 + x$ |
| $x^n$ | $n(n-1)e^x x^{n-2} + nx^{n-1} + x^n$ |
| $e^{-x}$ | |
| $e^x$ | |

2. For each of the following expressions $l$, determine whether $l$ is

Additive: $l(y_1 + y_2) = l(y_1) + l(y_2)$ for any two functions $y_1$ and $y_2$.
Homogeneous: $l(cy) = cl(y)$ for any constant $c$ and any function $y$.

Determine which $l$ are linear.
   (a) $l(y) \equiv x^2 y'' + xy' + (x^2 - n^2)y$
   (b) $l(y) \equiv xyy'' - 2y'$
   (c) $l(y) \equiv y'' - 2y^2$
3. The second order linear expression $l(y) \equiv a(x)y'' + b(x)y' + c(x)y$ satisfies

$$l(1) = x, \qquad l(x) = 0, \qquad l(e^x) = 0$$

Find $a(x)$.
4. Prove the following formula for multiplication of power series:

$$\left( \sum_{n=0}^{\infty} u_n x^n \right)\left( \sum_{n=0}^{\infty} v_n x^n \right) = \sum_{n=0}^{\infty} w_n x^n$$

where $w_n = \sum_{k=0}^{n} u_k v_{n-k} = \sum_{k=0}^{n} u_{n-k} v_k$

5. (Continuation) Using the formulas in Exercise 4, evaluate the products
   that occur in equation (2.2.3) as follows:

$$\left( \sum_{n=0}^{\infty} a_n x^n \right)\left( \sum_{n=0}^{\infty} (n+1)(n+2)y_{n+2} x^n \right) = \sum_{n=0}^{\infty} \alpha_n x^n$$

where $\alpha_n = \sum_{k=0}^{n} a_k(n - k + 1)(n - k + 2)y_{n-k+2}$

$$= \sum_{k=0}^{n} a_{n-k}(k + 1)(k + 2)y_{k+2}; \quad \left(\sum_{n=0}^{\infty} b_n x^n\right)\left(\sum_{n=0}^{\infty} (n + 1)y_{n+1}x^n\right)$$

$$= \sum_{n=0}^{\infty} \beta_n x^n \quad \text{where } \beta_n = \sum_{k=0}^{n} b_k(n - k + 1)y_{n-k+1}$$

$$= \sum_{k=0}^{n} b_{n-k}(k + 1)y_{k+1}; \quad \left(\sum_{n=0}^{\infty} c_n x^n\right)\left(\sum_{n=0}^{\infty} y_n x^n\right) = \sum_{n=0}^{\infty} \gamma_n x^n$$

where $\gamma_n = \sum_{k=0}^{n} c_k y_{n-k} = \sum_{k=0}^{n} c_{n-k} y_k$.

6. (Continuation) Show that, in the terminology of Exercise 5, equation (2.2.3) in the text can be written $\alpha_n + \beta_n + \gamma_n = 0, n = 0, 1, 2, \ldots$. Prove in turn that this system of equations leads to the system

$$y_{n+2} = -\frac{\sum_{l=2}^{n+1} (l(l-1)a_{n-l+2} + b_{n-l+1} + c_{n-l})y_l + (b_n + c_{n-1})y_1 + c_n y_0}{(n+1)(n+2)a_0}$$

for $n = 1, 2, \ldots$, $y_2 = -(b_0 y_1 + c_0 y_0)/2a_0$. Thus concludes rigorously that once $y_0, y_1$ are given, all subsequent $y_n$ are uniquely determined.
7. Find the unique polynomial solution to $y'' - xy' + 2y = 0$ that satisfies $y(0) = -1, y'(0) = 0$.
8. Use the method of undetermined coefficients to find the unique power series solution to $y'' - y = 0$ that satisfies $y(0) = y'(0) = 1$.
9. Given the second order linear differential expression

$$l(y) \equiv a(x)y'' + b(x)y' + c(x)y$$

let $y_1$ and $y_2$ be respectively solutions of the initial value problem $l(y) = 0$ and

$$y_1(x_0) = 1, \qquad y_1'(x_0) = 0$$
$$y_2(x_0) = 0, \qquad y_2'(x_0) = 1$$

which exist by Theorem 2.7.
(a) Show that $y_1$ and $y_2$ are linearly independent.
(b) If $y$ is a solution to the initial value problem $l(y) = 0, y(x_0) = a_1,$ $y'(x_0) = a_2$, show that the function

$$w(x) = y(x) - a_1 y_1(x) - a_2 y_2(x)$$

solves $l(w) = 0, w(x_0) = w'(x_0) = 0$. Conclude $w(x) = 0$.

The two mathematicians *Charles-François Sturm* (1803–1855) and *Joseph Liouville* (1809–1882) are generally credited with initiating the study that has subsequently developed into the theory of linear second order differential operators in $L^2$ spaces expounded in this book. It is accordingly often referred to as *Sturm-Liouville theory*. In a major paper published in 1836, Sturm found it convenient to convert the differential expression $a(x)y'' + b(x)y' + c(x)y$ to one of the form $(1/r(x))((p(x)y')' + q(x)y)$. If $p$, $q$, and $r$ are given, then obviously $a = p/r$, $b = p'/r$, and $c = q/r$.

10. Suppose $a(x)$, $b(x)$, and $c(x)$ are given functions. Determine functions $p(x)$, $q(x)$, and $r(x)$ so that

$$\frac{1}{r(x)}[(p(x)y')' + q(x)y] = a(x)y'' + b(x)y' + c(x)y$$

11. (Continuation) We shall refer to the expression on the left in Exercise 10 as the "Sturmian form" of the differential expression (also sometimes called "self-adjoint form"). Put each of the following in Sturmian form:
    (a) $(1 - x^2)y'' - 2xy'$
    (b) $y'' - xy'$
    (c) $xy'' + (1 - x)y'$
    (d) $x^2y'' + xy' + x^2y$

12. (Continuation) Show that, under the change of the independent variable,

$$z = \int_a^x \left(\frac{r(t)}{p(t)}\right)^{1/2} dt \qquad \text{(the lower limit } a \text{ is a constant)}$$

we have

$$\frac{1}{r(x)}\left(\frac{d}{dx}\left(p(x)\frac{dy}{dx}\right) + q(x)y\right) = \frac{d^2y}{dz^2} + \left(\frac{1}{2s}\frac{ds}{dz}\right)\frac{dy}{dz} + \frac{q}{r}y$$

where $s = rp$.

13. (Continuation) Starting with the expression on the right in Exercise 12 and making the substitution

$$y(z) = s^{-1/4}Y(z)$$

show that

$$\frac{1}{r}\left(\frac{d}{dx}\left(p\frac{dy}{dx}\right) + qy\right) = s^{-1/4}\left(\frac{d^2Y}{dz^2} - A(z)Y\right)$$

where

$$A(z) = \frac{1}{4s}\frac{d^2s}{dz^2} - 3\left(\frac{1}{4s}\frac{ds}{dz}\right)^2 - \frac{q}{r}$$

14. (Continuation) The combined change of variables

$$z = \int_a^x \left(\frac{r}{p}\right)^{1/2} dt, \qquad y = (rp)^{-1/4}Y$$

is called the *Liouville transformation*. Using the results of Exercises 12 and 13, show that under the Liouville transformation, the "eigenvalue problem"

$$\frac{1}{r}((py')' + qy) = \lambda y \qquad (\lambda \text{ constant})$$

becomes

$$Y'' - AY = \lambda Y$$

## 2.3 DIMENSION OF THE KERNEL; GENERAL SOLUTION; ABEL'S FORMULA

In Section 2.2 we proved this very important theorem:

**Theorem 2.9.** *Suppose $a(x)$, $b(x)$, $c(x)$ are analytic at $x = x_0$, and $a(x_0) \neq 0$. Let $y_0$, $y_1$ be any two fixed numbers. Then the equation*

$$a(x)y'' + b(x)y' + c(x)y = 0 \qquad (2.3.1)$$

*has one, and only one, solution $y(x)$ that satisfies*

$$y(x_0) = y_0, \qquad y'(x_0) = y_1 \qquad (2.3.2)$$

*This solution is analytic in the largest open disk, centered at $x_0$, in which both $b(x)/a(x)$ and $c(x)/a(x)$ are analytic.*

Recall that "analytic" means "expandable in a convergent power series." Virtually all the functions we have met in basic calculus, and most functions we will meet in this course, are analytic, except possibly at isolated points. Our proof of Theorem 2.9 showed formally that the power series coefficients of $y$ are uniquely determined by the differential equation (2.3.1) and the initial conditions (2.3.2). The proof of analyticity is given in Appendix 2.A. Our main purpose now is to draw some important practical conclusions from Theorem 2.9.

**Corollary 2.10.** *The kernel of a second order linear differential expression $l(y) \equiv a(x)y'' + b(x)y' + c(x)y$ is two-dimensional.*

Remember that the kernel of $l$ consists of all functions $y$ that satisfy $l(y) = 0$. Our assertion that the kernel is two-dimensional means this: there exist two linearly independent functions, say $f_1$ and $f_2$, in the kernel; and, given any such linearly independent pair in the kernel, every function $y$ in the kernel can be expressed $y = c_1 f_1 + c_2 f_2$ for suitable constants $c_1$ and $c_2$.

We are working at a point $x_0$ where $a(x)$, $b(x)$, $c(x)$ are analytic, and $a(x_0) \neq 0$. Our theorem tells us that the initial value problem $l(y) = 0$, $y(x_0) = 1$, $y'(x_0) = 0$ has a unique solution, say $f_1(x)$. Likewise, there is another unique function $f_2$ that satisfies $l(f_2) = 0$, $f_2(x_0) = 0$, $f_2'(x_0) = 1$. These functions $f_1$ and $f_2$ are in fact a basis for ker($l$).

First, their linear independence: suppose we had a relation $c_1 f_1(x) + c_2 f_2(x) = 0$ valid in some disk centered at $x_0$. Substitute $x = x_0$ to get $c_1 f_1(x_0) + c_2 f_2(x_0) = c_1 \cdot 1 + c_2 \cdot 0 = c_1 = 0$. As $c_1 = 0$, our relation becomes $c_2 f_2(x) = 0$, which becomes, on differentiation, $c_2 f_2'(x) = 0$. Put $x = x_0$ to get $c_2 = 0$. Thus the only relation $c_1 f_1(x) + c_2 f_2(x) = 0$ that can hold is the one with $c_1 = c_2 = 0$. By definition, $f_1$ and $f_2$ are linearly independent.

Second, we must show that $f_1$ and $f_2$ span ker($l$), which is to say, given $y \in \mathrm{ker}(l)$, we have $y = c_1 f_1 + c_2 f_2$ for suitable constants $c_1$ and $c_2$. Given $y \in \mathrm{ker}(l)$, let $c_1 = y(x_0)$, $c_2 = y'(x_0)$, and consider $w(x) = y(x) - c_1 f_1(x) - c_2 f_2(x)$. Then $w \in \mathrm{ker}(l)$ as ker($l$) is a subspace. Also $w(x_0) = y(x_0) - c_1 f_1(x_0) - c_2 f_2(x_0) = c_1 - c_1 \cdot 1 - c_2 \cdot 0 = 0$, and $w'(x_0) = y'(x_0) - c_1 f_1'(x_0) - c_2 f_2'(x_0) = c_2 - c_1 \cdot 0 - c_2 \cdot 1 = 0$. By Corollary 2.8 in the preceding section, $w = 0$, so $y = c_1 f_1 + c_2 f_2$. We have proved our assertion that $f_1, f_2$ constitute a basis for ker($l$).

Knowledge of ker($l$) will allow us to write down the "general" solution to the inhomogenous equation $l(y) = h$, just as in the first order case.

**Corollary 2.11.** *Let $l(y) \equiv a(x)y'' + b(x)y' + c(x)y$ be a second order linear differential expression, and let $f_1, f_2$ be a basis for ker($l$). Let $h(x)$ be a given function, and let $y_p$ be some solution to $l(y) = h$. Then every solution to $l(y) = h$ has the form $y = y_p + c_1 f_1 + c_2 f_2$ for suitable constants $c_1$ and $c_2$.*

The proof of this corollary is essentially the same as the proof of the corresponding result for first order expressions (Theorem 1.1 of Section 1.2): suppose $y$ solves $l(y) = h$. Then $y - y_p$ solves $l(y) = 0$ because $l(y - y_p) = l(y) - l(y_p) = h - h = 0$. This is another way of saying that $y - y_p$ belongs to ker($l$). But ker($l$) is spanned by $f_1$ and $f_2$. Hence $y - y_p = c_1 f_1 + c_2 f_2$ for suitable constants $c_1$ and $c_2$, which is what we wished to show.

Clearly, then, knowledge of ker($l$) is essential if we wish to solve the differential equation $l(y) = h$. For first order expressions, where the kernel is one-dimensional, we could simply write down a closed formula for a function that spanned the kernel (Section 1.2). For second order expressions, things aren't so simple, but we get some leverage on the problem via a remarkable result known as Abel's formula.

**Theorem 2.12.** *Suppose $f_1$ and $f_2$ are two functions in the kernel of $l(y) \equiv a(x)y'' + b(x)y' + c(x)y$. Let $W(f_1, f_2) = f_1 f_2' - f_1' f_2$ be the Wronskian* *of $f_1$ and $f_2$. Then*

$$W = c \exp\left(-\int_a^b \frac{b}{a} dx\right) \qquad (Abel's\ formula) \qquad (2.3.3)$$

*for a suitable constant c.*

Before deriving this formula, let us make some deductions from it. *Three deductions from Abel's formula:*

1. *If $f_1$ and $f_2$ belong to ker($l$), then either $W(f_1, f_2)$ is nowhere zero (linear independence), or $W$ is identically zero (linear dependence).*

Thus, when $f_1$ and $f_2$ both belong to ker($l$), their Wronskian offers a sharper test for linear dependence and independence than in the general case (Theorem 2.5 of Section 2.1). This deduction follows immediately from Abel's formula (2.3.3) because the exponential in (2.3.3) is never zero, so there are just two choices, $c \neq 0$ or $c = 0$.

2. *$W(f_1, f_2)$ is the same, up to a nonzero multiplicative constant, for any two functions $f_1, f_2$ that span ker($l$). We can therefore refer to $W$ as the Wronskian of $l$ ($W$ determined up to a nonzero multiplicative constant).*
3. *If we know one nonzero function $f_1$ in ker($l$), then we can find a second linearly independent one by solving the first order linear equation*

$$f_1 y' - f_1' y = W \qquad (2.3.4)$$

*where $W$ is the Wronskian of $l$.*

Deduction 3 is particularly important in applications. For example, a simple calculation shows that $e^{-x} = f_1(x)$ solves $l(y) \equiv y'' + 2y' + y = 0$. The Wronskian of $l$ is $W = c\exp(-\int(2/1)dx) = ce^{-2x}$. Put $c = 1$ for convenience (any nonzero $c$ will do), and solve $e^{-x}y' + e^{-x}y = e^{-2x}$. Using tricks we learned in Chapter 1, we multiply through this equation by $e^{2x}$ to get $e^x y' + e^x y = 1$. The left side is $(e^x y)'$, so our equation is $(e^x y)' = 1$, which has a solution $e^x y = x$ or $y = xe^{-x}$. Thus $f_2(x) = xe^{-x}$ is a second linearly independent function in ker($l$).

**Proof of Abel's formula (Theorem 2.12).** If we differentiate the equation $W = f_1 f_2' - f_1' f_2$ and use the product rule, we get $W' = (f_1 f_2'' + f_1' f_2') - (f_1' f_2' + f_1'' f_2) = f_1 f_2'' - f_1'' f_2$. Now by assumption we have

$$af_1'' + bf_1' + cf_1 = 0$$

and

$$af_2'' + bf_2' + cf_2 = 0$$

Multiply the first equation by $-f_2$, the second by $f_1$, and add:

$$a(f_1 f_2'' - f_1'' f_2) + b(f_1 f_2' - f_1' f_2) = 0$$

or $aW' + bW = 0$; which has the solution $W = c \exp(-\int (b/a)\, dx)$.

We note finally that in the application of deduction 3, where we determine a second linearly independent solution in the kernel by solving

$$f_1 y' - f_1' y = W \qquad\qquad (2.3.4)$$

where $f_1$ is a known function in the kernel and $W$ is the Wronskian, there is a nice trick that enables us to solve (2.3.4) explicitly. By the quotient rule

$$\left(\frac{y}{f_1}\right)' = \frac{f_1 y' - y f_1'}{f_1^2}$$

so if we divide (2.3.4) by $f_1^2$ we get $(y/f_1)' = W/f_1^2$. Thus

$$y = f_1(x) \int \frac{W(x)}{f_1(x)^2}\, dx$$

is a second linearly independent solution.

**Exercises**

1. By substituting $y = e^{rx}$ and solving for the constant $r$, find two functions that span the kernel of $l(y) \equiv y'' - 5y' + 6y$. By computing the Wronskian two ways, check Abel's formula.
2. Repeat Exercise 1 for $l(y) \equiv y'' - 3y' + 2y$.
3. By substituting $y = e^{rx}$ and solving for the constant $r$, find one function in the kernel of $l(y) \equiv y'' - 2y' + y$. Using the third deduction from Abel's formula, find another function in $\ker(l)$.
4. Let $l(y) \equiv (1 - x^2)y'' - 2xy' + 2y$.
   (a) Find the Wronskian of this expression (up to a nonzero multiplicative constant). Assume $-1 < x < 1$.
   (b) Noting that $y = x$ is a solution of $l(y) = 0$, find a first order equation which, if solved, would yield a second linearly independent solution.
5. (a) Show that the Wronskian of a differential expression in Sturmian form, $l(y) \equiv (1/r(x))((p(x)y')' + q(x)y)$, is $W = (\text{const})/p(x)$.
   (b) Using your solution to Exercise 11 of Section 2.2, find the Wronskian of the four differential expressions listed there.
6. By substituting $y = x^\alpha$ and solving for the constant $\alpha$, find two functions that span the kernel of $l(y) \equiv x^2 y'' + xy' - y$.
7. Using the method of Exercise 6, find two real functions that span the kernel of $l(y) \equiv x^2 y'' + xy' + y$. [Ans.: $f_1(x) = \cos(\ln x)$, $f_2(x) = \sin(\ln x)$.]

8. Let $g_1 = f_1 - f_2$, $g_2 = f_1 + f_2$. Prove that if $f_1$ and $f_2$ form a basis for the kernel of a second order linear differential expression $l$, then so do $g_1$ and $g_2$.

*Formal calculus of differential expressions.* Let $D$ be the operation of differentiation with respect to $x$, $Dy = dy/dx$, and let $M_f$ be the operation of multiplication by the function $f$, $M_f(y) \equiv fy$. Both $D$ and $M_f$ are linear. Note that $D$ and $M_f$ do not commute in general, because

$$DM_f(y) = D(fy) = fDy + (Df)y$$

while $M_f D(y) = fDy$. Thus $M_f D \neq DM_f$. The general first order expression $l(y) \equiv ay' + by$ can be written $l = M_a D + M_b$. The general second order linear expression $l(y) \equiv ay'' + by' + cy$ can be written $l = M_a D^2 + M_b D + M_c$. If we can factor a second order $l$ as the composition of two first orders, $l = l_1 l_2$, then any function in the kernel of $l_2$ will also be in $\ker(l)$, because if $l_2(y) = 0$, then $l(y) = l_1(l_2(y)) = l_1(0) = 0$. So if we can factor a second order expression, then we can find one function in its kernel.

9. Show that $DM_f = M_f D + M_{f'}$.
10. (Continuation) Let $p(x)$, $q(x)$ be two functions of $x$. Show that
    (a) $(M_p D + M_q)(D + M_x) = M_p D^2 + M_{xp+q} D + M_{p+xq}$.
    (b) There is a function $y = f(x)$ that lies in the kernel of the expression
    $l(y) \equiv py'' + (xp + q)y' + (p + xq)y$ whatever the choice of the functions $p = p(x)$, $q = q(x)$. Find $f$.
11. (Continuation) (a) Show that $l_1 = D + M_{x+1}$ commutes with $l_2 = D + M_x$; i.e., show that $l_1 l_2 = l_2 l_1$.
    (b) Find a basis for the kernel of
    $$\frac{d^2 y}{dx^2} + (2x + 1)\frac{dy}{dx} + (x^2 + x + 1)y$$

## 2.4 KERNEL OF CONSTANT-COEFFICIENT EXPRESSIONS

In Section 2.3 we established Abel's formula: if $f_1$ and $f_2$ belong to the kernel of $l(y) \equiv a(x)y'' + b(x)y' + c(x)y$, then $W(f_1, f_2) = c\exp(-\int (b/a)\,dx)$. We also drew three conclusions from this formula: (1) either $W(f_1, f_2)$ is nowhere zero ($f_1$ and $f_2$ linearly independent) or is identically zero ($f_1$ and $f_2$ linearly dependent); (2) $W(f_1, f_2)$ is the same (up to a nonzero multiplicative constant) for any two functions $f_1, f_2$ that span $\ker(l)$; and (3) if we know one nonzero function $f_1$ in $\ker(l)$ then we can find a second, $f_2$, by solving the first order equation $f_1 y' - f_1' y = W$ in the unknown $y$ ($y = f_2$). We shall apply these principles to describe explicitly the kernel of constant-coefficient expressions, that is, expressions $l(y) \equiv a(x)y'' + b(x)y' + c(x)y$ for which $a(x)$, $b(x)$, and $c(x)$ are (real) constants. Of course, $a \neq 0$.

We seek a solution of the form $y = e^{rx}$, $r$ a constant. Substituting, we get

$$l(e^{rx}) = (ar^2 + br + c)e^{rx}$$

The exponential is never zero, so $l(e^{rx}) = 0$ implies $ar^2 + br + c = 0$, which yields two solutions for the constant $r$ (in general), thus two functions in the kernel

$$f_1(x) = e^{r_1 x}, \qquad f_2(x) = e^{r_2 x}$$

where $r_1$ and $r_2$ are the two roots of $ar^2 + br + c = 0$. The quadratic formula tells us

$$r_1 = \frac{-b + \sqrt{b^2 - 4ac}}{2a}, \qquad r_2 = \frac{-b - \sqrt{b^2 - 4ac}}{2a}$$

We have $r_1 - r_2 = (\sqrt{b^2 - 4ac})/a$, so $r_1 \neq r_2$ except when $b^2 = 4ac$. If $r_1 \neq r_2$, we have two linearly independent solutions (Exercise 7, Section 2.1). We shall discuss the various cases individually:

$b^2 = 4ac$.  In this case our substitution gives us only one solution, $f_1(x) = e^{-(b/2a)x}$. To find a second solution, we use the third deduction from Abel's formula. We have $W = \text{const} \times \exp(-\int (b/a)\,dx) = \exp(-(b/a)x)$, where we have put $\text{const} = 1$ for convenience. Then solve $f_1 y' - f_1' y = W$, or

$$e^{-(b/2a)x}y' + (b/2a)e^{-(b/2a)x}y = e^{-(b/a)x}$$

Multiply through by $e^{(b/a)x}$ to get

$$e^{(b/2a)x}y' + (b/2a)e^{(b/2a)x}y = 1$$

Then recognize the left side as $(e^{(b/2a)x}y)'$, so our equation is $(e^{(b/2a)x}y)' = 1$, which has the solution $e^{(b/2a)x}y = x$, or $y = xe^{-(b/2a)x}$. So when $b^2 = 4ac$,

$$f_1(x) = e^{-(b/2a)x}, \qquad f_2(x) = xe^{-(b/2a)x}$$

form a basis for $\ker(l)$.

$b^2 - 4ac > 0$.  In this case $r_1 = (-b + \sqrt{b^2 - 4ac})/2a$ and $r_2 = (-b - \sqrt{b^2 - 4ac})/2a$ are distinct real numbers, so $f_1(x) = \exp(r_1 x)$, $f_2(x) = \exp(r_2 x)$ are real functions that span $\ker(l)$.

$b^2 - 4ac < 0$.  In this case our roots $r_1, r_2$ are distinct but complex. If we wish to deal in complex solutions, then we may take $f_1(x) = \exp(r_1 x)$, $f_2(x) = \exp(r_2 x)$ as spanning $\ker(l)$. But generally we will prefer real-valued solutions, and we may argue as follows. We have $r_1 = (-b + \sqrt{4ac - b^2}\,i)/2a$, $r_2 = (-b - \sqrt{4ac - b^2}\,i)/2a$ where $i = \sqrt{-1}$ and $\sqrt{4ac - b^2}$ is real. Let $s = b/2a$, $t = \sqrt{4ac - b^2}/2a$, both real, so $r_{1,2} = -s \pm it$. Then

$$e^{r_1 x} = e^{(-s + it)x} = e^{-sx}e^{itx} = e^{-sx}(\cos tx + i \sin tx)$$

and $\exp(r_2 x) = e^{-sx}(\cos tx - i \sin tx)$. In place of $\exp(r_1 x)$, $\exp(r_2 x)$ we now use

$$\frac{1}{2}(e^{r_1 x} + e^{r_2 x}) = e^{-(b/2a)x} \cos\left(\frac{\sqrt{4ac - b^2}}{2a}\right) x$$

$$\frac{1}{2i}(e^{r_1 x} - e^{r_2 x}) = e^{-(b/2a)x} \sin\left(\frac{\sqrt{4ac - b^2}}{2a}\right) x$$

These are two real functions that span ker($l$). Some examples follow.

*Find the kernel of* $l(y) \equiv y'' + y' - 2y$

Substituting $y = e^{rx}$ we get the equation $r^2 + r - 2 = 0$, or $(r+2)(r-1) = 0$, which has the roots $r_1 = -2$, $r_2 = 1$. Thus $e^{-2x}$, $e^x$ span ker($l$). Check:

$$l(e^{-2x}) = (e^{-2x})'' + (e^{-2x})' - 2e^{-2x} = 4e^{-2x} - 2e^{-2x} - 2e^{-2x} = 0$$

Likewise, $l(e^x) = 0$.

*Find the general solution of* $(d^2 y/dx^2) - 2\, dy/dx + y = 0$

Substitute $y = e^{rx}$ to get $r^2 - 2r + 1 = (r-1)^2 = 0$, which has the single root $r = 1$. Hence $e^x$ and $xe^x$ span the kernel, so the general solution is $y = c_1 e^x + c_2 xe^x$, $c_1$ and $c_2$ constants. Check (for $xe^x$):

$$\frac{d^2}{dx^2}(xe^x) - 2\frac{d}{dx}(xe^x) + xe^x = \frac{d}{dx}(e^x + xe^x) - 2\frac{d}{dx}(xe^x) + xe^x$$

$$= e^x - \frac{d}{dx}(xe^x) + xe^x$$

$$= e^x - (e^x + xe^x) + xe^x = 0$$

*Solve the initial value problem*

$$\frac{d^2 y}{dx^2} - 2\frac{dy}{dx} + 10y = 0, \qquad y(0) = 2,\ y'(0) = 5$$

Constant coefficients, so put $y = e^{rx}$ and get $r^2 - 2r + 10 = 0$. The roots are $\frac{1}{2}(+2 \pm \sqrt{4-40}) = 1 \pm 3i$. So the complex functions that span the kernel are

$$e^{(1+3i)x} = e^x e^{i(3x)} = e^x(\cos 3x + i \sin 3x)$$

$$e^{(1-3i)x} = e^x e^{i(-3x)} = e^x(\cos 3x - i \sin 3x)$$

Thus the general solution is $y = (c_1 \cos 3x + c_2 \sin 3x)e^x$. Then $y(0) = c_1$, so $c_1 = 2$. And

$$y'(x) = (-3c_1 \sin 3x + 3c_2 \cos 3x)e^x + (c_1 \cos 3x + c_2 \sin 3x)e^x$$

so $y'(0) = 3c_2 + c_1 = 5$, thus $c_2 = \frac{1}{3}(5 - 2) = 1$. The (necessarily unique) solution to the initial value problem is

$$y = (2 \cos 3x + \sin 3x)e^x$$

Note that in the examples we have derived the solution from first principles, rather than relying on the formulas already derived. When you do the exercises, follow the same pattern as in the examples so that you may reinforce your understanding of the method. Don't rely blindly on the formulas.

### Exercises

For each expression, find two linearly independent functions that span the kernel:

1. $l(y) \equiv y'' - y$
2. $l(y) \equiv y'' - cy, \quad c > 0$ (constant)
3. $l(y) \equiv y'' - 4y' + 3y$
4. $l(y) \equiv y'' - 4y' + 4y$
5. $l(y) \equiv y'' + y$
6. $l(y) \equiv y'' + cy, \quad c > 0$ (constant)
7. $l(y) \equiv y'' + 2y' + 2y$
8. $l(y) \equiv y'' - 2y' + 3y$
9. $l(y) \equiv y'' + y' + 5y$
10. $l(y) \equiv y'' + 4.0909657y' + 6.4340001y$
11. $l(y) \equiv y'' - iy$
    [Ans.: $\cosh(x/\sqrt{2}) \cos(x/\sqrt{2})$, $\sinh(x/\sqrt{2}) \sin(x/\sqrt{2})$]
    Solve the initial value problem:
12. $y'' - y = 0$, $y(0) = 0$, $y'(0) = 1$   (Ans.: $\sinh x$)
13. $y'' + \pi^2 y = 0$, $y(1) = -1$, $y'(1) = \pi$
14. Let $A$ and $B$ be two real constants, not both zero, and let $\alpha$ be a real nonzero number. Determine $R > 0$ and $\delta$, $-\pi < \delta \leqslant \pi$ so that

    $$A \cos \alpha x + B \sin \alpha x = R \cos(\alpha x - \delta)$$

    holds for all $x$. [*Ans.*: $R = +\sqrt{A^2 + B^2}$, $\sin \delta = B/R$, $\cos \delta = A/R$; $R$ is called the *magnitude* and $\delta$ the *phase angle*.]
15. Using the result of Exercise 14, show that the solution to Exercise 13 can be put in the form $y = \sqrt{2} \cos(\pi x + \pi/4)$.

In the following exercises, express the answer in two ways, the customary $A \cos \alpha x + B \sin \alpha x$, and in magnitude-phase angle form using Exercise 14.

16. $y'' + 2y' + 2y = 0$, $y(0) = 1$, $y'(0) = 0$
    [Ans.: $e^{-x}(\cos x + \sin x) = \sqrt{2} e^{-x} \cos(x - \pi/4)$]
17. $y'' - 2y' + 3y = 0$, $y(\pi/4\sqrt{2}) = y'(\pi/4\sqrt{2}) = (1/\sqrt{2})e^{\pi/4\sqrt{2}}$

## 2.5 THE CLASSICAL LINEAR OSCILLATOR

The theory of second order linear constant-coefficient equations enables us to solve the following classical mechanical problem: we consider "a particle of mass $m$ which, moving in a straight line, is attracted toward a fixed point in it, with a force proportional to its displacement $q$ from this point."

The setup is depicted in Figure 2.1. We shall use $x$ instead of $q$ to denote the distance of the particle of mass $m$ from the fixed point. We take $x$ positive to the right. The distance $x$ varies with time $t$. The velocity $v$ of the particle is $dx/dt$, and its acceleration is $a = d^2x/dt^2$. Newton's law tells us that force = mass × acceleration. According to our assumption, our particle is attracted toward the fixed point (labeled 0 in Figure 2.1) with a force proportional to $x$. Let us call the proportionality constant $k$ so that $F = -kx$, the minus sign because the force is in the negative $x$ direction when $x$ is positive. Applying Newton's law, $F = ma$, we get

$$-kx = m\frac{d^2x}{dt^2}$$

which leads to the second order constant-coefficient equation

$$\frac{d^2x}{dt^2} + \frac{k}{m}x = 0$$

As $k/m > 0$, we may set $\omega_0 = (k/m)^{1/2}$ (positive square root), so that our equation reads

$$\frac{d^2x}{dt^2} + \omega_0^2 x = 0 \tag{2.5.1}$$

which is the *equation of motion* of our mass particle.

Apply our standard method for solving constant-coefficient equations: substitute $x = e^{rt}$ and get the equation $r^2 + \omega_0^2 = 0$, which has the solutions $r = \pm\omega_0 i$ where $i = \sqrt{-1}$. Thus get two linearly independent solutions $x_1 = \exp(i\omega_0 t)$, $x_2 = \exp(-i\omega_0 t)$, and then take linear combinations $\cos\omega_0 t = (1/2)(x_1 + x_2)$, $\sin\omega_0 t = (1/2i)(x_1 - x_2)$ to get two real solutions

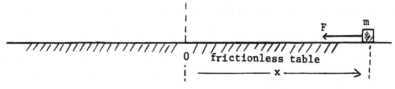

**Figure 2.1**  Setup for classical mechanical problem.

$\cos \omega_0 t$, $\sin \omega_0 t$. The general solution can then be put in magnitude-phase angle form (Exercise 14 of Section 2.4).

$$x = c_1 \cos \omega_0 t + c_2 \sin \omega_0 t = R \cos(\omega_0 t - \delta) \tag{2.5.2}$$

What units does $\omega_0$ have, say in the cgs system? From equation (2.5.1) we see that the product $\omega_0^2 x$ has the units of acceleration, which are cm/sec². Hence $\omega_0^2$ has units (sec)$^{-2}$, so $\omega_0$ has units (sec)$^{-1}$. Thus it is a kind of frequency. We call it the *angular frequency*.

Assume we start our mass particle in motion at time $t = 0$ from an initial position $x_0 = x(0)$ with initial velocity $v_0 = x'(0)$. So $x_0$ (cm) and $v_0$ (cm/sec) are prescribed in advance and may have any values we choose to give them. The constants $R$ and $\delta$ in the magnitude-phase angle form of our solution (2.5.2) are uniquely determined by $x_0$ and $v_0$ as follows. First, note that the velocity $v = x'(t)$ at any time $t$ is given by $v = x'(t) = -\omega_0 R \sin(\omega_0 t - \delta)$. Then

$$x_0 = x(0) = R \cos(-\delta) = R \cos \delta$$

$$v_0 = x'(0) = -\omega_0 R \sin(-\delta) = \omega_0 R \sin \delta$$

Hence

$$\left(\frac{x_0}{R}\right)^2 + \left(\frac{v_0}{\omega_0 R}\right)^2 = \cos^2 \delta + \sin^2 \delta = 1$$

which yields $R^2 = x_0^2 + (v_0/\omega_0)^2$, or $R = \sqrt{x_0^2 + (v_0/\omega_0)^2}$ (positive square root). Thus $R$ is determined. Then

$$\cos \delta = \frac{x_0}{R} = \frac{x_0}{\sqrt{x_0^2 + (v_0/\omega_0)^2}}, \qquad \sin \delta = \frac{v_0}{\omega_0 \sqrt{x_0^2 + (v_0/\omega_0)^2}}$$

determine the *phase angle* $\delta$. Then the complete solution to our mechanical problem is

$$x(t) = \sqrt{x_0^2 + (v_0/\omega_0)^2} \cos(\omega_0 t - \delta) \tag{2.5.3}$$

which specifies the position $x(t)$ of our particle at any time $t$ after $t = 0$.

The function $\cos(\omega_0 t - \delta)$ oscillates between $+1$ and $-1$. It will be $+1$ whenever $\omega_0 t - \delta$ is an integral multiple of $2\pi$, $\omega_0 t - \delta = 2n\pi$, or $t = (\delta + 2n\pi)/\omega_0$. If $\delta > 0$ the first maximum occurs at $t = \delta/\omega_0$.

The maximum distance of our mass particle from our fixed point 0 occurs when $|\cos(\omega_0 t - \delta)| = 1$. This maximum distance, $x_{max}$, is

$$x_{max} = \sqrt{x_0^2 + (v_0/\omega_0)^2} \tag{2.5.4}$$

The mass particle passes through the fixed point when $x(t) = 0$. This occurs when $\cos(\omega_0 t - \delta) = 0$, or $\omega_0 t - \delta = $ odd multiple of $\pi/2$, or $t = [(2n + 1)(\pi/2) + \delta]/\omega_0$.

The motion is periodic, repeating each time the quantity $\omega_0 t - \delta$ increases by $2\pi$ or when, for $t_1 < t_2$,

$$(\omega_0 t_2 - \delta) - (\omega_0 t_1 - \delta) = \omega_0(t_2 - t_1) = 2\pi$$

Thus the *period* $T_0 = t_2 - t_1$ of the motion is

$$T_0 = \frac{2\pi}{\omega_0} \quad \text{(sec)}$$

The *natural frequency* of the system, $v_0 = \omega_0/2\pi$, measures the number of oscillations per second; $v_0 = \sqrt{k/m}/2\pi$.

The work we do on the system when we move the particle initially to $x = x_0$ is given by $\int F\,dx$. (For a constant force, work is force times distance; for a nonconstant force, work $= \int (\text{force})\,dx$.) Thus, the work we do on the system, which equals the initial potential energy stored in the system, is given by

$$\int_0^{x_0} kx\,dx = \tfrac{1}{2}kx_0^2$$

which is taken positive. The initial kinetic energy is $\tfrac{1}{2}mv_0^2$, so the initial *total energy* is

$$E = \tfrac{1}{2}kx_0^2 + \tfrac{1}{2}mv_0^2 \tag{2.5.5}$$

Our system is assumed frictionless, so this energy must remain constant. That is, we must have

$$E = \tfrac{1}{2}k(x(t))^2 + \tfrac{1}{2}m(x'(t))^2 \qquad \text{independent of } t$$

and a simple calculation shows that is the case. From equations (2.5.4) and (2.5.5) we derive

$$x_{\max} = \frac{1}{\sqrt{2}\,\pi v_0} \sqrt{\frac{E}{m}} \tag{2.5.6}$$

relating the maximum range of the particle to three independent variables describing the system, its natural frequency $v_0$, the mass $m$ of the particle, and the total energy $E$ of the system. Note that the total energy $E$ can be assigned any positive value whatever (or zero).

Now if the natural frequency $v_0$ of our system is very large, say 1 million cycles per second, then the function $x(t)$ that describes the continuous motion would be of little practical value. A more practical quantity might be the probability that the particle, if observed by a strobe light, lies in the range $x$,

$x + dx$. Another way to put this is to ask what fraction of total time is spent in the range $x$, $x + dx$.

To find this quantity, we may compute the fraction of time the particle spends in the range $[0, x]$ and then differentiate. We have

$$\frac{x}{x_{max}} = \cos 2\pi v_0 t$$

where we have set $\delta = 0$ for convenience, as this clearly will not affect the probability. Let $u = x/x_{max}$, so we are working in the range $-1 \leqslant u \leqslant 1$. Then $u = \cos 2\pi v_0 t$, and our question becomes: What fraction of the time does the particle spend in the range $[0, u]$? In Figure 2.2 we show the graph of $u = \cos 2\pi v_0 t$ versus $2\pi v_0 t$ over one full cycle. The two shaded portions of the $2\pi v_0 t$ axis represent the time the particle spends in the range $[0, u]$. As these two shaded portions are exactly the same length, we may compute the length of one and multiply by 2. The total scaled time (the quantity $2\pi v_0 t$) over one cycle is $2\pi$, so the fraction we want is

Fraction of time spent by the particle in the range $[0, u]$ $\begin{cases} = \dfrac{2(\pi/2 - \text{arc} \cos u)}{2\pi} \\[2mm] = \dfrac{1}{2} - \dfrac{1}{\pi}\text{arc} \cos u \end{cases}$

The differential probability we seek is the derivative of this quantity. Remembering the formula

$$\frac{d}{du} \text{arc} \cos u = \frac{-1}{\sqrt{1 - u^2}}$$

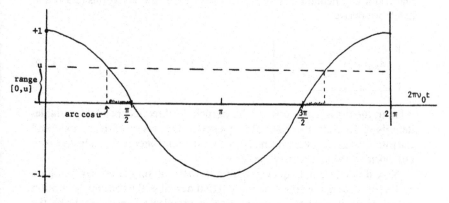

**Figure 2.2**   Graph of $u = \cos 2\pi v_0 t$ versus $2\pi v_0 t$. The shaded portion of the $2\pi v_0 t$ axis represents the time the particle spends in the range $[0, u]$.

we get

$$p(u) = \begin{cases} \dfrac{1}{\pi\sqrt{1 - u^2}}, & |u| < 1 \\ 0, & |u| > 1 \end{cases} \qquad u = x/x_{max} \qquad (2.5.7)$$

as the probability that the mass particle of our linear oscillator lies in the range $u, u + du$. In formula (2.5.7) we have included the information that the particle always lies in the range $|x| \leqslant x_{max}$, so the probability of finding it outside the range $|u| \leqslant 1$ is zero.

The probability of finding the particle in the range $0 \leqslant u \leqslant 1$ is 1/2, so

$$\int_0^1 p(u)\, du = \frac{1}{\pi} \int_0^1 \frac{du}{\sqrt{1 - u^2}} = \frac{1}{2}$$

as the reader can easily check using trigonometric substitution.

Note that $p(u) \to \infty$ as $u \to \pm 1$. But the integral is finite. The fact that $p(u) \to \infty$ as $u \to \pm 1$ reflects the fact that the particle spends most of its time

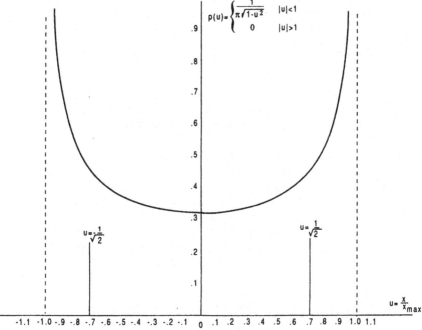

**Figure 2.3** Probability, $p(u)$, that the mass particle of the classical linear oscillator lies in the range $u, u + du$; $u = x/x_{max}$.

near the ends of its run. We can quantify this observation by asking, for what positive number $a$ does the particle spend half its time in the bands $-1 \leqslant u \leqslant -a$, $a \leqslant u \leqslant 1$? The number $a$ is determined by

$$\frac{1}{\pi} \int_a^1 \frac{du}{\sqrt{1-u^2}} = \frac{1}{4}$$

which yields $a = 1/\sqrt{2}$ (Exercise 3).

The graph of $p(u)$ is shown in Figure 2.3. Note that, after the scaling $u = x/x_{max}$, the probability density $p(u)$ is dependent of the natural frequency $v_0$, independent of the mass $m$, and independent of the total energy $E$.

### Exercises

1. Show that, for $x(t)$ given by equation (2.5.3), the quantity

    $$\tfrac{1}{2}k(x(t))^2 + \tfrac{1}{2}m(x'(t))^2$$

    is independent of $t$, and has the value $E = \tfrac{1}{2}k(x(0))^2 + \tfrac{1}{2}m(x'(0))^2$.
2. Derive equation (2.5.6).
3. Show that if $0 < a < 1$ and

    $$\frac{1}{\pi} \int_a^1 \frac{du}{\sqrt{1-u^2}} = \frac{1}{4}$$

    then $a = 1/\sqrt{2}$.

## 2.6 GUESSING A PARTICULAR SOLUTION TO A CONSTANT-COEFFICIENT EQUATION

For the constant-coefficient differential expression $l(y) \equiv ay'' + by' + cy$, we have determined the kernel explicitly. Say $f_1$ and $f_2$ are the two linearly independent functions spanning the kernel. We know that if we find any "particular" solution, $y_p$ say, to $l(y) = h$, then $y = y_p + c_1 f_1 + c_2 f_2$ is the general solution, where $c_1$ and $c_2$ are any constants.

For constant-coefficient expressions, we have a nonsystematic but often convenient procedure to find a particular solution. We call this procedure the *guessing method*, and we shall explain it by means of examples.

**Example 2.1.**  Find the general solutions to $l(y) \equiv y'' + y = e^x$ and the (unique) solution to the initial value problem $l(y) = e^x$, $y(0) = 0$, $y'(0) = 1$.

The kernel of $l(y) \equiv y'' + y$ is spanned by $f_1(x) = \sin x$, $f_2(x) = \cos x$. To find a particular solution, we guess $y_p = Ae^x$ and attempt to find the constant $A$.

$$y_p'' + y_p = Ae^x + Ae^x = 2Ae^x = e^x, \qquad \text{so } A = \tfrac{1}{2}$$

and our general solution is $y = c_1 \sin x + c_2 \cos x + \tfrac{1}{2}e^x$. To solve the initial value problem, we then fit the constants to the data.

$$0 = y(0) = c_1 \cdot 0 + c_2 \cdot 1 + \tfrac{1}{2}, \qquad c_2 = -\tfrac{1}{2}$$

$$y'(x) = c_1 \cos x - c_2 \sin x + \tfrac{1}{2}e^x$$

$$1 = y'(0) = c_1 + \tfrac{1}{2}, \qquad c_1 = \tfrac{1}{2}$$

So $y(x) = \tfrac{1}{2}(e^x + \sin x - \cos x)$ is the unique solution to $y'' + y = e^x$ that satisfies $y(0) = 0$, $y'(0) = 1$.

**Example 2.2.** Find the general solution $l(y) \equiv y'' + y' - 2y = x - 2x^3$, and solve the initial value problem $y'' + y' - 2y = x^3 - x$, $y(0) = \tfrac{7}{2}$, $y'(0) = 1$.

From a problem in the previous lecture, we know that the kernel of $l(y) \equiv y'' + y' - 2y$ is spanned by $e^x$ and $e^{-2x}$. To find a particular solution, we guess one of the form $y_p = Ax^3 + Bx^2 + Cx + D$ and try to determine the constants $A$, $B$, $C$, $D$.

$$y_p = Ax^3 + Bx^2 + Cx + D$$

$$y_p' = \qquad 3Ax^2 + 2Bx + C$$

$$y_p'' = \qquad\qquad 6Ax + 2B$$

$$y_p'' + y_p' - 2y_p = (6Ax + 2B) + (3Ax^2 + 2Bx + C) - 2(Ax^3 + Bx^2 + Cx + D)$$

$$= -2Ax^3 + (3A - 2B)x^2 + (6A + 2B - 2C)x + (2B + C - 2D)$$

so

$$-2A = -2, \qquad A = 1$$

$$3A - 2B = 0, \qquad B = \tfrac{3}{2}$$

$$6A + 2B - 2C = +1, \quad 6 + 3 - 2C = 1, \quad -2C = -8, \quad C = 4$$

$$2B + C - 2D = 0, \qquad 3 + 4 - 2D = 0, \qquad D = \tfrac{7}{2}$$

Our general solution is $y = c_1 e^x + c_2 e^{-2x} + x^3 + \tfrac{3}{2}x^2 + 4x + \tfrac{7}{2}$.

$$y(0) = c_1 + c_2 + \tfrac{7}{2} = \tfrac{7}{2}, \qquad c_1 + c_2 = 0$$

$$y'(x) = c_1 e^x - 2c_2 e^{-2x} + 3x^2 + 3x + 4$$

$$y'(0) = c_1 - 2c_2 + 4 = 1, \quad c_1 - 2c_2 = -3, \qquad c_1 = -1, \qquad c_2 = 1$$

so

$$y = e^{-2x} - e^x + x^3 + \tfrac{3}{2}x^2 + 4x + \tfrac{7}{2}$$

Reasonable guesses usually work. We must use a trick, however, when the guess is a solution of the homogeneous equation.

**Example 2.3.**   Find the general solution to $y'' - y = e^x$.

The kernel of $l(y) \equiv y'' - y$ is spanned by $e^x$ and $e^{-x}$, so if we try to guess $y_p = Ae^x$ then $l(y_p) = 0$ and we can never solve $l(y) = e^x$. The trick is to use instead $xe^x$:

$$y_p = Axe^x$$

$$y_p' = Ae^x + Axe^x$$

$$y_p'' = 2Ae^x + Axe^x$$

$$y_p'' - y_p = 2Ae^x + Axe^x - Axe^x = 2Ae^x = e^x \qquad \text{so } A = \tfrac{1}{2}$$

Hence

$$y = c_1 e^x + c_2 e^{-x} + \tfrac{1}{2}xe^x$$

is our general solution.

### Exercises

Do each of the first eight exercises below in two steps:
   (a) Use the method of guessing to find the general solution; then
   (b) Using your result from (a), find the unique solution to the initial value
       problem.

1. (a) $y'' + y' - 2y = 10 \sin x$   (Hint: guess $y_p = A \cos x + B \sin x$)
   (b) $y(0) = 1$, $y'(0) = 0$
2. (a) $y'' + y' - 2y = 2x$
   (b) $y(0) = y'(0) = 0$
3. (a) $y'' - 6y' + 9y = 27x^2$
   (b) $y(0) = 2$, $y'(0) = 4$
4. (a) $y'' + 4y = 5e^x$
   (b) $y(0) = 0$, $y'(0) = 1$
5. (a) $y'' + 4y = 4 \sin 2x$
   (b) $y(0) = y'(0) = 0$
6. (a) $y'' - 2y' + y = 2e^x$
   (b) $y(0) = 1$, $y'(0) = 0$
7. (a) $y'' + y' + y = e^{-(x/2)}$
   (b) $y(0) = y'(0) = 0$
8. (a) $y'' - 3y' + 2y = e^x + x$
   (b) $y(0) = \tfrac{3}{4}$, $y'(0) = \tfrac{1}{2}$
9. (a) $y'' - 2y' + y = 6xe^x$
   (b) $y(0) = y'(0) = 0$

10. (a) $y'' + y' - 2y = 3e^x$
    (b) $y(0) = y'(0) = 0$

11. In Theorem 2.7 of Section 2.2 we have proved that, under appropriate hypotheses on the coefficients of the second order linear differential expression $l(y) = a(x)y'' + b(x)y' + c(x)y$, the *homogeneous* initial value problem $l(y) = 0$, $y(x_0) = y_0$, $y'(x_0) = y_1$ is uniquely solvable. Prove that the *inhomogeneous* initial value problem $l(y) = h$, $y(x_0) = y_0$, $y'(x_0) = y_1$ is uniquely solvable. More precisely, prove the following: *Suppose $a(x)$, $b(x)$, $c(x)$ are analytic at $x = x_0$ and $a(x_0) \neq 0$. Let $y_0$, $y_1$ be any two fixed numbers, and let $h(x)$ be a given function such that the equation $ay'' + by' + cy = h$ has a solution $y = y_p$. Then the equation*

    $$a(x)y'' + b(x)y' + c(x)y = h(x)$$

    *has one, and only one, solution $y(x)$ that satisfies*

    $$y(x_0) = y_0, \qquad y'(x_0) = y_1$$

    (Hints: Start from Corollary 2.11 of Section 2.3, which guarantees a solution of the form $y = c_1 f_1 + c_2 f_2 + y_p$. Then the trick is to show that the constants $c_1$ and $c_2$ can be selected so that the initial conditions are satisfied. You will need to know that a certain $2 \times 2$ determinant is not zero.)

## 2.7 PARTICULAR SOLUTION BY VARIATION OF PARAMETERS

The method of guessing presented in Section 2.6, although often swift and convenient, yields a particular solution of an equation $l(y) = h$ only for constant-coefficient expressions $l$ and particularly simple $h$. We present now another method for finding a particular solution that, while more cumbersome, always works, at least in principle. It assumes that the kernel of $l$ is known.

Suppose $f_1$ and $f_2$ are two known functions that span $\ker(l)$, $l(y) \equiv ay'' + by' + cy$, where $a, b, c$ are analytic functions of $x$, not necessarily constants. The general function in $\ker(l)$ has the form $c_1 f_1 + c_2 f_2$ where $c_1$ and $c_2$ are constants. Let $h = h(x)$ be a known function. We shall attempt to find a particular solution $y = y(x)$ of $l(y) = h$ of the form

$$y = c_1(x)f_1(x) + c_2(x)f_2(x) \tag{2.7.1}$$

where $c_1$ and $c_2$ are now *functions* to be determined. Since we have replaced constant parameters by variable ones, the method is called *variation of parameters* or *variation of constants*.

Differentiate equation (2.7.1) twice, using the product rule, and substitute the results into the equation $l(y) = h$. As $f_1$ and $f_2$ belong to $\ker(l)$, we have $af_1'' + bf_1' + cf_1 = 0$, and similarly for $f_2$. Take advantage of these relations to simplify the resulting formula, getting finally

$$a(c_1'f_1' + 2c_1'f_1' + c_2''f_2 + 2c_2'f_2') + b(c_1'f_1 + c_2'f_2) = h \tag{2.7.2}$$

We seek *any* functions $c_1$ and $c_2$ that will solve this equation, so we can use any simplifying assumptions that will yield a solution. If we impose the condition $c_1'f_1 + c_2'f_2 = 0$ we eliminate the second term on the left side of (2.7.2). If we assume the equation $c_1'f_1 + c_2'f_2 = 0$, we must also assume the validity of the equation that results from differentiating it: $c_1''f_1 + c_1'f_1' + c_2''f_2 + c_2'f_2' = 0$. Using both these equations, we reduce equation (2.7.2) simply to $a(c_1'f_1' + c_2'f_2') = h$, or $c_1'f_1' + c_2'f_2' = h/a$. Thus our search for $c_1$ and $c_2$ will be successful if we can solve simultaneously

$$c_1'f_1 + c_2'f_2 = 0$$
$$c_1'f_1' + c_2'f_2' = \frac{h}{a} \tag{2.7.3}$$

But this $2 \times 2$ system of linear equations is easily solved in terms of the Wronskian determinant $W$,

$$W = W(f_1, f_2) = \begin{vmatrix} f_1 & f_2 \\ f_1' & f_2' \end{vmatrix} = f_1 f_2' - f_1' f_2 \tag{2.7.4}$$

We get $c_1' = -hf_2/aW$, $c_2' = hf_1/aW$. Hence $c_1 = -\int (hf_2/aW)\,dx$, $c_2 = \int (hf_1/aW)\,dx$. So, from equation (2.7.1), our desired particular solution is

$$y_p = -f_1(x) \int \frac{hf_2}{aW}\,dx + f_2(x) \int \frac{hf_1}{aW}\,dx \tag{2.7.5}$$

Compute the Wronskian $W$ from equation (2.7.4). If you use Abel's formula, $W = \text{const} \times \exp(-\int (b/a)\,dx)$, you must use the value of the constant appropriate to the choice of $f_1$ and $f_2$.

For example, consider the equation

$$y'' + y' - 2y = h(x)$$

In this case the kernel is spanned by $f_1(x) = e^x$ and $f_2(x) = e^{-2x}$. The Wronskian is

$$W = e^x(-2)e^{-2x} - e^x(e^{-2x}) = -3e^{-x}$$

so

$$y_p = +\frac{e^x}{3} \int \frac{h(x)e^{-2x}}{1 \cdot e^{-x}}\,dx - \frac{e^{-2x}}{3} \int \frac{h(x)e^x}{1 \cdot e^{-x}}\,dx$$

or

$$y_p = \frac{1}{3}\left[ e^x \int h(x)e^{-x}\,dx - e^{-2x}\int h(x)e^{2x}\,dx \right] \tag{2.7.6}$$

Suppose $h(x) = 1$. Then the guessing method gives us immediately $y_p = -\frac{1}{2}$. Let us do the calculation (2.7.6) to check:

$$y_p = \frac{1}{3}\left[ e^x \int e^{-x}\,dx - e^{-2x}\int e^{2x}\,dx \right]$$

$$= \tfrac{1}{3}[-(e^x)(e^{-x}) - (e^{-2x})\tfrac{1}{2}e^{2x}] = \tfrac{1}{3}[-1 - \tfrac{1}{2}] = -\tfrac{1}{2}$$

So it is a much longer route to the answer in this particular case. But this method has the very strong advantage that it always gives a closed form for the solution, namely equation (2.7.5), an expression which can even be used to give qualitative information on $y_p$ if $h(x)$ is given only numerically.

**Exercises**

In each of the following exercises, first find two real functions $f_1$, $f_2$ that span ker($l$). Then use variation of parameters to
  (a) Express a particular solution $y_p$ of $l(y) = h$ in integral form ($h(x)$ unspecified).
  (b) Evaluate the integrals in (a) for the given $h$.
  (c) Check your answer in (b) by substituting your $y_p$ into $l$.
  (d) Write down the general solution.
  (e) Solve the initial value problem $l(y) = h$, $y(0) = y'(0) = 0$.

1. $l(y) \equiv \dfrac{d^2y}{dx^2} + \dfrac{dy}{dx} - 2y,\qquad h(x) = x$

2. $l(y) \equiv \dfrac{d^2y}{dx^2} - 2\dfrac{dy}{dx} + 10y,\qquad h(x) = e^x\cos 3x$

## 2.8  THE KERNEL OF LEGENDRE'S DIFFERENTIAL EXPRESSION

If we know the kernel of a second order linear differential expression $l$, then we may use variation of parameters (Section 2.7) to solve—in principle at least—any equation $l(y) = h$. But everything hinges on knowledge of the kernel. For constant-coefficient expressions we have determined the kernel explicitly (Section 2.4); it is always spanned by elementary combinations of exponential and trigonometric functions. We turn now to the study of the kernel of some classical differential expressions with variable coefficients.

*Legendre's Expression* $l(y) \equiv (1 - x^2)y'' - 2xy' + n(n + 1)y$

Here $n$ is a nonnegative integer, $n = 0, 1, 2, \ldots$. Thus Legendre's expression depends on an integral parameter. Note that the coefficient of the leading term is $1 - x^2$, which equals zero at $x = \pm 1$, so we might anticipate some difficulties with solutions at $x = \pm 1$. Initially we may restrict our attention to the open interval $-1 < x < 1$. Noting that $(1 - x^2)y'' - 2xy' = ((1 - x^2)y')'$, we may write Legendre's expression in the Sturmian form* $l(y) \equiv ((1 - x^2)y')' + n(n + 1)y$. If $n = 0$, then $y = $ constant is a solution of $l(y) = 0$; for convenience we may take $y = 1$. So for $n = 0$ we may easily find a nonzero function in the kernel.

The general case: Following the method of Section 2.2 (undetermined coefficients) we seek a solution of $l(y) = 0$ in the form

$$y(x) = y_0 + y_1 x + y_2 x^2 + \cdots + y_k x^k + \cdots$$

where we know we may specify $y_0$ and $y_1$ arbitrarily. For the moment we leave them unspecified. We find

$$y' = y_1 + 2y_2 x + 3y_3 x^2 + 4y_4 x^3 + \cdots + ky_k x^{k-1} + \cdots$$

$$x^2 y' = \qquad\qquad y_1 x^2 + 2y_2 x^3 + 3y_3 x^4 + \cdots$$

so

$$(1 - x^2)y' = y_1 + 2y_2 x + (3y_3 - y_1)x^2 + (4y_4 - 2y_2)x^3 + \cdots$$
$$+ [(k + 1)y_{k+1} - (k - 1)y_{k-1}]x^k + \cdots$$

Thus

$$((1 - x^2)y')' = 2y_2 + 2(3y_3 - y_1)x + 3(4y_4 - 2y_2)x^2 + \cdots$$
$$+ k[(k + 1)y_{k+1} - (k - 1)y_{k-1}]x^{k-1} + \cdots$$

To satisfy $((1 - x^2)y')' + n(n + 1)y = 0$ we need to solve simultaneously the following system of equations:

$$2y_2 + n(n + 1)y_0 = 0$$
$$2(3y_3 - y_1) + n(n + 1)y_1 = 0$$
$$3(4y_4 - 2y_2) + n(n + 1)y_2 = 0$$
$$4(5y_5 - 3y_3) + n(n + 1)y_3 = 0$$
$$5(6y_6 - 4y_4) + n(n + 1)y_4 = 0$$
$$\vdots$$
$$(k + 1)[(k + 2)y_{k+2} - ky_k] + n(n + 1)y_k = 0$$
$$\vdots$$

*See Exercises 10 and 11 of Section 2.2.

where we have written out several terms to show the pattern. Note that the 1st, 3rd, 5th, etc. equations involve only the even-numbered coefficients, and the 2nd, 4th, 6th, etc. equations involve only the odd-numbered coefficients. As we may specify $y_0$ and $y_1$ independently, we can separate the equations into two groups:

$$2y_2 = -n(n + 1)y_0$$
$$3 \cdot 4y_4 = (2 \cdot 3 - n(n + 1))y_2$$
$$5 \cdot 6y_6 = (4 \cdot 5 - n(n + 1))y_4$$
$$\vdots \qquad\qquad \text{even group}$$
$$(2k - 1)(2k)y_{2k} = ((2k - 2)(2k - 1) - n(n + 1))y_{2k-2}$$
$$\vdots$$

and

$$2 \cdot 3y_3 = (2 - n(n + 1))y_1$$
$$4 \cdot 5y_5 = (3 \cdot 4 - n(n + 1))y_3$$
$$6 \cdot 7y_7 = (5 \cdot 6 - n(n + 1))y_5$$
$$\vdots \qquad\qquad \text{odd group}$$
$$(2k)(2k + 1)y_{2k+1} = ((2k - 1)(2k) - n(n + 1))y_{2k-1}$$
$$\vdots$$

Note further that if we set $y_0 = 0$ then all even coefficients necessarily vanish, and if we set $y_1 = 0$ then all odd coefficients necessarily vanish.

Now let us set about seeing if we can solve these equations to determine *some* nonzero function in the kernel of Legendre's expression for each $n = 0, 1, \ldots$. Start with $n = 0$. The first equation in the even group then shows that $y_2 = 0$; hence all subsequent even coefficients equal zero. We set $y_0 = 1$, and $y_1 = 0$ so that all odd terms vanish. Thus we find the solution $y = 1$ already derived.

$n = 1$. Then $n(n + 1) = 2$, so we see that $y_3$ and all subsequent odd terms equal zero. We set $y_1 = 1$, $y_0 = 0$, and find $y = x$ as a solution in this case.

$n = 2$. Here $n(n + 1) = 2 \cdot 3$ so $y_4 = y_6 = \cdots = 0$. We set $y_1 = 0$ to eliminate the odd terms. Then $2y_2 = -2 \cdot 3y_0$ so $y_2 = -3y_0$. We set $y_2 = 1$ to get $y = x^2 - \frac{1}{3}$ as a solution in this case.

$n = 3$. $n(n + 1) = 3 \cdot 4$ so $y_5 = y_7 = \cdots = 0$. Set $y_0 = 0$. Then $2 \cdot 3y_3 = (2 - 3 \cdot 4)y_1 = -10y_1$ so $3y_3 = -5y_1$. Set $y_3 = 1$ to get $y = x^3 - \frac{3}{5}x$.

We have carried the calculation far enough to see the pattern.

**Theorem 2.13.** *Legendre's equation* $(1-x^2)y'' - 2xy' + n(n+1)y = 0$ *has for each nonnegative integer n a polynomial solution* $P_n(x)$ *of degree n which for n even contains only even powers of x and for n odd, only odd powers.*

These are *Legendre's polynomials*. They are customarily normalized so that the leading coefficient is

$$\frac{(2n)!}{2^n(n!)^2} = \frac{(2n)(2n-1)(2n-2)(2n-3)\cdots(n+1)}{2^n n!}$$

Some values are given in Table 2.1.

So $P_0(x) = 1$, $P_1(x) = x$, $P_2(x) = \frac{3}{2}x^2 - \frac{1}{2} = \frac{1}{2}(3x^2 - 1)$ and $P_3(x) = \frac{1}{2}(5x^3 - 3x)$.

Note that these polynomial solutions are analytic in the entire plane, even though the leading coefficient of the operator is zero at $x = \pm 1$. Singularities at $x = \pm 1$ do show up in the second linearly independent solution in the kernel (Exercise 2).

**Exercises**

*Legendre's Equation* $(1 - x^2)y'' - 2xy' + n(n+1)y = 0$

1. Continuing the calculation in the lecture, compute $P_4$, $P_5$, and $P_6$ ($P_6(x) = \frac{1}{16}(231x^6 - 315x^4 + 105x^2 - 5)$).

2. (a) Show that with the appropriate choice of the multiplicative constant, the Wronskian of Legendre's expression is $W = 1/(1 - x^2)$ independent of $n$ (refer to Exercise 5 of Section 2.3).

   (b) Using the Wronskian computed in (a), derive the expression

   $$Q_n(x) = P_n(x) \int \frac{dx}{(1 - x^2)P_n^2(x)}$$

**Table 2.1**

| $n$ | $\dfrac{(2n)!}{2^n(n!)^2}$ |
|---|---|
| 0 | 1 |
| 1 | 1 |
| 2 | $\frac{3}{2}$ |
| 3 | $\frac{5}{2}$ |
| 4 | $\frac{35}{8}$ |
| 5 | $\frac{63}{8}$ |
| 6 | $\frac{231}{16}$ |

for a second linearly independent solution $Q_n(x)$ of Legendre's equation. Evaluate the integral for $n = 0$ and $n = 1$. [Ans.: $Q_0(x) = \frac{1}{2}\ln(1 + x/1 - x)$, $Q_1(x) = (x/2)\ln(1 + x/1 - x) - 1$.] The $Q_n(x)$, $n = 0, 1, 2, \ldots$, are called *Legendre functions of the second kind*.

## 2.9 THE KERNELS OF OTHER CLASSICAL EXPRESSIONS

*Hermite's Expression* $l(y) \equiv y'' - xy' + ny$, $n = 0, 1, 2, \ldots$

Hermite's expression is again really a family of expressions, one for each nonnegative integer $n$. The situation for Hermite's expression is very similar to that for Legendre's in that for each $n$ there is a polynomial of precise degree $n$ in $\ker(l)$. Note that the coefficient of $y''$ is 1, so all functions in the kernel are entire.

**Theorem 2.14.** *Hermite's equation $y'' - xy' + ny = 0$ has for each nonnegative integer $n$ a polynomial solution $H_n(x)$ of degree $n$ which for $n$ even contains only even powers of $x$ and for $n$ odd, only odd powers.*

The proof is left for Exercise 1. $H_n(x)$ is customarily normalized so that the coefficient of the highest power of $x$ is 1. These are *Hermite's polynomials*.

*Laguerre's Expression* $l(y) \equiv xy'' + (1 - x)y' + ny$, $n = 0, 1, 2, \ldots$

Again we have polynomial functions in the kernel:

**Theorem 2.15.** *Laguerre's equation $xy'' + (1 - x)y' + ny = 0$ has for each nonnegative integer $n$ a polynomial solution $L_n(x)$ of degree $n$.*

The proof is left for Exercise 3. $L_n(x)$ is customarily normalized so that the constant term is 1. These are *Laguerre's polynomials*.

*Bessel's Expression* $l(y) \equiv x^2 y'' + xy' + (x^2 - n^2)y$, $n = 0, 1, 2, \ldots$

The kernel of Bessel's expression contains new kinds of transcendental functions different from the functions we have met in calculus. For each nonnegative integer $n$ the kernel contains an entire function denoted $J_n(x)$ and called the *Bessel function of the first kind of order n*. The case $n = 0$ is treated in Exercise 5. Most of Chapter 6 is devoted to the general case.

### Exercises

*Hermite's Equation $y'' - xy' + ny = 0$*

1. Using the method of undetermined coefficients and separating the resulting system of equations for the coefficients into an even group and an odd

group, as in Section 2.8, prove Theorem 2.14: *Hermite's equation* $y'' - xy' + ny = 0$ *has for each nonnegative integer n a polynomial solution* $H_n(x)$ *of degree n which for n even contains only even powers of x and for n odd only odd powers.*

The $H_n(x)$ are *Hermite's polynomials*, usually normalized so that the coefficient of the highest power of $x$ is 1. Compute $H_0$ through $H_5$ $(H_5(x) = x^5 - 10x^3 + 15x)$.

2. (a) Show that with the appropriate choice of the multiplicative constant, the Wronskian of Hermite's expression is $W = \exp(x^2/2)$ independent of $n$.

   (b) Using the Wronskian computed in (a), derive the expression

$$H_n(x) \int \frac{1}{H_n^2(x)} e^{x^2/2}\, dx$$

   for a second linearly independent solution of Hermite's equation. Any multiple of that function by a nonzero constant is also a second linearly independent solution, and the functions

$$h_n(x) = n! H_n(x) \int \frac{1}{H_n^2(x)} e^{x^2/2}\, dx$$

   are usually taken as the second solution, where the factor $n!$ is included for convenience of normalization. The functions $h_n(x)$ are called *Hermite functions of the second kind.* Show that if we add the condition $h_0(0) = 0$ we may write

$$h_0(x) = \int_0^x e^{t^2/2}\, dt$$

*Laguerre's Equation* $xy'' + (1 - x)y' + ny = 0$

3. Prove the following stronger form of Theorem 2.15: *If n is a nonnegative integer, the only power series solution to Laguerre's equation* $xy'' + (1 - x)y' + ny = 0$ *is a polynomial of degree n.*

   These solutions, normalized so that the constant term is 1, are called *Laguerre's polynomials* and denoted $L_n(x)$. Compute $L_0$ through $L_3$.

4. Show that the Wronskian of Laguerre's expression can be taken to be $W = e^x/x$, and derive the expression

$$L_n(x) \int \frac{e^x}{xL_n^2(x)}\, dx$$

   for a second linearly independent solution of Laguerre's equation.

5. *Bessel's equation,* $x^2 y'' + xy' + (x^2 - n^2)y = 0$, depends on an integral

parameter $n = 0, 1, 2, \ldots$. Restrict yourself to the case $n = 0$, and derive by the method of undetermined coefficients the solution

$$J_0(x) = \sum_{k=0}^{\infty} \frac{(-1)^k}{(k!)^2} \left(\frac{x}{2}\right)^{2k}$$

to the Bessel equation of order zero, $xy'' + y' + xy = 0$. $J_0(x)$ is the *Bessel function of the first kind of order zero*. Do you think it is analytic everywhere, or does it have singularities?

## 2.10 DIRAC'S DELTA FUNCTION AND GREEN'S FUNCTIONS

### 2.10.1 Dirac's Delta Function

In his book the *Principles of Quantum Mechanics*, the physicist P. A. M. Dirac describes a quantity $\delta(x)$ which is to satisfy the two conditions

$$\int_{-\infty}^{\infty} \delta(x)\, dx = 1 \quad \text{and} \quad \delta(x) = 0 \quad \text{for } x \neq 0 \qquad (2.10.1)$$

If we think of $\delta(x)$ as a function, then these equations say that this function is zero everywhere except at $x = 0$, and at that point it has an infinity of sufficient strength so that its integral is 1. Within Lebesgue's theory of integration, no such function exists. A function zero everywhere except at $x = 0$ must have integral zero, no matter what the value at that one point. Dirac, who had introduced this concept originally in a 1926 paper, was well aware of the self-contradictory nature of his axioms. I quote from §15 of the fourth edition of his book:

$\delta(x)$ is not a function of $x$ according to the usual mathematical definition of a function, which requires a function to have a definite value for each point in its domain, but is something more general, which we may call an "improper function" to show up its difference from a function defined by the usual definition. Thus $\delta(x)$ is not a quantity which can be generally used in mathematical analysis like an ordinary function, but its use must be confined to certain simple types of expressions for which it is obvious that no inconsistency can arise.

The most important property of $\delta(x)$ is exemplified by the following equation:

$$\int_{-\infty}^{\infty} f(x)\delta(x)\, dx = f(0) \qquad (2.10.2)$$

where $f(x)$ is any continuous function of $x \ldots$.

Dirac set up a few basic rules of calculation for his delta function and used it systematically throughout his elegant and enormously influential book. Dirac's success with these new unorthodox methods posed a challenge for mathematicians, that of incorporating Dirac's ideas into a logically correct mathematical theory. The French mathematician Laurent Schwartz created his theory of "distributions" to reconcile Dirac's symbolic methods with the usual rules of calculus. Equation (2.10.2) provided the starting point. It gives a rule assigning a number to each continuous function $f(x)$, namely the number $f(0)$. This is an example of a "functional." Schwartz defined a distribution to be a special type of functional (the actual definition had been given earlier by the Russian mathematician Sobolev), studied them in depth, and then compiled and published his researches in a two-volume book. Though he wrote his book at length to explain his theory in complete detail, nonetheless it makes rather heavy going, requiring a knowledge of topological vector spaces. Subsequently, many other mathematicians, working from Schwartz's definitive book, sought to simplify his theory in a way that would still bring Dirac's idea within rigorous standards of mathematical proof. One of the more accessible of these approaches is that given by M. J. Lighthill in his book *Introduction to Fourier Analysis and Generalized Functions* (Cambridge University Press, 1958). Lighthill defines a generalized function to be a certain type of sequence of ordinary functions. For example, Dirac's $\delta(x)$ is defined in terms of the sequence $f_n(x) = (n/\pi)^{1/2} \exp(-nx^2)$ (see Exercise 1).

But Dirac's original formalism has maintained its appeal. While it is vitally important to know that Dirac's theory has a rigorous foundation, yet when it comes to practical calculations, no alternative formalism comes anywhere near Dirac's in elegance and simplicity. Quoting Dirac again:

> ... although an improper function does not itself have a well-defined value, when it occurs as a factor in an integrand the integral has a well-defined value. ... The use of improper functions thus does not involve any lack of rigour in the theory, but is merely a convenient notation, enabling us to express in a concise form certain relations which we could, if necessary, rewrite in a form not involving improper functions, but only in a cumbersome way which would tend to obscure the argument.

We are going to use the delta function to define and study the Green's function of a second order linear differential expression. Once we have derived a formula for the Green's function, we can then verify its properties by direct calculation, without further use of the delta function. So Dirac's original scheme fits our purpose, and we shall follow it.

Dirac's delta function has the following properties:

$$\delta(t - x) = 0, \qquad t \neq x$$

and

$$\int_{\alpha}^{\beta} \delta(t - x)\, dt = 1, \qquad \alpha < x < \beta \tag{2.10.3}$$

$$\int_{\alpha}^{\beta} \delta(t - x) f(t)\, dt = f(x), \qquad \alpha < x < \beta, f \text{ continuous} \tag{2.10.4}$$

$$\delta(-x) = \delta(x) \qquad (2.10.5)$$

$$x\delta(x) = 0 \qquad (2.10.6)$$

$$\delta(\alpha x) = \alpha^{-1}\delta(x), \qquad \alpha > 0 \qquad (2.10.7)$$

$$f(t)\delta(t - x) = f(x)\delta(t - x) \qquad (2.10.8)$$

We shall argue for the plausibility and consistency of these rules, as we cannot do a rigorous mathematical proof. We want the rules for the $\delta$ function to be consistent with the usual rules for the integral calculus, like integration by substitution, for example. Hence, if $t$ is a variable and $x$ a parameter, fixed independent of $t$, then, under the substitution $u = t - x$, $du = dt$, we have from (2.10.1),

$$1 = \int_{-\infty}^{\infty} \delta(u)\,du = \int_{-\infty}^{\infty} \delta(t - x)\,dt$$

Likewise from (2.10.1), $\delta(t - x) = 0$ when $t \neq x$. As we may disregard portions of the integral in which the integrand is zero, we get (2.10.3). Dirac justifies equation (2.10.2) as follows:

> The left-hand side of (2.10.2) can depend only on the values of $f(x)$ very close to the origin, so that we may replace $f(x)$ by its value at the origin, $f(0)$, without essential error.

Equation (2.10.4) then follows from (2.10.2) by substituting $u = t - x$, $du = dt$ as before (write (2.10.2) in terms of the dummy variable $u$).

Equation (2.10.5) asserts that $\delta(x)$ is even; it may be taken as an axiom. From (2.10.5) we deduce that $\delta(t - x) = \delta(-(t - x)) = \delta(x - t)$, so that (2.10.4) may also be written

$$\int_{\alpha}^{\beta} \delta(x - t)f(t)\,dt = f(x), \qquad (2.10.4')$$

$$\alpha < x < \beta, \ f(t) \text{ any continuous function of } t$$

Equation (2.10.6) comes from (2.10.2) with $f(x) = x$. Equation (2.10.7) states consistency with integration by substitution again, this time the substitution $u = \alpha x$, $du = \alpha\,dx$. And equation (2.10.8) comes from (2.10.4). That "proves" the properties (2.10.3) to (2.10.8).

We want our formalism to be consistent with another very important rule from calculus, namely the *fundamental theorem of calculus*, which says that if $f$ is a continuous function, then

$$\frac{d}{dx}\int_{\alpha}^{x} f(u)\,du = f(x)$$

Consider the integral

$$\int_\alpha^x \delta(t)\, dt$$

where $\alpha$ is a negative constant. Use (2.10.3) to evaluate this integral. (In (2.10.3), put $x = 0$. Then replace the upper limit $\beta$ by $x$.) According to (2.10.3), this integral will have the value 0 when $x < 0$, because the integrand is zero over the interval of integration, and will have the value 1 when $x > 0$. This function of $x$, which we shall denote $H(x)$, is often called *Heaviside's function*, after Oliver Heaviside who proposed another symbolic calculus before Dirac. Heaviside's function is pictured in Figure 2.4. The value of $H(x)$ at $x = 0$ doesn't matter, so we leave it undefined there. We now have the equation,

$$H(x) = \int_\alpha^x \delta(t)\, dt, \qquad \alpha < 0$$

so, if the fundamental theorem of calculus is to be maintained, we must have

$$\frac{d}{dx} H(x) = \delta(x) \tag{2.10.9}$$

Note that this equation makes no sense at all from the traditional point of view, because on the left we are differentiating a function, $H(x)$, which is not only nondifferentiable, but even discontinuous. But equation (2.10.9) is perfectly all right in Dirac's extended calculus. By translation, we get from (2.10.9)

$$\frac{d}{dx} H(x - t) = \delta(x - t) \tag{2.10.10}$$

where the differentiation, as indicated, is with respect to $x$, and $t$ is a fixed parameter.

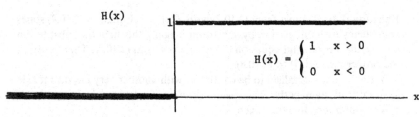

**Figure 2.4**   Heaviside's function, $H(x)$.

## 2.10.2 Green's Functions

In Corollary 2.11 in Section 2.3, we have spelled out a three-step procedure for solving $l(y) = h$ where $l(y) \equiv a(x)y'' + b(x)y' + c(x)y$.

*Step 1. Find the kernel of the differential expression l.* For our second order expression $l(y) \equiv ay'' + by' + cy$ we know that the kernel is two-dimensional (Corollary 2.10), so finding ker($l$) amounts to finding two linearly independent functions $f_1, f_2$ that satisfy $l(f_i) = 0$. If we find $f_1$, then Abel's formula will always give us $f_2$ (see Section 2.3). Hence it is a matter of determining one nonzero function in ker($l$). Undetermined coefficients has worked for all our variable-coefficient expressions (Sections 2.8 and 2.9). In the special case of constant coefficients, just substitute $y = e^{rx}$ (Section 2.4).

*Step 2. Find some solution $y_p$ to $l(y) = h$.* The function $y_p$ is called a particular solution. If you know the kernel of $l$ (Step 1), then variation of parameters (Section 2.7) will always yield a particular solution, at least in principle. For constant-coefficient expressions and simple elementary $h$, guessing is quickest (Section 2.6).

*Step 3. Then $y = y_p + c_1 f_1 + c_2 f_2$, where $c_1$ and $c_2$ are arbitrary constants, is the general solution to $l(y) = h$, in the sense that every solution has that form for suitable constants $c_1, c_2$.*

Green's functions give us another systematic method for Step 2, finding a particular solution.

**Definition 2.16.** A *Green's function* $G(x, t)$ for the second order linear differential expression $l(y) \equiv ay'' + by' + cy$ is any solution to

$$a(x)\frac{d^2}{dx^2} G(x, t) + b(x)\frac{d}{dx} G(x, t) + c(x)G(x, t) = \delta(x - t) \qquad (2.10.11)$$

Note that Green's function is a function of two variables, $x$ and $t$. We are using $x$ to symbolize the differentiation variable in the expression $l$; we can emphasize that by writing $l_x$. The variable $t$ is a parameter, independent of $x$. We can write (2.10.11) briefly as $l_x(G) = \delta(x - t)$.

A Green's function is by no means unique. If $G(x, t)$ is one solution to (2.10.11), and if $f_1, f_2$ span ker($l$), then $G(x, t) + c_1 f_1(x) + c_2 f_2(x)$ is also a solution for any choice of the constants $c_1$ and $c_2$. In fact, the "constants" $c_1$ and $c_2$ can depend on the parameter $t$, since $t$ is not involved in the differentiation. So, using Step 3, we can speak of the *general Green's function for $l$*

$$G(x, t) + c_1(t)f_1(x) + c_2(t)f_2(x) \qquad (2.10.12)$$

where $G(x, t)$ is some solution to (2.10.11); $f_1(x)$, $f_2(x)$ span ker($l$); and $c_1(t)$, $c_2(t)$ are arbitrary (reasonable) functions of $t$. If we add the two requirements $G(x_0, t) = 0$ and $(\partial/\partial x)(G(x_0, t) = 0$, we obtain a uniquely solvable initial value problem associated with (2.10.11). These two conditions determine the functions $c_1(t)$ and $c_2(t)$ in (2.10.12), so we may speak then of the unique Green's function for the initial value problem

$$l_x(G) = \delta(x - t), \qquad G(x_0, t) = \frac{\partial G}{\partial x}(x_0, t) = 0 \qquad (2.10.13)$$

**Theorem 2.17.** *If $G(x, t)$ is any Green's function for $l$, then*

$$y_p(x) = \int_\alpha^\beta G(x, t)h(t)\, dt, \qquad \alpha < x < \beta \qquad (2.10.14)$$

*is a particular solution to $l(y) = h$.*

*Proof.* $\alpha$ and $\beta$ denote constants chosen to include the range of interest of the variable $x$, as indicated. As the expression $l$ involves differentiation with respect to the variable $x$ whereas the integration in (2.10.14) is with respect to $t$, we shall take the reasonable view that we may interchange the order of differentiation and integration:

$$l_x(y_p) = l_x \int_\alpha^\beta G(x, t)h(t)\, dt = \int_\alpha^\beta l_x(G(x, t))h(t)\, dt$$

$$= \int_\alpha^\beta \delta(x - t)h(t)\, dt = h(x)$$

Which concludes the proof.

Theorem 2.17 explains the value of Green's function: if we know a Green's function for $l$, then we can write down the closed form (2.10.14) for a particular solution to $l(y) = h$ for any $h$.

Variation of parameters gives us an explicit formula for Green's function. According to formula (2.7.5) of Section 2.7, a solution to $l(y) \equiv a(x)y'' + b(x)y' + c(x)y = \delta(x - t)$ is

$$G(x, t) = -f_1(x) \int \frac{\delta(x - t)f_2(x)}{a(x)W(x)}\, dx + f_2(x) \int \frac{\delta(x - t)f_1(x)}{a(x)W(x)}\, dx$$

$$(2.10.15)$$

where $f_1$ and $f_2$ span ker($l$), and $W = f_1 f_2' - f_1' f_2$ is the Wronskian. The antiderivatives on the right of (2.10.15) are more conveniently expressed as definite integrals, using the fundamental theorem of calculus:

$$G(x, t) = -f_1(x) \int_\alpha^x \frac{\delta(u - t) f_2(u)}{a(u) W(u)} \, du + f_2(x) \int_\alpha^x \frac{\delta(u - t) f_1(u)}{a(u) W(u)} \, du$$

(2.10.16)

In equation (2.10.16) the constant $\alpha$ is so chosen that $\alpha < t$ over the range of interest of the parameter $t$. The singularity of the $\delta$ function occurs at $u = t$, so when $x < t$, $G(x, t) = 0$, and when $x > t$, each integral on the right of (2.10.16) equals the value of its respective integrand at $u = t$. We get

$$G(x, t) = \frac{f_1(t) f_2(x) - f_1(x) f_2(t)}{a(t) W(t)} H(x - t)$$

(2.10.17)

as a Green's function for the expression $l(y) \equiv a(x) y'' + b(x) y' + c(x) y$ where $f_1$ and $f_2$ span ker($l$), and $W = f_1 f_2' - f_1' f_2$ is the Wronskian.

For any pair of functions $c_1(t)$, $c_2(t)$, the function $G(x, t) + c_1(t) f_1(x) + c_2(t) f_2(x)$ is also a Green's function for $l$, where $G$ is given in (2.10.17). By appropriate choice of $c_1$ and $c_2$, we get the *unique Green's function*

$$G(x, t) = \frac{f_1(t) f_2(x) - f_1(x) f_2(t)}{a(t) W(t)} (H(x - t) - H(x_0 - t))$$

(2.10.18)

for the initial value problem $a(x) d^2 G/dx^2 + b(x) dG/dx + c(x) G = \delta(x - t)$, $G(x_0, t) = 0$, $(\partial G/\partial x)(x_0, t) = 0$. Remember that $H(u)$ is Heaviside's function: $H(u) = 1, u > 0$ and $H(u) = 0, u < 0$. Using (2.10.18), we may solve the general initial value problem in terms of the homogeneous initial value problem.

**Theorem 2.18.** Let $l(y) \equiv a(x) y'' + b(x) y' + c(x) y$ be a second order linear differential expression whose kernel is spanned by the functions $f_1(x)$ and $f_2(x)$, and let $W = f_1 f_2' - f_1' f_2$ be the Wronskian. Then, for any function $h$ and any numbers $y_0$ and $y_1$, the unique solution to the initial value problem

$$l(y) = h, \qquad y(x_0) = y_0, \, y'(x_0) = y_1$$

is

$$y = f(x) + \int_{-\infty}^{\infty} G(x, t) h(t) \, dt$$

(2.10.19)

where $f(x)$ is the unique solution to the homogeneous initial value problem $l(y) = 0$, $y(x_0) = y_0$, $y'(x_0) = y_1$, and $G(x, t)$ is the Green's function in (2.10.18).

The direct calculation in Exercise 12 provides the proof.

An example: consider again the equation $y'' + y' - 2y = h$ that we already discussed in Section 2.4. The kernel of $l(y) \equiv y'' + y' - 2y$ is spanned by $f_1(x) = e^x$, $f_2(x) = e^{-2x}$. We have $W = -3e^{-x}$. Equation (2.10.17) gives a Green's function for $l$ (remember that $a = 1$):

$$G(x, t) = \frac{e^t e^{-2x} - e^x e^{-2t}}{-3e^{-t}} H(x - t)$$

Note that $H(x - t) = 0$ when $t > x$ and $H(x - t) = 1$ when $t < x$. Hence the integral (2.10.14) cuts off at $t = x$, and we find

$$y_p = \frac{1}{3} \int_a^x (e^{2(t-x)} - e^{x-t})h(t)\, dt$$

as a particular solution to $l(y) = h$. If $h = 1$, this formula gives $y_p = -(1/3)(3/2 - (1/2)e^{2a}e^{-2x} - e^{-a}e^x)$. As the last two terms belong to the kernel, we can disregard them, and we get $y_p = -(1/2)$ as before.

Continuing with this example, suppose we want to solve the initial value problem $l(y) = h$, $y(0) = 0$, $y'(0) = 3$. We may use Theorem 2.18. First solve the homogeneous initial value problem: $y = c_1 e^x + c_2 e^{-2x}$, $y(0) = c_1 + c_2 = 0$, so $c_2 = -c_1$ and $y = c_1(e^x - e^{-2x})$. Then $y' = c_1(e^x + 2e^{-2x})$, $y'(0) = 3 = c_1(1 + 2)$, $c_1 = 1$. Hence $f(x) = e^x - e^{-2x}$ is the unique solution to the homogeneous initial value problem. Now use (2.10.19) and (2.10.18), remembering $x_0 = 0$;

$$y = e^x - e^{-2x} + \int_{-\infty}^{\infty} \frac{e^t e^{-2x} - e^x e^{-2t}}{-3e^{-t}} (H(x - t) - H(-t))h(t)\, dt.$$

Suppose $x > 0$. Then

$$H(x - t) - H(-t) = \begin{cases} 1 - 1 = 0, & t < 0 \\ 1 - 0 = 1, & 0 < t < x \\ 0 - 0 = 0, & x < t \end{cases}$$

Hence the integral goes from 0 to $x$, and we have

$$y = e^x - e^{-2x} - \frac{1}{3} \int_0^x (e^{2(t-x)} - e^{x-t})h(t)\, dt$$

as the unique solution to $y'' + y' - 2y = h$, $y(0) = 0$, $y'(0) = 3$. This also works for $x < 0$.

## Exercises

1. Let $f_n(x) = (n/\pi)^{1/2} \exp(-nx^2)$, $n = 1, 2, \ldots$.
   (a) Superimpose on the same graph plots of $f_n(x)$ versus $x$ for $n = 20$, 200, and 1900. (Use $x$-scale $-0.5 \leqslant x \leqslant 0.5$, $y$-scale $0 \leqslant y \leqslant 25$.)

(b) Compute

$$\int_{-\infty}^{\infty} f_n(x)\,dx \quad \text{and} \quad \lim_{n\to\infty} \int_{-\infty}^{\infty} f_n(x)\,dx$$

What is the area under each of the three graphs you have plotted in
(a)? (Hint: You may use the formula $\int_0^\infty \exp(-u^2)\,du = \sqrt{\pi}/2$,
which we will prove later.)

(c) Prove that, if $x \neq 0$, $\lim_{n\to\infty} f_n(x) = 0$. Prove also that $\lim_{n\to\infty} f_n(0)$
$= \infty$.

2. Evaluate:

(a) $\displaystyle\int_{-\infty}^{\infty} \delta(-x)\,dx$  (b) $\displaystyle\int_{-\infty}^{\infty} \delta(1-x)\,dx$  (c) $\displaystyle\int_{-\infty}^{\infty} \delta(2x)\,dx$

(d) $\displaystyle\int_{-\infty}^{\infty} x\delta(x-1)\,dx$  (e) $\displaystyle\int_{-\infty}^{\infty} \sin t\delta(x-t)\,dt$

3. What value would you assign to $\int_{-\infty}^{\infty} (\delta(x))^2\,dx$? Justify your answer.
4. Express the following functions as linear combinations of Heaviside
functions:

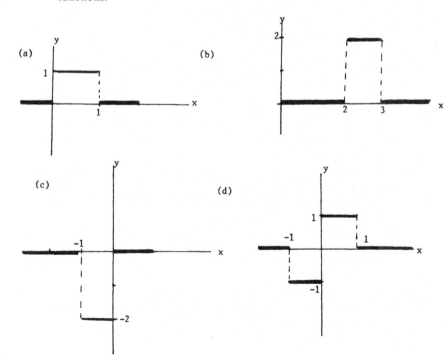

5. Express in terms of the Heaviside function

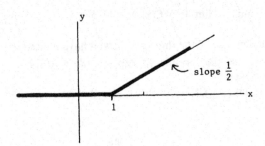

6. (a) Show that $(f(x)H(x-1))' = f'(x)H(x-1) + f(1)\delta(x-1)$.
   (b) Using (a), show that $(\sin \pi x H(x-1))' = \pi \cos \pi x H(x-1)$.
7. (a) If

$$G(x, t) = \frac{f(x)}{a(t)f(t)} H(x-t)$$

   show that

$$\frac{d}{dx} G(x, t) = \frac{f'(x)}{a(t)f(t)} H(x-t) + \frac{1}{a(t)} \delta(x-t)$$

   (b) Show that if $f(x)$ satisfies the equation $a(x)f'(x) + b(x)f(x) = 0$, then
   the function $G(x, t)$ from (a) satisfies the equation

$$a(x)\frac{d}{dx} G(x, t) + b(x)G(x, t) = \delta(x-t)$$

8. Find a particular solution to $d^2y/dx^2 + dy/dx - 2y = e^x$ by Green's
   function methods.
9. Verify that $f_1(x) = x$, $f_2(x) = x^2$ span the kernel of the expression
   $l(y) = x^2y'' - 2xy' + 2y$. Then
   (a) Find a Green's function for this expression [one answer:
   $G(x, t) = (x^2/t^3 - x/t^2)H(x-t)$].
   (b) Using the Green's function found in (a), write down an integral
   expression for a particular solution to $l(y) = h$. By solving the
   integral, find a particular solution to $x^2y'' - 2xy' + 2y = x^3$. Check
   your answer.
   (c) Find the general solution to $x^2y'' - 2xy' + 2y = x^3$.
10. (a) By substituting $y = x^a$ and solving for the constant $\alpha$, find two
    functions that span the kernel of $l(y) \equiv y'' - 6x^{-2}y$.
    (b) Using your answer in (a), find a Green's function for $l$.

(c) Using the Green's function found in (b), write down an integral expression for a particular solution to $l(y) = h$. By solving the integral, find a particular solution to $y'' - 6x^{-2}y = x \ln x$. Check your answer.

(d) Find the general solution to $y'' - 6x^{-2}y = x \ln x$.

11. Find a particular solution to $y'' - 2y' + y = 6xe^x$ by Green's function methods. Check your answer.

12. This exercise includes a proof of Theorem 2.18.

(a) Verify that the function $G(x, t)$ given in equation (2.10.18) satisfies

$$G(x_0, t) = \frac{\partial G}{\partial x}(x_0, t) = 0$$

(b) Verify that the function $y$ given in equation (2.10.19) satisfies

$$l(y) = h, \qquad y(x_0) = y_0, y'(x_0) = y_1$$

13. (a) Find the unique Green's function for the initial value problem

$$d^2y/dx^2 + dy/dx + y = \delta(x - t), y(0) = y'(0) = 0.$$

(b) Using the Green's function found in (a), express the unique solution $y$ to $y'' + y' + y = h$, $y(0) = y'(0) = 0$, in integral form.

(c) Evaluate your integral in (b) for $h(x) = e^{-x/2}$, and check your answer against that to Exercise 7 of Section 2.6.

## APPENDIX 2.A  SECOND ORDER LINEAR DIFFERENTIAL EQUATIONS IN THE COMPLEX DOMAIN

### 2.A.1  Ordinary Points

**Theorem 2.19.**  *Suppose $p(z)$ and $q(z)$ are analytic at $z = z_0$. Let $y_0, y_1$ be any two fixed complex numbers. Then the equation*

$$y'' + p(z)y' + q(z)y = 0 \qquad (2.A.1)$$

*has one, and only one, solution $y(z)$ that satisfies $y(z_0) = y_0$, $y'(z_0) = y_1$. This solution is analytic in the largest open disk centered at $z_0$ in which both $p(z)$ and $q(z)$ are analytic.*

In Theorem 2.7 of Section 2.2 we wrote our equation $a(x)y'' + b(x)y' + c(x)y = 0$, which we easily convert to the above form if we write $z$ instead of $x$ to emphasize the fact that we are working in the complex domain, and set $p(z) = b(z)/a(z)$, $q(z) = c(z)/a(z)$. Also, we lose no generality working at the point $z_0 = 0$, as replacing $z$ by $z - z_0$ we convert any result relative to the point $z = 0$ to one at the point $z = z_0$.

The formula for the coefficients of a power series solution to equation (2.A.1) has been derived in Section 2.2, Exercise 6. As $p(z)$ and $q(z)$ are assumed analytic at $z = 0$, we have power series expansions

$$p(z) = p_0 + p_1 z + p_2 z^2 + \cdots = \sum_{n=0}^{\infty} p_n z^n$$

$$q(z) = q_0 + q_1 z + q_2 z^2 + \cdots = \sum_{n=0}^{\infty} q_n z^n$$

both convergent in $|z| < R$, $R > 0$, where $R$ is the smaller of the radii of convergence of $p(z)$ and $q(z)$. In terms of the $p_n$ and $q_n$, the coefficients $y_n$ of the power series expansion of $y(z)$ are given by the formulas derived in Exercise 6 of Section 2.2,

$$y_{n+2} = -\frac{\sum_{l=2}^{n+1}(lp_{n-l+1}+q_{n-l})y_l+(p_n+q_{n-1})y_1+q_n y_0}{(n+1)(n+2)}, \qquad n \geqslant 1$$

$$y_2 = -\frac{p_0 y_1 + q_0 y_0}{2} \tag{2.A.2}$$

Once $y_0$ and $y_1$ are specified, all subsequent $y_n$ are determined by equation (2.A.2). To prove Theorem 2.19 we must show that the series for $y(z)$,

$$y(z) = y_0 + y_1 z + y_2 z^2 + \cdots = \sum_{n=0}^{\infty} y_n z^n$$

with coefficients determined by (2.A.2), always converges in $|z| < R$ independent of how $y_0$ and $y_1$ are specified.

We proceed as follows. Select a real number $S$ satisfying $0 < S < R$, choose $M > 0$ and $N > 0$, and consider the equation

$$Y''(z) - \frac{M}{(1 - z/S)} Y'(z) - \frac{N}{(1 - z/S)^2} Y(z) = 0 \tag{2.A.3}$$

One may easily check that $Y(z) = A(1 - z/S)^{-\lambda}$, where $A$ is any constant and $\lambda$ is the positive real number given by

$$2\lambda = ((1 - MS)^2 + 4NS^2)^{1/2} - (1 - MS)$$

is a solution to (2.A.3). The function $Y(z) = A(1 - z/S)^{-\lambda}$ is analytic in $|z| < S$ and, by the same analysis that we used to derive the recursion (2.A.2),

$$Y_{n+2} = +\frac{\sum_{l=2}^{n+1}(lP_{n-l+1}+Q_{n-l})Y_l+(P_n+Q_{n-1})Y_1+Q_n Y_0}{(n+1)(n+2)}, \qquad n \geqslant 1$$

where the $Y_n$ are the coefficients in the power series expansion of $Y(z)$ and the

$P_n$ and $Q_n$ are the coefficients in the power series expansion of the functions

$$P(z) = +\frac{M}{(1 - z/S)}, \qquad Q(z) = +\frac{N}{(1 - z/S)^2}$$

respectively.

As for the expansion coefficients of $Y(z)$ we have, by the binomial theorem,

$$Y(z) = A\left(1 - \frac{z}{S}\right)^{-\lambda} = A \sum_{n=0}^{\infty} \binom{-\lambda}{n}\left(-\frac{z}{S}\right)^n$$

where

$$\binom{-\lambda}{0} = 1$$

and

$$\binom{-\lambda}{n} = \frac{(-\lambda)(-\lambda-1)\cdots(-\lambda-n+1)}{n!}, \qquad n \geqslant 1$$

Noting that

$$\binom{-\lambda}{n} = (-1)^n \frac{\lambda(\lambda+1)\cdots(\lambda+n-1)}{n!}, \qquad n \geqslant 1$$

we have

$$Y(z) = A\left[1 + \frac{\lambda}{S}z + \sum_{n=2}^{\infty} \frac{\lambda(\lambda+1)\cdots(\lambda+n-1)}{n!S^n}z^n\right]$$

so

$$Y_0 = A, \quad Y_1 = \frac{A\lambda}{S}, \quad \text{and} \quad Y_n = \frac{\lambda(\lambda+1)\cdots(\lambda+n-1)}{n!S^n}, \qquad n \geqslant 2$$

Clearly, $Y_n > 0$ for all $n = 0, 1, 2, \ldots$. Furthermore, by taking $A$ large, we may make $Y_0$ and $Y_1$ as large as we please.

As the power series for $p(z)$ converges in $|z| < R$, it converges at $z = S$ because $0 < S < R$. Since $\Sigma p_n S^n$ converges, we have $p_n S^n \to 0$; thus there exists a constant $M > 0$ so that

$$|p_n| < \frac{M}{S^n}, \qquad n = 0, 1, 2, \ldots$$

We may make our earlier choice of $M$ agree with this $M$. A similar analysis shows that

$$|q_n| < \frac{N(n + 1)}{S^n}, \qquad n = 0, 1, 2, \ldots$$

for suitable $N > 0$, and we choose the original number $N$ to agree with this choice. The functions $P(z)$ and $Q(z)$ have the following power series expansions:

$$P(z) = M\left(1 - \frac{z}{S}\right)^{-1} = \sum_{n=0}^{\infty} \frac{M}{S^n} z^n$$

$$Q(z) = N\left(1 - \frac{z}{S}\right)^{-2} = NS\frac{d}{dz}\left(1 - \frac{z}{S}\right)^{-1} = \sum_{n=0}^{\infty} \frac{N(n+1)}{S^n} z^n$$

hence $P_n = M/S^n$, $Q_n = N(n+1)/S^n$.

Suppose now that $y_0$ and $y_1$ have been specified. We may then choose the constant $A$ so that

$$|y_0| < Y_0 = A \qquad \text{and} \qquad |y_1| < Y_1 = \frac{A\lambda}{S}$$

Then, from equations (2.A.2)

$$|y_2| = \tfrac{1}{2}|p_0 y_1 + q_0 y_0| \leqslant \tfrac{1}{2}(|p_0||y_1| + |q_0||y_0|)$$
$$< \tfrac{1}{2}(P_0 Y_1 + Q_0 Y_0) = Y_2$$

Next,

$$y_3 = -\frac{(2p_0 + q_1)y_2 + (p_1 + q_0)y_1 + q_1 y_0}{6}$$

so

$$|y_3| \leqslant \frac{(2P_0 + Q_1)Y_2 + (P_1 + Q_0)Y_1 + Q_1 Y_0}{6} = Y_3$$

One can see how this argument will proceed step by step to establish that

$$|y_n| < Y_n, \qquad n = 0, 1, 2, \dots$$

A rigorous proof can be given using mathematical induction, which we will cover in the body of the text later on. In any case, we can take it as proved that the coefficients of the power series expansion of our solution $y(z)$ are in absolute value less than the corresponding coefficients of $Y(z)$. As $Y(z) = A(1 - z/S)^{-\lambda}$ is analytic in $|z| < S$, its power series converges in this open disk. So, therefore, does the power series for $y(z)$. As this is true for every $S$ satisfying $0 < S < R$, the power series for $y(z)$ converges in $|z| < R$, and we have proved Theorem 2.19.

If $p(z)$ and $q(z)$ are analytic at $z = z_0$, we call $z_0$ an *ordinary point* for the differential equation $y'' + py' + qy = 0$. According to Theorem 2.19, at an ordinary point $z_0$ we may find one solution $f_1(z)$ such that

$$f_1(z_0) = 1, \qquad f_1'(z_0) = 0$$

and another solution $f_2(z)$ that satisfies

$$f_2(z_0) = 0, \qquad f_2'(z_0) = 1$$

Exactly the same argument used to prove Corollary 2.10 of Section 2.3 can be used here to prove that the functions $f_1(z)$, $f_2(z)$ are linearly independent and that any solution $y$ of $y'' + py' + qy = 0$ is a linear combination of $f_1$ and $f_2$, $y = af_1 + bf_2$ for (complex) constants $a$ and $b$. We thus have this corollary to Theorem 2.19.

**Corollary 2.20.** *At an ordinary point $z_0$, the kernel of the differential expression $y'' + p(z)y' + q(z)y$ is two-dimensional, and any function in the kernel is analytic in the largest open disk centered at $z_0$ in which both $p(z)$ and $q(z)$ are analytic. Moreover, the method of undetermined coefficients provides an effective algorithm for calculating the coefficients of the power series expansion of any function $y(z)$ in the kernel, once $y(z_0)$ and $y'(z_0)$ are specified.*

Of course, a function $y(z)$ in the kernel may be analytic in a much larger region than the common disk of convergence of the power series expansions of $p(z)$ and $q(z)$, but at least that much analyticity is guaranteed.

## 2.A.2 Singular Points

We say that the point $z_0$ is an *isolated singularity* of the function $f(z)$ if there is an $R > 0$ such that $f(z)$ is analytic and single valued in the annulus $0 < |z - z_0| < R$. At such a point, $f(z)$ has a Laurent series expansion

$$f(z) = \sum_{n=-\infty}^{\infty} a_n(z - z_0)^n$$

convergent at least in the same annulus $0 < |z - z_0| < R$. If there exists a positive integer $p$ such that $a_{-n} = 0$ for $n > p$ and $a_{-p} \neq 0$, then we say that $f$ has a *pole of order $p$ at $z_0$*. If $a_{-n} \neq 0$ for infinitely many positive $n$, then we call $z_0$ an *essential singularity*. For example,

$$f(z) = \frac{1}{z - i}$$

has a pole of order 1 at $z = i$, and

$$\exp\left(\frac{1}{z - 1}\right) = \sum_{n=0}^{\infty} \frac{1}{(n!)}(z - 1)^{-n}$$

has an essential singularity at $z = 1$. If $a_{-n} = 0$ for all $n \geq 1$, then $f(z)$ is actually analytic at $z_0$, in which case we refer to $z_0$ as an *apparent* or *removable* singularity.

If either of the functions $p(z)$, $q(z)$ has an isolated singularity at $z = z_0$, then we call $z_0$ a *singular point* for the differential equation $y'' + p(z)y' + q(z)y = 0$. In studying the behavior of such an equation at a singular point, we may take that point as $z_0 = 0$ because from any theorems that describe the facts relative to the point 0, we may get the general results relative to $z_0$ simply by substituting $z - z_0$ for $z$.

Suppose then that 0 is a singular point for the equation $y'' + p(z)y' + q(z)y = 0$. Thus $p(z)$ or $q(z)$ or both have an isolated singularity at $z = 0$ and are analytic in $0 < |z| < R$ for some $R > 0$. Note that there is no restriction on the type of singularity that $p$ and $q$ may have. They could both have poles of any order, or one or both could have an essential singularity at $z = 0$.

Select and fix a point $z_0$ in the annulus $0 < |z| < R$. Then $z_0$ is an ordinary point for our equation, so by Theorem 2.19 the solution space of $y'' + p(z)y' + q(z)y = 0$ in a neighborhood of $z_0$ is spanned by two functions $f_1(z)$ and $f_2(z)$ analytic at $z_0$. Select one from this linearly independent pair, say $f_1$, and continue it analytically once counterclockwise around the origin and return to the point $z_0$. At each stage of the analytic continuation, the continued function remains a solution of $y'' + py' + qy = 0$, and when we return to $z_0$ the analytic continuation of $f_1(z)$, call it $g_1(z)$, is also a solution of the equation at the ordinary point $z_0$. Hence, since $f_1$ and $f_2$ span the solution space at $z_0$ we have

$$g_1(z) = af_1(z) + bf_2(z) \qquad (2.A.4)$$

for suitable complex constants $a$ and $b$.

Now if $z = 0$ had been an apparent singularity only, then $f_1(z)$ would have been analytic and single valued in $|z| < R$ and we would have $g_1 = f_1$ so $a = 1, b = 0$. In general, however, with an actual singularity at $z = 0$, this will not happen, and (2.A.4) is the best we can say.

Continue $f_2(z)$ in the same way. Call $g_2(z)$ the continuation of $f_2(z)$, so that

$$g_2(z) = cf_1(z) + df_2(z) \qquad (2.A.5)$$

for suitable complex constants $c$ and $d$.

We can combine (2.A.4) and (2.A.5) into a matrix equation

$$\begin{bmatrix} g_1 \\ g_2 \end{bmatrix} = \begin{bmatrix} a & b \\ c & d \end{bmatrix} \begin{bmatrix} f_1 \\ f_2 \end{bmatrix}$$

Every solution $f(z)$ of our equation $y'' + py' + qy = 0$ at the ordinary point $z = z_0$ has the form $f = \alpha f_1 + \beta f_2$ for suitable complex constants $\alpha, \beta$. If we analytically continue such a function $f$ counterclockwise once around the origin, it becomes a function $g$ where $g = \alpha g_1 + \beta g_2$. We ask, is there such a function $f$, not the zero solution, which when analytically continued becomes a scalar multiple of itself, i.e., $g = \lambda f$? If this is the case, then we must have $\alpha g_1 + \beta g_2 = \lambda(\alpha f_1 + \beta f_2)$. Using (2.A.4) and (2.A.5), we get

$$(\alpha a + \beta c - \lambda \alpha)f_1 + (\alpha b + \beta d - \lambda \beta)f_2 = 0$$

As $f_1$ and $f_2$ are linearly independent, we must have the pair of equations

$$\alpha a + \beta c - \lambda \alpha = 0 \quad \text{and} \quad \alpha b + \beta d - \lambda \beta = 0$$

which can be put into matrix form

$$[\alpha, \beta]\begin{bmatrix} a & b \\ c & d \end{bmatrix} = \lambda[\alpha, \beta]$$

Thus $\lambda$ must be an eigenvalue for our matrix and $[\alpha, \beta]$ an eigenvector belonging to that eigenvalue.

The eigenvalues are the roots of the determinantal equation

$$\begin{bmatrix} a - \lambda & b \\ c & d - \lambda \end{bmatrix} = \lambda^2 - (a + d)\lambda + (ad - bc) = 0$$

which has either two distinct roots, say $\lambda_1$ and $\lambda_2$, or one root of multiplicity 2.

Let $[\alpha, \beta] \neq [0, 0]$ be an eigenvector belonging to the eigenvalue $\lambda_1$. Then $f = \alpha f_1 + \beta f_2$ has the property that, when it is continued once around the origin, it becomes $\lambda_1 f$.

Now $\lambda_1 \neq 0$ as our matrix is nonsingular. Let

$$\rho = \frac{\log \lambda_1}{2\pi i}$$

for some fixed value of the logarithm, and consider the function

$$z^\rho = e^{\rho \log z}$$

When $z^\rho$ is continued once around the origin, $\log z$ increases by $2\pi i$, hence $z^\rho$ is multiplied by

$$e^{\rho(2\pi i)} = e^{\log \lambda_1} = \lambda_1$$

Thus $f(z)/z^\rho$, when contained around the origin, returns to its original value. Thus $f(z)/z^\rho$ is *single valued* and analytic in $0 < |z| < R$ and so possesses a convergent Laurent series expansion

$$\frac{f(z)}{z^\rho} = \sum_{n=-\infty}^{\infty} a_n z^n, \quad 0 < |z| < R$$

We have proved:

**Theorem 2.21.** *At a singular point $z_0$, the differential equation $y'' + p(z)y' + q(z)y = 0$ always has one solution of the form*

$$y(z) = (z - z_0)^\rho \sum_{n=-\infty}^{\infty} a_n (z - z_0)^n \qquad (2.A.7)$$

*where $\rho$ is a (complex) constant, and the Laurent series converges in some annulus $0 < |z| < R$, $R > 0$.*

If the two roots, $\lambda_1$ and $\lambda_2$, are distinct, then our equation has another solution of the form (2.A.5) with a different $\rho$ and generally different Laurent series coefficients. If $\lambda_1 = \lambda_2$, then, unless our matrix is already a nonzero multiple of the identity matrix, there will be only one eigenvector, thus only one solution of the form (2.A.5). In this case there is a second solution involving $\log(z - z_0)$ which we can find either by using the Wronskian or by other methods. We shall go no further into the matter.

In any case, whether the point $z_0$ is ordinary or singular for the differential equation $y'' + p(z)y' + q(z)y = 0$, it always has two linearly independent solutions as described above. Note the absence of any restriction on the type of singularity $p(z)$ and $q(z)$ may have at $z = z_0$; our result is quite general in that respect. However, we have provided no algorithm for effectively calculating the $\rho$ and the $a_n$ in (2.A.7). Under special restrictions on the type of singularities of $p(z)$ and $q(z)$, the method of undetermined coefficients provides such an algorithm, and we turn to that matter now.

### 2.A.3  At a Regular Singular Point, the Method of Undetermined Coefficients Works

If our differential equation can be put in the form

$$z^2 y'' + z p_1(z) y' + q_1(z) y = 0 \qquad (2.A.8)$$

where $p_1(z)$ and $q_1(z)$ are *analytic* at $z = 0$, then 0 is called a *regular singular point* for the equation. In terms of the original notation, $y'' + p(z)y' + q(z)y = 0$, this is equivalent to requiring that $p(z)$ has, at $z = 0$, a pole of order 1 or less and $q(z)$ a pole of order 2 or less.

At any singular point, our equation has a solution of the form (2.A.7). At a regular singular point it turns out that the solution (2.A.7) has $a_{-n} = 0$ for all sufficiently large positive $n$. Hence, by changing the $\rho$, this solution can be written

$$y = z^\rho \sum_{n=0}^{\infty} a_n z^n \qquad (2.A.9)$$

The method of undetermined coefficients consists of substituting (2.A.9) into (2.A.8), then determining $\rho$ and the $a_n$ so that (2.A.9) is a solution to the differential equation (2.A.8). Our assertion here is that at a regular singular point this procedure will always produce a solution. We shall give no proof of this fact here, nor shall we do the substitution in general, as the results are somewhat complicated. In the text we deal only with Legendre's, Hermite's, Bessel's, and Laguerre's equations, and in these cases undetermined coefficients work (to get one solution) with $\rho = 0$. All these calculations are carried out in the text, and we shall go no further with the discussion here.

# 3
# Hilbert Space

This chapter presents the elementary theory of Hilbert space, with particular emphasis on the $L^2$ spaces. This material is contained in Sections 3.6 through 3.10. In teaching this subject over the years I found that students experienced substantially more difficulty with the Hilbert space theory than with the linear differential equations contained in Chapters 1 and 2, I suspect because the study of Hilbert space involves some new and abstract ideas, preparation in linear algebra notwithstanding. As the Hilbert space method is the main theme of the entire text, I wanted to explain it perspicuously, to put it across as effectively as I could. So I experimented with various ways to make the subject more accessible and finally settled on the present arrangement, which uses the wave equation to motivate Fourier series, which in turn exemplify orthogonal function expansions and provide a natural device to motivate the concept of an inner product. Hence Sections 3.1 through 3.5 serve a double purpose: (1) guiding the reader to the ensuing Hilbert space theory, and (2) expounding topics, namely the wave equation and Fourier series, very important in their own right.

## 3.1 THE VIBRATING WIRE

We consider now the classical problem of the vibrating wire. Our object is to analyze and predict the physical characteristics of the motion of a stretched

78

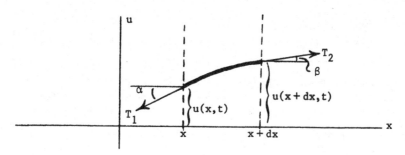

**Figure 3.1**

wire, like the strings of a violin, cello, guitar, harp, or piano, or a tightly stretched rubber band.

As our examples suggest, we will concern ourselves with very small lateral motions, although in Figure 3.1 we have exaggerated the displacement to make the figure clearer. We denote horizontal distance along the wire by $x$ and denote by $u(x, t)$ the vertical displacement of the wire at position $x$ and time $t$. The vertical displacement is measured from the equilibrium position given by $u = 0$. Suppose $T$ represents the tension of the wire (units of force) and $\rho$ its linear density, which we can measure by weighing the wire and dividing the weight by the stretched length.

In Figure 3.1 we have pictured an isolated infinitesimal piece of our wire. The symbols $T_1$ and $T_2$ represent the forces acting on the ends of this displaced segment. Under the reasonable assumption that the horizontal component of force is constant, we have $T = T_1 \cos \alpha = T_2 \cos \beta$, where $T$ is the (constant) tension in the wire. The net vertical force on this segment of wire is

$$dF = T_2 \sin \beta - T_1 \sin \alpha$$

Taking into account the just-derived relation between $T_1$ and $T_2$, we get $dF = T(\tan \beta - \tan \alpha)$ for the net vertical force. The expression $\tan \beta - \tan \alpha$ represents the change in slope as we go from $x$ to $x + dx$. Since $\partial u / \partial x$ equals the slope, we have $\tan \beta - \tan \alpha = d(\partial u / \partial x) = (\partial^2 u / \partial x^2) \, dx$. Thus the net vertical force on our segment of wire is $dF = T(\partial^2 u / \partial x^2) \, dx$.

We equate this net force to mass times acceleration. The mass of our segment of wire before displacement is $dm = \rho \, dx$, and this mass remains the same after displacement, even though it is distributed over a slightly longer stretched piece. The acceleration is $\partial^2 u / \partial t^2$. So Newton's law, force = mass × acceleration, yields

$$\frac{\partial^2 u}{\partial x^2} = \frac{\rho}{T} \frac{\partial^2 u}{\partial t^2} \qquad (3.1.1)$$

the partial differential equation governing the motion of our wire. This is called the *one-dimensional wave equation*. If we set $w^2 = T/\rho$, we may also write it

$$\frac{\partial^2 u}{\partial x^2} = \frac{1}{w^2} \frac{\partial^2 u}{\partial t^2}$$

where $w$ has the dimension of velocity (Exercise 1).

Let us examine the consequences of (3.1.1) and what it predicts for our physical system. Note first that in all our examples cited above, our wire is fixed at each end. Taking the coordinates of the ends of the string as $x = 0$ and $x = L$, we must therefore have

$$u(0, t) = u(L, t) = 0 \qquad (3.1.2)$$

for all times $t$. We attempt to solve equation (3.1.1) by the method of *separation of variables* by assuming that we can write the displacement $u(x, t)$ as the product of two functions, one depending solely on $x$ and the other on $t$:

$$u(x, t) = y(x)z(t)$$

Substituting into (3.1.1) we find

$$\frac{y''}{y} = \left(\frac{\rho}{T}\right)\frac{z''}{z}$$

Note that now the left-hand side depends only on $x$, the right only on $t$. Hence $y''/y$ in fact does not vary with $x$ (since the right side does not) and so is a constant $\lambda$. Thus we get two ordinary differential equations

$$\frac{d^2 y}{dx^2} = \lambda y \qquad \text{and} \qquad \frac{d^2 z}{dt^2} = \left(\frac{T}{\rho}\right)\lambda z$$

where the first equation must include the physical fact that the ends of the string are fixed: $y(0) = y(L) = 0$.

We look first at our $x$ system

$$\frac{d^2 y}{dx^2} = \lambda y, \qquad y(0) = y(L) = 0 \qquad (3.1.3)$$

This is a *boundary value problem.* By considering the general solution of $y'' - \lambda y = 0$ and attempting to fit this solution to the boundary conditions $y(0) = y(L) = 0$, we see that only for negative values of the parameter $\lambda$ do we have nontrivial solutions of (3.1.3) (refer to Exercises 5 and 6). When $\lambda < 0$ the general solution of $y'' = \lambda y$ is

$$y = A \cos \sqrt{-\lambda}x + B \sin \sqrt{-\lambda}x$$

The condition $y(0) = 0$ forces $A = 0$, and to have $y(L) = 0$ we must have $\sin \sqrt{-\lambda}L = 0$ or $\sqrt{-\lambda}L$ equal to an integral multiple of $\pi$:

$$\sqrt{-\lambda}L = n\pi \quad \text{or} \quad \lambda = -\frac{n^2\pi^2}{L^2}, \quad n = 0, 1, 2, \ldots$$

We may delete $n = 0$, since it corresponds to the zero solution, and so find that our system (3.1.3) has nonzero solutions only when

$$\lambda = -\frac{\pi^2}{L^2}, -4\frac{\pi^2}{L^2}, -9\frac{\pi^2}{L^2}, \ldots, -n^2\frac{\pi^2}{L^2}, \ldots$$

and the corresponding solutions are (up to multiplication by an arbitrary constant)

$$\sin\frac{\pi x}{L}, \sin\frac{2\pi x}{L}, \sin\frac{3\pi x}{L}, \ldots, \sin\frac{n\pi x}{L}, \ldots$$

Substituting the allowed values of $\lambda$ into the second equation, we get

$$\frac{d^2z}{dt^2} = -\frac{T}{\rho}n^2\frac{\pi^2}{L^2}z$$

which has the general solution

$$z = A_n \cos \sqrt{\frac{T}{\rho}}\frac{n\pi t}{L} + B_n \sin \sqrt{\frac{T}{\rho}}\frac{n\pi t}{L}, \quad n = 0, 1, 2, \ldots$$

We have therefore found an infinite family of solutions of our one-dimensional wave equation:

*The equation $\partial^2 u/\partial x^2 = (T/\rho)\partial^2 u/\partial t^2$ with boundary conditions $u(0, t) = u(L, t) = 0$ has the solutions*

$$u_1(x, t) = \sin\frac{\pi x}{L}\left(A_1 \cos \sqrt{\frac{T}{\rho}}\frac{\pi t}{L} + B_1 \sin \sqrt{\frac{T}{\rho}}\frac{\pi t}{L}\right)$$

$$u_2(x, t) = \sin\frac{2\pi x}{L}\left(A_2 \cos \sqrt{\frac{T}{\rho}}\frac{2\pi t}{L} + B_2 \sin \sqrt{\frac{T}{\rho}}\frac{2\pi t}{L}\right)$$

$$\vdots$$

$$u_n(x, t) = \sin\frac{n\pi x}{L}\left(A_n \cos \sqrt{\frac{T}{\rho}}\frac{n\pi t}{L} + B_n \sin \sqrt{\frac{T}{\rho}}\frac{n\pi t}{L}\right)$$

$$\vdots$$

For our linear one-dimensional wave equation, any linear combination of solutions is also a solution (refer to Exercise 2). Since we have an infinite family of solutions, we should consider the infinite sum $\Sigma u_n$. If we seize on a fixed time $t = t_0$ (a "freeze frame") and ask "What is the shape of our wire at that frozen moment in time?" we are led to consider a series of the form

$$u(x, t_0) = \sum_{n=1}^{\infty} c_n \sin \frac{n\pi x}{L}$$

where the $c_n$ are constants. This is a special kind of *Fourier series*.

## Exercises

1. Show that $\rho/T$ has dimensions $1/(\text{velocity})^2$, so $w = (T/\rho)^{1/2}$ has dimensions of velocity.

2. Show that the wave equation

   $$\frac{\partial^2 u}{\partial x^2} = \frac{1}{w^2} \frac{\partial^2 u}{\partial t^2}$$

   is *linear*; i.e., show that if $p(x, t)$ and $q(x, t)$ both satisfy the wave equation, then so does $u = ap + bq$ where $a$ and $b$ are any constants.

3. (a) Show that if $f(\cdot)$ is a function of one variable, then $u(x, t) = f(x - wt)$ solves the wave equation. Same for $f(x + wt)$.

   (b) In part (a) consider $f(z) = 1$ for $0 \leqslant z \leqslant 1$, $f(z) = 0$ elsewhere (the "unit pulse"). Supposing $w = 2$, sketch the solution $u(x, t) = f(x - wt)$ of the wave equation for $t = 0, 1, 2, 3$. Same for $f(x + wt)$. Does the name "traveling wave" describe these solutions?

   [Hints for part (a): Recall the following special case of the "chain rule" from partial differentiation. If $g(\cdot)$ is a function of one variable and $q = q(s, t)$ is a function of two variables, $s$ and $t$, then $h = g(q)$ can be considered a function of two variables, $s$ and $t$. Under these assumptions $\partial h/\partial s = g'(q) \, \partial q/\partial s$ and $\partial h/\partial t = g'(q) \, \partial q/\partial t$.]

4. Deduce from the results of Exercises 2 and 3a that if $f(\cdot)$ is a function of a single variable, then both $r(x, t) = f(x - wt) + f(x + wt)$ and $s(x, t) = f(x - wt) - f(x + wt)$ are solutions of the wave equation. Fix an integer $n \geqslant 1$. For the particular choice $f_n(z) = \frac{1}{2} \sin(n\pi z/L)$, show that the corresponding $r_n(x, t)$ satisfies the boundary conditions $r_n(0, t) = r_n(L, t) = 0$. For the particular choice $f_n(z) = \frac{1}{2} \cos(n\pi z/L)$, show that the corresponding $s_n(x, t)$ also satisfies the boundary conditions $s_n(0, t) = s_n(L, t) = 0$. Relate $r_n$ and $s_n$ to the solution $u_n$ given at the end of this section ($u_n = A_n r_n + B_n s_n$).

5. For $\lambda > 0$ find the general solution to $y'' = \lambda y$ and show that it must vanish identically in order to satisfy $y(0) = y(L) = 0$.

   [Hints for Exercise 5: Referring to Section 2.4, show first that when $\lambda > 0$ the equation $y'' - \lambda y = 0$ has the general solution $y = c_1 e^{\sqrt{\lambda}x} + c_2 e^{-\sqrt{\lambda}x}$.

Then show that the conditions $y(0)=0$, $y(L)=0$ yield two simultaneous linear equations in the unknowns $c_1$ and $c_2$. Compute the determinant of the coefficients, and using the fact that $e^t = 1$ only when $t = 0$, show that this determinant $\neq 0$. Conclude that $c_1 = c_2 = 0$.]

6. For $\lambda = 0$ find the general solution to $y'' = \lambda y$ and show that it must vanish identically in order to satisfy $y(0) = y(L) = 0$.

## 3.2   FOURIER SERIES

In our study of the wave equation we found ourselves led to the expansion of a function as a series of terms $c_n \sin(n\pi x/L)$. This is a special case of a *Fourier series*. The general Fourier series is written

$$f(x) = \frac{a_0}{2} + \sum_{n=1}^{\infty} (a_n \cos nx + b_n \sin nx) \tag{3.2.1}$$

which has the natural interval $-\pi \leqslant x \leqslant \pi$ owing to the periodicity of the trigonometric functions. We shall see later that a simple scaling converts our series to one appropriate to any given finite interval, such as $0 \leqslant x \leqslant L$. The $\frac{1}{2}$ in the constant term is included to make subsequent formulas simpler.

We set this problem: *Given $f(x)$, what are the constants in (3.2.1), and how well does the resulting series represent $f(x)$?*

Consider the following integrals, where $m$, $n$ are integers $\geqslant 0$.

$$I = \int_{-\pi}^{\pi} \cos mx \cos nx \, dx$$

$$II = \int_{-\pi}^{\pi} \sin mx \cos nx \, dx$$

$$III = \int_{-\pi}^{\pi} \sin mx \sin nx \, dx$$

Making use of the trigonometric formulas

$$\cos A \cos B = \tfrac{1}{2}(\cos(A - B) + \cos(A + B))$$

$$\sin A \cos B = \tfrac{1}{2}(\sin (A + B) + \sin(A - B))$$

$$\sin A \sin B = \tfrac{1}{2}(\cos(A - B) - \cos(A + B))$$

we find for $I$

$$I = \int_{-\pi}^{+\pi} \cos mx \cos nx \, dx = \frac{1}{2} \int_{-\pi}^{+\pi} \cos(m - n)x \, dx$$

$$+ \frac{1}{2} \int_{-\pi}^{+\pi} \cos(m + n)x \, dx$$

If $m=n=0$, we have $I=\frac{1}{2}\int_{-\pi}^{+\pi}dx+\frac{1}{2}\int_{-\pi}^{+\pi}dx=\pi+\pi=2\pi$. If one of $m$, $n$ is $\neq 0$, then the second integral will vanish, and the first will also be zero unless $m=n$, in which case it will have the value $\pi$. Summarizing

$$I = \int_{-\pi}^{+\pi} \cos mx \cos nx \; dx = \begin{cases} 2\pi & \text{if } m = n = 0 \\ \pi & \text{if } m = n \neq 0 \\ 0 & \text{if } m \neq n \end{cases}$$

Operating similarly with the other integrals, we find:

$$II = \int_{-\pi}^{+\pi} \sin mx \cos nx \; dx = 0 \quad \text{and} \quad III = \int_{-\pi}^{+\pi} \sin mx \sin nx \; dx$$

$$= \begin{cases} 0 & \text{if } m = n = 0 \\ \pi & \text{if } m = n \neq 0 \\ 0 & \text{if } m \neq n \end{cases}$$

To determine the coefficients in the series (3.2.1), multiply both sides by $\cos mx$ where $m$ is an integer $\geqslant 0$, and integrate both sides from $-\pi$ to $+\pi$. If we integrate the series on the right term by term, as seems only reasonable under the circumstances, and utilize formulas $I$ and $II$ just derived, we find

$$a_m = \frac{1}{\pi} \int_{-\pi}^{+\pi} f(x) \cos mx \; dx, \qquad m = 0, 1, 2, \ldots \tag{3.2.2}$$

Operating similarly with $\sin mx$, we get

$$b_m = \frac{1}{\pi} \int_{-\pi}^{+\pi} f(x) \sin mx \; dx, \qquad m = 1, 2, \ldots \tag{3.2.3}$$

While our derivation of formulas (3.2.2) and (3.2.3) utilized a reasonable but unjustified term-by-term integration, the formulas themselves provide a simple and neat way of computing the coefficients $a_n$ and $b_n$ from the given function $f(x)$. We may therefore take the point of view that we *start* from equations (3.2.2) and (3.2.3) and *then* ask: How well does the resulting series (3.2.1) represent the function $f(x)$?

Let us try some numerical experimentation. Consider the function $f(x) = x$. This function is odd (refer to Exercises 1 through 4) so, as $\cos nx$ is even, $x \cos nx$ is odd and

$$a_n = \frac{1}{\pi} \int_{-\pi}^{+\pi} x \cos nx \; dx = 0, \qquad n = 0, 1, 2, \ldots$$

To evaluate $b_n$ we use integration by parts:

$$b_n = \frac{1}{\pi} \int_{-\pi}^{+\pi} x \sin nx \, dx = \frac{2}{\pi} \int_0^{\pi} x \sin nx \, dx = \frac{2}{\pi} \left( \frac{-x}{n} \cos nx \bigg|_0^{\pi} + \frac{1}{n} \int_0^{\pi} \cos nx \, dx \right)$$

$$= \frac{-2\pi}{\pi n} \cos n\pi = (-1)^{n+1} \frac{2}{n}, \quad n = 1, 2, \ldots$$

Thus (3.2.1) becomes

$$x = 2 \sum_{n=1}^{\infty} \frac{(-1)^{n+1}}{n} \sin nx$$

$$= 2(\sin x - \tfrac{1}{2} \sin 2x + \tfrac{1}{3} \sin 3x - \tfrac{1}{4} \sin 4x + \tfrac{1}{5} \sin 5x - \cdots)$$

In Figure 3.2 we have plotted $x$ versus the sum of the first five terms of its Fourier series. We have generally good agreement.

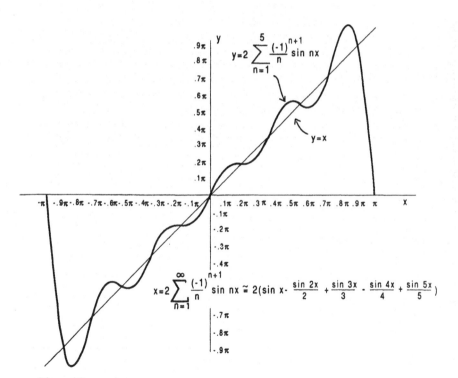

**Figure 3.2**

And we do notice some features of Fourier series in general and some features of the particular example.

*In general:*

1.  *The Fourier coefficients (3.2.2) and (3.2.3) depend only on the values of $f(x)$ in $-\pi \leqslant x \leqslant \pi$.*

2.  *The Fourier series is periodic of period $2\pi$;* that is, the series itself takes the same value at $x + 2n\pi$ as it does at $x$, for any $n = 0, \pm 1, \pm 2, \ldots$

We deduce that the Fourier series (3.2.1) cannot represent $f(x)$ for all $x$, but represents what we might call the "periodic extension" of $f(x)$: we take the graph of $f(x)$ in $-\pi \leqslant x \leqslant \pi$ and translate it to the intervals $-\pi + 2n\pi \leqslant x \leqslant +\pi + 2n\pi, n = \pm 1, \pm 2, \ldots$

Furthermore, the series will always have the same value at $+\pi$ as it does at $-\pi$ whether or not $f(\pi) = f(-\pi)$. Put another way, the series assigns the same value to $f(x)$ at $x = \pi$ as it does at $x = -\pi$, irrespective of the actual values of $f(x)$ there.

These observations suggest the following point of view, which we shall adopt. Remove the interval $-\pi \leqslant x \leqslant \pi$ from the real line and bend it in a circle, joining the point $+\pi$ to the point $-\pi$ so that they become one point (Figure 3.3). The circle has radius $+1$ (circumference $2\pi$) and forms a geometric entity that we call a *manifold without boundary*. There is no boundary because, as you traverse the circle, you never meet an obstacle. A Fourier series represents a single-valued function defined at every point of the circle. The fact that the series takes the same value at $x = +\pi$ as at $x = -\pi$ is now taken care of by the fact that these points are merged into the same point on the circle. Thus, in Fourier series we are concerned with functions defined

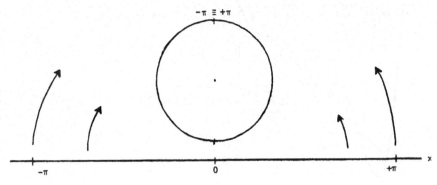

**Figure 3.3**

on the circle, and we shall keep that point of view in the foreground. We shall denote this circle of unit radius by $C$.

This point of view often has practical advantages. Suppose $g(x)$ is periodic, of period $2\pi$, and consider the integral

$$\int_{a-\pi}^{a+\pi} g(x)\,dx$$

As $g(x)$ is periodic of period $2\pi$, we may consider it as a function on the circle, and the first-displayed integral is an integral once around the circle, centered at point $a$ (modulo $2\pi$). As this integral is clearly independent of the starting point, we have

$$\int_{-\pi}^{+\pi} g(x)\,dx = \int_{a-\pi}^{a+\pi} g(x)\,dx \qquad \text{for any } a,\ g(x) \text{ periodic of period } 2\pi$$

In particular, in computing the Fourier coefficients (3.2.2) and (3.2.3), we may integrate over any interval of length $2\pi$ ($f(x)$ continued periodically).

*For the particular function* $f(x) = x$, for which $f(\pi) = \pi$ and $f(-\pi) = -\pi$, in order to define this function on the circle we must assign a single value at the amalgamated point $\pi \equiv -\pi$. We choose the average of the right- and left-hand limits, namely the value 0. Hence the function we are really dealing with is this one: $f(x) = x$, $-\pi < x < +\pi$, $f(\pm\pi) = 0$. In any case, no matter how we defined $f(x)$ at $+\pi \equiv -\pi$, the discrepancy between the right- and left-hand limits at that point would remain, and we see from Figure 3.2 that at the discontinuity the Fourier series converges to the average value of the left- and right-hand limits, a phenomenon that occurs in general. From the same figure we see also an "overshoot" at this discontinuity. This too occurs generally and is known as the *Gibbs phenomenon*.

We conclude this section with another example, $f(x) = \pi^2 - x^2$, for which $f(\pi) = f(-\pi) = 0$, so we may regard it already defined on the circle and needing no modification. As it is continuous there, we may expect better agreement. The function $f(x) = \pi^2 - x^2$ is even, so $f(x) \sin nx$ is odd, and all $b_n$'s are zero. The evaluation of the $a_n$'s requires a double integration by parts. We find

$$\pi^2 - x^2 = \frac{2\pi^2}{3} + 4 \sum_{n=1}^{\infty} \frac{(-1)^{n+1}}{n^2} \cos nx$$

$$\cong 6.5797 + 4 \cos x - \cos 2x + \tfrac{4}{9} \cos 3x$$

In Figure 3.4 we have plotted the four-term approximation against the function and find very good agreement.

These calculations were done on the Hewlett-Packard HP-15C, a relatively inexpensive programmable scientific electronic calculator. Exercises 5

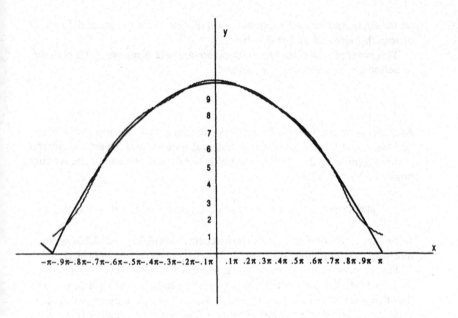

$$\pi^2 - x^2 = \frac{2\pi^2}{3} + 4\sum_{n=1}^{\infty} \frac{(-1)^{n+1}}{n^2} \cos nx \cong 6.5797 + 4 \cos x - \cos 2x + \frac{4}{9} \cos 3x$$

**Figure 3.4**

through 11, which ask you to do similar calculations, can also all be done on a hand-held calculator having capabilities comparable to those of the 15C. If you know a scientific programming language such as BASIC, C, FORTRAN, or PASCAL and have access to a computer, you may do the exercises on the computer. But no knowledge of a programming language or access to a computer is necessary to do any of these exercises. They are all within range of a good hand-held calculator. Sample programs for the HP-15C are provided in the appendix to this section.

### Exercises

*Even and Odd Functions*

1. An *even* function satisfies $f(-x) = f(x)$ and an *odd* function $f(-x) = -f(x)$. The function $\cos x$ is even, $\sin x$ is odd, and $e^x$ is neither

even nor odd. Classify the following functions:

(a) $x^3 - x$      (b) $\sin^2 x$      (c) $e^x \sin x$

(d) $x^4 - 2x^2 + 1$    (e) $e^{-x^2}$      (f) $(x^2 + 1) \sin x$

(g) $\sin 2x$      (h) $1/(x^4 + 1)$    (i) $\ln\left(\dfrac{1+x}{1-x}\right)$

(j) $|x|$      (k) $f(|x|)$, $f$ any function

2. (Continuation) (a) Describe the graph of an even function. Of an odd function.

(b) Complete and prove the following statements: even × even = ? even × odd = ? odd × odd = ?

(c) What is the only function that is both even and odd?

3. (Continuation) If $f(x)$ is any function, prove that then $f_e(x) = \frac{1}{2}[f(x) + f(-x)]$ is an even function and $f_o(x) = \frac{1}{2}[f(x) - f(-x)]$ is an odd function. Prove that any function is the sum of an even function and an odd function and that the even and odd parts are uniquely determined. Write $e^x$ as the sum of its even and odd parts.

4. (Continuation) (a) Prove that if $f(x)$ is even, then for any $a > 0$

$$\int_{-a}^{+a} f(x)\,dx = 2 \int_0^a f(x)\,dx$$

(b) Prove that if $f(x)$ is odd, then $\int_{-a}^{+a} f(x)\,dx = 0$.

*Calculating and Graphing Partial Sums of Fourier Series*

See Appendix 3.2.A for sample HP-15C programs.

5. Let $f(x) = 1$ for $0 < x < \pi$, $f(x) = -1$ for $-\pi < x < 0$, and $f(-\pi) = f(0) = f(\pi) = 0$. Find the Fourier series of $f(x)$, and graph the four-term partial sum against the function.

6. Find the Fourier series of $f(x) = x^2$, and graph the four-term partial sum against the function. Why is this so accurate?

7. Let $f(x) = \frac{1}{2}(-\pi - x)$ for $-\pi \leqslant x < 0$, $f(x) = \frac{1}{2}(\pi - x)$ for $0 < x \leqslant \pi$, and $f(0) = 0$. Find the Fourier series of $f(x)$ and graph the five-term partial sum against the function.

8. Find the Fourier series of $f(x) = |x|$, and graph the four-term partial sum against the function. [Ans.: $|x| = (\pi/2) - (4/\pi) \sum_{m=0}^{\infty} (2m+1)^{-2} \cos(2m+1)x$; count the constant as the first term.]

9. Let $f(x) = (2/\pi)x$ for $-\pi/2 \leqslant x \leqslant \pi/2$, $f(x) = (2/\pi)(\pi - x)$ for $\pi/2 \leqslant x \leqslant \pi$, and $f(x) = (2/\pi)(-\pi - x)$ for $-\pi \leqslant x \leqslant -\pi/2$. Find the Fourier series of $f(x)$ and graph the four-term partial sum against the function. [Ans.: $f(x) = (8/\pi^2) \sum_{m=0}^{\infty} (-1)^m (2m+1)^{-2} \sin(2m+1)x$.]

10. Find the Fourier series of $f(x) = x(\pi^2 - x^2)$ and graph the four-term

partial sum against the function. [Ans.: $x(\pi^2 - x^2) = 12 \sum_{n=1}^{\infty} (-1)^{n+1} n^{-3} \sin nx.$]

11. Find the Fourier series of $f(x) = e^{-|x|}$ and graph the four-term partial sum against the function.
    [Ans.: $\pi e^{-|x|} = (1 - e^{-\pi}) + 2 \sum_{n=1}^{\infty} (1 + (-1)^{n+1} e^{-\pi})(n^2 + 1)^{-1} \cos nx.$]

The following problems do not involve calculations.

12. Given $0 < a < \pi$, define $f(x) = 1$ for $-a < x < a$, $f(a) = f(-a) = \frac{1}{2}$, and $f(x) = 0$ for $-\pi \leqslant x < -a$ and $a < x \leqslant \pi$. Sketch the function $f(x)$ and find its Fourier series. From your answer deduce the formula

$$\sum_{n=1}^{\infty} \frac{\sin na}{na} = \frac{\pi - a}{2a}, \qquad 0 < a < \pi$$

From Exercise 7 derive the same formula.

13. Given $0 \leqslant a \leqslant \pi$, define $f(x) = 0$ for $-\pi \leqslant x \leqslant -a$, $f(x) = (1/a)(a + x)$ for $-a \leqslant x \leqslant 0$, $f(x) = (1/a)(a - x)$ for $0 \leqslant x \leqslant a$, and $f(x) = 0$ for $a \leqslant x \leqslant \pi$. Sketch the function $f(x)$ and find its Fourier series. From your answer deduce the formula

$$\frac{a(2\pi - a)}{4} = \sum_{n=1}^{\infty} \frac{1 - \cos na}{n^2}, \qquad 0 \leqslant a \leqslant \pi$$

From this, deduce the formula

$$\frac{\pi^2}{8} = \sum_{n=0}^{\infty} \frac{1}{(2n + 1)^2} = 1 + \frac{1}{3^2} + \frac{1}{5^2} + \frac{1}{7^2} + \cdots$$

From Exercise 9 deduce this last formula.

## Appendix 3.2.A   Calculating and Graphing Partial Sums of Fourier Series

This appendix provides some general suggestions and some sample HP-15C programs for the calculation and graphing Exercises 5 through 11.

I shall do two sample calculations for the series

$$x = 2 \sum_{n=1}^{\infty} \frac{(-1)^{n+1}}{n} \sin nx$$

$$\cong 2(\sin x - \tfrac{1}{2} \sin 2x + \tfrac{1}{3} \sin 3x - \tfrac{1}{4} \sin 4x + \tfrac{1}{5} \sin 5x)$$

derived in Section 3.2.

The objective is to compute the five-term sum on the right and graph it versus the function $y = x$. For graph paper I like 10 mm to the centimeter green-lined paper—it gives an attractive result. First, choose your scales so that your graph will fill the paper. In Figure 3.2 we have chosen equal scales for the $y$ axes of $0.1\pi$ per cm. Since we are plotting multiples of $\pi$, we shall divide the result of our calculation by $\pi$ and then plot the resulting number on the $y$ axis. Plot the original function, in this case $y = x$, first. Before you begin your calculation, be sure your machine is in the "RAD" mode (radian angle units).

*Sample Program #1*

In this case the HP-15C is programmed to do the calculation just as you would do it by hand.

*Keystroke*

| | |
|---|---|
| g P/R | clears all previous programs, and labels this program "A" |
| f CLEAR PRGM | |
| f LBL A | |
| RCL × 0 | multiplies entry in display by entry in storage register 0 |
| STO 1 | stores the result in register 1. This is "$x$" |
| SIN | $\sin x$ |
| LST x | retrieves $x$ |
| 2 × | $2x$ |
| SIN | $\sin 2x$ |
| 2 ÷ − | $\sin x - \frac{1}{2} \sin 2x$ |
| 3 RCL × 1 | $3x$ |
| SIN 3 ÷ + | adds $\frac{1}{3} \sin x$ |
| 4 RCL × 1 | $4x$ |
| SIN 4 ÷ − | subtracts $\frac{1}{4} \sin 4x$ |
| 5 RCL × 1 | $5x$ |
| SIN 5 ÷ + | adds $\frac{1}{5} \sin 5x$ |
| 2 × π ÷ | times $2/\pi$ |
| g RTN | return |
| g P/R | puts calculator back in run mode |

We will do our calculation in $x$ steps of $0.05\pi$, so enter $0.05\pi$ in storage register 0, enter an integer $k$ from 0 to 20 in the display (so $x = 0.05\pi k$), then key "fA." The table below gives the results. The graph is shown in Figure 3.2 in the text.

| $k$ | $x = 0.05\pi k$ | $\dfrac{2}{\pi}\left(\displaystyle\sum_{l=1}^{5}\dfrac{(-1)^{l+1}}{l}\sin lx\right)$ | $k$ | $x = 0.05\pi k$ | $\dfrac{2}{\pi}\left(\displaystyle\sum_{l=1}^{5}\dfrac{(-1)^{l+1}}{l}\sin lx\right)$ |
|---|---|---|---|---|---|
| 0 | 0 | 0 | 11 | 0.55 | 0.536 |
| 1 | $0.05\pi$ | 0.094 | 12 | $0.60\pi$ | 0.516 |
| 2 | $0.10\pi$ | 0.157 | 13 | $0.65\pi$ | 0.550 |
| 3 | $0.15\pi$ | 0.1798 | 14 | $0.70\pi$ | 0.662 |
| 4 | $0.20\pi$ | 0.1797 | 15 | $0.75\pi$ | 0.828 |
| 5 | $0.25\pi$ | 0.192 | 16 | $0.80\pi$ | 0.972 |
| 6 | $0.30\pi$ | 0.244 | 17 | $0.85\pi$ | 0.998 |
| 7 | $0.35\pi$ | 0.338 | 18 | $0.90\pi$ | 0.834 |
| 8 | $0.40\pi$ | 0.445 | 19 | $0.95\pi$ | 0.478 |
| 9 | $0.45\pi$ | 0.525 | 20 | $\pi$ | 0 |
| 10 | $0.50\pi$ | 0.552 | | | |

## Sample Program #2

In this program we take advantage of a counting function in the HP-15C, the "ISG" command (Increment and Skip if Greater than). Using this command, we can put a self-terminating loop in the program and so make the program sum as many terms of the series as we wish. This program is based on the formula

$$x \cong -\frac{2}{\pi}\sum_{l=1}^{n}\frac{(-1)^{l}}{l}\sin lx$$

*Keystroke*

| | |
|---|---|
| g  P/R | clears all previous programs, and labels |
| f  CLEAR PRGM | this program "A" |
| f  LBL A | |
| RCL × 0 | multiplies entry in display by entry in storage register 0 |
| STO 1 | stores the result in register 1. This is "$x$" |
| SIN | sin $x$ |
| CHS | $-\sin x$ |
| f LBL 1 | start of the loop |
| RCL  I | I contains 2.$pqr$01, $n = pqr =$ number of terms |
| g  INT | integral part of I = $l$ |
| STO .0 | puts $l$ in register .0 |
| RCL × 1 | $lx$ |
| SIN | sin $lx$ |
| π | |
| RLC × .0 | $(-1)^{l} = \cos l\pi$ |
| COS | |
| RCL ÷ .0 | $(-1)^{l}/l$ |

| | |
|---|---|
| × | $(-1)^l \sin lx/l$ |
| + | accumulates sum |
| f ISG I | increments integral part of I and compares with $n$ |
| GTO  1 | loops as long as $l \leqslant n$ |
| 2.00501 | $n = 5$ (this can be changed) |
| STO  I | returns I to original setting |
| CLR  x | |
| ↓ | |
| RCL × 2 | register 2 contains $-2/\pi$ |
| g  RTN | |
| g  P/R | puts calculator back in run mode |

When you run this program, enter 2.pqr01 in index register I, where $n = pqr$ is the number of terms of the series you wish to sum. For example, enter 2.00501 for 5 terms, 2.01001 for 10 terms, etc. Put this same number in the program as the sixth last command. Enter the $x$ step of the calculation in storage register 0. The larger $n$, the finer you would want this to be—for example, $0.05\pi$ when $n = 5$, $0.025\pi$ when $n = 10$, etc. Then enter an integer $k$ in display, so $x = 0.05\pi k$ and $x = 0.025\pi k$, respectively; then key "fA." The table below gives some results for $n = 10$.

| $k$ | $x = 0.025\pi k$ | $-\dfrac{2}{\pi} \sum\limits_{l=1}^{10} \dfrac{(-1)^l}{l} \sin lx$ |
|---|---|---|
| 0 | 0 | 0 |
| 1 | $0.025\pi$ | 0.003 |
| 2 | $0.05\pi$ | 0.020 |
| 3 | $0.075\pi$ | 0.056 |
| 4 | $0.10\pi$ | 0.105 |
| 5 | $0.125\pi$ | 0.151 |
| 6 | $0.150\pi$ | 0.180 |
| 7 | $0.175\pi$ | 0.190 |
| 8 | $0.20\pi$ | 0.190 |
| 9 | $0.225\pi$ | 0.196 |
| 10 | $0.25\pi$ | 0.220 |
| ⋮ | ⋮ | ⋮ |
| 38 | $0.95\pi$ | 0.854 |
| 39 | $0.975\pi$ | 0.481 |
| 40 | $\pi$ | 0 |

## 3.3   FOURIER SINE AND COSINE SERIES

In the examples worked out in the last section and in the exercises at the end of that section, we compared various functions with partial sums of their Fourier series and found generally good agreement. These examples give us confidence that for most reasonable functions $f(x)$, the Fourier series

$$\frac{a_0}{2} + \sum_{n=1}^{\infty} a_n \cos nx + b_n \sin nx \tag{3.3.1}$$

where

$$a_n = \frac{1}{\pi} \int_{-\pi}^{+\pi} f(x) \cos nx\, dx, \qquad n = 0, 1, 2, \ldots$$

$$b_n = \frac{1}{\pi} \int_{-\pi}^{+\pi} f(x) \sin nx\, dx, \qquad n = 1, 2, \ldots \tag{3.3.2}$$

represents $f(x)$ very well on the circle.

When we say "represents $f(x)$ very well," we mean that if the integer $N$ is taken large enough, then the $N$th partial sum of our Fourier series

$$\frac{a_0}{2} + \sum_{n=1}^{n=N} a_n \cos nx + b_n \sin nx \tag{3.3.3}$$

forms a good approximation to $f(x)$.

There are various ways to judge or measure the goodness of the approximation. One way is to compare the area discrepancy between the curves, sort of measuring the average discrepancy. This measure of convergence, whose details we shall make precise later, turns out to be a very natural concept for Fourier series. It will be a central concept for us.

Pointwise convergence, on the other hand, considers how the partial sum (3.3.3) behaves at individual fixed values of $x$ and asks the values of $x$ for which the partial sum tends to $f(x)$ as $N \to \infty$. The classical study of pointwise convergence yields the following conclusions:

(1) *Pointwise convergence is a local phenomenon.* In more detail: if (3.3.3) converges to $f(x)$ at each point $x$ in an interval $I$ of the circle as $N \to \infty$, then it will continue to converge to $f(x)$ at each point $x$ of an interval $I'$ properly inside $I$ no matter how the function $f(x)$ is changed outside $I$. Also, if we have convergence at a particular point $x = x_0$ and if we enclose this point in an interval $I$, however small, then the fact of convergence at $x = x_0$ will not be changed however we modify $f(x)$ outside $I$. Given that the Fourier coefficients $a_n$ and $b_n$ are determined from integrating $f(x)$ around the entire circle, the local nature of pointwise convergence is somewhat surprising. The rapidity of convergence will change, however.

(2) *If $f(x)$ is differentiable at each point of an interval I, then the Fourier series converges to $f(x)$ at each point interior to I.* Thus for most common functions we have pointwise convergence.

In the example $f(x) = x$ treated in Section 3.2 and the example $f(x) = 1$, $0 < x < \pi$, $f(x) = -1$, $-\pi < x < 0$ in Exercise 5 of that section, we dealt with functions that were discontinuous. The type of isolated discontinuity exhibited by these functions we shall call a simple discontinuity, without getting any more technical than that.

(3) *At a simple discontinuity, a Fourier series converges to the midpoint of the gap interval.*

Thus, at a simple discontinuity, a Fourier series "splits the difference" between the left and right values of the function.

Finally, in both the examples just cited, if we look at the last max/min on either side of the discontinuity, we see an example of the *Gibbs phenomenon.*

(4) *At a simple discontinuity, the difference between the last maximum and minimum of the partial sum (3.3.3) tends to approximately 1.09 times the length of the gap interval as $N \to \infty$.* Thus, at a discontinuity, a partial sum of a Fourier series will always overshoot by about 9%, however many terms are taken. In Exercise 5 of Section 3.2, the last peak of the four-term partial sum occurs at about $x = 0.12\pi$ and has a value of approximately 1.18, about 18% larger than the function value 1. As we take more terms, the amount of overshoot will come down toward 9%.

One can also make rigorous statements about the pointwise convergence of term-by-term differentiated and integrated Fourier series, but we shall treat these matters formally only in the exercises. We shall not subsequently be concerned with pointwise convergence, but rather with "mean-square convergence" alluded to earlier, which we shall introduce shortly.

For the remainder of this section we shall deal with pragmatic aspects of Fourier series expansions.

Suppose we are given a function $f(x)$ defined on $[0, \pi]$ which we wish to represent by a Fourier series. If we extend the definition of $f(x)$ to the full interval $-\pi \leqslant x \leqslant +\pi$, i.e., to the circle, we may deal with the problem in the framework of our theory already developed. There are two natural ways to extend $f$, as an odd function and as an even function:

*Odd extension*: in $-\pi < x < 0$ define $f(x) = -f(-x)$.
*Even extension*: in $-\pi < x < 0$ define $f(x) = f(-x)$.

We picture these extensions in Figure 3.5. The odd extension will be an odd function on $-\pi \leqslant x \leqslant \pi$, so its Fourier series will contain only terms $\sin nx$; and the even extension is an even function on $-\pi \leqslant x \leqslant \pi$, so its expansion contains only terms $\cos nx$. Thus we speak of the *Fourier sine series* for $f(x)$ on $0 \leqslant x \leqslant \pi$, meaning the Fourier series of its odd extension, and likewise

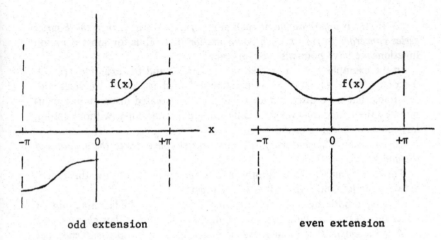

odd extension                                    even extension

**Figure 3.5**

speak of the *Fourier cosine series* of $f(x)$ meaning the Fourier series of its even extension.

For example, suppose $f(x) = x$, $0 \leqslant x < \pi$. Its even extension is $f_e(x) = |x|$, and the Fourier series for $|x|$ is the Fourier cosine series for $x$. Thus

$$x = \frac{\pi}{2} - \frac{4}{\pi} \sum_{n=0}^{\infty} \frac{\cos(2n+1)x}{(2n+1)^2}, \qquad \begin{array}{l} \text{Fourier cosine series for} \\ f(x) = x \text{ on } 0 \leqslant x \leqslant \pi \end{array}$$

This series was computed in Exercise 8 of Section 3.2, and you may refer to the graph you did there to see how well the four-term approximation fits the function.

On the other hand, the odd extension of $f(x) = x$ is given by the same formula, $f(x) = x$. Thus we have (refer to Section 3.2)

$$x = 2 \sum_{n=1}^{\infty} \frac{(-1)^{n+1}}{n} \sin nx \qquad \begin{array}{l} \text{Fourier sine series for} \\ f(x) = x \text{ on } 0 \leqslant x < \pi \end{array}$$

Observe that the four-term partial sum of the Fourier cosine series of $f(x) = x$ seems to be a better approximation than the four-term partial sum of its Fourier sine series. This is reasonable, because the odd extension contains a discontinuity (at $\pi \equiv -\pi$) and the even extension does not.

In some cases, one of these series may be dramatically better than the other. For example, the function $f(x) = 1$ has for odd extension the function

$f(x) = 1, 0 < x < \pi; f(x) = -1, -\pi < x < 0$, which yields (Exercise 5 of Section 3.2)

$$1 = \frac{4}{\pi} \sum_{n=0}^{\infty} \frac{1}{2n+1} \sin(2n+1)x \qquad \begin{array}{l} \text{Fourier sine series for} \\ f(x) = 1 \text{ on } 0 < x < \pi \end{array}$$

On the other hand, the even extension of $f(x) = 1$ is the function $f(x) = 1$, $-\pi < x < +\pi$, which has the Fourier series

$$1 = 1 \qquad \begin{array}{l} \text{Fourier cosine series for} \\ f(x) = 1 \text{ on } 0 < x < \pi \end{array}$$

so in this case the one-term partial sum fits the function exactly.

**Exercises**

1. (a) Find the Fourier series of $f(x) = 1$. Of $f(x) = \cos x$. Of $f(x) = \sin x$. Of $f(x) = 1 + \cos x + \sin x + \cos 2x + \sin 2x$.
   (b) Find the Fourier series of $\cos^2 x$. [Ans.: $\cos^2 x = \frac{1}{2} + \frac{1}{2}\cos 2x$.] Find the Fourier series of $\sin^2 x$. Of $\sin x \cos x$.

2. (Continuation) We may use Euler's formula in the form $\cos x = \frac{1}{2}(e^{ix} + e^{-ix})$ to express powers of the cosine as a sum of terms of the form $\cos nx$. For example,

$$\cos^3 x = \frac{1}{8}(e^{ix} + e^{-ix})^3 = \frac{1}{8}(e^{3ix} + 3e^{2ix}e^{-ix} + 3e^{ix}e^{-2ix} + e^{-3ix})$$

$$= \frac{1}{4}\left(\frac{e^{3ix} + e^{-3ix}}{2}\right) + \frac{3}{4}\left(\frac{e^{ix} + e^{-ix}}{2}\right)$$

$$= \frac{1}{4}\cos 3x + \frac{3}{4}\cos x$$

which gives the Fourier series for $\cos^3 x$. Find the Fourier series for $\cos^4 x$. For $\sin^4 x$.

3. (a) Using the formula $x^2 = \pi^2 - (\pi^2 - x^2)$ and the Fourier series for $\pi^2 - x^2$ given in Section 3.2, derive again the Fourier series for $f(x) = x^2$ (Exercise 6 of Section 3.2).
   (b) Using the formula $x = \frac{1}{2}(d/dx)(x^2)$ and differentiating the Fourier series for $x^2$ term by term, derive again the Fourier series for $f(x) = x$ (Section 3.2).
   (c) Using the formula $x^3 = 3\int_0^x t^2 \, dt$ and integrating the Fourier series for $t^2$ term by term, derive again the Fourier series for $f(x) = x(\pi^2 - x^2)$ (Exercise 10 of Section 3.2). By combining this answer with the answer to (b), derive the Fourier series for $f(x) = x^3$.

4. Let $f(x)$ be defined on $0 \leqslant x \leqslant \pi$. Show that the coefficients of its Fourier sine series, $f(x) = \Sigma_{n=1}^{\infty} b_n \sin nx$, are given by

$$b_n = \frac{2}{\pi} \int_0^\pi f(x) \sin nx \, dx$$

Show that the coefficients of its Fourier cosine series $f(x) = (a_0/2) + \Sigma_{n=1}^{\infty} a_n \cos nx$ are given by

$$a_n = \frac{2}{\pi} \int_0^\pi f(x) \cos nx \, dx$$

5. Let $f(x) = (2/\pi)x$ for $0 \leqslant x \leqslant \pi/2$; $f(x) = (2/\pi)(\pi - x)$ for $\pi/2 \leqslant x \leqslant \pi$. Sketch the even extension of $f(x)$ to $-\pi \leqslant x \leqslant \pi$. Find the Fourier cosine series for $f(x)$. [Hints: $\cos(n\pi/2) = 0$ when $n$ is odd. If $n = 2m$, then $\cos(n\pi/2) = \cos m\pi = (-1)^m$. The Fourier cosine series is

$$f(x) = \frac{1}{2} - \frac{4}{\pi^2} \sum_{m=0}^{\infty} \frac{1}{(2m+1)^2} \cos 2(2m+1)x]$$

6. (Continuation) Sketch the odd extension to $-\pi \leqslant x \leqslant \pi$ of the function $f(x)$ of Exercise 5. Find the Fourier sine series of $f(x)$. (Check with Exercise 9 of Section 3.2.)

7. Find the Fourier cosine series for $\sin x$. From your answer deduce the formula

$$\frac{1}{2} = \sum_{n=1}^{\infty} \frac{1}{4n^2 - 1}$$

Find the Fourier sine series for $\sin x$.

8. Find the Fourier sine series for $\cos x$. From your answer deduce the formula

$$\frac{\pi}{8\sqrt{2}} = \sum_{n=0}^{\infty} \frac{(-1)^n(2n+1)}{4(2n+1)^2 - 1} = \frac{1}{3} - \frac{3}{35} + \frac{5}{99} - \frac{7}{195} + \cdots$$

Find the Fourier cosine series for $\cos x$.

9. The following quotation is from Article 225 of Joseph Fourier's *The Analytical Theory of Heat*, translated by Alexander Freeman (Dover, 1955; originally published in 1822).

> We may remark that we have arrived at three different developments for $x$, namely,
>
> $$\frac{1}{2}x = \sin x - \frac{1}{2}\sin 2x + \frac{1}{3}\sin 3x - \frac{1}{4}\sin 4x + \frac{1}{5}\sin 5x - \& \text{ c.},$$
>
> $$\frac{1}{2}x = \frac{2}{\pi}\sin x - \frac{2}{3^2\pi}\sin 3x + \frac{2}{5^2\pi}\sin 5x - \& \text{ c.} \quad \text{(Art. 181)},$$

$$\frac{1}{2} x = \frac{1}{4} \pi - \frac{2}{\pi} \cos x - \frac{2}{3^2 \pi} \cos 3x - \frac{2}{5^2 \pi} \cos 5x - \text{\& c.}$$

It must be remarked that these three values of $x/2$ ought not to be considered as equal; with reference to all possible values of $x$, the three preceding developments have a common value only when the variable $x$ is included between 0 and $\pi/2$.

Thus, Fourier claims there are three different Fourier series that, when restricted to the subinterval $0 < x < \pi/2$, represent the function $x/2$ there. In full detail, either justify Fourier's assertion or refute it. [Hint: Refer to earlier exercises.]

## 3.4 FOURIER SERIES OVER OTHER INTERVALS

### 3.4.1 Functions Defined on $-L \leqslant x \leqslant L$

Consider a function $f(x)$ defined on an interval $-L \leqslant x \leqslant L$ instead of the interval $-\pi \leqslant x \leqslant \pi$ ($L > 0$). We make the change of variable $z = \pi x/L$, $x = zL/\pi$ and set $g(z) = f(x)$. Then $g(z)$ is defined on $-\pi \leqslant z \leqslant \pi$ and has a conventional Fourier series

$$g(z) = \frac{a_0}{2} + \sum_{n=1}^{\infty} a_n \cos nz + b_n \sin nz$$

$$a_n = \frac{1}{\pi} \int_{-\pi}^{+\pi} g(z) \cos nz \, dz, \qquad b_n = \frac{1}{\pi} \int_{-\pi}^{+\pi} g(z) \sin nz \, dz$$

Now substitute $z = \pi x/L$ to get *the Fourier series for $f(x)$ on $-L \leqslant x \leqslant L$ (with $+L$ and $-L$ identified)*:

$$f(x) = \frac{a_0}{2} + \sum_{n=1}^{\infty} a_n \cos \frac{n\pi x}{L} + b_n \sin \frac{n\pi x}{L} \tag{3.4.1}$$

$$a_n = \frac{1}{L} \int_{-L}^{L} f(x) \cos \frac{n\pi x}{L} dx, \qquad b_n = \frac{1}{L} \int_{-L}^{L} f(x) \sin \frac{n\pi x}{L} dx \tag{3.5.1}$$

For example, consider the function $f(x) = x$, $-L < x < L$, and $f(L) = f(-L) = 0$. As $f(x)$ is odd, all the $a_n$'s are zero. For the $b_n$'s we get (integrating by parts):

$$b_n = \frac{1}{L} \int_{-L}^{+L} x \sin \frac{n\pi x}{L} dx = \frac{2}{L} \int_0^L x \sin \frac{n\pi x}{L} dx$$

$$= \frac{2}{L} \left[ \frac{-xL}{n\pi} \cos \frac{n\pi x}{L} \Big|_0^L + \frac{L}{n} \int_0^L \cos \frac{n\pi x}{L} dx \right]$$

$$u = x, \qquad dv = \sin \frac{n\pi x}{L}$$

$$du = dx, \qquad v = \frac{L}{n\pi} \cos \frac{n\pi x}{L}$$

$$= \frac{2}{L} \left[ \frac{-L^2}{n\pi} \cos n\pi + \frac{L}{n\pi} \frac{L}{n\pi} \sin \frac{n\pi x}{L} \Big|_0^L \right] = \frac{2L(-1)^{n+1}}{n\pi}, \qquad n = 1, 2, \ldots$$

So our Fourier series is

$$x = \frac{2L}{\pi} \sum_{n=1}^{\infty} \frac{(-1)^{n+1}}{n} \sin \frac{n\pi x}{L}, \qquad -L < x < +L$$

If $L = \pi$, this reduces to the series derived in Section 3.1.

If $f(x)$ is defined only on the half-interval $0 \leqslant x \leqslant L$, we may extend it as either an odd function or an even function to the symmetric interval $-L \leqslant x \leqslant +L$. The odd extension is defined by $f(x) = -f(-x)$, $-L \leqslant x \leqslant 0$. This function, being odd, has all the $a_n$'s in its Fourier expansion equal zero, and we speak of its *Fourier sine series in* $0 \leqslant x \leqslant L$:

$$f(x) = \sum_{n=1}^{\infty} b_n \sin \frac{n\pi x}{L}, \qquad b_n = \frac{2}{L} \int_0^L f(x) \sin \frac{n\pi x}{L} dx, \qquad n = 1, 2, \ldots$$

$$(3.4.3)$$

The expansion of $u(x, t_0)$ given at the end of Section 3.1 is a Fourier sine series.

The even extension of $f(x)$ is defined by $f(x) = f(-x)$ on $-L \leqslant x \leqslant 0$ and has all its $b_n$'s zero. We speak of the *Fourier cosine series of* $f(x)$ *on* $0 \leqslant x \leqslant L$:

$$f(x) = \frac{a_0}{2} + \sum_{n=1}^{\infty} a_n \cos \frac{n\pi x}{L}, \qquad a_n = \frac{2}{L} \int_0^L f(x) \cos \frac{n\pi x}{L} dx, \qquad n = 0, 1, 2, \ldots$$

$$(3.4.4)$$

### 3.4.2   Functions Defined on $a \leqslant x \leqslant b$

The Fourier series of a function defined on a general finite interval is obtained by translating the interval to one symmetric about the origin and using the expansion just derived. Suppose that $f(x)$ is defined on the interval $a \leqslant x \leqslant b$. The midpoint of this interval is $(a+b)/2$, so we make the change of variable $z = x - (a+b)/2$. As $x$ ranges over the interval $a \leqslant x \leqslant b$, the variable $z$ ranges over the interval $-L \leqslant z \leqslant L$, where $L = (b-a)/2$. For the function $g(z) = f(x)$, we have the expansion

$$g(z) = \frac{u_0}{2} + \sum_{n=1}^{\infty} u_n \cos \frac{n\pi z}{L} + v_n \sin \frac{n\pi z}{L}$$

$$u_n = \frac{1}{L} \int_{-L}^{+L} g(z) \cos \frac{n\pi z}{L} dz, \qquad v_n = \frac{1}{L} \int_{-L}^{+L} g(z) \sin \frac{n\pi z}{L} dz \qquad (3.4.5)$$

where we have used different letters for the Fourier coefficients for reasons that will become apparent in a minute. Now

$$\frac{z}{L} = \frac{2x}{b-a} - \frac{a+b}{b-a}$$

Set $c = (a+b)/(b-a)$ so $z/L = 2x/(b-a) - c$. Changing variables in equations (3.4.5), we get

$$f(x) = \frac{u_0}{2} + \sum_{n=1}^{\infty} u_n \cos\left(\frac{2n\pi x}{b-a} - n\pi c\right) + v_n \sin\left(\frac{2n\pi x}{b-a} - n\pi c\right)$$

where

$$u_n = \frac{2}{b-a} \int_a^b f(x) \cos\left(\frac{2n\pi x}{b-a} - n\pi c\right) dx,$$

$$v_n = \frac{2}{b-a} \int_a^b f(x) \sin\left(\frac{2n\pi x}{b-a} - n\pi c\right) dx$$

Set

$$a_n = \frac{2}{b-a} \int_a^b f(x) \cos\frac{2n\pi x}{b-a} dx, \qquad b_n = \frac{2}{b-a} \int_a^b f(x) \sin\frac{2n\pi x}{b-a} dx$$

and use trig identities to derive

$$u_n = a_n \cos n\pi c + b_n \sin n\pi c, \qquad v_n = b_n \cos n\pi c - a_n \sin n\pi c$$

so the general term of our Fourier series is

$$(a_n \cos n\pi c + b_n \sin n\pi c)\left(\cos\frac{2n\pi x}{b-a}\cos n\pi c + \sin\frac{2n\pi x}{b-a}\sin n\pi c\right)$$

$$+ (b_n \cos n\pi c - a_n \sin n\pi c)\left(\sin\frac{2n\pi x}{b-a}\cos n\pi c - \cos\frac{2n\pi x}{b-a}\sin n\pi c\right)$$

$$= a_n \cos\frac{2n\pi x}{b-a} + b_n \sin\frac{2n\pi x}{b-a}$$

As $a_0 = u_0$, we have finally our general Fourier series:

$$\begin{cases} f(x) = \dfrac{a_0}{2} + \displaystyle\sum_{n=1}^{\infty} a_n \cos\dfrac{2n\pi x}{b-a} + b_n \sin\dfrac{2n\pi x}{b-a}, & a \leqslant x \leqslant b, a \equiv b \\[2ex] a_n = \dfrac{2}{b-a} \displaystyle\int_a^b f(x) \cos\dfrac{2n\pi x}{b-a} dx, & n = 0, 1, 2, \ldots \\[2ex] b_n = \dfrac{2}{b-a} \displaystyle\int_a^b f(x) \sin\dfrac{2n\pi x}{b-a} dx, & n = 1, 2, \ldots \end{cases} \qquad (3.4.6)$$

The coefficients $a_n$ and $b_n$ in (3.4.6) depend only on the values of $f(x)$ in the interval $a \leqslant x \leqslant b$. Moreover, the series in (3.4.6) takes the same value at $x = a$ as it does at $x = b$ (Exercise 3), so assigns the same value to $f(x)$ at $x = a$ as it does at $x = b$ irrespective of the actual values of $f(x)$ there. These facts suggest that we might better view the Fourier series in (3.4.6) as representing a function defined on the circle that we obtain as follows: remove the interval $a \leqslant x \leqslant b$ from the real line, bend it round, and join (identify) the point $x = a$ to the point $x = b$. This circle has circumference $b - a$ and radius $r = (b - a)/2\pi$. We shall also refer to it as a *one-dimensional manifold without boundary*; the unit circle we constructed in Section 3.2 is the special case with $r = 1$. The series in (3.4.6) depends only on the radius $(b - a)/2\pi$ of the circle (or the circumference $b - a$), not on the individual points $x = a$ and $x = b$, which reinforces our view that the circle is the proper "home" for this series.

### Exercises

1. Let $L > 0$, and let $f(x) = 1$ for $0 < x < L$; $f(x) = -1$ for $-L < x < 0$, and $f(-L) = f(0) = f(L) = 0$. Find the Fourier series of $f(x)$. Check that your answer agrees with that to Exercise 5 of Section 3.2 when $L = \pi$.
2. Show that the Fourier series for $f(x) = x$ on $0 \leqslant x \leqslant 2\pi$ differs from the Fourier series for $f(x) = x$ on $-\pi \leqslant x \leqslant \pi$. How is this consistent?
3. Show that $\cos(2n\pi a/(b-a)) = \cos(2n\pi b/(b-a))$. Same for sine.

## 3.5 THE VIBRATING WIRE, REVISITED

In Section 3.1 we derived the wave equation $\partial^2 u/\partial x^2 = (1/w^2)\,\partial^2 u/\partial t^2$ that governed the behavior of our vibrating wire. The particular problem we considered had the wire clamped at both ends, $x = 0$ and $x = L$, which implied the boundary conditions $u(0, t) = u(L, t) = 0$ for the vertical displacement $u(x, t)$ of our wire. Separation of variables led us to the solution

$$u(x, t) = \sum_{n=1}^{\infty} \left( A_n \cos \frac{n\pi wt}{L} + B_n \sin \frac{n\pi wt}{L} \right) \sin \frac{n\pi x}{L} \qquad (3.5.1)$$

and we are now in a position to determine the constants $A_n$ and $B_n$ and so completely determine the solution.

Put $t = 0$. The function $u(x, 0)$ describes the configuration of our sketched wire at time $t = 0$. We begin our observations at time $t = 0$ and describe $u(x, 0)$ as the *initial position* of the wire. With $t = 0$ equation (3.5.1) becomes

$$u(x, 0) = \sum_{n=1}^{\infty} A_n \sin \frac{n\pi x}{L}$$

which is a Fourier sine series for $u(x, 0)$. The coefficients $A_n$ are given by equation (3.4.3) of Section 3.4:

$$A_n = \frac{2}{L} \int_0^L u(x, 0) \sin \frac{n\pi x}{L} dx, \qquad n = 1, 2, \ldots \tag{3.5.2}$$

In our study of the classical linear oscillator (Section 2.5), we found the motion determined by initial position and initial velocity, and it is reasonable to pursue the same idea here. The velocity of our string is $\partial u/\partial t$. Differentiating (3.5.1) term by term and setting $t = 0$, we get

$$\frac{\partial u}{\partial t}(x, 0) = \sum_{n=1}^{\infty} \frac{n\pi w}{L} B_n \sin \frac{n\pi x}{L}$$

another Fourier sine series, in this instance for the *initial velocity* $(\partial u/\partial t)(x, 0)$. Using formula (3.4.3) again, we find

$$B_n = \frac{2}{n\pi w} \int_0^L \frac{\partial u}{\partial t}(x, 0) \sin \frac{n\pi x}{L} dx, \qquad n = 1, 2, \ldots \tag{3.5.3}$$

Combining formulas (3.5.1), (3.5.2), and (3.5.3), we have

*The motion of the vibrating wire is completely determined by its initial position $u(x, 0)$ and its initial velocity $(\partial u/\partial t)(x, 0)$. The motion is described by equation (3.5.1), the constants $A_n$ being determined by the initial position according to equation (3.5.2) and the constants $B_n$ being determined by the initial velocity according to equation (3.5.3).*

Let us now look in detail at this solution. Suppose we begin with initial velocity 0 so that $B_n \equiv 0$ and with initial position $u(x, 0) = \sin(\pi x/L)$, a simple bowlike initial configuration of our wire. Then

$$A_n = \frac{2}{L} \int_0^L \sin \frac{\pi x}{L} \sin \frac{n\pi x}{L} dx = \frac{2}{L} \int_0^\pi \sin z \, (\sin nz) \frac{L}{\pi} dz,$$

$$z = \frac{\pi x}{L}, \, dz = \frac{\pi}{L} dx$$

$$A_n = \frac{2}{\pi} \int_0^\pi \sin z \sin nz \, dz = \frac{1}{\pi} \int_{-\pi}^\pi \sin z \sin nz \, dz$$

where in the last step we have taken advantage of the fact that $\sin z \sin nz$ is even. Referring to the calculations done in Section 3.2, we see that this integral is zero unless $n = 1$, in which case it has the value $\pi$. So $A_1 = 1$, and $A_n = 0$ when $n \geqslant 2$. Thus our solution for these initial conditions is simply

$$u(x, t) = \cos \frac{\pi w t}{L} \sin \frac{\pi x}{L}$$

In Figure 3.6a we have plotted this function as a function of $x$ for various times $t$. The wire flexes back and forth across its equilibrium position. At each instant in time the wire has the same bowlike shape: $\text{const} \times \sin(\pi x/L)$ and oscillates between the configurations $+\sin(\pi x/L)$ and $-\sin(\pi x/L)$. The *period* $P$ of the oscillation is given by $(\pi w/L)P = 2\pi$ or $P = 2L/w = (2L\sqrt{\rho})/\sqrt{T}$. The *frequency* $v$, which is the number of complete oscillations per second, is $1/P$; thus

$$v = \frac{\sqrt{T}}{2L\sqrt{\rho}} = \frac{w}{2L} \tag{3.5.4}$$

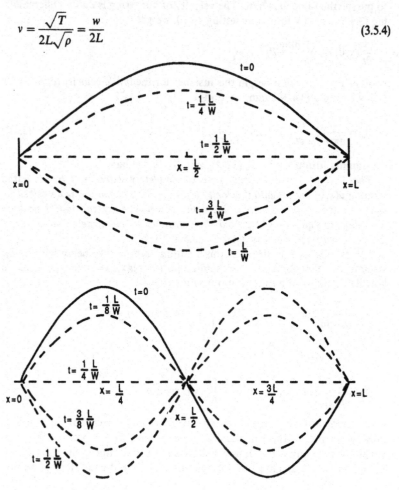

**Figure 3.6** (a) The fundamental $u = \cos(\pi wt/L)\sin(\pi x/L)$. (b) The first harmonic $u = \cos(2\pi wt/L)\sin(2\pi x/L)$.

The tone, or pitch, of the sound given off by the vibrating wire is determined by the frequency $v$. For example, "middle A" in American Standard pitch corresponds to 440 cycles/sec.

The *wavelength* $\lambda$ associated with a pure mode of the vibrating wire is the distance $x$ for the (periodically continued) solution to repeat itself. Thus the wavelength $\lambda$ for the mode $u = \cos(\pi w t/L) \sin(\pi x/L)$ is determined by $\pi\lambda/L = 2\pi$, so

$$\lambda = 2L \tag{3.5.5}$$

In Exercise 1 of Section 3.1 we saw that the quantity $w$ in the wave equation $\partial^2 u/\partial x^2 = (1/w^2)\partial^2 u/\partial t^2$ has the dimensions of velocity, and in Exercise 3 of that section we interpreted $w$ as the velocity of a traveling-wave solution $u = f(x - wt)$. For the wavelength $\lambda = 2L$ and the frequency $v = w/2L$ we have the equation

$$(\text{wavelength}) \times (\text{frequency}) = \lambda v = w \tag{3.5.6}$$

We shall refer to $w$ as the *wave velocity*.

Formula (3.5.4) tells us how the frequency varies with tension $T$, linear density $\rho$, and length $L$. The tighter we string the wire, the higher the frequency. We might have guessed that qualitative correlation from ordinary experience. But (3.5.4) gives us the quantitative relationship: $v$ increases as $\sqrt{T}$. Likewise $v$ decreases with increasing $\rho$: the bass strings on any instrument are always the thickest. Again (3.5.4) gives us the precise quantitative relationship: $v$ goes as $\rho^{-1/2}$ and $v$ goes as $1/L$, so if we halve the length, we double the frequency.

For example a 10 mil diameter (0.01 inch) brass wire weights 0.00029 lb/ft. If the tension applied is 100 lb, then

$$w = \sqrt{\frac{T}{\rho}} = \sqrt{\frac{100}{2.9 \times 10^{-4}}} = 5.87 \times 10^2 \text{ ft/sec}$$

so if $L = 2$ feet, the frequency is

$$v = \frac{5.87 \times 10^2}{4} \cong 147 \text{ cycles/sec}$$

which is approximately D an octave below middle A.

The solution we have been considering here is $u(x, t) = \cos(\pi w t/L) \sin(\pi x/L)$, which is the $n = 1$ term of the series (3.5.1). Look now at the $n = 2$ term, considered separately as a solution corresponding to initial position $u(x, 0) = \sin(2\pi x/L)$ and initial velocity $(\partial u/\partial t)(x, 0) = 0$.

$$u(x, t) = \cos\frac{2\pi w t}{L} \sin\frac{2\pi x}{L}$$

In Figure 3.6b we show a picture of the wire at various times $t$. In this motion the midpoint of the wire stays fixed. This point is called a *node*. At any fixed instant of time the wire has the shape const $\times \sin(2\pi x/L)$. At $t=0$ the wire is high to the left of the midpoint and low to the right. As time increases the left drops down, the right rises up, and at $t=L/4w$ it passes through its equilibrium position $u=0$. Then the left drops down and the right continues to rise, until at $t=L/2w$ its shape is the reflection through the midpoint position of its initial shape. Then at $t=L/w$ the wire has returned to its original shape. Therefore the period $P_2$ of this mode of vibration is $L/w$ and the frequency $v_2$ is $w/L$, which is double the frequency of the $n=1$ mode. The wavelength for the $n=2$ mode, $\lambda_2$, is given by $2\pi\lambda_2/L=2\pi$, so $\lambda_2=L$, half the wavelength of the $n=1$ mode. We have $\lambda_2 v_2=w$, so the equation (wavelength) $\times$ (frequency) = wave velocity is maintained.

In general, the $n$th term

$$\cos\frac{n\pi wt}{L}\sin\frac{n\pi x}{L} \tag{3.5.7}$$

corresponds to a mode of vibration with $n$ nodes (the zeros of $\sin(n\pi x/L)$), with frequency $v_n=nv_1$ and wavelength $\lambda_n=\lambda_1/n$ where $v_1=w/2L$ and $\lambda_1=2L$. The $n=1$ mode is called the *fundamental*, and the mode (3.5.7) for $n \geqslant 2$ the $(n-1)st$ harmonic. Thus equation (3.5.1) exhibits the motion of the wire as the sum of the fundamental mode plus the higher harmonics. We say we have done a *harmonic analysis* of the motion of the wire.

For the wavelength $\lambda_n$ and the frequency $v_n$ of the $(n-1)$st harmonic we have $\lambda_n v_n=w$, so the equation (wavelength) $\times$ (frequency) = wave velocity is maintained.

Our analysis has shown that with initial position $u(x,0)=\sin(\pi x/L)$ and initial velocity 0, $(\partial u/\partial t)(x,0)=0$, the motion of the wire is given by $u(x,t)=\cos(\pi wt/L)\sin(\pi x/L)$, so it vibrates only in the fundamental mode. It accordingly emits the pure tone of the fundamental with no harmonics. In practice, such special initial conditions are difficult to achieve. A more realistic initial condition is this: The wire is pulled from equilibrium at a point $x=x_0$, then released with zero velocity (Figure 3.7). This approximates what

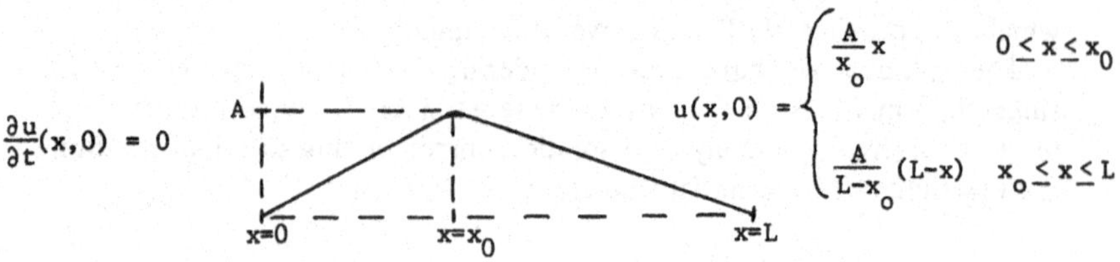

$$\frac{\partial u}{\partial t}(x,0)=0$$

$$u(x,0)=\begin{cases} \dfrac{A}{x_0}x & 0 \leq x \leq x_0 \\[2ex] \dfrac{A}{L-x_0}(L-x) & x_0 \leq x \leq L \end{cases}$$

**Figure 3.7** The plucked wire ($A > 0$).

happens when a guitar string is plucked, for example. With these initial conditions, the sound emitted is a blend of the fundamental and all the harmonics. By selecting the point $x_0$ where the wire is plucked, one can arrange to suppress some of the harmonics. It is especially desirable to suppress the 7th harmonic, as it is dissonant against the fundamental (Exercise 2).

**Exercises**

1. A standard piano has 88 wires, the 49th from the bass end, the middle A, being tuned to a frequency of 440 cycles/sec. There are 12 tones to an octave (an octave represents a doubling of the frequency). Thus the 49th wire is tuned to 440, the 61st to 880, the 73rd to 1760, and the 85th to 3520. The tension is approximately 160 pounds.
   (a) If the ratio of the frequencies of successive tones is constant, find the frequencies of the tones from middle A to A above middle A: A, B♭, B, C, C♯, D, E♭, E, F, F♯, G, A♭, A.
   (b) What is the frequency of C below middle A? What is the frequency of the 88th note?
   (c) If the 88th wire is approximately 3 inches long, what is its density in pounds per foot?
   (d) If, between octaves, the ratio of lengths is 1.875/1, what is the ratio of linear densities? Of radii (assuming the same material)?

2. A wire is plucked at the point $x = x_0$ (see Figure 3.7). Solve for the Fourier coefficients, and thus find the subsequent motion of the wire (equation (3.5.1)). Show that if the wire is plucked at its midpoint, then the 1st, 3rd, 5th, 7th, ... harmonics are all absent. Where should the wire be plucked to maintain the first six harmonics, but eliminate the seventh?

3. Consider a wire stretched between two supports, one at $x = 0$ and the other at $x = L$. Suppose the support at $x = 0$ is fixed, but the support at $x = L$ moves up and down with frequency $v$. This system is described by the wave equation

$$\frac{\partial^2 u}{\partial x^2} = \frac{1}{w^2}\frac{\partial^2 u}{\partial t^2}, \qquad w = \sqrt{\frac{T}{\rho}} \tag{1}$$

subject to the boundary conditions

$$u(0, t) = 0, \qquad u(L, t) = A \sin 2\pi v t \tag{2}$$

where the notation is that introduced in this section and $A$ is a small positive constant.
   (a) Show that if $p(x, t)$ is any solution of (1) and (2), and if $u_0(x, t)$ is any solution of (1) and $u(0, t) = u(L, t) = 0$ (clamped supports), then $u = u_0 + p$ is also a solution of (1) and (2).

(b) Suppose

$$p(x,t) = f(x) \sin 2\pi vt$$

satisfies (1) and (2). Find $f(x)$ and $p(x,t)$.

(c) What happens to your solution $p(x,t)$ as $v \to nw/2L$, for any fixed $n = 1, 2, \ldots$? Explain on physical grounds. When the wire is driven this way, what are its resonant frequencies?

4. We consider the two-dimensional wave equation

$$\frac{\partial^2 u}{\partial x^2} + \frac{\partial^2 u}{\partial y^2} = \frac{\rho}{T}\frac{\partial^2 u}{\partial t^2}$$

where $u(x,y,t)$ is the vertical displacement of a stretched membrane (a drumhead, for example) at position $(x,y)$ and time $t$, $\rho$ is the density of the membrane (g/cm$^2$), and $T$ is the tension per unit length. Derive this equation.

5. (Continuation). By making a change of variables $x = r \cos\theta$, $y = r \sin\theta$ in the equation of Exercise 4, derive the two-dimensional wave equation in polar coordinates:

$$\frac{\partial^2 u}{\partial r^2} + \frac{1}{r}\frac{\partial u}{\partial r} + \frac{1}{r^2}\frac{\partial^2 u}{\partial \theta^2} = \frac{\rho}{T}\frac{\partial^2 u}{\partial t^2}$$

Write this equation in the special case that $u(r,\theta,t)$ does not depend on $\theta$ (circular symmetry).

## 3.6  THE INNER PRODUCT

Recall what we mean by a "function space" or "vector space of functions." (Refer to Section 2.1, especially Exercise 3.) A set $V$ of functions is a *function space* when the following two conditions are satisfied:

1.  If the functions $f$ and $g$ belong to $V$, then so does their sum $f + g$.
2.  If the function $f$ belongs to $V$ then so does $cf$ for any scalar $c$.

We can deal with real-valued functions and real scalars or complex-valued functions and complex scalars. The following very important definition covers both cases.

**Definition 3.1.**  An *inner product* on a (real or complex) function space $V$ is a scalar-valued expression $\langle f, g \rangle$, defined for all $f$, $g$ in $V$, satisfying the following three conditions:

1.  *Linear in the first variable, conjugate linear in the second*: For any scalars $a, b$ and any functions $f, g$, and $h$ it is true that

$$\langle af + bg, h \rangle = a\langle f, h \rangle + b\langle g, h \rangle$$

and

$$\langle f, ag + bh \rangle = \bar{a}\langle f, g \rangle + \bar{b}\langle f, h \rangle$$

2. *Hermitian symmetric:* $\langle f, g \rangle = \overline{\langle g, f \rangle}$ for all $f$, $g$ in $V$.
3. *Positive definite:* $\langle f, f \rangle \geqslant 0$ for all $f$ in $V$, and $\langle f, f \rangle = 0$ implies $f = 0$.

The function space $V$ together with its inner product is called an *inner product space*.

In the case of real-valued functions and real scalars, complex conjugation has no effect, $\bar{a} = a$ for all scalars $a$. Hence, in the real case, condition 1 states that the inner product is linear in each variable separately, and condition 2 says simply that $\langle f, g \rangle = \langle g, f \rangle$ for all $f$, $g$ in $V$. With this understanding, Definition 3.1 covers both the real and complex cases. For convenience, we shall abbreviate condition 1, "linear in the first variable, conjugate linear in the second," as simply "conjugate bilinear." Thus the three key properties that characterize an inner product are these: conjugate bilinear, Hermitian symmetric, and positive definite. These three axioms are very important; memorize them.

The conjugate bilinearity applies to single terms, $\langle af, g \rangle = a\langle f, g \rangle$ and $\langle f, ag \rangle = \bar{a}\langle f, g \rangle$. It extends as well to any number of terms. For example, if we have three terms, $a_1 f_1 + a_2 f_2 + a_3 f_3$, then we may group the first two as one function, $(a_1 f_1 + a_2 f_2) + a_3 f_3$ and apply the bilinearity property (1) of Definition 3.1 thus:

$$\langle a_1 f_1 + a_2 f_2 + a_3 f_3, g \rangle = \langle (a_1 f_1 + a_2 f_2) + a_3 f_3, g \rangle$$
$$= \langle (a_1 f_1 + a_2 f_2), g \rangle + \langle a_3 f_3, g \rangle$$

Then apply bilinearity again:

$$= a_1 \langle f_1, g \rangle + a_2 \langle f_2, g \rangle + a_3 \langle f_3, g \rangle$$

Extending this to $n$ terms, we have

$$\langle a_1 f_1 + a_2 f_2 + \cdots + a_n f_n, g \rangle = a_1 \langle f_1, g \rangle + a_2 \langle f_2, g \rangle + \cdots + a_n \langle f_n, g \rangle$$

or, more succinctly,

$$\left\langle \sum_{i=1}^{n} a_i f_i, g \right\rangle = \sum_{i=1}^{n} a_i \langle f_i, g \rangle$$

In the second variable,

$$\left\langle f, \sum_{i=1}^{n} a_i g_i \right\rangle = \sum_{i=1}^{n} \bar{a}_i \langle f, g_i \rangle$$

Bilinearity even has an extension to infinitely many terms; we will cover this aspect in Theorem 3.12.

Definition 3.1 characterizes an inner product abstractly by axioms or properties. Let's take a look at an example.

Consider real-valued functions defined on the circle $C$, $-\pi \leqslant x \leqslant \pi$, $-\pi \equiv \pi$ (we have already dealt with the circle $C$ in our study of Fourier series). For any pair of functions $f, g$ on $C$, define

$$\langle f, g \rangle = \frac{1}{\pi} \int_{-\pi}^{\pi} f(x)g(x)\, dx \tag{3.6.1}$$

Thus the formula (3.6.1) assigns to every pair of functions a real number. It is therefore "scalar valued" as required in Definition 3.1. Does it possess the three properties conjugate bilinear, Hermitian symmetric, and positive definite?

As for the conjugate bilinearity (which becomes simply bilinearity since our functions are real valued), we have

$$\langle af + bg, h \rangle = \frac{1}{\pi} \int_{-\pi}^{\pi} (af + bg)h \, dx = \frac{1}{\pi} \int_{-\pi}^{\pi} (afh + bgh)\, dx$$

$$= \frac{a}{\pi} \int_{-\pi}^{\pi} fh \, dx + \frac{b}{\pi} \int_{-\pi}^{\pi} gh \, dx = a\langle f, h \rangle + b\langle g, h \rangle$$

The argument for the second variable is the same. Thus we easily verify, using nothing more than the simple fact from calculus that the definite integral of a sum is the sum of the integrals, that formula (3.6.1) defines an expression that is linear in each variable separately.

As for the Hermitian symmetry, which becomes simply symmetry for real-valued functions, the proof is even easier;

$$\langle f, g \rangle = \frac{1}{\pi} \int_{-\pi}^{\pi} fg \, dx = \frac{1}{\pi} \int_{-\pi}^{\pi} gf \, dx = \langle g, f \rangle$$

As for positive definiteness, the first part, namely $\langle f, f \rangle \geqslant 0$, is easy because

$$\langle f, f \rangle = \frac{1}{\pi} \int_{-\pi}^{\pi} (f(x))^2 \, dx$$

and the definite integral of a nonnegative function is nonnegative ($(f(x))^2 \geqslant 0$ because $f(x)$ is real). The second part, $\langle f, f \rangle = 0$ implies $f = 0$, involves a subtle point. The equation $\langle f, f \rangle = 0$ means $\int_{-\pi}^{\pi} (f(x))^2 \, dx = 0$. Now recall the meaning of the definite integral $\int_{a}^{b} g(x) \, dx$: it is the signed area between the graph of $y = g(x)$ and the $x$ axis, the area above the $x$ axis counted positively, the area below negatively. As $y = (f(x))^2 \geqslant 0$, the equation $\int_{-\pi}^{\pi} (f(x))^2 \, dx = 0$ means that there is zero area between the graph of $y = (f(x))^2$ and the $x$ axis. Thus $(f(x))^2$, and so $f(x)$, has to be "essentially" zero (zero "almost everywhere" in the picturesque language of Lebesgue theory). But it does not necessarily mean that $f(x)$ has to be zero at every point. For example, the function $f(x) = 0$ when $x \neq 0$, $f(0) = 1$, has $\langle f, f \rangle = (1/\pi) \int_{-\pi}^{\pi} (f(x))^2 \, dx = 0$. In fact, any function $f(x)$ that is zero except at a finite number of points has a zero integral. On the other hand, if $f(x)$ is continuous and $\int_{-\pi}^{\pi} (f(x))^2 \, dx = 0$, then $f(x) = 0$ for every $x$ in $C$ (Exercise 4). What we are really dealing with here is a new kind of function equality different from those we have studied in Section 2.1. We can say that functions $f$ and $g$ on $C$ are "essentially equal" if $\int_{-\pi}^{\pi} |f(x) - g(x)|^2 \, dx = 0$ and that $f$ is essentially zero if $\int_{-\pi}^{\pi} |f(x)|^2 \, dx = 0$. If we adopt this understanding, which is enough for the present, then our proof of positive definiteness is complete. And we have proved that formula (3.6.1) defines an inner product.

Formula (3.6.1) really streamlines the discussion of our classical Fourier series and even enables us to get some new results. Given a function $f(x)$ defined on the circle $C$, its Fourier series is (Section 3.2)

$$f(x) = a_0 \left( \frac{1}{2} \right) + \sum_{n=1}^{\infty} a_n \cos nx + b_n \sin nx \qquad (3.6.2)$$

where

$$a_n = \frac{1}{\pi} \int_{-\pi}^{\pi} f(x) \cos nx \, dx, \qquad n = 0, 1, 2, \ldots$$

$$b_n = \frac{1}{\pi} \int_{-\pi}^{\pi} f(x) \sin nx \, dx, \qquad n = 1, 2, \ldots \qquad (3.6.3)$$

First, using the new notation we may express the equations (3.6.3) for the Fourier series coefficients $a_n$ and $b_n$ in a more compact form,

$$a_n = \langle f(x), \cos nx \rangle, \qquad n = 0, 1, 2, \ldots$$

$$b_n = \langle f(x), \sin nx \rangle, \qquad n = 1, 2, \ldots \qquad (3.6.3')$$

Second, using our new terminology, we can express the "orthogonality" relations derived in Chapter 2 more neatly. These relations now become

$$\langle \cos mx, \cos nx \rangle = \begin{cases} 2 & \text{if } m = n = 0 \\ 1 & \text{if } m = n \neq 0 \\ 0 & \text{if } m \neq n \end{cases}$$

$$\langle \sin mx, \sin nx \rangle = \begin{cases} 1 & m = n \\ 0 & m \neq n \end{cases} \quad m, n \geqslant 1$$

$$\langle \cos mx, \sin nx \rangle = 0, \qquad \text{all } m, n$$

Third, our new notation permits us to derive the formulas for the Fourier coefficients, equations (3.6.3') or (3.6.3), in a very transparent way. Recall the problem as stated at the beginning of Section 3.2: *Given $f(x)$, what are the constants $c_n$, $d_n$ in the following series?*

$$f(x) = \frac{1}{2}c_0 + \sum_{n=1}^{\infty} c_n \cos nx + d_n \sin nx \qquad (3.6.4)$$

We shall work on $c_0$ first. Take the inner product of both sides of the series with the function $\cos 0x = 1$:

$$\langle f(x), 1 \rangle = \langle \tfrac{1}{2}c_0 + c_1 \cos x + d_1 \sin x + c_2 \cos 2x + d_2 \sin 2x + \cdots, 1 \rangle$$

We have shown that the inner product is bilinear for any *finite* number of terms. Here we have an infinite number of terms. The bilinearity does in fact extend to an infinite series of terms, and we shall justify that fact in Theorem 3.12. Take it for granted now. Using this bilinearity, we get

$$\langle f(x), 1 \rangle = \langle \tfrac{1}{2}c_0 + c_1 \cos x + d_1 \sin x + c_2 \cos 2x + \cdots, 1 \rangle$$
$$= c_0 \langle \tfrac{1}{2}, 1 \rangle + c_1 \langle \cos x, 1 \rangle + d_1 \langle \sin x, 1 \rangle + c_2 \langle \cos 2x, 1 \rangle + \cdots$$

Now $\langle \cos x, 1 \rangle = \langle \cos 1x, \cos 0x \rangle = 0$ from our first orthogonality relation with $m = 1$, $n = 0$. And $\langle \sin x, 1 \rangle = \langle \sin 1x, \cos 0x \rangle = 0$ also. The next term is likewise zero; $\langle \cos 2x, 1 \rangle = \langle \cos 2x, \cos 0x \rangle = 0$. In fact, our orthogonality relations show that all terms, save the first, are zero. As for the first term, we have

$$\langle \tfrac{1}{2}, 1 \rangle = \frac{1}{\pi} \int_{-\pi}^{\pi} \frac{1}{2} \cdot 1 \, dx = \frac{1}{2\pi} \int_{-\pi}^{\pi} dx = 1$$

Thus we get $c_0 = \langle f(x), 1 \rangle = \langle f(x), \cos 0x \rangle = a_0$.

The pattern is the same for all coefficients. For example, to find $c_1$ we take the inner product of both sides of (3.6.4) with $\cos x$. We get

$$\langle f(x), \cos x \rangle = c_0 \langle \tfrac{1}{2}, \cos x \rangle + c_1 \langle \cos x, \cos x \rangle$$
$$+ d_1 \langle \sin x, \cos x \rangle + c_2 \langle \cos 2x, \cos x \rangle + \cdots$$

Now $\langle \tfrac{1}{2}, \cos x \rangle = \tfrac{1}{2} \langle 1, \cos x \rangle = \tfrac{1}{2} \langle \cos 0x, \cos 1x \rangle = 0$ by our first orthogonality relation with $m = 0$, $n = 1$. Similarly, all the other inner products equal zero except for $\langle \cos x, \cos x \rangle = 1$. Then we find $c_1 = \langle f(x), \cos x \rangle = a_1$.

In this way we derive again the formulas (3.6.3') or (3.6.3) for the Fourier coefficients. Our new inner product notation emphasizes the properties that are essential for the kind of calculation we have just done.

As a fourth, and final, illustration of the usefulness of our new notation, we shall derive the Parseval equality for classical Fourier series. Start with the Fourier series (3.6.2),

$$f(x) = \tfrac{1}{2}a_0 + a_1 \cos x + b_1 \sin x + a_2 \cos 2x + b_2 \sin 2x + \cdots$$

then take the inner product of both sides with the function $f(x)$,

$$\langle f, f \rangle = \langle \tfrac{1}{2}a_0 + a_1 \cos x + b_1 \sin x + a_2 \cos 2x + b_2 \sin 2x + \cdots, f(x) \rangle$$

Here again we will use a fact that has yet to be justified, namely that the bilinearity of the inner product, which we know to hold for finite sums, is valid for such infinite sums as well. Taking this fact for granted now, we get

$$\langle f, f \rangle = a_0 \langle \tfrac{1}{2}, f(x) \rangle + a_1 \langle \cos x, f(x) \rangle + b_1 \langle \sin x, f(x) \rangle$$
$$+ a_2 \langle \cos 2x, f(x) \rangle + \cdots$$

Now use symmetry to compute the various inner products:

$$\langle \tfrac{1}{2}, f(x) \rangle = \tfrac{1}{2} \langle 1, f(x) \rangle = \tfrac{1}{2} \langle f(x), 1 \rangle = \tfrac{1}{2} \langle f(x), \cos 0x \rangle = \tfrac{1}{2}a_0$$

$$\langle \cos x, f(x) \rangle = \langle f(x), \cos x \rangle = a_1$$

$$\langle \sin x, f(x) \rangle = \langle f(x), \sin x \rangle = b_1 \quad \text{etc.}$$

Inserting these values, we get *Parseval's equality*,

$$\langle f, f \rangle = \tfrac{1}{2}a_0^2 + a_1^2 + b_1^2 + a_2^2 + b_2^2 + \cdots$$

or, in "sigma" form,

$$\boxed{\langle f, f \rangle = \frac{1}{2}a_0^2 + \sum_{n=1}^{\infty} a_n^2 + b_n^2}$$

Parseval's equality is a numerical series that derives from a Fourier series. Using examples of Fourier series that we have already computed, we can derive many interesting numerical series.

For example, for the function $f(x) = x$ we have $\langle f, f \rangle = (1/\pi)$ $\int_{-\pi}^{\pi} x^2 \, dx = 2\pi^2/3$. In Section 3.2 we found the Fourier coefficients of $f(x) = x$: $a_n = \langle x, \cos nx \rangle = 0$ for all $n$, and $b_n = \langle x, \sin nx \rangle = 2(-1)^{n+1}/n$. Hence $b_n^2 = 4/n^2$, so Parseval's equality becomes

$$\frac{2\pi^2}{3} = 4 \sum_{n=1}^{\infty} \frac{1}{n^2}$$

or

$$\frac{\pi^2}{6} = \sum_{n=1}^{\infty} \frac{1}{n^2} = 1 + \frac{1}{4} + \frac{1}{9} + \frac{1}{16} + \cdots$$

In calculus we proved that the series on the right converges (the usual proof involves the integral test). Now, thanks to the Parseval equality, we know its sum. Note that every finite partial sum of the series on the right is a rational number, yet the sum of the infinite series is $\pi^2/6$, which is not rational. We have $\pi^2/6 = 1.644934067\ldots$. Summing the first 50 terms of the series on the right we get 1.625132736, which involves about a 1.2% error.

Look back now at equation (3.6.1), which was

$$\langle f, g \rangle = \frac{1}{\pi} \int_{-\pi}^{\pi} f(x)g(x)\, dx$$

This equation defines the inner product in the example we have just finished discussing, namely the *real-valued* functions on the unit circle $C$. If we allow *complex-valued* functions, this formula no longer works. The expression it defines fails to satisfy the criteria of Definition 3.1: it is not conjugate linear in the second variable, not Hermitian symmetric, and not positive definite. If we deal with complex-valued functions on $C$, then the appropriate definition of the inner product is

$$\langle f, g \rangle = \frac{1}{\pi} \int_{-\pi}^{\pi} f(x)\overline{g(x)}\, dx \qquad (3.6.5)$$

In Exercises 5, 6, and 7 you are asked to verify that the expression defined by equation (3.6.5) satisfies the criteria of Definition 3.1: conjugate bilinear, Hermitian symmetric, and positive definite.

In doing these exercises, bear in mind the definition of complex conjugation and the rules for working with it. If $z = x + iy$ is a complex number ($x$ and $y$ real), its complex conjugate $\bar{z}$ is defined by $\bar{z} = x - iy$. The complex conjugate of a sum (or difference) is the sum (or difference) of the complex conjugates: $\overline{z_1 \pm z_2} = \bar{z}_1 \pm \bar{z}_2$. Similarly for product and quotient: $\overline{z_1 z_2} = \bar{z}_1 \bar{z}_2$ and $\overline{(z_1/z_2)} = \bar{z}_1/\bar{z}_2$. Also $\bar{\bar{z}} = z$ for any complex $z$. Moreover, $\bar{z} = z$ if, and only if, $z$ is real. Also $|z|^2 = z\bar{z} = x^2 + y^2$, so $|z| = 0 \Leftrightarrow z = 0$.

If $z$ is a complex number, then both $x = (1/2)(z + \bar{z})$ and $y = (1/2i)(z - \bar{z})$ are real (check that $\bar{x} = x$ and $\bar{y} = y$). And $z = x + iy$. The same trick works for a complex-valued function $f(x)$ of a real variable $x$: $f = g + ih$ where both $g = (1/2)(f + \bar{f})$ and $h = (1/2i)(f - \bar{f})$ are real-valued. One can use this latter decomposition to prove that

$$\overline{\int_a^b f(x)\, dx} = \int_a^b \overline{f(x)}\, dx$$

for a complex-valued function $f(x)$ of a real variable $x$ (Exercise 6).

Note that equation (3.6.5) reduces to equation (3.6.1) when we deal with real functions because then $\overline{g(x)} = g(x)$. Hence we might as well use equation (3.6.5) as it covers both the real and complex cases.

Equation (3.6.5) provides a particular example of an inner product, one that is familiar to us because we used it in our work on Fourier series. There are many other examples of inner products. All the ones we will use in this book are covered by the general formula

$$\langle f, g \rangle = \int_a^b f(x)\overline{g(x)}w(x)\,dx \tag{3.6.6}$$

where $[a, b]$ is a finite or infinite interval, and $w(x)$ is an analytic function that is strictly positive in the open interval $a < x < b$.

In Exercise 8 you will prove that the expression defined by formula (3.6.6) is conjugate bilinear and Hermitian symmetric. You will also prove part of positive definiteness, namely that $\langle f, f \rangle \geq 0$. As for the other part, namely $\langle f, f \rangle = 0 \Rightarrow f$ is "essentially zero," that argument goes as follows. Suppose that $\langle f, f \rangle = 0$, which means

$$\int_a^b |f(x)|^2 w(x)\,dx = 0$$

Consider a proper closed subinterval $[c, d]$ of $[a, b]$. As the analytic function $w(x)$ is strictly positive on the open interval $a < x < b$, it will be bounded away from 0 on $[c, d]$; that is, there exists a positive constant $m$ such that $w(x) \geq m > 0$, $c \leq x \leq d$. Then

$$\int_c^d |f(x)|^2\,dx = \frac{1}{m}\int_c^d |f(x)|^2 m\,dx \leq \frac{1}{m}\int_c^d |f(x)|^2 w(x)\,dx$$

$$\leq \frac{1}{m}\int_a^b |f(x)|^2 w(x)\,dx = \frac{1}{m}\cdot 0 = 0$$

Hence the integral of $|f(x)|^2$ over every proper subinterval of $[a, b]$ is zero. From this it follows that

$$\int_a^b |f(x)|^2\,dx = 0$$

Hence the area between the graph of $y = |f(x)|^2$ and the $x$ axis is zero, which is what we mean by saying that $f$ is essentially zero.

### Exercises

1. For each of the following expressions, determine whether it is (1) bilinear, (2) symmetric, and (3) positive definite. (Either prove that it has the property, or show by specific example that it does not.) Which are inner products? (Real-valued functions and real scalars.)

   (a) $\langle f, g \rangle = \int_{-1}^{+1} f(x)g(x)x\,dx$  　[Dealing here with functions defined on $-1 \leq x \leq +1$.

   (b) $\langle f, g \rangle = \int_{-1}^{+1} f(x)g(x)x^2\,dx$  　$H(x)$ is Heaviside's function; $H(x) = 1$ when $x > 0$,

   (c) $\langle f, g \rangle = \int_{-1}^{+1} f(x)g(x)H(x)\,dx$  　$H(x) = 0$ when $x < 0$.]

(d) $\langle f, g \rangle = \int_0^{+1} f(x) g(x)\, dx$ \hspace{2em} (e) $\langle f, g \rangle = \int_{-\pi}^{+\pi} f(x) g(x)^2\, dx$

(f) $\langle f, g \rangle = \int_{-\pi}^{+\pi} f(x)^2 g(x)^2\, dx$ \hspace{2em} (g) $\langle f, g \rangle = \int_0^{\infty} f(x) g(x) e^{-x}\, dx$

2. In each of the following, sketch a graph of the function $y = f(x)$; then, from the meaning of the definite integral as a signed area, evaluate, without integrating, the definite integral $\int_a^b f(x)\, dx$.

(a) $\displaystyle\int_0^2 f(x)\, dx$ \hspace{1em} where $f(x) = \begin{cases} 2x, & 0 \leqslant x \leqslant \frac{1}{2} \\ 2 - 2x, & \frac{1}{2} \leqslant x \leqslant 1 \\ -1, & 1 < x \leqslant 2 \end{cases}$

$\Bigg[$ Soln. graph:

area $\triangle = \frac{1}{2}(1)(1) = \frac{1}{2}$ \hspace{1em} $\displaystyle\int_0^2 f\, dx = \frac{1}{2} - 1 = \frac{-1}{2} \Bigg]$

area $\square = (1)(1) = 1$

(b) $\displaystyle\int_0^2 f(x)\, dx$ \hspace{1em} where $f(x) = \begin{cases} 1, & 0 \leqslant x < 1 \\ 0, & x = 1 \\ -1, & 1 < x \leqslant 2 \end{cases}$

(c) $\displaystyle\int_0^2 (f(x))^2\, dx$, \hspace{1em} $f(x)$ same as (b).

(d) $\displaystyle\int_0^2 f(x)\, dx$, \hspace{1em} $f(x) = \begin{cases} x & 0 \leqslant x < 1 \\ 0 & x = 1 \\ 2 - x, & 1 < x \leqslant 2 \end{cases}$

3. (Continuation) (a) Sketch the graph of the function $g(x)$ defined on $[a, b]$ by

$$g(x) = \begin{cases} m, & r \leqslant x \leqslant s \\ 0, & \text{otherwise} \end{cases}$$

where $m > 0$ and $a \leqslant r < s \leqslant b$. Evaluate $\int_a^b g(x)\, dx$.

(b) Prove that if a function $f(x)$ satisfies $f(x) \geqslant g(x)$, $a \leqslant x \leqslant b$, where $g(x)$ is the function defined in (a), then $\int_a^b f(x)\, dx > 0$.

4. (Continuation) (a) Prove that if $f(x)$ is continuous and nonnegative on $[a, b]$ and

$$\int_a^b f(x)\, dx = 0$$

then $f(x) = 0$ for every $x$, $a \leqslant x \leqslant b$.

(b) From your result in (a) deduce that if $f(x)$ is continuous on $[a, b]$ and $\int_a^b (f(x))^2 \, dx = 0$ then $f(x) = 0$ for every $x$, $a \leqslant x \leqslant b$.

[Hint for part (a): Suppose for some $x_0$, $a \leqslant x_0 \leqslant b$, $f(x_0) = c > 0$. Using the continuity, show that there must then exist $r$, $s$ such that $a \leqslant r < s \leqslant b$, $r \leqslant x_0 \leqslant s$, and $f(x) \geqslant c/2$ for $r \leqslant x \leqslant s$. Then use Exercise 3.]

(Exercises 5, 6, and 7, taken together, prove that the expression defined by equation (3.6.5) is an inner product.)

5. Show that the expression $\langle f, g \rangle$ defined by equation (3.6.5) is conjugate bilinear.

6. (Continuation) (a) Let $f(x)$ be a complex-valued function of a real variable $x$. Let $g = (1/2)(f + \bar{f})$ and $h = (1/2i)(f - \bar{f})$. Show that $g$ and $h$ are real-valued and that $f = g + ih$.

(b) Using the result in part (a), show that

$$\overline{\int_a^b f(x) \, dx} = \int_a^b \overline{f(x)} \, dx$$

(c) Using the result in part (b), show that the expression $\langle f, g \rangle$ defined by equation (3.6.5) is Hermitian symmetric.

7. (Continuation) (a) Show that the expression defined by equation (3.6.5) has the property that $\langle f, f \rangle \geqslant 0$.

(b) Show that $\langle f, f \rangle = 0$ implies that $\int_{-\pi}^{\pi} |f|^2 \, dx = 0$, thus that $|f|$ is essentially zero.

8. Let $[a, b]$ be a finite or infinite interval and $w(x)$ an analytic function such that $w(x) > 0$ in the open interval $a < x < b$. Show formally (disregarding questions of convergence of the integrals involved) that the formula

$$\langle f, g \rangle = \int_a^b f(x)\overline{g(x)}w(x) \, dx$$

defines an inner product, i.e., satisfies the three conditions of Definition 3.1.

9. Given that the function $f(x)$ has the Fourier series

$$f(x) = \sum_{n=0}^{\infty} \frac{1}{n^2 + 1} \cos nx$$

find $\int_{-\pi}^{+\pi} f(x) \sin^2 x \, dx = \pi \langle f, \sin^2 x \rangle$. [Hint: $2 \sin^2 x = 1 - \cos 2x = \cos 0x - \cos 2x$.]

10. (a) Set $\phi_n = e^{inx}$, $-\pi \leqslant x \leqslant \pi$, $n = 0, \pm 1, \pm 2, \ldots$. Show that with respect to the inner product

$$\langle f, g \rangle = \frac{1}{\pi} \int_{-\pi}^{+\pi} f(x)\overline{g(x)} \, dx$$

the functions $\phi_n$ satisfy $\langle \phi_n, \phi_m \rangle = 0$ when $n \neq m$ and $\langle \phi_n, \phi_n \rangle = 2$, $n = 0, \pm 1, \pm 2, \ldots$.

(b) By taking the inner product of both sides of the "Fourier series"

$$f = \sum_{n=-\infty}^{n=+\infty} c_n \phi_n$$

with $\phi_m$, determine the coefficients $c_m$.

(c) Compute the explicit series in (b) for $f(x) = x$. By pairing off the $+n$ and $-n$ terms of your series, derive again the ordinary Fourier for $f(x) = x$ (Section 3.2).

11. Determine the Parseval equality corresponding to the Fourier series of the function of Exercise 5, Section 3.2, and use it to find the sum of the series

$$\sum_{n=0}^{\infty} \frac{1}{(2n+1)^2}$$

Compare this value with the 30-term partial sum of the series.

12. Determine the Parseval equality corresponding to the Fourier series of $f(x) = x^2$ (Exercise 6 of Section 3.2), and use it to find the sum of the series

$$\sum_{n=1}^{\infty} \frac{1}{n^4}$$

Using a hand calculator, sum the first 30 terms of this series, and use this number to find an approximate value for $\pi$.

13. Determine the Parseval equality corresponding to the Fourier series of $f(x) = |x|$ (Exercise 8 of Section 3.2), and use it to sum the series

$$\sum_{n=0}^{\infty} \frac{1}{(2n+1)^4}$$

14. Using Parseval's equality, show that the Fourier coefficients

(1) $a_n = \frac{1}{\pi} \int_{-\pi}^{+\pi} f(x) \cos nx \, dx$    and    (2) $b_n = \frac{1}{\pi} \int_{-\pi}^{+\pi} f(x) \sin nx \, dx$

of any (reasonable) function $f(x)$ tend to zero as $n \to \infty$.

15. (Continuation) Do you think there exists a function $f(x)$ for which $a_n = 0$, $b_n = 1/\sqrt{n}$? Why or why not?

## 3.7  SCHWARZ'S INEQUALITY

We begin by repeating the key definition from the previous lecture:

**Definition 3.2.**  An *inner product* on a (real or complex) function space $V$ is a scalar-valued expression $\langle f, g \rangle$ defined for all $f$, $g$ in $V$ that is *conjugate* *bilinear* (linear in the first variable, conjugate linear in the second):

$$\langle af + bg, h \rangle = a\langle f, h \rangle + b\langle g, h \rangle$$

and

$$\langle f, ag + bh \rangle = \bar{a}\langle f, g \rangle + \bar{b}\langle f, h \rangle$$

*Hermitian symmetric*:

$$\langle f, g \rangle = \overline{\langle g, f \rangle}$$

and *positive definite*:

$$\langle f, f \rangle \geqslant 0 \text{ for all } f \text{ in } V, \text{ and } \langle f, f \rangle = 0 \text{ implies } f = 0$$

The function space $V$ together with its inner product is called an *inner product space*.

Note particularly that this definition describes an inner product totally in terms of its three properties, not by any formula or expression. In deriving additional facts about inner products, we shall argue directly from these key properties: conjugate bilinear, Hermitian symmetric, and positive definite. Adopting this abstract point of view—this is part of the Hilbert space method—is very important for us as we shall have to deal with various different integral formulas all of which share these three basic properties. Thus our derived results will hold automatically in all the special cases.

This step of abstraction is somewhat akin to that which we took in high school when we learned to manipulate letters in algebraic expressions in place of numbers, a skill whose usefulness we no longer question. The ability to deal with an abstract inner product will prove similarly useful.

We note again that when we are dealing with real-valued functions and real scalars, then the operation of complex conjugation has no effect, and the three properties listed in Definition 3.2 reduce to those considered in Section 3.6: bilinear, symmetric, and positive definite. Thus Definition 3.2 covers both the real and complex cases.

It is common in physics to use an inner product that is conjugate linear in the first variable and linear in the second. Definition 3.2, which makes the inner product linear in its first variable and conjugate linear in its second, accords with standard mathematical practice. There is no material difference.

A *Hilbert space* is an inner product space with one additional property, *metric completeness*, which adds nothing to the practical calculations that will concern us. This additional axiom is discussed in Section 3.10.

We have come across the three properties in Definition 3.2 in early courses that used vectors. In vector algebra the *dot product* is often defined this way: If $A$ and $B$ are vectors in ordinary two-dimensional space and $\theta$ is the acute

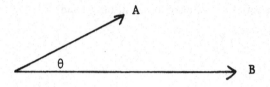

**Figure 3.8**

angle between them, then

$$A \cdot B = \text{(length of } A) \times \text{(length of } B) \times \cos\theta \qquad (3.7.1)$$

The situation is depicted in Figure 3.8. As the expression on the right of (3.7.1) does not change when we interchange $A$ and $B$, we have $A \cdot B = B \cdot A$, so the dot product is symmetric. Also, when $B = A$, then $\theta = 0$, $\cos\theta = 1$, and (3.7.1) becomes $A \cdot A = \text{(length of } A)^2$. Hence $A \cdot A > 0$ when $A \neq 0$, so the dot product is also positive definite. As for the bilinearity, when $a > 0$, the length of $aA = a \times$ length of $A$, which translates to $(aA) \cdot B = a(A \cdot B)$. When $a < 0$, then $aA$ has the opposite direction to $A$, and the new acute angle between $aA$ and $B$ is $\pi - \theta$. As $\cos(\pi - \theta) = -\cos\theta$, the equation $(aA) \cdot B = a(A \cdot B)$ still holds when $a < 0$. As it obviously holds when $a = 0$, we have $(aA) \cdot B = a(A \cdot B)$ for all real scalars $a$. Then, using symmetry, we easily deduce that $A \cdot (bB) = b(A \cdot B)$ for all real scalars $b$. To complete the proof that $A \cdot B$ is bilinear, we need to show that $(A_1 + A_2) \cdot B = A_1 \cdot B + A_2 \cdot B$, and this is easily done from a figure, using the parallelogram law that defines vector addition. This is left as an exercise.

So our abstract inner product of Definition 3.2 has the same three key properties as the familiar dot product from elementary vector algebra. It is then natural to carry over to general inner product spaces the language and concepts used in vector algebra, such as angle, orthogonality, and distance. When transferred to function spaces, these geometrical ideas will guide our solution to problems in analysis, differential equations, and quantum physics. They are part of the language of Hilbert space.

For example, if we put $A = B$ in equation (3.7.1), then $\theta = 0$ and we have for an ordinary vector $A$, length of $A = (A \cdot A)^{1/2}$. We shall carry this idea over to a general inner product space to define the "length" or "norm" (the term we shall use) of a function in an inner product space.

**Definition 3.3.** If $f$ is a function in an inner product space, then the *norm*, $\|f\|$, of $f$ (with respect to the given inner product) is defined by $\|f\| = \langle f, f \rangle^{1/2}$ (positive square root).

Since an inner product is positive definite, we have $\langle f, f \rangle \geqslant 0$, so we can always take the square root. Note also that $\|f\| > 0$ unless $f = 0$.

Look at the formula for the norm in case our inner product is that considered earlier, $\langle f, g \rangle = (1/\pi) \int_{-\pi}^{\pi} f\bar{g}\, dx$. Then

$$\|f\| = \left[ \frac{1}{\pi} \int_{-\pi}^{\pi} |f(x)|^2\, dx \right]^{1/2}$$

So, for this particular inner product, we measure the size of a function this way: first we square it, then integrate to get an average or mean, then take the square root. Thus arises the term "root-mean-square metric" to describe this concept of length. We shall shorten the name to mean-square metric when we discuss it in the next lecture.

Let us now bring over the concept of angle $\theta$ into our abstract inner product space. Equation (3.7.1) gives us a natural starting point. If we solve for $\cos \theta$ in equation (3.7.1) we get $\cos \theta = A \cdot B/(\text{length of } A)(\text{length of } B)$. Thus (sticking to real functions and real scalars for simplicity) we could try to define $\cos \theta$ in an abstract inner product space by

$$\cos \theta = \frac{\langle f, g \rangle}{\|f\| \, \|g\|} \tag{3.7.2}$$

In order for this to make sense, however, the expression $\langle f, g \rangle / \|f\| \, \|g\|$ must lie between $-1$ and $1$. Otherwise equation (3.7.2) is nonsense, because $\cos \theta$ always lies between $-1$ and $1$.

That brings us to the main result of this section, which is to prove exactly that $-1 \leqslant \langle f, g \rangle / \|f\| \, \|g\| \leqslant 1$, in an equivalent form.

**Theorem 3.4.** *For any functions $f$ and $g$ in an inner product space, we have*

$$|\langle f, g \rangle| \leqslant \|f\| \, \|g\| \qquad \text{(Schwarz's inequality)}$$

*Proof.* We shall deal with the real case first. The proof must rely on our three basic properties: bilinear, symmetric, and positive definite. Thus our proof is "abstract."

Now if either $f$ or $g$ is zero, both sides of Schwarz's inequality are zero and it becomes $0 = 0$, which is obviously true. So we may assume $f \neq 0$ and $g \neq 0$.

Fix the nonzero functions $f$ and $g$, and set $h = f + \lambda g$ where $\lambda$ is real. By positive definiteness we have always $\langle f + \lambda g, \ f + \lambda g \rangle = \langle h, h \rangle \geqslant 0$. Using bilinearity and symmetry, we find

$$\langle f + \lambda g, f + \lambda g \rangle = \langle f, f \rangle + 2\lambda \langle f, g \rangle + \lambda^2 \langle g, g \rangle \geqslant 0$$

The expression on the right is a second degree polynomial in $\lambda$ with constant coefficients and our inequality shows that this polynomial is nonnegative for all real $\lambda$, $-\infty < \lambda < \infty$. In other words, if we put

$$p(\lambda) = \langle f,f \rangle + 2\lambda \langle f,g \rangle + \lambda^2 \langle g,g \rangle$$

and graph $p(\lambda)$ versus $\lambda$, the graph lies wholly above the $\lambda$ axis because $p(\lambda) = \langle h,h \rangle \geqslant 0$ for all $\lambda$.

Let us find the point at which our polynomial $p(\lambda)$ has its minimum. We differentiate and set the derivative equal to zero:

$$\frac{d}{d\lambda} p(\lambda) = 2\langle f,g \rangle + 2\lambda \langle g,g \rangle = 0$$

Thus the minimum occurs at $\lambda_0 = -\langle f,g \rangle / \langle g,g \rangle$. As $p(\lambda)$ is everywhere nonnegative, it is nonnegative in particular at $\lambda = \lambda_0$: $p(\lambda_0) \geqslant 0$. Substituting, we get

$$p(\lambda_0) = \langle f,f \rangle + 2\left( \frac{-\langle f,g \rangle}{\langle g,g \rangle} \right) \langle f,g \rangle + \frac{\langle f,g \rangle^2}{\langle g,g \rangle^2} \langle g,g \rangle \geqslant 0$$

or $\langle f,f \rangle - \langle f,g \rangle^2 / \langle g,g \rangle \geqslant 0$, which we may put in the equivalent form $\langle f,g \rangle^2 \leqslant \langle f,f \rangle \langle g,g \rangle$. Now take the square root of both sides of this last inequality. The function $\sqrt{x}$ is monotone increasing for positive $x$, so preserves the inequality. And $\sqrt{x^2} = |x|$ for all real $x$, so we get our result. That proves Schwarz's inequality in the real case.

For the general complex case, we expand $f + \lambda g$ just as before, where now $\lambda$ is a complex variable. We get

$$\langle f + \lambda g\, f + \lambda g \rangle = \langle f,f \rangle + \lambda \langle g,f \rangle + \bar{\lambda} \langle f,g \rangle + \lambda \bar{\lambda} \langle g,g \rangle \geqslant 0 \qquad (3.7.3)$$

This expression is $\geqslant 0$ for any value of the complex variable $\lambda$, so in particular it is $\geqslant 0$ for $\lambda = \lambda_0 = -\langle f,g \rangle / \langle g,g \rangle$. Substituting, we derive again $|\langle f,g \rangle| \leqslant \|f\| \, \|g\|$. That completes the proof of Theorem 3.4.

So, in a general (real) inner product space, we may define the angle between two functions by $\cos \theta = \langle f,g \rangle / \|f\| \, \|g\|$. Of course, the mental picture depicted in Figure 3.8 no longer applies, because functions are not arrows. Nonetheless, all the formal properties still hold. We have transplanted the geometric concept of angle to function space. This will enable us to use geometric ideas to solve problems in analysis.

Let us see what Schwarz's inequality states for the familiar inner product $\langle f,g \rangle = (1/\pi) \int_{-\pi}^{\pi} f\bar{g}\, dx$. We find (real scalars)

$$\left| \frac{1}{\pi} \int_{-\pi}^{\pi} f(x)g(x)\, dx \right| \leqslant \left[ \frac{1}{\pi} \int_{-\pi}^{\pi} f(x)^2\, dx \right]^{1/2} \left[ \frac{1}{\pi} \int_{-\pi}^{+\pi} g(x)^2\, dx \right]^{1/2}$$

We can apply this inequality equally well to the functions $|f|$ and $|g|$ in place of $f$ and $g$. If we do this and cancel the $\pi$'s, we get

$$\int_{-\pi}^{\pi} |f(x)| \, |g(x)| \, dx \leqslant \left[ \int_{-\pi}^{\pi} |f(x)|^2 \, dx \right]^{1/2} \left[ \int_{-\pi}^{\pi} |g(x)|^2 \, dx \right]^{1/2}$$

an important integral equality that we have derived by purely "abstract" methods.

Schwarz's inequality enables us to prove a familiar-looking inequality for the norm.

**Theorem 3.5.** *If $V$ is an inner product space, with inner product $\langle f, g \rangle$ and associated norm $\|f\| = \langle f, f \rangle^{1/2}$, then for any $f$, $g$ in $V$*

$\|f + g\| \leqslant \|f\| + \|g\|$     *(triangle inequality)*

*Proof.* We begin by computing $\|f + g\|^2$:

$$\|f + g\|^2 = \langle f + g, f + g \rangle = \langle f, f \rangle + \langle f, g \rangle + \langle g, f \rangle + \langle g, g \rangle$$

$$= \|f\|^2 + \langle f, g \rangle + \overline{\langle f, g \rangle} + \|g\|^2$$

For any complex number $z$, $z + \bar{z} = 2 \operatorname{Re} z \leqslant |z|$, so $\langle f, g \rangle + \overline{\langle f, g \rangle} = 2 \operatorname{Re} \langle f, g \rangle \leqslant 2|\langle f, g \rangle| \leqslant 2\|f\| \, \|g\|$. In the last step we used Schwarz's inequality. Thus we get

$$\|f + g\|^2 \leqslant \|f\|^2 + 2\|f\| \, \|g\| + \|g\|^2 = (\|f\| + \|g\|)^2$$

and, taking the square root of both sides, we get the triangle inequality. Thus, Theorem 3.5 is proved.

Look again at the concept of angle in the general inner product space. We have defined the angle $\theta$ between two elements (functions) $f$ and $g$ in an inner product space by the formula $\cos \theta = \langle f, g \rangle / \|f\| \, \|g\|$. In particular, we can specify when $f$ and $g$ have a right angle between them, i.e., when $\theta = \pi/2$ radians (90°). This is so exactly when $\cos \theta = 0$, or when $\langle f, g \rangle = 0$, which leads to the following very important definition.

**Definition 3.6.** Functions $f$ and $g$ in an inner product space are called *orthogonal* when $\langle f, g \rangle = 0$. A finite or infinite family of functions $f_1, f_2, \ldots,$ $f_n, \ldots$ is called an *orthogonal family* if $\langle f_i, f_j \rangle = 0$ when $i \neq j$.

For example, the functions $\cos mx$ and $\cos nx$ are orthogonal with respect to the inner product $\langle f, g \rangle = (1/\pi) \int_{-\pi}^{\pi} f \bar{g} \, dx$ when $m \neq n$. With respect to the same inner product, $\cos mx$ is orthogonal to $\sin nx$ for all $m$ and $n$. And $\sin mx$ is orthogonal to $\sin nx$ when $m \neq n$. So, according to Definition 3.6, the functions

$$\tfrac{1}{2}, \cos x, \sin x, \cos 2x, \sin 2x, \ldots, \cos nx, \sin nx, \ldots$$

form an orthogonal family with respect to the inner product $\langle f, g \rangle = (1/\pi) \int_{-\pi}^{\pi} f\bar{g}\, dx$. Hence, we can regard our Fourier series

$$f(x) = a_0(1/2) + a_1 \cos x + b_1 \sin x + a_2 \cos 2x + b_2 \sin 2x + \cdots$$

as an orthogonal expansion exactly comparable to that we had in linear algebra for a vector $A = (a_1, a_2, a_3)$,

$$A = a_1 I_1 + a_2 I_2 + a_3 I_3$$

where the $I_n$ are the orthogonal unit vectors $I_1 = (1, 0, 0)$, $I_2 = (0, 1, 0)$, and $I_3 = (0, 0, 1)$. Thus a transplanted geometric concept affords us a new perspective on classical Fourier series.

## Exercises

1. Using only the three abstract properties of an inner product, conjugate bilinear, Hermitian symmetric, and positive definite,
   (a) Prove that $\langle 0, g \rangle = 0$ for all $g$. [Two solutions: (1) $\langle 0, g \rangle = \langle 0 + 0, g \rangle$ $= \langle 0, g \rangle + \langle 0, g \rangle$. Subtract $\langle 0, g \rangle$ from both sides to get $\langle 0, g \rangle = 0$. (2) If $f$ is any function then $0 = 0 \cdot f$, where the zero on the left is the zero function and the zero on the right is the zero scalar. Then $\langle 0, g \rangle = \langle 0 \cdot f, g \rangle = 0 \langle f, g \rangle = 0$.]
   (b) Prove that $\langle f, 0 \rangle = 0$ for all $f$.
   (c) Prove that if $f$ and $g$ are each orthogonal to $h$, then $af + bg$ is also orthogonal to $h$ for any (real or complex) constants $a$ and $b$. Generalize.
   (d) Prove that if $a$ is any real or complex number, then $\langle af, af \rangle = |a|^2 \langle f, f \rangle$ and $\|af\| = |a|\, \|f\|$.
2. Reprove Schwarz's inequality (in the real case) by arguing that since the polynomial $p(\lambda) = \langle f, f \rangle + 2\lambda \langle f, g \rangle + \lambda^2 \langle g, g \rangle$ is everywhere nonnegative, it cannot have two distinct real roots, so that its "$b^2 - 4ac$" must be $\leq 0$.
3. (K. Mertes)  Re-prove Schwarz's inequality in the complex case by solving simultaneously the equations

$$\frac{\partial}{\partial \lambda} \langle f + \lambda g, f + \lambda g \rangle = 0 \quad \text{and} \quad \frac{\partial}{\partial \bar{\lambda}} \langle f + \lambda g, f + \lambda g \rangle = 0$$

and substituting the resulting value of $\lambda$ into equation (3.7.3) of the text.
4. In the following steps, prove that *equality holds in Schwarz's inequality, i.e., $|\langle f, g \rangle| = \|f\|\, \|g\|$, when and only when $f$ and $g$ are linearly dependent* (real functions and real scalars).
   (a) In our proof of Schwarz's inequality, we have tacitly assumed that

$g \neq 0$ as we have divided by $\langle g,g \rangle$. Show that if either $f = 0$ or $g = 0$ (or both) then $|\langle f,g \rangle| = \|f\| \, \|g\|$.

(b) From (a) we may assume $f \neq 0$ and $g \neq 0$. Prove that if $af + bg = 0$ for nonzero constants $a$ and $b$, then (by substitution) $|\langle f,g \rangle| = \|f\| \, \|g\|$.

(c) Suppose $f \neq 0$ and $g \neq 0$ and that $|\langle f,g \rangle| = \|f\| \, \|g\|$. By examining the text proof of Schwarz's inequality, prove that then $af + bg = 0$. Find the constants $a$ and $b$.

(Exactly the same result holds in the complex case.)

5. (a) Write Schwarz's inequality for the inner product $\langle f,g \rangle = \int_a^b f(x)g(x)\, dx$ (real case).

(b) By substituting $|f|$ for $f$ and $|g|$ for $g$ in your answer to (a), derive

$$\left( \int_a^b |f| \, |g| \, dx \right)^2 \leqslant \left( \int_a^b |f|^2 \, dx \right) \left( \int_a^b |g|^2 \, dx \right)$$

(c) From this, derive $\int_a^b |g| \, dx \leqslant (b - a)^{1/2} (\int_a^b |g|^2 \, dx)^{1/2}$.

(d) Using (c), prove that a function that is square-integrable on a finite interval is integrable there; i.e., $\int_a^b |g|^2 \, dx < \infty \Rightarrow \int_a^b |g| \, dx < \infty$.

6. Derive the inequality

$$\int_0^\infty |f(x)|e^{-x} \, dx \leqslant \frac{1}{\sqrt{2}} \left[ \int_0^\infty |f(x)|^2 \, dx \right]^{1/2}$$

7. (a) Show that if $[a,b]$ is a finite interval and $\int_a^b |f(x)|^2 \, dx = 0$, then $\int_a^b |f(x)| \, dx = 0$.

(b) Assuming the validity of the equation $\int_{-\infty}^\infty |g(x)| \, dx = \sum_{n=-\infty}^\infty \int_n^{n+1} |g(x)| \, dx$, use your result in (a) to prove that

$$\int_{-\infty}^\infty |f(x)|^2 \, dx = 0 \qquad \text{implies} \qquad \int_{-\infty}^\infty |f(x)| \, dx = 0$$

[Hint for part (a): Use a result from Exercise 5.]

8. Write Schwarz's inequality for the inner product

$$\langle f,g \rangle = \int_a^b f(x)\overline{g(x)}w(x)\, dx \quad w(x) \text{ real and positive}$$

9. Fix an integer $n \geqslant 1$. The set of all $n$-tuples $(a_1, a_2, \ldots, a_n)$ of real numbers under componentwise addition and scalar multiplication:

$$(a_1, a_2, \ldots, a_n) + (b_1, b_2, \ldots, b_n) = (a_1 + b_1, a_2 + b_2, \ldots, a_n + b_n)$$

$$c(a_1, a_2, \ldots, a_n) = (ca_1, ca_2, \ldots, ca_n)$$

forms a vector space which we call $\mathbf{R}^n$.

(a) If $A=(a_1,a_2,\ldots,a_n)$ and $B=(b_1,b_2,\ldots,b_n)$ show that the expression

$$\langle A,B \rangle = a_1 b_1 + a_2 b_2 + \cdots + a_n b_n$$

defines an inner product on $\mathbf{R}^n$; i.e., show that is bilinear, symmetric, and positive definite.

(b) Show that $U_1=(1,0,\ldots,0)$, $U_2=(0,1,0,\ldots,0)$, $\ldots$, $U_n=(0,0,\ldots,0,1)$ form an orthogonal family in $\mathbf{R}^n$, and find the "Fourier series" of $A=(a_1,a_2,\ldots,a_n)$ with respect to $U_1,U_2,\ldots,U_n$.

(c) Interpret Schwarz's inequality in this context, and derive from it the inequality

$$(|b_1|+|b_2|+\cdots+|b_n|)^2 \leqslant n(b_1{}^2+b_2{}^2+\cdots b_n{}^2)$$

10. (Continuation) Prove that if the series $\Sigma_{i=1}^{\infty} a_i{}^2$ and $\Sigma_{i=1}^{\infty} b_i{}^2$ both converge, then so do $\Sigma_{i=1}^{\infty} |a_i| |b_i|$ and $\Sigma_{i=1}^{\infty} a_i b_i$.

11. Prove that $f(x)=1$, $g(x)=2x^2-1$ are orthogonal with respect to the inner product

$$\langle f,g \rangle = \int_{-1}^{1} f(x)g(x)\frac{dx}{\sqrt{1-x^2}}$$

but not with respect to the inner product

$$\langle f,g \rangle = \int_{-1}^{1} f(x)g(x)\,dx$$

12. Show that the functions $p_0=1$, $p_1=x$, $p_2=3x^2-1$, and $p_3=5x^3-3x$ form an orthogonal family with respect to the inner product $\langle f,g \rangle = \int_{-1}^{+1} f(x)g(x)\,dx$. [Take advantage of even-odd considerations to evaluate the integrals.]

13. Using the inner product $\langle f,g \rangle = \int_{0}^{\infty} f(x)g(x)e^{-x}\,dx$,
(a) Find the positive constant $a$ so that $f(x)=ax$ has norm 1.
(b) Find the constant $a$ so that $a-x$ is orthogonal to $x$.

14. Prove that if $\langle f,g \rangle=0$ then $\|f+g\|^2=\|f\|^2+\|g\|^2$. Prove that if $\langle \phi_i,\phi_j \rangle=0$, $i \neq j$, then $\|\phi_1+\phi_2+\cdots+\phi_n\|^2 = \|\phi_1\|^2+\|\phi_2\|^2+\cdots+\|\phi_n\|^2$.

15. Prove that if $f_1, f_2, \ldots, f_n$ are pairwise orthogonal and nonzero, then they are linearly independent.

16. Let $M$ be a collection of functions in an inner product space $V$. Denote by $M^{\perp}$ the following set of functions in $V$: $f$ belongs to $M^{\perp}$ if and only if $\langle f,m \rangle=0$ for every function $m$ in $M$. Show that $M^{\perp}$ is always a subspace of $V$; i.e., show that if $f_1$ and $f_2$ belong to $M^{\perp}$, so does $af_1+bf_2$ for any constants $a$ and $b$. Show that always $M \cap M^{\perp} = 0$ and $M \subseteq M^{\perp\perp}$. [$M^{\perp}$ is called the *orthocomplement* of $M$.]

17. (a) Let $m$ be a fixed nonzero function in an inner product space $V$. Show that for any $f$ in $V$ there are a constant $a$ and a function $n$ such that both

$$f = am + n \qquad \text{and} \qquad \langle m, n \rangle = 0$$

Find $a$ and $n$ in terms of $f$ and $m$. [Ans.: $a = \langle f, m \rangle / \langle m, m \rangle$, $n = f - am$.]

   (b) Let $m_1, m_2, \ldots, m_l$ be fixed nonzero orthogonal functions in an inner product space $V$. Show that for any $f$ in $V$ there are constants $a_1, a_2, \ldots, a_l$ and a function $n$ such that both

$$f = a_1 m_1 + a_2 m_2 + \cdots + a_l m_l + n$$

and

$$\langle m_i, n \rangle = 0, \qquad 1 \leqslant i \leqslant l$$

Find $a_i$ in terms of $f$ and $m_i$.

18. (Continuation) In Exercise 17b take $V$ to be the space of functions $f(x)$ on $-\pi \leqslant x \leqslant \pi$ with inner product

$$\langle f, g \rangle = \frac{1}{\pi} \int_{-\pi}^{\pi} f(x) g(x) \, dx$$

and take the functions $m_k = \sin kx$. Find the constants $a_k$ explicitly.

## 3.8 THE MEAN-SQUARE METRIC; ORTHOGONAL BASES

In the previous section we proved Schwarz's inequality for an abstract inner product, and we were accordingly able to carry over to abstract inner product spaces the concept of angle $\theta$ by means of the formula $\cos \theta = \langle f, g \rangle / \|f\| \, \|g\|$.

In this lecture we shall show that another important geometric concept derives from the inner product: *distance*. This is the "mean-square" distance between functions that we have alluded to earlier.

The formula for distance in an inner product space uses the norm defined earlier: $\|f\| = \langle f, f \rangle^{1/2}$.

**Definition 3.7.** The distance $d(f, g)$ between two functions $f$ and $g$ in an inner product space is defined by

$$d(f, g) = \|f - g\|$$

We call this the intrinsic or *mean-square* metric.

**Theorem 3.8.** *The mean-square metric in an inner product space has the following properties:*

1.  $d(f,g) \geqslant 0$; $d(f,g) = 0$ when and only when $f = g$.
2.  $d(f,g) = d(g,f)$.
3.  $d(f,h) \leqslant d(f,g) + d(g,h)$.

Before getting into a proof, let us consider the meaning of each of these assertions. The first says that the distance between two unequal functions is strictly positive and that the distance between a function and itself is always zero. The second says that the distance between $f$ and $g$ equals the distance between $g$ and $f$. Clearly, these are properties we expect any decent concept of "distance" to have. The third says that if we go directly from $f$ to $h$, then this is always shorter than going via an intermediate point $g$. We illustrate this in Figure 3.9. Thus we might interpret 3.8(3) as asserting that "the shortest distance between two points (functions) is a straight line."

These three properties follow directly from what we have already done. Item 1 is an immediate consequence of positive definiteness. Item 2 follows from Exercise 1d of the previous section: $\|f-g\| = \|-(g-f)\| = |-1| \, \|g-f\| = \|g-f\|$. And item 3 follows from the triangle inequality:

$$\|f - h\| = \|f - g + g - h\| \leqslant \|f - g\| + \|g - h\|$$

which concludes the proof.

Consider the space $V$ of functions on the circle $C$, $-\pi \leqslant x \leqslant \pi$, $-\pi \equiv \pi$, with the inner product

$$\langle f,g \rangle = \frac{1}{\pi} \int_{-\pi}^{+\pi} f(x)g(x)\, dx$$

What is the interpretation of our mean square metric in this case?

$$\|f - g\|^2 = \langle f - g, f - g \rangle = \frac{1}{\pi} \int_{-\pi}^{+\pi} |f(x) - g(x)|^2\, dx$$

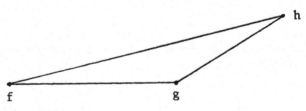

**Figure 3.9**

so

$$\|f - g\| = \left[\frac{1}{\pi} \int_{-\pi}^{+\pi} |f(x) - g(x)|^2 \, dx\right]^{1/2}$$

Thus the distance between the functions $f(x)$ and $g(x)$ is measured by integrating the square of the ordinate between the functions, dividing by $\pi$, and taking the square root of the result. Thus the distance between $f$ and $g$ measures what we might call the root-mean-square area between the functions—thus the term "mean-square metric."

Look back now to our study in Section 3.2 of Fourier series of a function $f(x)$, whose partial sum we write as

$$S_N(x) = \frac{a_0}{2} + \sum_{n=1}^{N} a_n \cos nx + b_n \sin nx$$

In Section 3.2 we compared $S_4$ and $S_5$ with $f(x)$ for various specific functions and subsequently raised the question of how we might best measure the difference between $f$ and the partial sum of its Fourier series $S_N$. In Section 3.3 we discussed pointwise convergence, where we look at the difference $|f(x) - S_N(x)|$ at individual points $x$. The mean-square distance now gives us a new conceptually different measure of the discrepancy between a function $f(x)$ and the $N$th partial sum of its Fourier series, $S_N(x)$:

$$\|f - S_N\| = \left[\frac{1}{\pi} \int_{-\pi}^{+\pi} |f(x) - S_N(x)|^2 \, dx\right]^{1/2}$$

Rather than comparing $f(x)$ and $S_N(x)$ at individual points $x$, we are now looking at the whole picture at once, as it were. The measure of the discrepancy between $f$ and $S_N$ is the root mean square of the area between their graphs. It is with respect to *this* metric that the two questions

1. How well does $S_N$ approximate $f$?
2. Does $S_N$ converge to $f$ as $N \to \infty$?

have their most natural and definitive answer.

**Definition 3.9.** Let $V$ be a real or complex inner product space. Suppose we are given in $V$ a sequence $\phi_1, \phi_2, \ldots, \phi_n, \ldots$ of orthogonal nonzero functions, i.e., $\langle \phi_m, \phi_n \rangle = 0$ when $m \neq n$. Given $f$ in $V$, the numbers

$$c_n = \frac{\langle f, \phi_n \rangle}{\|\phi_n\|^2} = \frac{\langle f, \phi_n \rangle}{\langle \phi_n, \phi_n \rangle}, \qquad n = 1, 2, \ldots$$

are called the *Fourier coefficients* of $f$ with respect to the orthogonal sequence $\{\phi_n\}$.

**Theorem 3.10.** *Let V and $\{\phi_n\}$ be as in the preceding definition. Then for any f in V and any real or complex numbers $a_1, a_2, \ldots, a_n$ we have*

$$\|f - (a_1\phi_1 + a_2\phi_2 + \cdots + a_n\phi_n)\|^2$$

$$= \|f - (c_1\phi_1 + c_2\phi_2 + \cdots + c_n\phi_n)\|^2 + \sum_{i=1}^{n} |c_i - a_i|^2 \|\phi_i\|^2$$

*where the $c_i$ are the Fourier coefficients of f with respect to the $\phi_i$.*

*Proof.* Write

$$f - (a_1\phi_1 + a_2\phi_2 + \cdots + a_n\phi_n) = u + v$$

where $u = f - (c_1\phi_1 + c_2\phi_2 + \cdots + c_n\phi_n)$ and $v = (c_1 - a_1)\phi_1 + \cdots + (c_n - a_n)\phi_n$. I claim that $u$ is orthogonal to $v$, i.e., $\langle u, v \rangle = 0$. As $v$ is a linear combination of the $\phi_i$, it will be enough to show that $\langle u, \phi_i \rangle = 0$, $1 \leqslant i \leqslant n$. Remembering that the $\phi_i$ are orthogonal, we get

$$\langle u, \phi_i \rangle = \langle f, \phi_i \rangle - c_i\langle \phi_i, \phi_i \rangle = \langle f, \phi_i \rangle - \frac{\langle f, \phi_i \rangle}{\|\phi_i\|^2}\langle \phi_i, \phi_i \rangle = 0$$

because the $c_i$ were chosen to be the Fourier coefficients of $f$ with respect to the $\phi_i$: $c_i = \langle f, \phi_i \rangle / \|\varphi_i\|^2 = \langle f, \phi_i \rangle / \langle \phi_i, \phi_i \rangle$. Since $u \perp v$, we have $\|u + v\|^2 = \|u\|^2 + \|v\|^2$. Using again the fact that the $\phi_i$ are orthogonal, we have $\|v\|^2 = \sum_{i=1}^{n} |c_i - a_i|^2 \|\phi_i\|^2$. Putting these facts together, we have our result.

The theorem has very important consequences for us. Consider the term on the far right,

$$\sum_{i=1}^{n} |c_i - a_i|^2 \|\phi_i\|^2$$

This term will be positive unless $a_i = c_i$, $i = 1, 2, \ldots, n$. Thus

$$\|f - (a_1\phi_1 + a_2\phi_2 + \cdots + a_n\phi_n)\|^2 \geqslant \|f - (c_1\phi_1 + c_2\phi_2 + \cdots + c_n\phi_n)\|^2$$

with equality holding only when every $a_i$ equals the Fourier coefficient of $f$ with respect to $\phi_i$. We have this interpretation:

*Among all possible linear combinations $a_1\phi_1 + a_2\phi_2 + \cdots + a_n\phi_n$, the best mean-square approximation to f is obtained when every $a_i$ is chosen to be the Fourier coefficient of f with respect to $\phi_i$.*

Consider, for example, the inner product space $V$ of functions on the circle $C$, $-\pi \leqslant x \leqslant \pi$, $-\pi \equiv \pi$, with inner product

$$\langle f, g \rangle = \frac{1}{\pi} \int_{-\pi}^{+\pi} f(x)g(x)\,dx$$

The trigonometric system

$\frac{1}{2}, \cos x, \sin x, \cos 2x, \sin 2x, \ldots, \cos nx, \sin nx, \ldots$

is an orthogonal family in $V$. We first show that the Fourier coefficients of $f$ with respect to this family, in the sense of the definition just given, are the same as those we computed previously in Section 3.2. First, the norms

$$\left\|\frac{1}{2}\right\|^2 = \frac{1}{\pi}\int_{-\pi}^{+\pi}\left(\frac{1}{2}\right)^2 dx = \frac{1}{4\pi}\int_{-\pi}^{+\pi} dx = \frac{2\pi}{4\pi} = \frac{1}{2}$$

$$\|\cos nx\|^2 = \frac{1}{\pi}\int_{-\pi}^{+\pi}\cos^2 nx\, dx = 1,$$

$$\|\sin nx\|^2 = \frac{1}{\pi}\int_{-\pi}^{+\pi}\sin^2 nx\, dx = 1$$

Hence, the first coefficient is

$$\frac{\langle f, \frac{1}{2}\rangle}{\|\frac{1}{2}\|^2} = \frac{1}{\frac{1}{2}}\frac{1}{\pi}\int_{-\pi}^{+\pi}\frac{f(x)}{2}\, dx = \frac{1}{\pi}\int_{-\pi}^{+\pi} f(x)\, dx = a_0$$

and the others are

$$\frac{\langle f, \cos nx\rangle}{\|\cos nx\|^2} = \frac{1}{1}\frac{1}{\pi}\int_{-\pi}^{+\pi} f(x)\cos nx\, dx = a_n$$

$$\frac{\langle f, \sin nx\rangle}{\|\sin nx\|^2} = \frac{1}{1}\frac{1}{\pi}\int_{-\pi}^{+\pi} f(x)\sin nx\, dx = b_n$$

Thus our theorem gives, in this special case, this result:

*Among all possible linear combinations of the functions $\frac{1}{2}, \cos x, \sin x, \cos 2x, \sin 2x, \ldots, \cos nx, \sin nx$, the best mean-square approximation to a function $f$ is obtained when every coefficient is chosen to be the corresponding (trigonometric) Fourier coefficient of $f$.*

In Section 3.2 we compared $f(x)$ and $S_4(x)$,

$$S_4(x) = \frac{a_0}{2} + a_1\cos x + b_1\sin x + a_2\cos 2x + b_2\sin 2x$$

$$+ a_3\cos 3x + b_3\sin 3x + a_4\cos 4x + b_4\sin 4x$$

for various functions $f(x)$ by graphing $S_4$ against the function. If we review these graphs, we can see the graph of $S_4(x)$ "wobbles" across the graph of $f(x)$ so as to minimize the mean-square area between the graphs.

In particular, for the function $f(x) = x$ we found

$$x \cong 2\sin x - \sin 2x + \tfrac{2}{3}\sin 3x - \tfrac{1}{2}\sin 4x + \tfrac{2}{5}\sin 5x$$

If we were to alter, in any manner whatsoever, any of the coefficients on the right, we would only worsen the mean-square discrepancy. We get the best possible mean-square approximation when we choose the Fourier coefficients. This justifies the use of Fourier coefficients without making any appeal to infinite series.

Turn back again now to the general case as described in Definition 3.9 and Theorem 3.10: given an orthogonal sequence of functions $\phi_1, \phi_2, \ldots, \phi_m \ldots$ in an inner product space $V$, and given a function $f$ in $V$, then the linear combination $c_1\phi_1 + c_2\phi_2 + \cdots + c_n\phi_n$, where the $c_i$ are the Fourier coefficients of $f$ with respect to the $\phi_i$, $c_i = \langle f, \phi_i \rangle / \langle \phi_i, \phi_i \rangle$, offers the best square approximation to $f$ in the sense that the mean-square distance between $f$ and $c_1\phi_1 + \cdots + c_n\phi_n$,

$$\|f - (c_1\phi_1 + c_2\phi_2 + \cdots + c_n\phi_n)\| \tag{3.8.1}$$

is an absolute minimum.

What happens to the approximation (3.8.1) when $n \to \infty$?

**Definition 3.11.** Let $\phi_1, \phi_2, \ldots, \phi_m \ldots$ be a fixed orthogonal sequence of nonzero functions in a real or complex inner product space $V$. Given $f \in V$, if, for some sequence of scalars $a_i$, $\|f - \Sigma_{i=1}^n a_i\phi_i\| \to 0$ as $n \to \infty$, we say that the series $\Sigma_{i=1}^\infty a_i\phi_i$ converges to $f$ in the mean-square metric (of $V$), and write

$$f = \sum_{i=1}^\infty a_i\phi_i$$

We call this the *general Fourier series* or *general Fourier expansion* of $f$ (with respect to the given orthogonal family $\{\phi_i\}$).

**Theorem 3.12.** Let $\phi_1, \phi_2, \ldots, \phi_m \ldots$ be an orthogonal family of nonzero functions in an inner product space $V$. Suppose that a function $f$ in $V$ has a general Fourier expansion with respect to the $\phi_i$:

$$f = \sum_{i=1}^\infty a_i\phi_i$$

*Then:*

1. *The $a_i$ are the Fourier coefficients of $f$ with respect to the $\phi_i$, $a_i = c_i = \langle f, \phi_i \rangle / \langle \phi_i, \phi_i \rangle$. Hence, in any general Fourier series, the coefficients are uniquely determined and must be the Fourier coefficients as given in Definition 3.9.*
2. *Given any $g \in V$, we have*

$$\langle f, g \rangle = \sum_{i=1}^\infty a_i \langle \phi_i, g \rangle$$

*where the series of real or complex numbers on the right converges in the ordinary calculus sense to the quantity on the left. Hence, in any general Fourier series, we may take the inner product term by term and will always obtain a valid numerical series.*

*Proof.* (1) We shall give two proofs of this important fact. The first proof starts with the equality of Theorem 3.10:

$$\|f - (a_1\phi_1 + \cdots + a_n\phi_n)\|^2 = \|f - (c_1\phi_1 + \cdots + c_n\phi_n)\|^2$$
$$+ \sum_{i=1}^{n} |c_i - a_i|^2 \|\phi_i\|^2 \tag{3.8.2}$$

In this expression the $c_i$ are the first $n$ Fourier coefficients of $f$ with respect to the $\phi_i$, and the $a_i$ are the first $n$ coefficients in the Fourier expansion of $f$. Our assumption in the theorem, namely that $f = \Sigma\, a_i\phi_i$, means exactly that the left side of equation (3.8.2) goes to zero as $n \to \infty$. (Refer to Definition 3.11.) Hence the right side of equation (3.8.2) must also go to zero as $n \to \infty$. The right side is the sum of two terms, both $\geqslant 0$. As each term is $\leqslant$ right side, each term must go to zero as $n \to \infty$. But the term $\Sigma^n |c_i - a_i|^2 \|a_i\|^2$ *increases* with $n$. The only way it can go to zero as $n \to \infty$ is to be identically zero, which means $c_i = a_i$, $i = 1, 2, \ldots$, which was to be proved. The second proof will be based on 2, which we turn to now.

Proof of (2). We are given

$$f = \sum_{i=1}^{\infty} a_i\phi_i$$

in the mean-square sense, i.e.,

$$\left\| f - \sum_{i=1}^{n} a_i\phi_i \right\| \to 0 \qquad \text{as } n \to \infty$$

We wish to prove that the numerical series in 2 converges, i.e.,

$$\left| \langle f, g \rangle - \sum_{i=1}^{n} a_i\langle \phi_i, g \rangle \right| \to 0 \qquad \text{as } n \to \infty$$

where $|\cdot|$ is the usual absolute value. But

$$\left| \langle f, g \rangle - \sum_{i=1}^{n} a_i\langle \phi_i, g \rangle \right| = \left| \left\langle f - \sum_{i=1}^{n} a_i\phi_i, g \right\rangle \right|$$
$$\leqslant \left\| f - \sum_{i=1}^{n} a_i\phi_i \right\| \cdot \|g\| \tag{3.8.3}$$

where we have used the finite additivity of the inner product, then the Schwarz inequality. By assumption, the right side of equation (3.8.3) tends to zero as $n \to \infty$; therefore so does the left side, which is what we wanted to prove.

Now, using part 2 of Theorem 3.12, we can give a second proof of the uniqueness of Fourier coefficients. Item 2 asserts that you can take the inner product term by term of a Fourier expansion with any element. So take the inner product of both sides of $f = \Sigma a_i \phi_i$ with $\phi_m$ to get $\langle f, \phi_m \rangle = \Sigma_i a_i \langle \phi_i, \phi_m \rangle$ ($m$ fixed). But $\langle \phi_i, \phi_m \rangle = 0$ when $i \neq m$. Hence, $\langle f, \phi_m \rangle = a_m \langle \phi_m, \phi_m \rangle$, or $a_m = \langle f, \phi_m \rangle / \langle \phi_m, \phi_m \rangle = c_m$, which is what we wanted to show. Theorem 3.12 is proved.

Part 2 of Theorem 3.12 says that the bilinearity of the inner product that we previously had established for finite sums holds for infinite series as well.

Part 1 of Theorem 3.12, the uniqueness, holds for finite sums as well. There are two ways we can look at this. If we have

$$f = \sum_{i=1}^{n} a_i \phi_i$$

so that $f$ is *equal* to a finite linear combination of the $\phi_i$, then obviously this finite sum is the best possible mean-square approximation to $f$, so by the italicized corollary to Theorem 3.10, the $a_i$ must be the Fourier coefficients. Or we can look at the finite sum $f = \Sigma^n a_i \phi_i$ as an infinite series $f = \Sigma^\infty a_i \phi_i$ with $a_i = 0$ for $i > n$, and apply Theorem 3.12 directly.

Suppose now that $c_i, i = 1, 2, \ldots$, are the Fourier coefficients of a function $f$ with respect to an orthogonal family $\phi_i, i = 1, 2, \ldots, c_i = \langle f, \phi_i \rangle / \langle \phi_i, \phi_i \rangle$. All this in an inner product space $V$. Then, as one can easily check, $u = f - \Sigma^n c_i \phi_i$ is orthogonal to every $\phi_i$, $1 \leqslant i \leqslant n$, so $u$ is orthogonal to $v = \Sigma^n c_i \phi_i$. Thus $\|u + v\|^2 = \|u\|^2 + \|v\|^2$, or $\|u\|^2 = \|u + v\|^2 - \|v\|^2$. As $u + v = f$, we get

$$\left\| f - \sum_{i=1}^{n} c_i \phi_i \right\|^2 = \|f\|^2 - \sum_{i=1}^{n} |c_i|^2 \|\phi_i\|^2 \tag{3.8.4}$$

The expression on the left of equation (3.8.4) is the (square of) the mean-square error we make when we approximate $f$ by the first $n$ terms of its Fourier expansion. The right side of equation (3.8.4) is easily computable in most cases and thus affords a means of computing the mean-square error (see Exercises 17, 18, and 19).

Equation (3.8.4) also yields the following.

**Theorem 3.13.** *Always* $\Sigma_{i=1}^{\infty} |c_i|^2 \|\phi_i\|^2 \leqslant \|f\|^2$, *with equality holding if and only if* $f = \Sigma_{i=1}^{\infty} c_i \phi_i$.

The inequality in Theorem 3.13 is called *Bessel's inequality*, and the equality, when it holds, is called *Parseval's relation*.

**Definition 3.14.** If $\phi_i$, $i = 1, 2, \ldots$, is an orthogonal family of nonzero functions in an inner product space $V$ such that every $f$ in $V$ has a general Fourier expansion,

$$f = \sum_{i=1}^{\infty} c_i \phi_i$$

then $\{\phi_i\}$ is called an *orthogonal basis* of $V$.

According to Theorem 3.13, $\{\phi_i\}$ is an orthogonal basis of $V$ if, and only if, the Parseval relation

$$\|f\|^2 = \sum_{i=1}^{\infty} |c_i|^2 \|\phi_i\|^2$$

which involves an ordinary numerical series with nonnegative terms, holds for every $f$ in $V$.

**Exercises**

1. Given vectors $A$ and $B$ in ordinary three-dimensional space, the vector from $A$ to $B$ is $B - A$. Correspondingly, given functions $f$ and $g$ in an inner product space, the "line segment" from $f$ to $g$ is $g - f$. Thus in Figure 3.9 the three sides of the triangle shown are $g - f$, $h - g$, and $f - h$. Show that if the angle at $g$ is a right angle, then the Pythagorean theorem holds: the square of the hypotenuse equals the sum of the squares of the other two sides, i.e., $\|f - h\|^2 = \|h - g\|^2 + \|g - f\|^2$.

2. For the inner product $\langle f, g \rangle = (1/\pi) \int_{-\pi}^{+\pi} f(x)g(x)\, dx$, find the mean square distance between
   (a) $f(x) = 1$ and $g(x) = \sin x$   [Ans.: $\sqrt{3}$]
   (b) $f(x) = \sin x$ and $g(x) = \cos x$
   (c) $f(x) = \sin x$ and $g(x) = \sin 2x$

3. For the inner product $\langle f, g \rangle = \int_0^{\infty} f(x)g(x)e^{-x}\, dx$, find the mean square distance between
   (a) $f(x) = 1$ and $g(x) = x$
   (b) $f(x) = \sin x$ and $g(x) = \cos x$

4. Let $a$ and $b$ be real constants $< \frac{1}{2}$. Show that with respect to the mean-square metric derived from the inner product $\langle f, g \rangle = \int_0^{\infty} f(x)g(x)e^{-x}\, dx$ the distance between $e^{ax}$ and $e^{bx}$ is

$$\frac{\sqrt{2}|a - b|}{[(1 - 2a)(1 - 2b)(1 - a - b)]^{1/2}}$$

5. Find the distance between the functions $\sin \pi x$ and $\cos \pi x$ with respect to the mean-square metric derived from the inner product
   (a) $\langle f, g \rangle = \int_{-\frac{1}{2}}^{+\frac{1}{2}} f(x)g(x)\, dx$
   (b) $\langle f, g \rangle = \int_{-\frac{1}{2}}^{+\frac{1}{2}} f(x)g(x)(1/\sqrt{1 - x^2})\, dx$   [Ans.: $\sqrt{\pi}$]

6. For the space $\mathbf{R}^3$ of all triples $(x_1, x_2, x_3)$ of real numbers, the ordinary inner product is

$$\langle A, B \rangle = a_1 b_1 + a_2 b_2 + a_3 b_3$$

where $A = (a_1, a_2, a_3)$ and $B = (b_1, b_2, b_3)$. Show that in this case our mean-square metric is the ordinary Euclidean distance.

7. Consider the inner product space of functions on $0 \leqslant x \leqslant 1$ with inner product $\langle f, g \rangle = \int_0^1 f(x) g(x) \, dx$. For each integer $n = 1, 2, \ldots$ define a function $f_n(x)$ by

$$f_n(x) = \begin{cases} 0, & x = 0 \\ n, & 0 < x < 1/n \\ 0, & 1/n \leqslant x \leqslant 1 \end{cases}$$

(a) Sketch the graph of $f_n$.
(b) Show that $\| f_n \| \to \infty$ as $n \to \infty$.
(c) Show that for each fixed $x$ in $[0, 1]$, $f_n(x) \to 0$ as $n \to \infty$.

8. With respect to the inner product in Exercise 7, show that if $|f(x)| \leqslant M$, $0 \leqslant x \leqslant 1$, then $\| f \| \leqslant M$.

9. Given orthogonal nonzero functions $\phi_1, \phi_2, \ldots, \phi_n$ in an inner product space, the subspace $M$ spanned by the $\phi_i$ consists of all (real or complex) linear combinations $a_1 \phi_1 + a_2 \phi_2 + \cdots + a_n \phi_n$. The line segment from a function $f$ to a general point on $M$ is $f - (a_1 \phi_1 + a_2 \phi_2 + \cdots + a_n \phi_n)$. For what selection of the constants $a_i$ is this line segment orthogonal to $M$?

10. In Exercise 12 of Section 3.7 we showed that the functions $p_0 = 1$, $p_1 = x$, $p_2 = 3x^2 - 1$, and $p_3 = 5x^3 - 3x$ form an orthogonal family with respect to the inner product $\langle f, g \rangle = \int_{-1}^{+1} f(x) g(x) \, dx$. Find the linear combination of $p_0, p_1, p_2, p_3$ that offers the best mean-square approximation to $f(x) = x^4$. Graph carefully on graph paper the approximation against the function.

11. If $\langle f, g \rangle = 0$, then $d(f, g) = (\| f \|^2 + \| g \|^2)^{1/2}$.

12. (Tschebyscheff polynomials of the first kind)   Consider the inner product space $V$ consisting of sufficiently smooth real functions on $-1 \leqslant x \leqslant 1$ with the inner product

$$\langle f, g \rangle = \int_{-1}^{+1} f(x) g(x) \frac{dx}{\sqrt{1 - x^2}}$$

(a) Let $x = \cos \theta$, $T_n(x) = \cos n\theta$, $n = 0, 1, 2, \ldots$. Prove that the functions $T_n(x)$ form an orthogonal family in $V$.
(b) Find $\| T_0(x) \|$. Find $\| T_n(x) \|$, $n \geqslant 1$. [$\| T_0 \|^2 = \pi$, $\| T_n \|^2 = \pi/2$, $n \geqslant 1$.]
(c) Compute $T_n(x)$ for $n = 0, 1, 2, 3, 4$. [$T_0 = 1$, $T_1 = x$, $T_2 = 2x^2 - 1$, $T_3 = 4x^3 - 3x$, $T_4 = 8x^4 - 8x^2 + 1$.]

(d) Using the answers to (c), show by simple polynomial addition that $x^4 = \frac{3}{8}T_0 + \frac{1}{2}T_2 + \frac{1}{8}T_4$. Then, paying attention to Theorem 3.12 (the uniqueness part) write down the Fourier coefficients of $x^4$ with respect to $T_0$, $T_1$, $T_2$, and $T_3$. Find the linear combination of $T_0$, $T_1$, $T_2$, and $T_3$ that offers the best mean-square approximation to $x^4$. Graph carefully on graph paper the approximation against the function $-1 \leqslant x \leqslant 1$.

13. (Continuation) (a) Prove that $T_{n+1} = 2xT_n - T_{n-1}$.
    (b) Starting with $T_0 = 1$, $T_1 = x$, and using the recursion in (a), compute again $T_2$, $T_3$, and $T_4$.

14. (Continuation) Consider the following statement: "$T_n(x)$ is a polynomial of degree $n$ containing only even powers of $x$ when $n$ is even, only odd powers when $n$ is odd." Call this statement $S(n)$.
    (a) Check, using $T_0$ through $T_4$ already computed, that $S(0)$, $S(1)$, $S(2)$, $S(3)$, and $S(4)$ are true.
    (b) Suppose $n$ is even. Show that if $S(j)$ is true for $0 \leqslant j \leqslant n$, then $S(n+1)$ is true. [Hint: Use the recursion $T_{n+1} = 2xT_n - T_{n-1}$ from Exercise 13.]
    (c) Suppose $n$ is odd. Show that if $S(j)$ is true for $0 \leqslant j \leqslant n$, then $S(n+1)$ is true.
    (d) Conclude from (b) and (c) that if $S(j)$ is true for $0 \leqslant j \leqslant n$, then $S(n+1)$ is true no matter what the parity of $n$.
    (e) You know from (a) that $S(j)$ is true for $0 \leqslant j \leqslant 4$. Using (d), what can you deduce about $S(5)$? $S(6)$? $S(7)$? $S(n)$ for all $n$?
    This method is called *mathematical induction*.

15. (Continuation) Consider the following statement: "$x^n$ is a linear combination of $T_0, T_1, T_2, \ldots, T_n$." Call the statement $S(n)$. Show that $S(0)$ through $S(4)$ are true, and prove by mathematical induction that $S(n)$ is true for all $n$ (no need to separate even and odd $n$ in this case). Using this, prove that $\langle x^m, T_n \rangle = 0$ when $m \leqslant n-1$.

16. Prove that the Tschebyscheff polynomial $T_n(x) = \cos n\theta$, $x = \cos \theta$, is a solution of the differential equation

$$(1 - x^2)y'' - xy' + n^2 y = 0$$

[Hint: Show first that $(1-x^2)y'' - xy' = (1-x^2)^{1/2}((1-x^2)^{1/2}y')'$. Then, using $dy/dx = (dy/d\theta)(d\theta/dx)$, show that

$$(1 - x^2)^{1/2} \frac{d}{dx}\left((1 - x^2)^{1/2} \frac{dy}{dx}\right) = \frac{d^2 y}{d\theta^2}]$$

17. (a) Show that with respect to the inner product $\langle f, g \rangle = (1/\pi) \int_{-\pi}^{+\pi} f(x)g(x)\,dx$ we have $\|1/2\|^2 = 1/2$ and $\|\cos nx\|^2 = \|\sin nx\|^2 = 1$, $n = 1, 2, \ldots$.

(b) Show that

$$\frac{1}{\pi} \int_{-\pi}^{+\pi} \left| f(x) - \left( \frac{a_0}{2} + \sum_{n=1}^{N} a_n \cos nx + b_n \sin nx \right) \right|^2 dx$$

$$= \frac{1}{\pi} \int_{-\pi}^{+\pi} |f(x)|^2 dx - \frac{a_0^2}{2} - \sum_{n=1}^{N} (a_n^2 + b_n^2)$$

[Hint: Specialize equation (3.8.4).]

18. (Continuation)   Compute the mean-square error we made in Section 3.2 when we approximated $f(x) = x$ by the first five terms of its Fourier series.
19. (Continuation)   Compute the mean-square error we made in Exercise 10 of Section 3.2 when we approximated $f(x) = x(\pi^2 - x^2)$ by the first four terms of its Fourier series.

## 3.9   $L^2$ SPACES

In the previous sections we have introduced the concept of inner product (via the motivating example of trigonometric Fourier series) (Section 3.6), have defined an inner product space and proved that the Schwarz inequality is valid in any inner product space (Section 3.7), and have shown that any inner product space carries its own intrinsic distance function, the mean-square metric so-called (Section 3.8).

In this lecture we shall study a very important class of inner product spaces, the $L^2$ spaces. We shall also discuss bases in $L^2$ spaces.

**Definition 3.15.**   Let $a$ and $b$ satisfy $-\infty \leqslant a < b \leqslant +\infty$, and let $w(x)$ be a real-valued function on the interval $a \leqslant x \leqslant b$ such that $w(x) > 0$ in the interior $a < x < b$. We denote by

$$L^2(a, b; w)$$

the set of all real (resp. complex) valued functions on $a \leqslant x \leqslant b$ that satisfy

$$\int_a^b |f(x)|^2 w(x) \, dx < \infty \qquad (3.9.1)$$

We call $L^2(a, b; w)$ the space of all real (resp. complex) square integrable functions with respect to the weight $w$ on $(a, b)$.

The use of the term "space" in Definition 3.15 will be justified subsequently in Theorem 3.16.

Note the peculiar nature of the condition (3.9.1) that defines the class $L^2(a, b; w)$. We don't require the integral in (3.9.1) to have any particular *value*; we require only that it be *finite*. So this comes under what we used to call in calculus "improper" integrals. The issue here is to decide whether or not the integral displayed in (3.9.1) is convergent or divergent.

Now, to be technically correct, we should restrict our attention to functions that are Lebesgue measurable and should interpret all integrals as Lebesgue integrals. However, this aspect of the mathematical theory does not affect our applications, so we shall ignore it.

Another comment on condition (3.9.1). In equation (3.9.1) the term underneath the integral sign, $|f(x)|^2 w(x)$, is $\geq 0$. Hence the integral itself will be $\geq 0$, so it is either finite and nonnegative or it is $+\infty$. We are dealing here with what is called *absolute convergence*. And we have a general principle that absolute convergence implies convergence. Thus, if the integral in (3.9.1) converges, so does the integral with the absolute value sign removed.

We can reach the same conclusion if we use the familiar inequality from calculus

$$\left| \int_a^b g(x)\,dx \right| \leq \int_a^b |g(x)|\,dx$$

If we apply this to (3.9.1), remember $|w(x)| = w(x)$ because $w(x) \geq 0$, and remember as well that $|z^2| = |z|^2$ for any real or complex number $z$, we get

$$\left| \int_a^b f(x)^2 w(x)\,dx \right| \leq \int_a^b |f(x)|^2 w(x)\,dx$$

If the integral on the right is finite, so will be the integral inside the absolute value signs on the left. The qualitative reason behind this is that if the integral of the absolute value of a function converges, then if we remove the absolute value signs, we will allow cancellation between positive and negative values of the function, which can only improve convergence. That is also the qualitative reason behind the principle that absolute convergence implies convergence.

Note also that (3.9.1) involves the square, $|f(x)|^2$, and not $|f(x)|$ itself. The convergence of the one integral does not necessarily imply convergence of the other. (It will though if the interval $(a, b)$ is finite; refer to Exercise 5d of Section 3.7.) We are interested in the integral of $|f(x)|^2$.

Let us look at two typical examples.

**Example 3.1:** $a = -\pi, b = \pi, w(x) = 1/\pi$. In this case our condition for square integrability is this: $(1/\pi) \int_{-\pi}^{+\pi} |f(x)|^2\,dx < \infty$. As the $1/\pi$ clearly does not influence the convergence or divergence of the integral, we shall often omit it, and shall also write $L^2(-\pi, \pi)$ in place of $L^2(-\pi, \pi; 1/\pi)$.

Clearly, if $f(x)$ is bounded, $|f(x)| \leqslant M$, $-\pi \leqslant x \leqslant \pi$, then $f(x)$ will satisfy this square integrability condition:

$$\int_{-\pi}^{+\pi} |f(x)|^2 \, dx \leqslant \int_{-\pi}^{+\pi} M^2 \, dx = M^2 \int_{-\pi}^{+\pi} dx = 2\pi M^2 < \infty$$

Hence $L^2(-\pi, \pi)$ contains every bounded function on the interval $-\pi \leqslant x \leqslant \pi$ no matter how many discontinuities it may have or how wildly it may behave. Obviously, this includes every example we considered in Section 3.2.

For the interval $(-\pi, \pi)$ (in fact, for any finite interval), the condition of square integrability becomes a restriction only when we deal with functions that tend to infinity at some point or points in $-\pi \leqslant x \leqslant \pi$. For example, suppose $f(x) = 1/|x|^{1/2}$, $x \neq 0$, and $f(0) = 0$. Does this function belong to $L^2(-\pi, \pi)$? Here we must deal with an improper integral by the methods we learned in calculus. Fix a number $a > 0$, and compute

$$\int_{-\pi}^{-a} \left(\frac{1}{|x|^{1/2}}\right)^2 dx + \int_{a}^{\pi} \left(\frac{1}{|x|^{1/2}}\right)^2 dx = 2\int_{a}^{\pi} \left(\frac{1}{|x|^{1/2}}\right)^2 dx$$

Then evaluate the limit of this quantity, as $a$ tends to zero through positive values: $a \to 0+$. If the limit is finite, then the function is square integrable; otherwise not. For our example, we have

$$2\int_{a}^{\pi} \left(\frac{1}{|x|^{1/2}}\right)^2 dx = 2\int_{a}^{\pi} \frac{dx}{x} = 2(\ln \pi - \ln a) \to +\infty \text{ as } a \to 0+.$$

Hence $f(x) = 1/|x|^{1/2}$ is not square integrable on $-\pi \leqslant x \leqslant \pi$, so does not belong to $L^2(-\pi, \pi)$. On the other hand, $f(x) = 1/|x|^{1/3}$, although singular at $x = 0$, is square integrable, so does belong to $L^2(-\pi, \pi)$. (Exercise 1.)

Thus on $-\pi \leqslant x \leqslant \pi$, the square integrability condition allows all sorts of complicated functions, including those that have one or many singularities, as long as they are not too strong. The same conclusion holds for any finite interval, for example $(-1, 1)$, $(0, 1)$, etc.

**Example 3.2:** $a = 0$, $b = \infty$, $w(x) = e^{-x}$. In this case our condition for square integrability is this: $\int_0^\infty |f(x)|^2 e^{-x} \, dx < \infty$. This condition involves the same sort of restriction on the singularities of $f(x)$ as we had in the previous case and adds a new requirement that was absent for finite intervals, namely $|f(x)|$ cannot increase too rapidly as $x \to \infty$. For example, suppose $f(x) = e^{x/2}$. To decide whether or not this function belongs to $L^2(0, \infty; e^{-x})$ we must decide whether or not the integral

$$\int_0^\infty |f(x)|^2 e^{-x} \, dx$$

is convergent, for this particular $f$.

Remember how this is done. We first select a number $a > 0$, then evaluate (or estimate) the integral

$$\int_0^a |f(x)|^2 e^{-x}\, dx$$

then evaluate the limit as $a \to +\infty$. Here is how the calculation goes for $f(x) = e^{x/2}$.

$$\int_0^a |e^{x/2}|^2 e^{-x}\, dx = \int_0^a e^x e^{-x}\, dx = \int_0^a dx = a \to +\infty \qquad \text{as } a \to +\infty$$

Hence our integral diverges, so the function $f(x) = e^{x/2}$ does *not* belong to $L^2(0, \infty; e^{-x})$. On the other hand, the functions $1, x, x^2, \ldots$ all do belong (Exercise 6). The functions $f(x) = x^n$ for $n \geqslant 1$ go to $+\infty$ as $x \to +\infty$, but not so rapidly as to make the integral $\int_0^\infty |f(x)|^2 e^{-x}\, dx$ divergent.

Other examples are dealt with in the exercises. There are two main techniques used in doing these exercises: (1) evaluation of improper integrals by the methods just illustrated in Examples 1 and 2, and (2) evaluation of limits such as $\lim x^n e^{-x}$ as $x \to +\infty$, $\lim x \ln x$ as $x \to 0+$, etc. Both these things are normally covered in basic calculus but sometimes not stressed. Hence, doing these exercises may encourage a little review of these techniques.

Note that in deciding whether or not a given function belongs to $L^2$, we need not always evaluate $\int_a^b |f(x)|^2 w(x)\, dx$ exactly, as it is only a question of whether this integral is finite or infinite. For example, if $|f(x)|^2 \leqslant g(x)$ and $\int_a^b g(x)w(x)\, dx$ is finite, then so is $\int_a^b |f|^2 w\, dx$. By choosing $g$ adroitly, it is often possible to simplify the work (Exercises 8 and 10b).

Another useful remark: in deciding when $\int_a^\infty |f(x)|^2 w(x)\, dx$ is finite for a function $f(x)$ that has no singularities, we need only consider the behavior of $f(x)$ as $x \to \infty$.

By and large, the main thing we should stress about the spaces $L^2(a, b; w)$ is how broad they are. The *only* requirement a function $f(x)$ must pass in order to qualify for membership in $L^2(a, b; w)$ is that the integral

$$\int_a^b |f(x)|^2 w(x)\, dx$$

be finite (other than the tacit requirement of Lebesgue measurability that won't concern us).

Having concluded our discussion of the definition of the $L^2$ spaces, we turn now to their two principal properties: they are vector spaces and, in a natural way, inner product spaces.

**Theorem 3.16.** $L^2(a, b; w)$ *is a vector space; that is, if the functions $f$ and $g$ belong to $L^2(a, b; w)$, then so do $f + g$ and $cf$ for any (real or complex) scalar $c$.*

*Proof.* The assertion about $cf$ follows easily, because

$$\int_a^b |cf(x)|^2 w(x)\, dx = |c|^2 \int_a^b |f(x)|^2 w(x)\, dx$$

As the integral on the right is finite, so is that on the left.

To prove that $f + g$ belongs to $L^2$ we must show that the integral

$$\int_a^b |f(x) + g(x)|^2 w(x)\, dx$$

is finite. Using the triangle inequality, $|f+g| \leqslant |f| + |g|$, we have

$$|f(x) + g(x)|^2 \leqslant (|f(x)| + |g(x)|)^2$$
$$= |f(x)|^2 + 2|f(x)|\,|g(x)| + |g(x)|^2$$

Hence

$$\int_a^b |f(x) + g(x)|^2 w(x)\, dx \leqslant \int_a^b |f(x)|^2 w(x)\, dx$$
$$+ 2\int_a^b |f(x)|\,|g(x)|w(x)\, dx + \int_a^b |g(x)|^2 w(x)\, dx$$

Now the first and last integrals on the right are finite by virtue of the assumption that $f$ and $g$ belong to $L^2$. So it boils down to proving the finiteness of the integral

$$\int_a^b |f(x)|\,|g(x)|w(x)\, dx$$

At this point it is tempting to try to complete the proof by using the Schwarz inequality. However, our derivation of the Schwarz inequality *assumed* we had an inner product defined on a vector space of functions. At this point in our argument, we don't know that we are dealing with a vector space of functions. Indeed, that is exactly what we are trying to prove, namely that if $f$ and $g$ belong to $L^2(a, b; w)$, then so does their sum $f + g$. So using the Schwarz inequality here would involve us in circular reasoning.

We use a different idea as follows. We use the fact that the square of any real number is $\geqslant 0$. Hence

$$(|f(x)| - |g(x)|)^2 = |f(x)|^2 - 2|f(x)|\,|g(x)| + |g(x)|^2 \geqslant 0$$

which leads to the inequality

$$2|f(x)|\,|g(x)| \leqslant |f(x)|^2 + |g(x)|^2$$

which leads immediately to our result, because then

$$2 \int_a^b |f(x)|\, |g(x)| w(x)\, dx \leqslant \int_a^b |f(x)|^2 w(x)\, dx + \int_a^b |g(x)|^2 w(x)\, dx \tag{3.9.2}$$

and the two integrals on the right are finite. Hence, we have our result: $L^2(a, b; w)$ is a vector space of functions.

**Theorem 3.17.** *The formula*

$$\langle f, g \rangle = \int_a^b f(x)\overline{g(x)}w(x)\, dx$$

*defines an inner product on* $L^2(a, b; w)$.

*Proof.* All that we need to show is that the displayed integral is finite. Once this is established, the three defining properties of an inner product—conjugate bilinear, Hermitian symmetric, and positive definite—follow just as in Section 3.6.

To prove that the above integral is finite, we can use the principle that absolute convergence implies convergence and deal instead with the integral of $|f(x)\overline{g(x)}|$. Using the fact that $|\bar{z}| = |z|$ for any complex number $z$, we see that we can use the integral

$$\int_a^b |f(x)|\, |g(x)| w(x)\, dx$$

But, as we have assumed that $f$ and $g$ belong to $L^2$, formula (3.9.2) directly above shows that this integral is finite.

Or we may argue directly, as follows, again using formula (3.9.2) above:

$$\left| \int_a^b f\bar{g}w\, dx \right| \leqslant \int_a^b |f|\, |g| w\, dx \leqslant \frac{1}{2} \left[ \int_a^b |f|^2 w\, dx + \int_a^b |g|^2 w\, dx \right]$$

Either way our result is proved: $L^2(a, b; w)$ is an inner product with the inner product as given in Theorem 3.17.

These $L^2$ spaces are the primary spaces that we will use henceforth. They are all Hilbert spaces, and we shall often refer to them as such. The additional property that makes them Hilbert spaces, the property of metric completeness, will be explained in Section 3.10 that follows. We won't be concerned with this additional property ourselves—for us the key feature is the inner product.

Knowing now that $L^2(a, b; w)$ is an inner product space, we know that all the properties and concepts associated with an inner product are available to us. For example, we have the *norm* $\|f\| = \langle f, f \rangle^{1/2}$ (Definition 3.3), which in our $L^2$ space has the form

$$\|f\| = \left[ \int_a^b |f(x)|^2 w(x)\, dx \right]^{1/2}$$

and the mean-square metric derived from this norm (Section 3.8), which here has the form

$$d(f,g) = \|f - g\| = \left[ \int_a^b |f(x) - g(x)|^2 w(x)\, dx \right]^{1/2}$$

Moreover, Schwarz's inequality, $|\langle f, g \rangle| \leqslant \|f\|\, \|g\|$ (Section 3.7) holds, which says here that

$$\left| \int_a^b f(x)\overline{g(x)}w(x)\, dx \right| \leqslant \left[ \int_a^b |f(x)|^2 w(x)\, dx \right]^{1/2} \left[ \int_a^b |g(x)|^2 w(x)\, dx \right]^{1/2}$$

Likewise we can speak of an orthogonal family $\phi_1, \phi_2, \ldots, \phi_m \ldots$ in $L^2(a, b; w)$, meaning

$$\langle \phi_i, \phi_j \rangle = \int_a^b \phi_i(x)\overline{\phi_j(x)}w(x)\, dx = 0, \qquad i \neq j$$

the *Fourier coefficients* $c_i$ of an $f$ in $L^2(a, b; w)$ with respect to an orthogonal family of nonzero functions $\{\phi_i\}$,

$$c_i = \frac{\langle f, \phi_i \rangle}{\langle \phi_i, \phi_i \rangle} = \frac{\int_a^b f(x)\overline{\phi_i(x)}w(x)\, dx}{\int_a^b |\phi_i(x)|^2 w(x)\, dx}$$

and the *general Fourier expansion* of an $f$ in $L^2(a, b; w)$,

$$f(x) = \sum_{i=1}^{\infty} c_i \phi_i(x)$$

meaning

$$\left\| f - \sum_{i=1}^{n} c_i \phi_i \right\| = \left[ \int_a^b \left| f(x) - \sum_{i=1}^{n} c_i \phi_i(x) \right|^2 w(x)\, dx \right]^{1/2} \to 0$$

as $n \to \infty$.

Especially important is the notion of *orthogonal basis* of $L^2(a, b; w)$ (Definition 3.14): the orthogonal family $\{\phi_i\}$ is an orthogonal basis of $L^2(a, b; w)$ if *every* $f$ in $L^2(a, b; w)$ has a general Fourier expansion with respect to the $\phi_i$,

$$f(x) = \sum_{i=1}^{\infty} c_i \phi_i(x)$$

convergent in the mean-square sense as defined in the preceding paragraph. We are already familiar with one orthogonal basis.

**Theorem 3.18.** *The trigonometric system*

$\frac{1}{2}$, cos $x$, sin $x$, cos $2x$, sin $2x$, ..., cos $nx$, sin $nx$, ...

*forms an orthogonal basis of $L^2(C)$, where $C$ is the unit circle, $-\pi \leqslant x \leqslant \pi$, $-\pi \equiv \pi$.*

We shall not prove this theorem in this book.

The theorem asserts that for *any* function $f(x)$ in $L^2(C)$, we have

$$\int_{-\pi}^{\pi} \left| f(x) - \left( \frac{1}{2} a_0 + \sum_{n=1}^{N} a_n \cos nx + b_n \sin nx \right) \right|^2 dx \to 0 \qquad \text{as } N \to \infty$$

(3.9.3)

where the $a_n$ and $b_n$ are the trigonometric Fourier coefficients of $f(x)$. It is the broadness of this assertion that is so remarkable. It says that given any function $f(x)$ in $L^2(C)$, no matter how nondifferentiable, discontinuous, or wild in various other ways, we may, by taking $N$ sufficiently large, make the mean-square area between the graph of $f(x)$ and that of $S_N(x)$, the $N$th partial sum of the Fourier series of $f$, as small as we please. If taken in this sense, the equation

$$f(x) = \frac{a_0}{2} + \sum_{n=1}^{\infty} a_n \cos nx + b_n \sin nx$$

(3.9.4)

then holds for *every* function $f(x)$ in $L^2(C)$. We don't have to assume that $f(x)$ is differentiable, or continuous, or anything else. All that is required is that the integral $\int_{-\pi}^{\pi} |f(x)|^2 dx$ be finite. But note that the equality in (3.9.4) has the meaning of (3.9.3), equality in the mean-square metric. If we let $g(x)$ represent the function on the right side of (3.9.4), then the equality in (3.9.4) means

$$\int_{-\pi}^{\pi} |f(x) - g(x)|^2 dx = 0$$

what we have called in Section 3.6 "essential equality." As was pointed out in Section 3.6, it is possible for functions to be essentially equal, yet have different values at a finite number of points. So it is possible that the series on the right of (3.9.4) may be pointwise convergent and have, at a finite number of points, values different from those of $f(x)$.

Joseph Fourier published his *Théorie Analytique de la Chaleur* in 1822, reporting on work from approximately 1807. Fourier's book has been translated into English by Alexander Freeman (*The Analytical Theory of Heat*, Dover, 1955). Section VI of Chapter III of this work is entitled "Development of an Arbitrary Function in Trigonometric Series." In Article

233 of this section, Fourier writes:

> Thus we obtain the following equation (p), which serves to develop any function whatever in a series formed of sines and cosines of multiple arcs:
>
> $$\pi F(x) = \frac{1}{2} \int F(x)\, dx + \cos x \int F(x) \cos x\, dx + \cos 2x \int F(x) \cos 2x\, dx + \&c.$$
>
> $$+ \sin x \int F(x) \sin x\, dx + \sin 2x \int F(x) \sin 2x\, dx + \&c. \qquad \text{(p)}$$

Slightly earlier he asserted:

> The integrals must be taken from $x = -\pi$ to $x = \pi$.

Then in Article 235 he goes on to say:

> The series arranged according to sines or cosines of multiple arcs are always convergent; that is to say, on giving to the variable any value whatever that is not imaginary, the sum of the terms converges more and more to a single fixed limit, which is the value of the developed function.

Fourier appears to be speaking of pointwise convergence, and subsequent analysis has not borne out his assertions; pointwise convergence of Fourier series is quite erratic. If, however, we replace pointwise convergence by mean-square convergence, then a sweeping statement like Fourier's *is* true: the Fourier series of any $L^2$ function $f$ whatever converges to $f$ in the mean-square metric.

Furthermore, our Hilbert space method sets the theory of Fourier series in a natural general setting: orthogonal function expansion in inner product spaces. Fourier series provide but one example of many interesting such expansions.

## Exercises

1. Let $f(x) = 1/|x|^r$, $x \neq 0$, and $f(0) = 0$.
   (a) Suppose $0 < r < \frac{1}{2}$. By evaluating the integral $\int_a^\pi (x^{-r})^2\, dx$ for $a > 0$ and computing its limit as $a \to 0+$, find $\|f\| = [(1/\pi) \int_{-\pi}^{+\pi} |f(x)|^2\, dx]^{1/2}$. Thus show that $1/|x|^r$ belongs to $L^2(-\pi, +\pi)$ when $0 < r < \frac{1}{2}$. Show further that $\|f\| \to \infty$ as $r \to \frac{1}{2}$.
   (b) Suppose $r > \frac{1}{2}$. Show then that $\|f\| = \infty$, so that $1/|x|^r$ does *not* belong to $L^2(-\pi, +\pi)$ when $r > \frac{1}{2}$.

2. Let $f(x) = 1/x^r$, $x \geqslant 1$.
   (a) Suppose $r > \frac{1}{2}$. By evaluating the integral $\int_1^M (x^{-r})^2\, dx$ for $M > 1$ and computing the limit as $M \to \infty$, find $\|f\| = [\int_1^\infty |f(x)|^2\, dx]^{1/2}$. Thus

show that $1/x^r$ belongs to $L^2(1, \infty)$ when $r > \frac{1}{2}$. Show further that $\|f\| \to \infty$ as $r \to \frac{1}{2}$.

(b) Suppose $0 < r \leqslant \frac{1}{2}$. Show then that $\|f\| = [\int_1^\infty |f(x)|^2 \, dx]^{1/2} = \infty$ so that $1/x^r$ does *not* belong to $L^2(1, \infty)$ when $0 < r \leqslant \frac{1}{2}$.

3. Use your knowledge of Exercises 1 and 2 to decide: For which positive values of $r$ does $1/x^r$ belong to $L^2(0, 1)$? $L^2(1, \infty)$? $L^2(0, \infty)$?

4. By evaluating the integrals

$$\int_0^{1-a} \frac{dx}{\sqrt{1 - x^2}} \quad \text{and} \quad \int_0^{1-a} \frac{dx}{1 - x^2}$$

for small positive $a$, then computing the limits as $a \to 1-$, show that $f(x) = 1/(1 - x^2)^{1/4}$ belongs to $L^2(-1, +1)$ but not to $L^2(-1, +1;$ $1/\sqrt{1 - x^2})$.

5. (a) Use integration by parts to show $\int \ln x \, dx = x(\ln x - 1) + c$, and $\int (\ln x)^2 \, dx = x((\ln x)^2 - 2 \ln x + 2) + c$.

(b) By evaluating the integral $\int_a^1 (\ln x)^2 \, dx$ and computing the limit as $a \to 0+$, find $\|\ln x\| = [\int_0^1 (\ln x)^2 \, dx]^{1/2}$. Thus show that $\ln x$ belongs to $L^2(0, 1)$.

(c) Using similar limit calculations, compute the $L^2(-1, +1)$ norms $\|\ln(1 + x)\| = [\int_{-1}^{+1} \ln^2(1 + x) \, dx]^{1/2}$ and $\|\ln(1 - x)\| = [\int_{-1}^{+1} \ln^2(1 - x) \, dx]^{1/2}$. Thus show that both $\ln(1 + x)$ and $\ln(1 - x)$ belong to $L^2(-1, +1)$. Using the fact that $L^2(-1, +1)$ is a vector space, show without further calculation that $Q_0(x) = (1/2) \ln((1 + x)/(1 - x))$ belongs to $L^2(-1, +1)$. Using the triangle inequality, estimate $\|Q_0(x)\|$.

(d) Does $Q_0'(x)$ belong to $L^2(-1, +1)$?

6. Compute the norm of $x^n$, $n = 0, 1, 2, \ldots,$ in $L^2(0, \infty; e^{-x})$. Compute $\langle x^n, x^m \rangle$ (for the inner product $\langle f, g \rangle = \int_0^\infty f(x)\overline{g(x)}e^{-x} \, dx$).

7. Let $c_n = \|x^n\|^2$ in $L^2(-\infty, +\infty; e^{-x^2/2})$. Evaluate $c_n$ in the following steps:

(a) $c_0 = \int_{-\infty}^{+\infty} e^{-x^2/2} \, dx$. The following trick is probably the easiest way to evaluate $c_0$. First, by iterated integration, $\iint_{-\infty}^{+\infty} e^{-x^2/2 - y^2/2} \, dx \, dy = (\int_{-\infty}^{+\infty} e^{-x^2/2} \, dx)(\int_{-\infty}^{+\infty} e^{-y^2/2} \, dy) = c_0^2$. Now transform the double integral into polar coordinates

$$\iint_{-\infty}^{+\infty} \exp\left(-\frac{x^2 + y^2}{2}\right) dx \, dy = \int_0^\infty r \exp\left(-\frac{r^2}{2}\right) dr \int_0^{2\pi} d\theta = 2\pi$$

Thus conclude $c_0 = \sqrt{2\pi}$.

(b) Prove by integration by parts that $c_n = (2n - 1)c_{n-1}$, $n \geqslant 1$. From this deduce $c_1 = c_0 = \sqrt{2\pi}$, $c_2 = 3\sqrt{2\pi}$, $c_3 = 5 \cdot 3\sqrt{2\pi}, \ldots,$ $c_n = (2n - 1)(2n - 3) \cdots 5 \cdot 3\sqrt{2\pi}$.

(c) Show that the formula in (b) can also be written

$$c_n = \frac{(2n)!}{n!2^n} \sqrt{2\pi}$$

8. Prove that $\sin x$ belongs to $L^2(0, \infty; e^{-x})$. [Hint: $|\sin x| \leqslant 1$.]
9. Let $V$ be a vector space of functions. Suppose that the function $f_1$ belongs to $V$ and $f_2$ does not. Show that the linear combination $c_1 f_1 + c_2 f_2$ belongs to $V$ only if $c_2 = 0$ ($c_1$ and $c_2$ are constants).
10. (a) Does the function $f(x) = \exp((x-1)^2/2)$ belong to $L^2(-\infty, +\infty; \exp(-x^2/2))$? [Ans.: No, as

$$\int_{-\infty}^{+\infty} f(x)^2 e^{-x^2/2} \, dx = \int_{-\infty}^{+\infty} e^{x^2/2 - 2x + 1} \, dx$$

and this integral obviously diverges because $x^2/2 - 2x + 1 = x^2(1/2 - 2/x + 1/x^2) \to \infty$ as $x \to \pm\infty$.]
  (b) Consider the function $g(x) = \int_0^x \exp(t^2/2) \, dt$. By sketching the function $\exp(t^2/2)$ and interpreting the integral as the area under this graph between $t = 0$ and $t = x$, argue that

$$g(x) > \int_{x-1}^{x} e^{t^2/2} \, dt > \exp\left((x-1)^2/2\right) \qquad \text{if } x > 1$$

Discuss whether or not $g(x)$ belongs to $L^2(-\infty, +\infty; \exp(-x^2/2))$.
11. Denote by $l^2(\mathbf{R})$ (resp. $l^2(\mathbf{C})$) the set of all real (resp. complex) sequences $U = (u_1, u_2, \ldots, u_n, \ldots)$ such that

$$\sum_{i=1}^{\infty} |u_i|^2 < \infty$$

  (a) Given two $l^2$ sequences, $A = (a_1, a_2, \ldots)$ and $B = (b_1, b_2, \ldots)$, prove that the series $\Sigma |a_i| |b_i|$ converges. From this deduce that the expression

$$\langle A, B \rangle = \sum_{i=1}^{\infty} a_i \bar{b}_i$$

  is finite valued.
  (b) Prove that $l^2$ is a vector space under componentwise operations.
  (c) Prove that the expression $\langle A, B \rangle$ given in (a) is an inner product on $l^2$.
12. Consider the complex Hilbert space $L^2(C)$ (where $C$ is the unit circle $-\pi \leqslant x \leqslant \pi$, $-\pi \equiv \pi$) with the inner product

$$\langle f, g \rangle = \frac{1}{2\pi} \int_{-\pi}^{\pi} f(x)\overline{g(x)} \, dx$$

  (a) Show that the $e^{inx}$, $n = 0, \pm 1, \pm 2, \ldots$, form an orthogonal family in this Hilbert space. What is $\|e^{inx}\|$?

(b) Given that the $e^{inx}$, $n = 0, \pm 1, \pm 2, \ldots$, form an orthogonal *basis* for $L^2(C)$, write down the general Fourier series and Parseval relation for $f$.

(c) Prove that $\int_{-\pi}^{\pi} f(x)e^{-inx}\, dx \to 0$ as $n \to \pm \infty$ for any $f \in L^2(C)$. [Hint: Use the Parseval relation in (b).]

(Exercise 12 is essentially the same as Exercise 10 of Section 3.6.)

## 3.10   HILBERT SPACE

We have had a gradual progression in this chapter. We started with the wave equation (we'll return to it later); the wave equation led us to Fourier series; Fourier series in turn suggested the idea of an inner product (which we were careful to define abstractly by its three properties: conjugate bilinear, Hermitian symmetric, and positive definite); and finally we ended with a study of the very important $L^2$ spaces. The fact that the trigonometric system forms an orthogonal basis of $L^2(C)$, where $C$ is the unit circle, links the classical theory of Fourier series with our $L^2$ concept.

This section will provide a review, a summing up, and a more general overview. In this section I will tell you what a Hilbert space is, exactly, leaving nothing out. That will be the overview, and, in explaining this abstract definition, I shall draw on the material from this chapter and from your linear algebra course, as review and summing up.

Hilbert space is defined abstractly—by "axioms" or "properties." It has been my experience that when you start to list axioms and to bring abstract material into an undergraduate course, students' eyes glaze over pretty rapidly. If you find this material a little spongy, don't fret. The concrete examples from earlier sections, as well as familiar specific examples from your linear algebra course, will be woven into the explanation, so there will be something to fix your ideas on. In subsequent chapters we'll be using these concepts again and again, so you'll have plenty of opportunity to look at them from this side, then that, and to refer back to this section.

I'll begin with the definition, then follow with a discursive explanation of all the terms involved.

**Definition 3.19.**   A *Hilbert space* is a separable real or complex inner product space that is complete in the metric derived from its inner product.

### 3.10.1   Vector Spaces

A Hilbert space is, first of all, a vector space. Now the word "vector" conjures up a mental picture of an arrow pointing in a fixed direction. But we have already used the term vector in a more general sense, when we referred to a "vector space of functions." A function certainly isn't an arrow pointing in a fixed direction. But functions do share two key properties with elementary vectors: functions can be added and can be multiplied by scalars. It is just these two key properties that define the mathematical term *vector space*: A vector space is a collection $V$ of objects, or symbols for objects, together with a way of combining any two elements, say $v$ and $w$ of $V$, to get a third element, always denoted $v + w$, in $V$, and another law that associates to each (real or complex) scalar $\alpha$ and each $v$ in $V$ another element, denoted $\alpha v$ in $V$. The first operation is called addition, the second, scalar multiplication. But—and this is important—the "addition" and "scalar multiplication" are described by axioms and not by specific calculational procedures.

What are these axioms? Well, they are the same axioms you had in your linear algebra course: associativity, distributivity, etc. I won't list them again here. These axioms are just the commonsense rules we use routinely when we add 3-tuples $(a_1, a_2, a_3)$ and multiply them by scalars. They are the very same rules we use instinctively when we add functions, $f(x) + g(x)$, and multiply *them* by scalars.

One of these axioms does perhaps merit comment, and that is closure, which says that if $v$ and $w$ belong to $V$, then so does $v + w$. For the specific example of $n$-tuples, closure is obvious: if $v = (v_1, v_2, \ldots, v_n)$ is an $n$-tuple and $w = (w_1, w_2, \ldots, w_n)$ is another, then so is their term-by-term sum $v + w = (v_1 + w_1, v_2 + w_2, \ldots, v_n + w_n)$. The proof of closure of $L^2(a, b; w)$ is not so easy. We did it in Section 3.9. It amounts to proving that if $\int |f|^2 w\, dx$ and $\int |g|^2 w\, dx$ are both finite, so is $\int |f + g|^2 w\, dx$. Here is an analogous problem that you can use to test your understanding of Theorem 3.16. Let $l^2$ denote the set of $\infty$-tuples $(a_1, a_2, \ldots)$ that are square summable: $\Sigma |a_i|^2 < \infty$. We shall use $l^2(\mathbf{R})$ if all the entries are restricted to be real numbers and $l^2(\mathbf{C})$ if we allow complex entries. Question: Is $l^2$ closed under term-by-term addition? Refer to Exercise 11 of Section 3.9.

So closure can be a little tricky in some examples. The other axioms are generally automatic in most situations. But don't lose sight of the fact that it's the axioms that count. The mathematical term "vector space" refers to an axiomatic system; nothing is said about the specific nature of the objects we are adding and multiplying by scalars.

This open-endedness of the definition has the advantage that any concept you formulate for the abstract vector space then applies to all the specific models, like the space $\mathbf{R}^n$ of $n$-tuples (from linear algebra), the space $L^2(C)$, consisting of functions square integrable on the unit circle, etc. One very important such concept is that of *dimension*. We have already discussed this in Section 2.1. The spaces $\mathbf{R}^n$ have finite dimension $n$. But the $L^2$ spaces we consider in this book all have infinite dimension. Exercise 1 has you show this for $L^2(-1, 1)$. A Hilbert space can be finite- or infinite-dimensional, but most of the interesting ones for analysis are infinite-dimensional.

In summary, a Hilbert space is, first of all, a real or complex abstract vector space, defined by the same axioms as in your linear algebra course, but not restricted to finite dimensions.

### 3.10.2  Inner Product

Let the symbol $V$ stand for an abstract vector space, either real or complex scalars. An *inner product* on $V$ is a rule for assigning to each pair $v$, $w$ of elements of $V$ a scalar $\langle v, w \rangle$ so that the two-variable function $\langle \cdot, \cdot \rangle$ is conjugate bilinear, Hermitian symmetric, and positive definite. One can refer back to Section 3.6 for a definition of these terms. That is all we mean by an abstract inner product space: a vector space with an inner product on it.

Despite their abstract nature, inner product spaces are rich in geometric structure. An abstract inner product satisfies Schwarz's inequality (Section 3.7), which allows us then to bring in the concept of angle, especially right angle (orthogonality), and a measure of distance, $d(v, w) = \|v - w\|$, where $\|x\| = \langle x, x \rangle^{1/2}$ (see Section 3.8). When we specialize to the space $V = \mathbf{R}^3$ with the usual inner product, our abstract notions of angle and distance then become exactly our familiar real-world angle and distance (see Exercise 6 of Section 3.8). When we specialize to the spaces $L^2(a, b; w)$ with their inner product $\langle f, g \rangle = \int_a^b f\bar{g}w \, dx$, we have then transplanted to these function spaces the concepts and language of familiar geometry.

The intrinsic or "mean-square" metric $d(v, w) = \|v - w\|$ deserves some additional comment. With a measure of distance, we can then say when two elements of our inner product space are "close" and when a sequence $v_1, v_2, \ldots$ of elements *converges to an element* $v$, namely when $d(v_i, v) = \|v_i - v\| \to 0$ as $i \to \infty$. This leads us to the term "separable."

### 3.10.3  Separable

**Definition 3.20.** Let $V$ be an abstract inner product space, $M$ a subspace of $V$. We say that $M$ is *dense* in $V$ if, given any $v \in V$ and any positive real number $\varepsilon$, however small, we can find an $m \in M$ such that $\|v - m\| < \varepsilon$.

Put into words, Definition 3.20 says that $M$ is dense in $V$ when we can find elements of $M$ as close as we wish to any element of $V$.

**Definition 3.21.** An inner product space $V$ is *separable* if it contains a sequence of elements $m_1, m_2, \ldots$ that span a dense subspace of $V$.

Recall what we mean by "span." Given the sequence $m_1, m_2, \ldots$, we form all possible finite linear combinations, $\alpha_1 m_{i(1)} + \alpha_2 m_{i(2)} + \cdots + \alpha_k m_{i(k)}$, where the $\alpha_p$ are any scalars and the $m_{i(p)}$ any vectors from our sequence. The set of those vectors in $V$ that can be represented this way for some set of scalars $\alpha$ and some selection of elements $m$ from our sequence form a subspace of $V$ called the subspace *spanned* by the sequence $m_1, m_2, \ldots$.

Definition 3.21 is trivially satisfied for any finite-dimensional vector space, because if $V$ is finite-dimensional, then it has a finite basis $m_1, m_2, \ldots, m_k$. Then the subspace spanned by the $m$'s is $V$ itself, so is certainly dense in $V$.

So "separable" is a restriction only for infinite-dimensional inner product spaces. It rules out very large pathological spaces of little interest in applied mathematics. All our $L^2$ spaces are separable, but we won't prove this here. There is a fact about separable inner product spaces that is instructive to consider here:

**Theorem 3.22.** *Any separable inner product space, whether complete or not, has an orthogonal basis.*

*Proof.* Forget about the word "complete" in the theorem; we'll get to that in a moment. I've included that phrase just to emphasize that completeness isn't necessary; separability is enough by itself.

Any finite-dimensional space is separable, so as a special case the theorem says that any finite-dimensional inner product space has an orthogonal basis. You may remember that result from your linear algebra course. It is proved by the Gram-Schmidt process. Call the space $V$, and let's say its dimension is $n$. So $V$ has a basis of $n$ elements, say $v_1, v_2, \ldots, v_n$. This means that the $v_i$ are linearly independent and span $V$. Further, any other linearly independent spanning set in $V$ will have exactly $n$ elements—this is the uniqueness of dimension.

From $v_1, v_2, \ldots, v_n$ we wish to construct an *orthogonal* basis of $V$, say $e_1, e_2, \ldots, e_n$. Orthogonal means this: $e_i \neq 0$, $1 \leqslant i \leqslant n$, and $\langle e_i, e_j \rangle = 0$ when $i \neq j$. The Gram-Schmidt process for constructing the $e_i$'s goes as follows: start by setting $e_1 = v_1$. We know $v_1 \neq 0$, so this is permissible. Then set

$$e_2 = v_2 - \frac{\langle v_2, e_1 \rangle}{\langle e_1, e_1 \rangle} e_1$$

First, $e_2 \neq 0$ because $v_2$ and $e_1 = v_1$ are linearly independent. Next, $\langle e_2, e_1 \rangle = 0$, as a direct calculation shows. Finally, it is easy to see that $\{e_1, e_2\}$ span the same subspace as $\langle v_1, v_2 \rangle$.

Next, set

$$e_3 = v_3 - \frac{\langle v_3, e_2 \rangle}{\langle e_2, e_2 \rangle} e_2 - \frac{\langle v_3, e_1 \rangle}{\langle e_1, e_1 \rangle} e_1$$

The same three conclusions follow: (1) $e_3 \neq 0$ lest we induce a dependency relation among $v_1, v_2, v_3$; (2) $\langle e_3, e_2 \rangle = \langle e_3, e_1 \rangle = 0$; and (3) $\{e_1, e_2, e_3\}$ span the same subspace as $\{v_1, v_2, v_3\}$.

Continue. At the $k$th stage, you will have constructed $k$ orthogonal vectors $e_1, e_2, \ldots, e_k$ that span the same subspace as $v_1, v_2, \ldots, v_k$. When $k = n$, we reach our desired orthogonal basis. That completes the review of the proof from your linear algebra course for the finite-dimensional case.

How does the proof work in the infinite-dimensional separable case? Essentially the same way. It's helpful to prepare the way with an intermediate result.

**Lemma 3.23.** *Let $e_1, e_2, \ldots$ be an orthogonal family in an inner product space V. Then the $e_i$, $1 \leqslant i < \infty$, form an orthogonal basis of V if, and only if, they span a dense subspace of V.*

*Proof.* We have to prove the implication in both directions. First, suppose that the $e_i$ form an orthogonal basis of $V$. Then, by definition of orthogonal basis (Definition 3.14), given any $x \in V$ we have $x = \sum_{i=1}^{\infty} c_i e_i$ where $c_i = \langle x, e_i \rangle / \langle e_i, e_i \rangle$ are the (scalar) Fourier coefficients of $x$ with respect to the $e_i$. To say that $x$ equals this infinite series means exactly that $\|x - \sum_{i=1}^{N} c_i e_i\| \to 0$ as $N \to \infty$ (Definition 3.11). Thus, given any $\varepsilon > 0$, however small, we can find an integer $N$ so that $\|x - \sum_{i=1}^{N} c_i e_i\| < \varepsilon$. As the finite sum $\sum_{i=1}^{N} c_i e_i$ clearly belongs to the subspace spanned by the $e_i$, and as $x$ is chosen arbitrarily in $V$, we have proved our result.

To complete the proof of Lemma 3.23, we need to prove the implication in the other direction. Suppose that the orthogonal family $e_1, e_2, \ldots$ spans a dense subspace of $V$. What does this mean? It means that, given any $x \in V$, and given any positive number $\varepsilon$, however small, we can find a finite linear combination of the $e_i$'s that approximates $x$ to within $\varepsilon$. By throwing in some zero coefficients, we may take this finite linear combination to include $e_1$ through $e_N$ for some $N$. Hence we have

$$\|x - (\alpha_1 e_1 + \alpha_2 e_2 + \cdots + \alpha_N e_N)\| < \varepsilon$$

Now we know that if we replace each scalar $\alpha_i$ by the Fourier coefficient $c_i = \langle x, e_i \rangle / \langle e_i, e_i \rangle$, we only improve the approximation (Theorem 3.10 and remarks following it). Hence,

$$\left\| x - \sum_{i=1}^{N} c_i e_i \right\| < \varepsilon \qquad \text{where } c_i = \frac{\langle x, e_i \rangle}{\langle e_i, e_i \rangle}$$

This holds for arbitrary positive $\varepsilon$ and for each $x \in V$. By Definition 3.11, $x = \sum_{i=1}^{\infty} c_i e_i$ for every $x \in V$, which says that the $e_i$ form an orthogonal basis of $V$. That completes the proof of Lemma 3.23.

We are now in a position to complete the proof of Theorem 3.22. We are assuming that our inner product space $V$ is separable, so contains a sequence $m_1, m_2, \ldots$ that spans a dense subspace. If some of the $m_i$'s are zero, we may delete them—the remaining nonzero terms will span the same dense subspace. So we may assume $m_i \neq 0, i = 1, 2, \ldots$. Having cast out all the zero terms, look at $m_2$ and ask: Is $m_2$ dependent on $m_1$? If so, we can delete it, because the remaining $m_i$'s will span the same dense subspace. If $m_1$ and $m_2$ are independent, then keep $m_2$. Now examine $m_3$ and ask: Is $m_3$ dependent on the terms that precede it? If yes, discard it. If no, keep it. Continue in this manner. You will produce a new sequence that spans the same dense subspace and is such that any finite number of its terms are linearly independent. So, in substance, we may assume that our original sequence $m_1, m_2, \ldots$ has the property not only that it spans a dense subspace but also that any finite subset of it is linearly independent.

Now apply Gram-Schmidt to $m_1, m_2, \ldots$ just as we did for the finite-dimensional case. At the $k$th stage, you will have constructed $k$ orthogonal vectors $e_1, e_2, \ldots, e_k$ that span the same subspace as $m_1, m_2, \ldots, m_k$. Continue. There results an orthogonal sequence $e_1, e_2, \ldots$ such that, for every positive integer $k$, the first $k$ $e_i$'s span the same subspace as the first $k$ $m_i$'s. Hence the full sequence of $e_i$'s spans the same dense subspace as the $m_i$'s. By Lemma 3.23, the $e_i$'s constitute an orthogonal basis of $V$. That finishes the proof of Theorem 3.22 and finishes as well our discussion of the concept of "separability."

### 3.10.4 Complete

The notion of completeness is defined in terms of what is called a Cauchy sequence.

**Definition 3.24.** A sequence $v_n, n = 1, 2, \ldots$, of elements in an inner product space is called a *Cauchy sequence* if it has this property: given any positive number $\varepsilon$, however small, we can find an integer $N$ (dependent on $\varepsilon$ in general) so that $\|v_n - v_m\| < \varepsilon$ whenever both $m, n \geqslant N$.

We might describe a Cauchy sequence as a sequence whose terms eventually cluster together.

**Definition 3.25.** An inner product space $V$ is said to be *complete* if, given any Cauchy sequence $v_n, n = 1, 2, \ldots$, of elements from $V$, there is a $v \in V$ such that the sequence $v_n$ converges to $v$.

So the meaning of completeness is this: every Cauchy sequence in $V$ must converge to an element of $v$. I shall illustrate the idea behind completeness using the rational numbers and the real numbers. The number $\sqrt{2}$ is not rational; that is, $\sqrt{2} \neq m/n$ for any integers $m$ and $n$ (Exercise 6). The sequence of rational numbers defined recursively by $a_{n+1} = (a_n^2 + 2)/2a_n$, $a_1 = 1$, converges to $\sqrt{2}$ (Exercise 7). Since this sequence converges, it is Cauchy (Exercise 8). But it is therefore a Cauchy sequence of rational numbers that converges to a nonrational. Thus the rational numbers are not complete. There is a "hole" in the rational number line at the point $x = \sqrt{2}$. Thus we might say that a space is complete when it has no holes.

That finishes our discussion of the definition of Hilbert space (Definition 3.19). The concept of Hilbert space is abstract, so it is not only a "thing" in itself, it is also an organizing principle—a way of thinking about other things. Thus, when we have the Hilbert space concept in mind, we can regard the various spaces $\mathbf{R}^3$, $\mathbf{R}^n$, $l^2$, $L^2(C)$, $L^2(-1,1)$, $L^2(-\infty, \infty; e^{-x^2/2})$, etc. as all instances, or examples, of one thing, namely Hilbert space. The concept of Hilbert space provides a framework that enables us to distinguish the main features that all these spaces have in common. Further, when we speak of (trigonometric) Fourier series or—looking ahead—of series of Legendre polynomials, or series of Hermite polynomials, or even the three vectors $e_1 = (1, 0, 0)$, $e_2 = (0, 1, 0)$, and $e_3 = (0, 0, 1)$, we can recognize all these as instances of the same general concept: each one is an orthogonal basis of its respective Hilbert space.

## Exercises

1. Show that, for any positive integer $n$, the functions $1, x, x^2, \ldots, x^n$ are linearly independent in $L^2(-1, 1)$. What is the vector space dimension of $L^2(-1, 1)$? Explain.
2. Which of the following formulas define inner products on the real vector space $\mathbf{R}^3$ of all real 3-tuples? Prove your answer.
   (a) $\langle (x_1, x_2, x_3), (y_1, y_2, y_3) \rangle = 2x_1 y_1 + x_2 y_2 + x_3 y_3$
   (b) $\langle (x_1, x_2, x_3), (y_1, y_2, y_3) \rangle = x_1 y_2 - x_2 y_1 + x_3 y_3$
   (c) $\langle (x_1, x_2, x_3), (y_1, y_2, y_3) \rangle = x_1 y_2 + x_2 y_1 + x_3 y_3$
3. A subspace $M$ of an inner product space $V$ is said to be *closed* if whenever we have a sequence $m_1, m_2, \ldots$ of elements of $M$ that converges to a $v \in V$, then $v \in M$. Suppose $M$ is closed and dense in $V$. What conclusions can you draw?
4. As in the text, apply the Gram-Schmidt procedure to the three vectors $v_1 = (1, 1, 0)$, $v_2 = (1, 0, -1)$, and $v_3 = (0, 1, 0)$. (Ordinary inner product.)
5. The functions $1, x, x^2, x^3, \ldots$ are linearly independent in $L^2(-1, 1)$. Apply the Gram-Schmidt procedure to the first four of these functions. Compare your answer to the first four Legendre polynomials (Section 2.8). (The inner product in $L^2(-1, 1)$ is $\langle f, g \rangle = \int_{-1}^{1} f\bar{g}\, dx$.)

6. (a) Show that if we have an equation $\sqrt{2} = m/n$ for integers $m$ and $n$, then we must necessarily also have an equation $\sqrt{2} = r/s$ where $r$ and $s$ are *positive* integers that have only 1 as a common integral divisor.

   (b) Prove that if the equation $2s^2 = r^2$ holds for integers $r$ and $s$, then both $r$ and $s$ must be divisible by 2.

   (c) Using (a) and (b), prove that $\sqrt{2} = m/n$ is impossible for integral $m$ and $n$.

7. Let $a_1 = 1$, $a_{n+1} = (a_n^2 + 2)/2a_n$. Compute $a_2$, $a_3$, and $a_4$ as fractions. [Ans.: $a_4 = 577/408$.] Compute $(a_4)^2$ on your calculator. Assuming $\lim a_n = a$ exists, prove that $a = \sqrt{2}$. [Hint: $\lim a_n = \lim a_{n+1}$.]

8. Suppose that the sequence $v_n$, $n = 1, 2, \ldots$, of elements from an inner product space $V$ converges to $v \in V$. Prove that $v_n$, $n = 1, 2, \ldots$, is a Cauchy sequence. [Hint: Write $v_n - v_m = (v_n - v) + (v - v_m)$ and use the triangle inequality, Theorem 3.5.]

9. Let $\phi_n$, $n = 1, 2, \ldots$, be an orthogonal family in a Hilbert space $H$. Prove: if $a_n$, $n = 1, 2, \ldots$, is any sequence of scalars such that the numerical series $\sum_{i=1}^{\infty} |a_i|^2 \langle \phi_i, \phi_i \rangle$ converges, then the series $\sum_{i=1}^{\infty} a_i \phi_i$ converges to an element of $H$. [Hint: Let $g_n = \sum_{i=1}^{n} a_i \phi_i$ and, by computing $\|g_n - g_m\|^2$, $n > m$, show that $\{g_n\}$ is Cauchy.]

# 4

# Linear Second Order Differential Operators in $L^2$ Spaces and Their Eigenvalues and Eigenfunctions

## 4.1 COMPATIBILITY

A *linear operator* $L$ in a Hilbert space $H = L^2(a, b; r)$ is a mapping, or assignment, that assigns to each function $f$ in a certain subspace $D$ of $H$, called the *domain* of the operator, another function $L(f)$ in $H$ so that the following rules are satisfied:

$$L(f + g) = L(f) + L(g) \qquad \text{for all } f, g \text{ in the domain } D$$

and

$$L(cf) = cL(f) \qquad \text{for all constants } c \text{ and all } f \text{ in } D$$

A *differential operator* in Hilbert space is a special kind of linear operator. A differential operator is made up of two parts: (1) a linear differential expression like the ones we have studied in Chapter 2 and (2) an independently specified domain $D$. Both parts are equally important. We may symbolize this makeup with the suggestive, if somewhat unorthodox, equation

Differential operator in Hilbert space = differential expression + domain

The domain of a differential operator consists of functions that satisfy two stipulations: (1) a compatibility condition and (2) a boundary condition. The boundary condition is usually, but not always, present. However, the

**157**

compatibility condition is always needed. In this section we explain compatibility.

Consider the following example: Begin with the Hilbert space $L^2(0,1)$, which consists of all functions $f(x)$ that satisfy the square-integrability condition

$$\int_0^1 |f(x)|^2 \, dx < \infty$$

For simplicity, we shall restrict our attention to real-valued functions. Suppose we wish to construct a differential operator $L$ in the Hilbert space $L^2(0,1)$ which has as its "law of action" the linear differential expression $l(y) \equiv -y''$. This means we want to define a Hilbert space operator $L$ such that for every function $y$ in the domain of $L$ we have $L(y) = -y''$. Look at the function $y = \sqrt{x}$. This function belongs to $L^2(0,1)$ because

$$\int_0^1 y(x)^2 \, dx = \int_0^1 (\sqrt{x})^2 \, dx = \int_0^1 x \, dx = \frac{x^2}{2}\Big|_0^1 = \frac{1}{2} < \infty$$

But when we apply our differential expression $l(y) \equiv -y''$ to $y = \sqrt{x}$ we get

$$l(\sqrt{x}) = -x^{1/2''} = -\tfrac{1}{2}(x^{-1/2})' = \tfrac{1}{4}x^{-3/2}$$

and this function does *not* belong to $L^2(0,1)$:

$$\int_0^1 \left[\frac{1}{4}x^{-3/2}\right]^2 dx = \frac{1}{16}\int_0^1 x^{-3} \, dx = \frac{1}{16}\lim_{a\to 0+}\int_a^1 x^{-3} \, dx$$

$$= \frac{1}{16}\lim_{a\to 0+}\left(-\frac{1}{2}\right)x^{-2}\Big|_a^1 = \frac{1}{16}\lim_{a\to 0+}\left(-\frac{1}{2}+\frac{1}{a^2}\right)$$

$$= \frac{1}{32}\lim_{a\to 0+}\left(\frac{2}{a^2}-1\right) = +\infty$$

Hence the differential expression $l(y) \equiv -y''$, when it acts on the function $y = \sqrt{x}$ which belongs to $L^2(0,1)$, produces a function $l(y) = \tfrac{1}{4}x^{-3/2}$ which does *not* belong to $L^2(0,1)$. Our objective is to construct a linear operator $L$ on the Hilbert space $L^2(0,1)$ which takes *all* its values in the same Hilbert space. Hence we must exclude the function $y = \sqrt{x}$, and functions like it, from the domain of our operator. This leads us to the precise definition of "compatibility."

First, recall the definition of the Hilbert space $L^2(a,b;r)$. Here $a \leqslant x \leqslant b$ is a finite or infinite interval, $r(x)$ is a function that is positive on the interior $a < x < b$ (we are using $r$ here instead of $w$ as in Section 3.9), and $L^2(a,b;r)$

consists of all functions $f(x)$ that satisfy the square-integrability condition

$$\int_a^b |f(x)|^2 r(x)\, dx < \infty$$

**Definition 4.1.** Suppose we are given a Hilbert space $L^2(a, b; r)$ and a second order linear differential expression $l(y) \equiv A(x)y'' + B(x)y' + C(x)y$, where $A$, $B$, $C$ are analytic functions of $x$ and $A(x) \neq 0$, $a < x < b$. We say that a function $y(x)$ in $L^2(a, b; r)$ is *compatible with $l$* in $L^2(a, b; r)$ if $l(y)$ also belongs to $L^2(a, b; r)$.

If we follow the common mathematical usage of using the symbol "$\in$" as an abbreviation for "belongs to," then we may state Definition 4.1 in the form: *$y$ is compatible with $l$ in $L^2(a, b; r)$ when both $y \in L^2(a, b; r)$ and $l(y) \in L^2(a, b; r)$.*

Or we may write it out in full: *$y$ is compatible with $l$ in $L^2(a, b; r)$ when both*

$$\int_a^b |y|^2 r(x)\, dx < \infty \qquad and \qquad \int_a^b |l(y)|^2 r(x)\, dx < \infty$$

Here is a second example. Consider the Hilbert space $L^2(0, 1; x)$ that consists of functions $f(x)$ that satisfy $\int_0^1 |f(x)|^2 x\, dx < \infty$. We are concerned in this case with the differential expression

$$l(y) \equiv -\frac{1}{x}\frac{d}{dx}\left(x\frac{dy}{dx}\right)$$

which we have written in Sturmian form (Section 2.2, Exercises 10 and 11). We ask, for what values of the constants $a_0$, $a_1$, $a_2$ is the polynomial $y(x) = a_0 + a_1 x + a_2 x^2$ compatible with $l$ in $L^2(0, 1; x)$?

Because $y$ is a polynomial, it is obvious that the integral $\int_0^1 |y|^2 x\, dx$ is finite. We don't have to compute this integral. So for any values of the constants $a_0$, $a_1$, $a_2$ we have $y \in L^2(0, 1; x)$.

Now do the calculation $l(y) = -(1/x)(xy')'$ for the polynomial $y = a_0 + a_1 x + a_2 x^2$. You get $y' = a_1 + 2a_2 x$, $xy' = a_1 x + 2a_2 x^2$, $(xy')' = a_1 + 4a_2 x$, $l(y) = -(a_1/x + 4a_2)$. Hence

$$\int_0^1 |l(y)|^2 x\, dx = \int_0^1 \left(\frac{a_1^2}{x^2} + \frac{8a_1 a_2}{x} + 16a_2^2\right) x\, dx$$

$$= \int_0^1 \left(\frac{a_1^2}{x} + 8a_1 a_2 + 16a_2^2 x\right) dx$$

$$= a_1^2 \int_0^1 \frac{dx}{x} + 8a_1 a_2 \int_0^1 dx + 16a_2^2 \int_0^1 x\, dx$$

The second and third integrals are finite, but the first is infinite:

$$\int_0^1 \frac{dx}{x} = \lim_{a \to 0+} \int_a^1 \frac{dx}{x} = \lim_{a \to 0+} (-\ln a) = \lim_{a \to 0+} \ln\left(\frac{1}{a}\right) = +\infty$$

Hence, if $a_1 \neq 0$, then $l(y)$ does *not* belong to $L^2(0, 1; x)$; thus in that case $y = a_0 + a_1 x + a_2 x^2$ is not compatible with $l(y) = -(1/x)(xy')'$ in $L^2(0, 1; x)$.

On the other hand, if $a_1 = 0$, then $\int_0^1 |l(y)|^2 x\, dx = 16a_2^2 \int_0^1 x\, dx$, which is finite. Thus we have answered our question: *the function $y = a_0 + a_1 x + a_2 x^2$ is compatible with the differential expression $l(y) \equiv -(1/x)(xy')'$ in $L^2(0, 1; x)$ if, and only if, $a_1 = 0$.*

Our third example brings us to the matter of differentiability. In Section 3.9 we stressed the "democratic" nature of the Hilbert spaces $L^2(a, b; r)$. The *only* test a function has to pass to belong to this space is the test of square-integrability. Hence the space $L^2(a, b; r)$ contains many nondifferentiable and discontinuous functions, because there are many such functions for which the integral $\int_a^b |f(x)|^2 r(x)\, dx$ is finite. But when we write a differential expression $l(y) \equiv Ay'' + By' + Cy$, we are implying that the function $y$ is smooth enough to have a second derivative. So when we construct a differential operator in Hilbert space based on such a differential expression, we would expect to have to impose conditions that limit the domain of that operator to sufficiently smooth functions. A thorough discussion of such conditions would get us into a lot of fussy mathematics that would divert us from our main purpose, which is to lay out the broad lines of the Hilbert space method and apply it to physical problems. The bulk of the details of these smoothness conditions is relegated to the optional Appendix 4.A and a few optional paragraphs at the end of this section. But there is one very important aspect of smoothness that is directly related to compatibility that we do need to take up here.

Consider the differential expression $l(y) \equiv -y''$ in $L^2(-1, 1)$. Suppose we wish to apply this expression to the function $f(x)$ defined by

$$f(x) = \begin{cases} 0, & -1 \leqslant x \leqslant 0 \\ x, & 0 \leqslant x \leqslant 1 \end{cases} \tag{4.1.1}$$

Question: Is $y = f(x)$ compatible with $l(y) \equiv -y''$ in $L^2(-1, 1)$? Answering this question means answering two questions: (1) does $f(x)$ itself belong to $L^2(-1, 1)$? and (2) does $l(f) = -f''$ belong to $L^2(-1, 1)$?

As to the first question, we note that $|f(x)| \leqslant 1$ on $-1 \leqslant x \leqslant 1$, so also

$|f(x)|^2 \leqslant 1$ there. Thus

$$\int_{-1}^{1} |f(x)|^2 \, dx \leqslant \int_{-1}^{1} 1 \, dx = 2 < \infty \tag{4.1.2}$$

Hence $f(x) \in L^2(-1, 1)$. Note that while we could have computed the exact value of the integral in (4.1.2), that calculation would have represented wasted work, as the exact value of the integral is not of interest here. The only information needed is whether the integral (4.1.2) is finite or infinite. Keep that viewpoint in mind. It can save you a lot of work, because the question of whether a function belongs to a particular $L^2$ space occurs repeatedly throughout the book.

Now look at the second question: does $l(f) = -f''$ belong to $L^2(-1, 1)$? Start by computing $f'(x)$, the first derivative of $f$. From (4.1.1) we get

$$f'(x) = \begin{cases} 0, & -1 \leqslant x < 0 \\ 1, & 0 < x \leqslant 1 \end{cases} \tag{4.1.3}$$

As $f(x)$ is not differentiable at $x = 0$, $f'(x)$ is not defined there. We have $f'(x) = H(x)$, Heaviside's function (Section 2.10). And our question now becomes: does $l(f) = -(f'(x))' = -H'(x)$ belong to $L^2(-1, 1)$? Note that $H(x)$ is not even continuous at $x = 0$, let alone differentiable there. So the equation $l(f) = -H'(x)$ falls outside the framework of traditional calculus because it calls for the differentiation of a discontinuous function. But Dirac's formalism does give us a value for $H'(x)$, namely $H'(x) = \delta(x)$, the delta function (Section 2.10, equation 2.10.9)). So $l(f) = -\delta(x)$, and our question "does $l(f) = -f''$ belong to $L^2(-1, 1)$?" leads to this two-part question: *Is it possible to assign a value to*

$$\int_{-1}^{1} (\delta(x))^2 \, dx$$

*that is consistent with the other rules for the $\delta$ function and, if so, what is that value?*

Indeed, it is possible to assign a consistent value to that integral, and, in fact,

$$\int_{-1}^{1} (\delta(x))^2 \, dx = \infty \tag{4.1.4}$$

Hence the answer to our original question is "no"; $l(f) = -f''$ is not square integrable over $-1 \leqslant x \leqslant 1$, hence the function $f(x)$ of (4.1.1) is not compatible with $l(y) = -y''$.

I shall give two separate arguments to support (4.1.4). The first is based on Lighthill's idea of defining the $\delta$ function in terms of the sequence $f_n(x) = (n/\pi)^{1/2} \exp(-nx^2)$, $n = 1, 2, \ldots$. Exercise 1 of Section 2.10 showed that

$$\lim_{n \to \infty} \int_{-\infty}^{\infty} f_n(x)\,dx = 1 = \int_{-\infty}^{\infty} \delta(x)\,dx$$

so one is led to define

$$\int_{-\infty}^{\infty} (\delta(x))^2\,dx = \lim_{n \to \infty} \int_{-\infty}^{\infty} (f_n(x))^2\,dx$$

As $(f_n(x))^2 = (n/\pi)\exp(-2nx^2)$, we have

$$\int_{-\infty}^{\infty} (f_n(x))^2\,dx = \frac{2n}{\pi} \int_0^{\infty} e^{-2nx^2}\,dx = \frac{(2n)^{1/2}}{\pi} \int_0^{\infty} e^{-u^2}\,du$$

$$= \left(\frac{n}{2\pi}\right)^{1/2} \to \infty \qquad \text{as } n \to \infty$$

Hence (4.1.4).

The second derivation of (4.1.4) starts from a formula in Dirac's book, *Principles of Quantum Mechanics* (4th ed., Oxford 1958), namely his formula (10) in §15. Dirac writes

$$\int_{-\infty}^{\infty} \delta(a - x)\,dx\,\delta(b - x) = \delta(a - b) \tag{4.1.5}$$

and does not exclude the possibility that $a = b$. If we put $a = b = 0$ in (4.1.5) and use $\delta(-x) = \delta(x)$ (see Section 2.10) we get

$$\int_{-\infty}^{\infty} \delta(x)\,dx\,\delta(x) = \delta(0) \tag{4.1.6}$$

As $\delta(0)$ must be infinite, we get (4.1.4). In connection with his formula (10) (our formula (4.1.5)), Dirac connects it with the equation $\int f(x)\delta(x-a)\,dx = f(a)$, which is number (4) in his listing (also number (2.10.4) in our Section 2.10). He says:

> Equation (10) would be given by an application of (4) with $f(x) = \delta(x - b)$. We have here an illustration of the fact that we may often use an improper function as though it were an ordinary continuous function, without getting a wrong result.

In summary: We raised the question of whether the function $f(x)$ defined in (4.1.1) is compatible with the differential expression $l(y) \equiv -y''$ in

$L^2(-1, 1)$. We found that $f' = H(x)$, so that $l(f) = -f'' = -H'(x) = -\delta(x)$. As $\delta(x)$ does not belong to $L^2(-1, 1)$, the answer to our question is "no," the function $f(x)$ is not compatible with $l(y) = -y''$ in $L^2(-1, 1)$.

The argument generalizes. Suppose we are given a function $f(x)$ such that $f'(x)$ shows a discontinuity like that of the Heaviside function $H(x)$. Then the second derivative, $f''$, will involve a delta function which is not square integrable. So for $f$ to be compatible it must at least be true that $f'$ (therefore also $f$) be continuous. We state our conclusion formally.

**Theorem 4.2.** *A necessary condition that a function $y(x)$ be compatible with any given linear second order differential expression in $L^2(a, b; r)$ is that both $y(x)$ and $y'(x)$ be continuous at each point of the interval $(a, b)$, excluding the boundary points $x = a$ and $x = b$.*

Our "proof" of Theorem 4.2 used Dirac's delta function. One can also justify Theorem 4.2 by methods that meet conventional standards of mathematical rigor. At the end of this section I sketch one such argument that requires a knowledge of the (graduate level) Lebesgue theory of integration, particularly "absolutely continuous" functions. That discussion is optional, inasmuch as it calls on material well beyond the prerequisites for this course.

We turn now to the applications of Theorem 4.2.

In Chapter 3 we dealt with the unit circle $C$ constructed by bending the interval $-\pi \leqslant x \leqslant \pi$ into a circle and then identifying the end points $-\pi$ and $\pi$. This is an example of a *one-dimensional manifold without boundary*. There is no boundary because, as you go around the circle, you encounter no obstruction. We write $L^2(C)$ to designate the Hilbert space consisting of square-integrable functions on the circle $C$. The necessary condition in Theorem 4.2 for the compatibility of a function on $C$ with any differential expression in $L^2(C)$ then applies to every point of $C$ because there are no boundary points to exclude.

Our fourth example illustrates the latter point. Consider the function $f(x) = \cos \mu x$ as a function defined on the unit circle $C$, which, as just remarked, we take as the interval $-\pi \leqslant x \leqslant \pi$ with the points $-\pi$ and $\pi$ identified, $-\pi \equiv \pi$. For what values of the constant $\mu$ is $f(x) = \cos \mu x$ compatible in $L^2(C)$? To give a complete answer we would need to know the explicit form of the differential expression involved. But the necessary conditions in Theorem 4.2 must be fulfilled in any case. Put another way, unless $f$ and $f'$ are continuous, $f$ cannot be compatible with any second order linear differential expression.

For what values of the constant $\mu$ is $f(x) = \cos \mu x$ continuous on the unit circle $C$? The function is obviously continuous at each point of the open interval $-\pi < x < \pi$. If $f(x)$ is to remain continuous on the circle $C$ that we obtain by bending the interval $[-\pi, \pi]$ and identifying the point $-\pi$ with the point $\pi$, then we must have $f(-\pi) = f(\pi)$ or $\cos \mu(-\pi) = \cos \mu\pi$. But this is evidently true because $\cos x$ is an even function, $\cos(-x) = \cos x$. Hence $f(x) = \cos \mu x$ is continuous on $C$ for all values of the constant $\mu$.

For what values of the constant $\mu$ is $f'(x) = -\sin \mu x$ continuous on $C$? In order for $f'$ to be continuous on $C$ we must have $f'(-\pi) = f'(\pi)$, or $-\mu \sin \mu(-\pi) = -\mu \sin \mu\pi$. This is clearly true when $\mu = 0$. When $\mu \neq 0$, we get $\sin \mu(-\pi) = \sin \mu\pi$. As $\sin(-x) = -\sin x$, this leads to $2\sin \mu\pi = 0$ or $\sin \mu\pi = 0$. Thus $\mu\pi$ must be an integral multiple of $\pi$, or $\mu = 0, \pm 1, \pm 2, \ldots$. Thus Theorem 4.2 has led us to this conclusion: *If the function $f(x) = \cos \mu x$ is to be compatible in $L^2(C)$ with any given second order linear differential expression, then $\mu$ must be an integer.*

**Theorem 4.3.** *Given any Hilbert space $L^2(a, b; r)$ and any linear differential expression $l$, the class of functions in $L^2(a, b; r)$, compatible with $l$, forms a subspace of $L^2(a, b; r)$.*

*Proof.* Remember the definition of subspace: closed under sum and closed under scalar multiplication. So we have to show that if $f$ and $g$ are compatible with $l$ then so are $f + g$ and $cf$ for any constant $c$.

To show that $f + g$ is compatible with $l$ we have to prove that $l(f + g) \in L^2(a, b; r)$. But by the linearity of $l$, $l(f + g) = l(f) + l(g)$. And, since $f$ and $g$ are assumed compatible with $l$, both $l(f)$ and $l(g)$ belong to $L^2(a, b; r)$. In Section 3.9 we proved that $L^2(a, b; r)$ is a vector space, which means it contains along with any pair of functions $u, v$ their sum $u + v$ and contains $cu$ for any constant $c$. As $L^2(a, b; r)$ contains both $l(f)$ and $l(g)$, it contains their sum $l(f) + l(g) = l(f + g)$, which shows that $f + g$ is compatible with $l$. In a similar way, using $l(cf) = cl(f)$, we prove that $cf$ is compatible with $l$. Thus the functions compatible with $l$ form a subspace of $L^2(a, b; r)$.

Summarizing this section: The basic "equation" is differential operator = differential expression + domain. The domain and the differential expression are equally important parts of the definition of a differential operator; both must be explicitly given. The domain can never be larger than the subspace of compatible functions and might be smaller if further restricted by boundary conditions. Boundary conditions, if used, must define a subspace of functions—see Exercise 10.

All the definitions of the various differential operators that we will use in this book will be put in the general framework of the following typical example:

Let $L$ be the operator in $L^2(0, 1)$ defined by

Differential expression $l(y) \equiv -y''$
Domain$(L)$ = compatible functions $y$ that satisfy $y(0) = y(1) = 0$

When you read such a definition, you are to understand the following.

1.  When the function $y$ belongs to the domain of the operator, which in the example written above means that $y$ is compatible with $l(y) \equiv -y''$ *and* satisfies the boundary conditions $y(0) = y(1) = 0$, then the action of the operator $L$ is given by $L(y) = l(y)$.
2.  If $y$ does not belong to the domain, then $L(y)$ is *not* defined, even though $l(y)$ might make perfectly good sense.

For example, suppose you are asked to find the kernel of a differential operator $L$. The kernel of $L$ is the subspace consisting of all solutions of the equation $L(y) = 0$. When we write the symbol $L(y)$, we shall invariably mean that the function $y$ belongs to the domain of the operator $L$. Hence solving $L(y) = 0$ means finding all functions *in the domain of* $L$ that satisfy $L(y) = 0$. In practice, such problems are done this way: First, solve $l(y) = 0$ where $l$ is the second order linear differential expression of $L$. As we know from Chapter 2, ker($l$) is always two-dimensional, hence the general solution to $l(y) = 0$ will have the form $y = c_1 f_1 + c_2 f_2$, where $f_1$ and $f_2$ are linearly independent solutions and $c_1$ and $c_2$ are arbitrary constants. Second, determine the values of these constants for which the function $y = c_1 f_1 + c_2 f_2$ lies in the domain of $L$. To belong to the domain of $L$, the function $y$ must be compatible with $l$ and must satisfy the boundary conditions. Exercises 11 and 12 give practice in this kind of calculation.

I shall close this section with the promised sketch of a rigorously based line of reasoning to establish Theorem 4.2. This part is optional.

Consider the simplest possible differential expression, $l(y) \equiv y'$. What is the broadest conceivable domain of functions on which this expression makes sense? Answer: the class of functions $f(x)$ having the property that knowledge of $f'(x)$ determines $f(x)$ up to an additive constant. Why no larger class than this? Because the entire theory of differential equations (Chapters 1 and 2) is based on the tacit principle that if $f' = g'$, then $f = g + \text{constant}$. Or, equivalently, that the *only* solutions to $y' = 0$ are $y = \text{constant}$. Without this principle, virtually no theorem on differential equations remains true. So, without this principle, we have no theory of differential equations, nor any theory of differential operators, for that matter.

Within Lebesgue's theory of integration, the phrase "$f'(x)$ determines $f(x)$ up to an additive constant" is taken to mean that $f(x)$ is differentiable "almost everywhere" and $f(x)$ is the Lebesgue integral of its derivative $f'(x)$. This is a well-defined class of functions in the Lebesgue theory. Such functions are called *absolutely continuous*. In the Appendix 4.A to this chapter we touch briefly on absolutely continuous functions, but to get further into this graduate-level concept, one needs to consult an appropriate text such as R. L. Wheeden and A. Zygmund, *Measure and Integral*, (Dekker, 1977) Chapter 7.

The function $f(x)$ in (4.1.1) is absolutely continuous because $f(x)$ is the integral of its derivative $f'(x) = H(x)$,

$$f(x) = \int_{-1}^{x} H(t)\, dt$$

as you can easily check. Note that while $f$ is absolutely continuous, it is not differentiable at $x = 0$.

The Heaviside function $H(x)$ is not absolutely continuous because $H'(x) = 0$, except at $x = 0$, and $H(x)$ is not the integral of the zero function. A more famous and dramatic example of a function that is not absolutely continuous is the Cantor-Lebesgue function, $g(x)$. This function is defined on $-1 \leqslant x \leqslant 1$, continuous at every point of that interval, monotone increasing in the sense that $-1 \leqslant x_1 \leqslant x_2 \leqslant 1 \Rightarrow g(x_1) \leqslant g(x_2)$ such that $g(-1) = 0$ and $g(1) = 1$ (so that as $x$ goes from $-1$ to $1$, $g(x)$ rises continuously from 0 to 1), but such that $g'(x) = 0$ "almost everywhere." Lebesgue's theory cannot distinguish between a function that is zero almost everywhere and one that is zero at every point; hence the Lebesgue integral of $g'(x)$ is zero, so $g(x)$ is not the Lebesgue integral of its derivative.

Getting back to our original differential expression $l(y) \equiv y'$, the answer to our question is: the absolutely continuous functions constitute the largest class of functions on which this expression makes sense. We can generalize this: *the domain of any nth order linear differential expression, no matter what the overall context, can be no larger than the class of those functions $f(x)$ whose $(n-1)$st derivatives, $f^{(n-1)}(x)$, are absolutely continuous*. Theorem 4.2 follows from this principle (take $n = 2$), because an absolutely continuous function is continuous.

## Exercises

1. Consider the expression $l(y) \equiv -y''$ in $L^2(0,1)$. By computing $l(y)$ for each of the following functions $y$, then testing $l(y)$ for square-integrability, decide in each case whether the function $y$ is compatible with $l$.
   (a) $y = x^0 = 1$     (c) $y = x$     (e) $y = x^{3/2}$
   (b) $y = x^{2/3}$      (d) $y = x^{5/4}$    (f) $y = x^{7/4}$
2. (Continuation) Let $r$ be a constant $\geqslant 0$. Show that $y = x^r$ is compatible with the expression $l$ of Exercise 1 in the following cases and only in these cases: $r = 0$, $r = 1$, $r > \frac{3}{2}$.
3. (Continuation) Let $S$ be the subspace of $L^2(0,1)$ consisting of functions that are compatible with the expression $l$ of Exercise 1.
   (a) Show that $S$ contains all powers $x^n$, $n = 0, 1, 2, \ldots$.
   (b) Using (a) and the fact that $S$ is a subspace, show without further calculation that $S$ contains all polynomials.

(c) Give several examples of functions in $L^2(0, 1)$ that do *not* lie in $S$ (thus verifying that $S$ is a *proper* subspace of $L^2(0, 1)$).

4. (a) Show that $y = x(1 - \ln x)$ is bounded on $0 \leqslant x \leqslant 1$, thus in $L^2(0, 1)$. Same for $y = x^2(\frac{1}{2} - \ln x)$.

   (b) Which, if either, of the functions in (a) is compatible with the expression $l(y) \equiv -d^2y/dx^2$ in $L^2(0, 1)$?

   [Hints for 4a: Compute the local max and local min by solving $y' = 0$. Compute also $y(0) = \lim_{x \to 0+} y(x)$ and $y(1)$.]

5. $Q_0(x) = \frac{1}{2} \ln((1 + x)/(1 - x))$ belongs to $L^2(-1, +1)$ (Section 3.9, Exercise 5c). Show that this function is compatible with the expression

$$l(y) \equiv -\frac{d}{dx}\left((1 - x^2)\frac{dy}{dx}\right) \quad \text{in } L^2(-1, +1)$$

6. Let $n$ be an integer, and let $l$ denote the differential expression

$$l(y) \equiv -\frac{1}{x}\frac{d}{dx}\left(x\frac{dy}{dx}\right) + \frac{n^2}{x^2}y \quad \text{in } L^2(0, 1; x)$$

Consider the quadratic polynomial $y = a_0 + a_1 x + a_2 x^2$, which obviously belongs to $L^2(0, 1; x)$.

   (a) Prove that, if $n = 0$, then $y$ is compatible with $l$ if, and only if, $a_1 = 0$.

   (b) Prove that, if $n = \pm 1$, then $y$ is compatible with $l$ if, and only if, $a_0 = 0$.

   (c) Prove that if $|n| \geqslant 2$, then $y$ is compatible with $l$ if, and only if, both $a_0 = 0$ and $a_1 = 0$.

7. Which of the following functions satisfy the necessary condition for compatibility stated in Theorem 4.2?

   (a) $f(x) = |x|$ on $-1 \leqslant x \leqslant +1$

   (b) $f(x) = \begin{cases} -x^2, & -1 \leqslant x \leqslant 0 \\ x^2, & 0 \leqslant x \leqslant 1 \end{cases} \quad \text{on } -1 \leqslant x \leqslant 1$

8. Let $C$ be the circle obtained by bending the interval $0 \leqslant x \leqslant 2\pi$ and identifying the point $x = 0$ with the point $x = 2\pi$. For what values of the constants $a$ and $b$ will the following functions satisfy on $C$ the necessary condition for compatibility stated in Theorem 4.2?

   (a) $f(x) = \sin ax$            (c) $f(x) = a + bx$

   (b) $f(x) = \sin ax + \cos ax$    (d) $f(x) = ae^{-x} + be^x$

9. Make a circle $C$ by bending the interval $-1 \leqslant x \leqslant 1$ and identifying the point $x = -1$ with the point $x = 1$. For what value of the constant $a$ will the function $f(x) = \cos ax$ satisfy on $C$ the necessary condition for compatibility stated in Theorem 4.2?

10. For each of the following sets $D$ of functions defined on $0 \leqslant x \leqslant 1$, determine (1) whether it is closed under sum and (2) whether it is closed

under multiplication by scalars. Which are subspaces?
(a) $D = \{f : f(0) = f(1) = 2\}$
(b) $D = \{f : f(0) = f(1) = 0\}$
(c) $D = \{f : f'(0) = f'(1) = 1\}$
(d) $D = \{f : f(0) = f'(0),\ f(1) = f'(1)\}$
(e) $D = \{f : f(0) = f(1),\ f'(0) = f'(1)\}$
(f) $D = \{f : f(0),\ f'(0),\ f(1),\ f'(1)$ are all finite$\}$
11. Show that the operator $L$ defined in $L^2(0, 1)$ by

Differential expression $l(y) \equiv -d^2y/dx^2$
Domain($L$) = compatible functions $y$ that satisfy $y(0) = 0$, $y(1) = 0$

has kernel 0. (That is, $L(y) = 0$ implies $y = 0$.)
12. (Continuation)   Show that the operator $M$ defined in $L^2(0, 1)$ by

Differential expression $m(y) \equiv -d^2y/dx^2$
Domain($M$) = compatible functions $y$ that satisfy $y'(0) = 0$, $y'(1) = 0$

has a one-dimensional kernel. Find a convenient function that spans the kernel.

## 4.2   EIGENVALUES AND EIGENFUNCTIONS

Let us look again at the boundary value problem that occurred in our study of the vibrating wire of length 1 clamped at both ends:

$$-\frac{d^2y}{dx^2} = \lambda y, \qquad y(0) = y(1) = 0$$

In Hilbert space theory, we incorporate the boundary conditions into the definition of the operator and rephrase the problem this way:

Let $L$ be the operator in $L^2(0, 1)$ defined by

Differential expression $l(y) \equiv -d^2y/dx^2$
Domain($L$) = compatible functions $y$ that satisfy $y(0) = y(1) = 0$

Then for what values of $\lambda$ does the equation

$$L(y) = \lambda y \tag{4.2.1}$$

possess nonzero solutions $y$ in domain($L$)? We are thus asking for the nonzero functions $y$ in the domain of $L$ that are mapped by $L$ to scalar multiples of themselves. In Chapter 3 we solved this problem and found that equation (4.2.1) has nonzero solutions only when $\lambda = n^2\pi^2$, $n = 1, 2, \ldots$, and the respective solutions $y$ are $y = \sin n\pi x$, $n = 1, 2, \ldots$.

The proper general setting here is provided by the following key definition:

**Definition 4.4.** Let $H = L^2(a, b; r)$ be a Hilbert space and $L$ a linear operator whose domain is a subspace $D$ of $H$. (Hence $L$ is defined on $D$, and only there, and for each $f \in D$, $L(f) \in H$.) The scalar $\lambda$ is said to be an *eigenvalue* of $L$ if there is a nonzero function $f$ in the domain of $L$ such that

$$L(f) = \lambda f$$

Any nonzero function $f$ in the domain of $L$ that satisfies this equation is called an *eigenfunction belonging to the eigenvalue* $\lambda$. The collection of all eigenfunctions belonging to the given eigenvalue $\lambda$, together with the zero function, form a subspace of $D$ called the *eigenspace* belonging to $\lambda$. The dimension of the eigenspace belonging to $\lambda$ is the *multiplicity* of $\lambda$. Eigenvalues of multiplicity 1 are called *simple*.

Using this language, we can restate our earlier result: *The operator $L$ on $L^2(0, 1)$ defined by $l(y) \equiv -y''$, domain($L$) = compatible functions that satisfy $y(0) = y(1) = 0$ has the eigenvalues $n^2 \pi^2$, $n = 1, 2, \ldots$ each of multiplicity 1. The corresponding eigenfunctions are $\sin n\pi x$, $n = 1, 2, \ldots$.*

The set of eigenvalues of an operator is called its *spectrum* (more technical mathematical usage would say discrete or point spectrum), and the finding of its spectrum is its *spectral analysis*. We shall soon see the enormous benefits to be derived from the knowledge of the spectrum of an operator. Indeed, Schrödinger, cofounder with Heisenberg of quantum mechanics, asserted that the existence of eigenvalues of a certain operator provided for him the true explanation of the discrete nature of the emission and absorption lines of atomic radiation.

We shall carry through the spectral analysis of another example, the operator on $L^2(-1, 1)$ defined by

$$m(y) \equiv -y'',$$

domain($M$) = compatible functions that satisfy $y(-1) = y(1) = 0$

Notice that $M$ has the same law of action as our previously studied operator $L$ but is nonetheless a different operator as it acts in a different space.

The equation $-y'' = \lambda y$, equivalently $y'' + \lambda y = 0$, has the general solution

$$y = c_1 e^{\sqrt{-\lambda} x} + c_2 e^{-\sqrt{-\lambda} x} \qquad \text{when } \lambda < 0$$

$$y = c_1 x + c_2 \qquad \text{when } \lambda = 0$$

$$y = c_1 \cos \sqrt{\lambda} x + c_2 \sin \sqrt{\lambda} x \qquad \text{when } \lambda > 0$$

All three functions lie in $L^2(-1, 1)$ and have continuous second derivatives bounded on $[-1, +1]$, so the compatibility condition is clearly satisfied. We need only concern ourselves with the boundary conditions $y(-1) = 0$, $y(1) = 0$. When does $y'' + \lambda y = 0$ have solutions that satisfy these conditions?

When $\lambda < 0$, applying the boundary conditions to the general solution displayed above we get the following two equations in the unknowns $c_1$ and $c_2$:

$$y(-1) = c_1 e^{a(-1)} + c_2 e^{-a(-1)} = 0$$
$$y(1) = c_1 e^a + c_2 e^{-a} = 0 \tag{4.2.2}$$

where we have set $a = \sqrt{-\lambda}$ for convenience. As $\lambda < 0$, it follows that $a > 0$. This homogeneous system of two equations in two unknowns $c_1$ and $c_2$ always has the solution $c_1 = c_2 = 0$. It will have a nonzero solution if and only if the determinant of its coefficients is zero. Compute this determinant:

$$\begin{vmatrix} e^{-a} & e^a \\ e^a & e^{-a} \end{vmatrix} = e^{-2a} - e^{2a}$$

Were this zero, $e^{-2a} = e^{2a}$, we would then have $e^{4a} = 1$, which necessitates $4a = 0$, as the only solution to $e^x = 1$ is $x = 0$. But $a \neq 0$, so this is impossible and the equation $M(y) = \lambda y$ has only the solution $y = 0$ when $\lambda < 0$. Thus our operator $M$ has no negative eigenvalues.

When $\lambda = 0$, we get the following equations:

$$y(-1) = c_1(-1) + c_2 = 0$$
$$y(1) = c_1 + c_2 = 0 \tag{4.2.3}$$

which have only the solution $c_1 = c_2 = 0$. Thus $\lambda = 0$ is not an eigenvalue.

When $\lambda > 0$, we get

$$y(-1) = c_1 \cos \sqrt{\lambda}\,(-1) + c_2 \sin \sqrt{\lambda}\,(-1) = 0$$
$$y(1) = c_1 \cos \sqrt{\lambda} + c_2 \sin \sqrt{\lambda} = 0 \tag{4.2.4}$$

The determinant of this system is

$$\begin{vmatrix} \cos \sqrt{\lambda} & -\sin \sqrt{\lambda} \\ \cos \sqrt{\lambda} & \sin \sqrt{\lambda} \end{vmatrix} = 2 \sin \sqrt{\lambda} \cos \sqrt{\lambda} = \sin 2\sqrt{\lambda}$$

Our system of equations in the unknown $c_1$ and $c_2$ will have a nonzero solution when and only when $\sin 2\sqrt{\lambda} = 0$. This latter equation holds exactly when $2\sqrt{\lambda}$ is an integral multiple of $\pi$:

$$2\sqrt{\lambda} = n\pi \quad \text{or} \quad 4\lambda = n^2\pi^2, \qquad n = 1, 2, \ldots$$

Thus our operator has an infinite sequence of positive eigenvalues $\pi^2/4$, $\pi^2$, $9\pi^2/4$, $25\pi^2/4, \ldots, n^2\pi^2/4, \ldots$. To find the eigenfunctions and the multiplicities we must solve the original pair of equations in the case $\lambda = n^2\pi^2/4$,

$\sqrt{\lambda} = n\pi/2$:

$$c_1 \cos \frac{n\pi}{2} - c_2 \sin \frac{n\pi}{2} = 0$$

$$c_1 \cos \frac{n\pi}{2} + c_2 \sin \frac{n\pi}{2} = 0$$

When $n$ is even, $n/2$ is an integer, so $\sin n\pi/2 = 0$, $\cos n\pi/2 = \pm 1$. Then $c_1 = 0$ and $c_2$ is arbitrary—we may take $c_2 = 1$. Thus for even $n$ the eigenspace corresponding to $n^2\pi^2/4$ is spanned by the single function $y_n = \sin(n\pi x/2)$. Thus each of these eigenvalues has multiplicity 1, and any nonzero multiple of $\sin(n\pi x/2)$ is an eigenfunction belonging to $n^2\pi^2/4$.

When $n$ is odd, $\cos(n\pi/2) = 0$ and $\sin(n\pi/2) = \pm 1$. Hence in this case we must have $c_2 = 0$. We may take $c_1 = 1$ and find the eigenfunction $y_n = \cos(n\pi x/2)$. Thus the eigenvalues for $M$ occur in two separate sequences: (I) those with $n = 2m$ and (II) those with $n = 2m + 1$:

(I) Eigenvalues $\pi^2, 4\pi^2, 9\pi^2, \ldots, n^2\pi^2, \ldots$ with corresponding eigenfunctions $\sin \pi x$, $\sin 2\pi x$, $\sin 3\pi x, \ldots, \sin n\pi x, \ldots$.

(II) Eigenvalues $\pi^2/4, 9\pi^2/4, 25\pi^2/4, \ldots, (2n+1)^2\pi^2/4, \ldots$ with corresponding eigenfunctions $\cos(\pi x/2)$, $\cos(3\pi x/2)$, $\cos(5\pi x/2), \ldots,$ $\cos((2n+1)\pi x/2), \ldots$.

That completes the spectral analysis of our operator $M$.

You will follow the same procedure in the exercises at the end of this section as we have followed in the example just completed. Let us look in more detail at the steps in this procedure:

*Step 1. Solve the second order ordinary differential equation $l(y) - \lambda y = 0$ in terms of the parameter $\lambda$.*

To solve this differential equation, use the methods of Chapter 2. There will always be two linearly independent solutions, say $f_1(x)$ and $f_2(x)$, each depending on the parameter $\lambda$. In writing out these solutions in the previous example, we have tacitly assumed that $\lambda$ is real. In doing this we have anticipated a result, to be proved shortly, that for the kind of operators we consider, namely *Hermitian* operators, the eigenvalues are real. You will also assume $\lambda$ real in the exercises.

In writing out the solution to the general differential equation we have also taken, in the example, the functions $f_1$ and $f_2$ as real and have assumed as well that the constants $c_1$ and $c_2$ in the general solution $c_1 f_1 + c_2 f_2$ are real. This is mainly for convenience in the applications. You will make the same assumptions in the exercises.

Sticking to real functions generally necessitates using different functions when $\lambda > 0$ and when $\lambda < 0$. In our example we had to use exponential functions when $\lambda < 0$ and trigonometric functions when $\lambda > 0$. When we develop some additional theory, we shall be able to rule out the possibility of negative eigenvalues right from the start for many operators. For the present, we will deal with these calculations case by case, from first principles.

*Step 2. Determine the values of the parameter $\lambda$ for which the general solution $y = c_1 f_1 + c_2 f_2$ found in Step 1 belongs to the domain of L and is not zero.*

In practical terms, the values of $\lambda$ are found this way: apply the boundary conditions that are part of the specification of domain. You will get in general a system of two equations in the two unknowns $c_1$ and $c_2$:

$$\mu_{11}c_1 + \mu_{12}c_2 = 0$$
$$\mu_{21}c_1 + \mu_{22}c_2 = 0 \qquad\qquad (4.2.5)$$

where the coefficients $\mu_{ij}$ depend on $\lambda$. Equations (4.2.2), (4.2.3), and (4.2.4) of the previous example are typical. Equations (4.2.5) will have a nonzero solution if and only if the determinant

$$\begin{vmatrix} \mu_{11} & \mu_{12} \\ \mu_{21} & \mu_{22} \end{vmatrix} = \mu_{11}\mu_{22} - \mu_{12}\mu_{21} \qquad\qquad (4.2.6)$$

is zero. Equation (4.2.6) gives the condition that determines the value of $\lambda$. These are the eigenvalues.

*Step 3. Substitute the eigenvalues $\lambda$ into equation (4.2.5) and solve for the corresponding $c_1$ and $c_2$.*

When you substitute a particular eigenvalue $\lambda$ into equations (4.2.5), there are two mutually exclusive possibilities:

1.  All four $\mu_{ij}$ are zero. Then *any* pair $(c_1, c_2)$ is a solution to (4.2.5). As $(c_1, c_2) = c_1(1, 0) + c_2(0, 1)$, we can take $c_1 = 1$, $c_2 = 0$ as one basic solution and $c_1 = 0$, $c_2 = 1$ as the other. In the first case we get $y = f_1(x)$, in the second case $y = f_2(x)$. So there are two linearly independent eigenfunctions, $f_1$ and $f_2$, that span the eigenspace belonging to this $\lambda$. If $a$ and $b$ are constants not both 0, then $y = af_1 + bf_2$ is an eigenfunction belonging to $\lambda$, and every eigenfunction has this form. We could use other pairs of linearly independent eigenfunctions to span the same eigenspace, for example, $f_1 + f_2$ and $f_1 - f_2$. Thus, in case 1 the multiplicity is two.
2.  Not all four $\mu_{ij}$ are zero. Then one row (at least) of the matrix has some entry not zero. Say it is the first row. If $\mu_{11} \neq 0$ and $\mu_{12} = 0$, then $c_1 = 0$ but $c_2$ can have any value. Give $c_2$ some convenient nonzero value, say $c_2 = 1$. Then $y = f_2$ spans the eigenspace belonging to $\lambda$. If $\mu_{11} = 0$ and

$\mu_{12} \neq 0$, then $y = f_1$ spans. If both $\mu_{11} \neq 0$ and $\mu_{12} \neq 0$, then the constants $c_1$ and $c_2$ satisfy $c_2 = -(\mu_{11}/\mu_{12})c_1$, so we can put $c_1 = \mu_{12}$, whence $c_2 = -\mu_{11}$. Then $y = \mu_{12}f_1 - \mu_{11}f_2$ spans the eigenspace belonging to $\lambda$. Thus, in case 2 the multiplicity is one.

For a second order linear differential operator, an eigenvalue has multiplicity either one or two; those are the only possibilities.

One final note: While all the operators that we study in this chapter do have eigenvalues—in fact, each of the several operators that we examine in detail has enough eigenvalues that the corresponding eigenfunctions form an orthogonal basis of the underlying Hilbert space—nonetheless many Hilbert space operators don't have any eigenvalues at all. For one simple example, see Exercise 7. Refer also to Section 8.5. Whether or not any given differential operator, or class of operators, does have eigenvalues has to be determined in each case, and the proof is generally a major enterprise.

### Exercises

1. Let $L$ be the operator in $L^2(-\pi, +\pi)$ defined by

   $l(y) \equiv -y''$,

   domain$(L)$ = compatible functions that satisfy $y'(-\pi) = y'(+\pi) = 0$

   (a) Show that the equation $L(y) = \lambda y$ has for $\lambda < 0$ only the solution $y = 0$ ($y \in$ domain$(L)$ understood). Thus conclude that $L$ has no negative eigenvalues.
   (b) Show that $\lambda = 0$ *is* an eigenvalue of $L$ of multiplicity one. Find a conveniently normalized eigenfunction.
   (c) Show that the positive eigenvalues of $L$ are $\lambda = n^2/4, n = 1, 2, \ldots$. Show also that for $n$ even the eigenfunctions are $\cos(nx/2), n = 2, 4, 6, \ldots$, and that for $n$ odd the eigenfunctions are $\sin(nx/2), n = 1, 3, 5, \ldots$. All eigenvalues are simple (multiplicity 1).

2. Determine the eigenvalues, their multiplicities, and the corresponding eigenfunctions for the operator $L$ whose differential expression is

   $$l(y) \equiv -\frac{d^2y}{dx^2}$$

   and whose domain is the subspace of compatible functions $y$ in $L^2(0, 1)$ that satisfy
   (a) $y'(0) = 0, y'(1) = 0$
   (b) $y(0) = 0, y'(1) = 0$
   (c) $y'(0) = 0, y(1) = 0$

3. (Continuation) Same as Exercise 2 except with boundary conditions $y(0) = y(1), y'(0) = y'(1)$. What are the multiplicities in this case?

4. (Continuation) Same as Exercise 2 except with boundary conditions $y(0) = y'(0)$, $y(1) = y'(1)$. Show that in this case there are exactly one negative eigenvalue and an infinite sequence of positive eigenvalues, all of multiplicity 1. Find them and their eigenfunctions.

5. Determine the eigenvalues, their multiplicities, and the corresponding eigenfunctions for the operator $L$ in $L^2(0, \pi)$ with differential expression $l(y) \equiv -y'' - y$ and whose domain is the subspace of compatible functions $y$ in $L^2(0, \pi)$ that satisfy $y(0) = 0$, $y(\pi) = 0$.

6. Consider the operator $L$ in $L^2(0, \infty)$ with differential expression $l(y) \equiv -y''$ and domain consisting of the entire subspace of functions in $L^2(0, \infty)$ that are compatible with $l$. (Thus domain($L$) consists of all functions $y$ such that both $y$ and $l(y) = -y''$ belong to $L^2(0, \infty)$.) Show that every $\lambda < 0$ is a simple eigenvalue for $L$ and find the corresponding eigenfunction. Show also that no $\lambda \geq 0$ is an eigenvalue.

7. (Continuation) Show that if we define an operator $M$ on $L^2(0, \infty)$ the same as in Exercise 6 except that the domain is further restricted by the condition $y(0) = 0$, then $M$ has no eigenvalues.

The definitions of eigenvalue, eigenfunction, and eigenspace (Definition 4.4) apply to any linear operator on any Hilbert space—in fact, on any vector space as the inner product is not used. In particular, Definition 4.4 applies to the familiar vector space $\mathbf{R}^3$ of your linear algebra course, although in that case we say *eigenvector* rather than eigenfunction.

The following definitions and terminology are also standard. If $L$ is a linear operator whose domain is a subspace of the vector space $V$, then the *kernel* of $L$, denoted ker($L$), consists of all $f$ in the domain of $L$ such that $L(f) = 0$. Always $0 \in$ ker($L$). If $0$ is the only vector (or function) in ker($L$), we say that $L$ has *zero kernel* and write ker($L$) = 0; otherwise say that $L$ has a *nontrivial kernel*. For us, the symbol $I$ will denote the *identity operator*: $If = f$ for all $f$. If $\lambda$ is a scalar, $\lambda I$ is the operation of multiplication by the scalar $\lambda$: $\lambda I(f) = \lambda f$. If $L$ is a linear operator, then the domain of $L - \lambda I$ is the same as the domain of $L$ and, for any $f$ in that domain, $(L - \lambda I)(f) = L(f) - \lambda f$.

8. Prove: The linear operator $L$ has zero kernel if, and only if, $L$ is one-to-one (or "injective"): $L(f_1) = L(f_2) \Rightarrow f_1 = f_2$.

9. Prove: $\lambda$ is an eigenvalue for $L$ if, and only if, $L - \lambda I$ has nontrivial kernel. Moreover, if $\lambda$ is an eigenvalue, then ker($L - \lambda I$) is the eigenspace belonging to $\lambda$.

10. Let $L$ be a linear operator on $\mathbf{R}^3$, the space of all real 3-tuples. Prove: $\lambda$ is an eigenvalue for $L$ if, and only if, det($L - \lambda I$) = 0, where det means determinant.

11. (Continuation) Suppose that the matrix of $L$ with respect to the standard basis $e_1 = (1, 0, 0)$, $e_2 = (0, 1, 0)$, and $e_3 = (0, 0, 1)$ is

$$L = \begin{bmatrix} 2 & 0 & 0 \\ 1 & -1 & -2 \\ -1 & 0 & 1 \end{bmatrix}$$

Show that $L$ has three real eigenvalues, $\lambda_1 < \lambda_2 < \lambda_3$; find them and their corresponding eigenvectors $v_1, v_2, v_3$.

12. (Continuation) What is the matrix of $L$ with respect to the basis $v_1, v_2, v_3$?

## 4.3 HERMITIAN OPERATORS

While we shall frame our basic Definition 4.5 below in terms of a general inner product space $\{V, \langle \cdot, \cdot \rangle\}$, most applications will be for the case where $V$ is a real or complex Hilbert space, $V = L^2(a, b; r)$, with inner product $\langle f, g \rangle = \int_a^b f(x)\overline{g(x)}r(x)\,dx$ where $r(x) > 0$, $a < x < b$.

Our forthcoming Definition 4.5 also involves a linear operator $L$ on $V$, which for us will ordinarily be a second order differential operator determined by two pieces of information:

1. A differential expression $l(y) \equiv A(x)y'' + B(x)y' + C(x)y$ which gives the law of action of the operator $L$

2. A subspace of $V$, the *domain* of $L$ which is the set of functions to which $l$ is to be applied

Now, the definition:

**Definition 4.5.** Let $\{V, \langle \cdot, \cdot \rangle\}$ be a real or complex inner product space, $L$ a linear operator with domain a subspace of $V$.

(a) We say that $L$ is *Hermitian* if

$$\langle L(f), g \rangle = \langle f, L(g) \rangle \qquad \text{for all } f, g \text{ in the domain of } L$$

(b) We say that $L$ is *positive semidefinite* if it is Hermitian, and

$$\langle L(f), f \rangle \geq 0 \qquad \text{for all } f \text{ in the domain of } L$$

(c) We say that $L$ is *positive definite* if it is positive semidefinite, and

$$\langle L(f), f \rangle = 0 \qquad \text{implies } f = 0 \text{ for } f \text{ in the domain of } L$$

Note particularly that, in the definition of Hermitian operator, the condition $\langle L(f), g \rangle = \langle f, L(g) \rangle$ holds for those, and only those, $f, g$ that belong to the domain of $L$. Thus, whether or not $L$ is Hermitian will depend crucially on its independently specified domain. It happens frequently that for two operators with the same differential expression but different domains one will be Hermitian and the other not. This fact provides further emphasis for the point that operators with different domains are different operators, even though they have the same differential expression.

Why is Definition 4.5 so important? Its importance lies in the consequences it has for the eigenvalues and eigenfunctions of $L$.

**Theorem 4.6.** *Let* $\{V, \langle \cdot, \cdot \rangle\}$ *be an inner product space and $L$ a linear operator on $V$. If $L$ is Hermitian, then*

(a) *All eigenvalues of $L$ are real.*

(b) *If $\phi_1$ and $\phi_2$ are eigenfunctions of $L$ belonging to the eigenvalues $\lambda_1$ and $\lambda_2$, respectively, and if $\lambda_1 \neq \lambda_2$, then $\langle \phi_1, \phi_2 \rangle = 0$. Briefly, eigenfunctions belonging to different eigenvalues are orthogonal.*

The inner product space $V$ in Theorem 4.6 may be either real or complex. If it is real, then of course item (a) is automatic.

Let us turn now to the proof of Theorem 4.6, part (a). To say that $\lambda$ is an eigenvalue of $L$ is to say that there exists a *nonzero* function $\phi$ in the domain of $L$ such that $L(\phi) = \lambda \phi$. Supposing this, we are to show that $\lambda$ must be real. We shall prove this fact in its equivalent form, that $\lambda$ be its own complex conjugate: $\lambda = \bar{\lambda}$. We know that $\|\phi\|^2 = \langle \phi, \phi \rangle > 0$. Then, using the fact that $L$ is Hermitian, we get

$$\lambda \langle \phi, \phi \rangle = \langle \lambda \phi, \phi \rangle = \langle L(\phi), \phi \rangle = \langle \phi, L(\phi) \rangle = \langle \phi, \lambda \phi \rangle = \bar{\lambda} \langle \phi, \phi \rangle$$

Dividing by $\|\phi\|^2$, we get $\lambda = \bar{\lambda}$, so $\lambda$ is real.

In many of the computations in Section 4.2, we ignored the possibility of complex eigenvalues. The result just proved justifies that omission.

We turn now to the proof of (b). We are given nonzero functions $\phi_1$ and $\phi_2$ in domain($L$) such that $L(\phi_1) = \lambda_1 \phi_1$, $L(\phi_2) = \lambda_2 \phi_2$, and $\lambda_1 \neq \lambda_2$. Be careful to note where in the following computation we use the Hermitian character of $L$. We will also use the fact just proved, that $\lambda_1$ and $\lambda_2$ are real.

$$\lambda_1 \langle \phi_1, \phi_2 \rangle = \langle \lambda_1 \phi_1, \phi_2 \rangle = \langle L(\phi_1), \phi_2 \rangle = \langle \phi_1, L(\phi_2) \rangle$$
$$= \langle \phi_1, \lambda_2 \phi_2 \rangle = \lambda_2 \langle \phi_1, \phi_2 \rangle$$

Hence $(\lambda_1 - \lambda_2) \langle \phi_1, \phi_2 \rangle = 0$. As $\lambda_1 - \lambda_2 \neq 0$ by assumption, we must have $\langle \phi_1, \phi_2 \rangle = 0$. That completes our proof.

**Theorem 4.7.** *Under the same hypotheses as in Theorem 4.6, if $L$ is also positive definite (resp. positive semidefinite) then all eigenvalues of $L$ are positive (resp. nonnegative).*

*Proof.* Suppose $\lambda$ is an eigenvalue for $L$ so that $L(\phi) = \lambda\phi$ for some $\phi \neq 0$ in the domain of $L$. Then

$$\lambda\langle \phi, \phi \rangle = \langle \lambda\phi, \phi \rangle = \langle L(\phi), \phi \rangle > 0$$

on the assumption that $L$ is positive definite ($\geq 0$ if $L$ is positive semidefinite). As $\langle \phi, \phi \rangle = \|\phi\|^2 > 0$ we may divide by this positive quantity to get $\lambda > 0$ in the positive definite case, $\geq 0$ in the positive semidefinite case. That completes the proof.

We summarize the very important Theorems 4.6 and 4.7:

---

If $L$ is a Hermitian operator on an inner product space, then the eigenvalues of $L$ are real, and eigenfunctions belonging to different eigenvalues are orthogonal.

If $L$ is in addition positive definite (resp. positive semidefinite), then the eigenvalues of $L$ are positive (resp. nonnegative).

---

We shall turn now to one example to illustrate the preceding material. Our example uses integration by parts and is therefore quite typical, as integration by parts is the standard method for deciding whether or not a given differential operator is Hermitian.

Consider the operator $L$ in $L^2(0, 1)$ defined as follows:

Differential expression $l(y) \equiv -y''$
Domain$(L)$ = compatible functions $y$ such that $y(0) = y(1) = 0$

In analyzing this example we shall need Theorem 4.62 from Appendix 4.A, which is restated below as Theorem 4.8. You are not responsible for the proof.

**Theorem 4.8.** *Let $(a, b)$ be a finite interval. If $y(x)$ belongs to $L^2(a, b)$ and is compatible with $l(y) \equiv -y''$, then $y(a)$, $y'(a)$, $y(b)$, and $y'(b)$ are all finite.*

We wish to prove that our operator $L$ is Hermitian, i.e., that $\langle L(f), g \rangle = \langle f, L(g) \rangle$ for all $f$, $g$ in domain$(L)$. The inner product in $L^2(0, 1)$ is

$$\langle f, g \rangle = \int_0^1 f(x)\overline{g(x)}\, dx$$

We begin with one integration by parts on the expression $\langle L(f), g \rangle$:

$$\langle L(f), g \rangle = \int_0^1 L(f)\bar{g}\, dx = -\int_0^1 f''\bar{g}\, dx = -\int_0^1 \bar{g}\, d(f') \tag{4.3.1}$$

The integration by parts is now referred to the usual format by setting $u = \bar{g}$, $dv = df'$. Then the standard formula,

$$\int_a^b u \, dv = uv|_a^b - \int_a^b v \, du$$

yields

$$\int_0^1 \bar{g} \, d(f') = \bar{g} f'|_0^1 - \int_0^1 f' \, d\bar{g} = \bar{g}(1)f'(1) - \bar{g}(0)f'(0) - \int_0^1 f'\bar{g}' \, dx \qquad (4.3.2)$$

Introducing a minus sign and combining (4.3.2) with (4.3.1), we get

$$\langle L(f), g \rangle = f'(0)\bar{g}(0) - f'(1)\bar{g}(1) + \int_0^1 f'\bar{g}' \, dx \qquad (4.3.3)$$

valid for all $f, g$ in domain($L$).

Now examine the integrated portion of equation (4.3.3), namely the term $f'(0)\bar{g}(0) - f'(1)\bar{g}(1)$. By definition, our operator $L$ has for domain the set of compatible functions $y$ in $L^2(0, 1)$ such that $y(0) = y(1) = 0$. By Theorem 4.8, $y'(0)$ and $y'(1)$ are finite. As the functions $f$ and $g$ in equation (4.3.3) belong to the domain of $L$, we have

$$f'(0)\bar{g}(0) - f'(1)\bar{g}(1) = (\text{finite}) \times 0 - (\text{finite}) \times 0 = 0 - 0 = 0$$

Hence, equation (4.3.3) becomes

$$\langle L(f), g \rangle = \int_0^1 f'\bar{g}' \, dx, \qquad f, g \in \text{domain}(L) \qquad (4.3.4)$$

Now, in equation (4.3.4), we may interchange $f$ and $g$ because equation (4.3.4) holds for any functions $f, g$ in the domain of $L$. If we do this, then take the complex conjugate of both sides of the resulting equation, and finally use the Hermitian symmetry of the inner product, $\langle x, y \rangle = \overline{\langle y, x \rangle}$, we get

$$\langle f, L(g) \rangle = \overline{\langle L(g), f \rangle} = \overline{\int_0^1 g'\bar{f}' \, dx} = \int_0^1 f'\bar{g}' \, dx \qquad (4.3.5)$$

Comparing (4.3.4) and (4.3.5), we conclude that $\langle L(f), g \rangle = \langle f, L(g) \rangle$ for all $f$ and $g$ in the domain of $L$. Hence our operator $L$ is Hermitian.

We may also use equation (4.3.4) to deduce that $L$ is positive definite. Putting $g = f$ in equation (4.3.4), we get

$$\langle L(f), f \rangle = \int_0^1 f'\bar{f}' \, dx = \int_0^1 |f'|^2 \, dx \qquad (4.3.6)$$

As $|f'|^2$ is $\geq 0$, so is its integral, thus $\langle L(f), f \rangle \geq 0$ for all $f$ in the domain of $L$. This proves that $L$ is positive semidefinite (refer to Definition 4.5). To prove the stronger statement that it is positive definite, we must show that $\langle L(f), f \rangle = 0$ only when $f = 0$. But, if $\langle L(f), f \rangle = 0$, then

$$\int_0^1 |f'|^2 \, dx = 0$$

As the integrand is everywhere $\geq 0$, no cancellation is possible, and the only way the integral can be zero is to have $|f'|^2 = 0$, whence $f' = 0$ (function equality). Thus $f = $ constant. But $f$ belongs to the domain of $L$, hence $f(0) = f(1) = 0$. As $f$ is constant, we must have $f = 0$. We have shown that $\langle L(f), f \rangle = 0$ implies $f = 0$, so $L$ is positive definite.

That completes the analysis of the operator $L$ of our example. Let's review that analysis, because there is more information to be gleaned from it, and the analysis can be recast in a more compact form.

Look at equation (4.3.6). The function $f$ is compatible with $l(y) \equiv -y''$, so $l(f) = L(f) \in L^2(0, 1)$. Hence the inner product $\langle L(f), f \rangle$ on the left of (4.3.6) is finite; thus so is the integral on the right. We conclude: *if $f$ is compatible with $l(y) \equiv -y''$ on $L^2(0, 1)$, then $f' \in L^2(0, 1)$*. Hence we may write (4.3.4) in the form $\langle L(f), g \rangle = \langle f', g' \rangle$ for any $f, g \in \mathrm{domain}(L)$. This slick notation enables us to streamline our proof that $L$ is Hermitian positive definite: interchange $f$ and $g$ to get $\langle L(g), f \rangle = \langle g', f' \rangle$, so (taking the complex conjugate of both sides) $\langle f, L(g) \rangle = \langle f', g' \rangle = \langle L(f), g \rangle$, which proves that $L$ is Hermitian. Then, putting $g = f$, $\langle L(f), f \rangle = \langle f', f' \rangle = \|f'\|^2 \geq 0$, so $L$ is positive semidefinite. Finally, if $\langle L(f), f \rangle = 0$, then $\|f'\| = 0 \Rightarrow f' = 0 \Rightarrow f = $ constant. And that constant must be 0 as $f(0) = f(1) = 0$. Thus $L$ is positive definite. Same proof, but cleaner.

When, in Section 3.1, we computed the eigenvalues of $L$, we ignored the possibility of complex eigenvalues. That omission is now justified; our operator $L$ is Hermitian, so all its eigenvalues are real. We also went through some detailed calculations to rule out the cases $\lambda < 0$ and $\lambda = 0$. Now we know that these calculations were unnecessary; our operator $L$ is positive definite, so all its eigenvalues are positive. Thus, Theorems 4.6 and 4.7 provide very practical and useful information.

In Section 3.1 (see also Section 4.2), we found that the eigenvalues of $L$ are $\lambda_n = n^2 \pi^2$, $n = 1, 2, \ldots$ with corresponding eigenfunctions $\phi_n = \sin n\pi x$. Theorem 4.6(b) tells us that when $m \neq n$, the eigenfunctions $\phi_m$ and $\phi_n$ are orthogonal. Hence

$$\langle \sin m\pi x, \sin n\pi x \rangle = \int_0^1 \sin m\pi x \sin n\pi x \, dx = 0,$$

$$m \neq n, \ m, n \geq 1 \tag{4.3.7}$$

The important point here is that we know the integral in equation (4.3.7) is zero *without having to compute it*. The vanishing of this integral is guaranteed because it represents the inner product of eigenfunctions for distinct eigen-values of a Hermitian operator.

Now it is easy enough to check equation (4.3.7) by a direct integration, but the point is that we don't have to. And we shall soon see examples where a direct integration to verify orthogonality gets to be quite involved. As we get to those cases, we will gain an increased appreciation of Theorem 4.6.

A final note on terminology. Operators that we call Hermitian are called by many authors "symmetric." So be aware of that when you consult other works. The term "selfadjoint" means something stronger than Hermitian; every selfadjoint operator is Hermitian, but not vice versa. We'll discuss selfadjoint operators briefly in the next section.

## Exercises

1. Let $L$ be the operator in $L^2(-\pi, \pi)$ defined by

   Differential expression $l(y) \equiv -d^2y/dx^2$
   Domain($L$)=compatible functions $y$ that satisfy $y'(-\pi)=0$, $y'(\pi)=0$

   (a) Prove that $L$ is Hermitian. From this fact, what can you infer about the eigenvalues of $L$?
   (b) Prove that $L$ is positive semidefinite. What does this imply about the eigenvalues?
   (c) Is $L$ positive definite? If so, what additional information does this give you about the eigenvalues?
   (d) Referring to Exercise 1 of Section 4.2, deduce without any further calculation that

   $$\int_{-\pi}^{+\pi} \cos\frac{mx}{2} \cos\frac{nx}{2}\, dx = 0, \qquad m \neq n, \text{ both even}$$

   $$\int_{-\pi}^{+\pi} \sin\frac{mx}{2} \cos\frac{nx}{2}\, dx = 0; \qquad m \text{ odd}, n \text{ even}$$

   $$\int_{-\pi}^{+\pi} \sin\frac{mx}{2} \sin\frac{nx}{2}\, dx = 0, \qquad m \neq n, \text{ both odd}.$$

2. Let $L$ be the operator in $L^2(0, 1)$ defined by

   Differential expression $l(y) \equiv -d^2y/dx^2$
   Domain($L$)=compatible functions $y$ that satisfy $y(0)=0$, $y'(1)=0$

   (a) Prove that $L$ is Hermitian. From this fact, what can you infer about the eigenvalues of $L$?

(b) Prove that $L$ is positive semidefinite. What does this imply about the eigenvalues?

(c) Is $L$ positive definite? If so, what additional information does this give you about the eigenvalues?

(d) Referring to Exercise 2b of Section 4.2, deduce, without any further calculation, that

$$\int_0^1 \sin\frac{2m+1}{2}\pi x \sin\frac{2n+1}{2}\pi x\, dx = 0, \qquad m \neq n$$

3. Let $L$ be the operator in $L^2(0,1)$ defined by

Differential expression $l(y) \equiv -d^2y/dx^2$
Domain($L$) = compatible functions $y$ that satisfy $y'(0)=0$, $y(1)=0$

(a) Prove that $L$ is Hermitian. From this fact, what can you infer about the eigenvalues of $L$?

(b) Prove that $L$ is positive semidefinite. What does this imply about the eigenvalues?

(c) Is $L$ positive definite? If so, what additional information does this give you about the eigenvalues?

(d) Referring to Exercise 2c of Section 4.2, deduce, without any further calculation, that

$$\int_0^1 \cos\frac{2m+1}{2}\pi x \cos\frac{2n+1}{2}\pi x\, dx = 0, \qquad n \neq m$$

4. Let $L$ be the operator in $L^2(0,1)$ defined by

Differential expression $l(y) \equiv -d^2y/dx^2$
Domain($L$) = compatible functions $y$ that satisfy $y(0)=y(1)$, $y'(0)=y'(1)$

(a) Prove that $L$ is Hermitian. From this fact, what can you infer about the eigenvalues of $L$?

(b) Prove that $L$ is positive semidefinite. What does this imply about the eigenvalues?

(c) Is $L$ positive definite? If so, what additional information does this give you about the eigenvalues?

(d) Referring to Exercise 3 of Section 4.2, deduce, without any further calculation, that

$$\int_0^1 \cos 2n\pi x\, dx = 0, \qquad \int_0^1 \sin 2n\pi x\, dx = 0, \qquad n \neq 0$$

and

$$\left.\begin{array}{l} \displaystyle\int_0^1 \cos 2m\pi x \cos 2n\pi x \, dx = 0 \\[2mm] \displaystyle\int_0^1 \cos 2m\pi x \sin 2n\pi x \, dx = 0 \\[2mm] \displaystyle\int_0^1 \sin 2m\pi x \sin 2n\pi x \, dx = 0 \end{array}\right\} \quad m \neq n, \; m,n \geq 1$$

(e) Does Theorem 4.6(b) allow you to say anything about the following integral?

$$\int_0^1 \cos 2n\pi x \sin 2n\pi x \, dx$$

5. Let $L$ be the operator in $L^2(0, 1)$ defined by

Differential expression $l(y) \equiv -d^2y/dx^2$
Domain($L$) = compatible functions $y$ that satisfy $y(0) = y'(0)$, $y(1) = y'(1)$

(a) Prove that $L$ is Hermitian; i.e., prove that $\langle L(f), g \rangle = \langle f, L(g) \rangle$ for all $f, g$ in the domain of $L$.

(b) In Exercise 4 of Section 4.2 we showed that $\lambda = -1$ is an eigenvalue of $L$ with corresponding eigenfunction $y = e^x$ and that $\lambda_n = n^2\pi^2$, $n = 1, 2, \ldots$, are all simple eigenvalues with corresponding eigenfunctions $y_n = \sin n\pi x + n\pi \cos n\pi x$. Deduce, *without further calculation*, that

$$\int_0^1 (\sin n\pi x + n\pi \cos n\pi x)e^x \, dx = 0, \qquad n = 1, 2, \ldots$$

(c) Verify the orthogonality relation in (b) by direct integration.

6. Consider the Hilbert space $L^2(-1, 1; e^{2x})$ which carries the inner product

$$\langle f, g \rangle = \int_{-1}^{+1} f(x)\overline{g(x)}e^{2x} \, dx$$

Let $L$ be the operator in $L^2(-1, 1; e^{2x})$ defined by

Differential expression $l(y) \equiv -y'' - 2y'$
Domain($L$) = compatible functions $y$ that satisfy $y(-1) = e^2y(1)$, $y'(-1) = e^2y'(1)$

Prove that $L$ is Hermitian. [Hint: Note that $L(y) = -e^{-2x}(e^{2x}y')'$. Refer to Exercise 10 of Section 2.2.]

7. Let $\{V, \langle \cdot, \cdot \rangle\}$ be an inner product space. You are given no integral formula for the abstract inner product $\langle f, g \rangle$, so, in doing this problem, you must use only the three properties of the inner product: conjugate bilinear, Hermitian, and positive definite.

   (a) The *identity operator* $I$ on $V$ has domain equal to all of $V$ and maps each element of $V$ to itself: $I(f) = f$ for every $f$ in $V$. Prove that $I$ is Hermitian and positive definite.

   (b) Let $\alpha$ be a constant. The operator $M_\alpha$, of *multiplication by* $\alpha$, has domain equal to all of $V$ and maps each element of $V$ to $\alpha$ times itself: $M_\alpha(f) = \alpha f$ for all $f$ in $V$. Prove that $M_\alpha$ is Hermitian if, and only if, $\alpha$ is real. For what $\alpha$ is $M_\alpha$ positive definite?

   (c) What can you say about the eigenvalues and eigenfunctions of $I$ and $M_\alpha$? (Usually $M_\alpha$ is written $\alpha I$.)

8. Let $\{V, \langle \cdot, \cdot \rangle\}$ be an inner product space. You are given no integral formula for the abstract inner product $\langle f, g \rangle$, so, in doing this problem, you must use only the three properties of the inner product: conjugate bilinear, Hermitian, and positive definite.

   (a) Suppose $L_1$ is a Hermitian operator with domain $D_1$ and $L_2$ a Hermitian operator with domain $D_2$. Let $D = D_1 \cap D_2$, so that $D$ consists of elements in both $D_1$ and $D_2$. Define the sum $L_1 + L_2 = L$ as the operator with domain $D$ and law of action

   $$L(f) = L_1(f) + L_2(f) \qquad \text{for all } f \text{ in } D$$

   Prove that $L$ is Hermitian. Prove that if $L_1$ and $L_2$ are positive semidefinite, so is $L$. When is $L$ positive definite?

   (b) If $\alpha$ is a constant, when is $\alpha L_1$ Hermitian?

   (c) If $\alpha$ is a constant, when is $L_1 + \alpha I$ Hermitian?

9. Let $V$ be the Hilbert space $L^2(a, b; r)$ consisting of all functions that are square integrable over the interval $(a, b)$ with respect to the positive weight function $r(x)$. Let $m(x)$ be a given (fixed) function defined on $(a, b)$, and denote by $D$ the subset of functions $f(x)$ in $V$ such that the product $m(x)f(x)$ also belongs to $V$.

   (a) Using the fact that $V$ is a vector space, prove that $D$ is a subspace of $V$. That is, prove that if $f_1$ and $f_2$ belong to $D$, so do $f_1 + f_2$ and $cf_1$ for any constant $c$.

   (b) Define an operator $M$ on $V$ with domain $D$ by the rule $M(f) = mf$ for all $f$ in $D$. That is the operation of multiplication by $m$. Prove that, if $m(x)$ is real valued, then $M$ is Hermitian.

10. Consider the space $\mathbf{C}^3$ of all 3-tuples $(\alpha_1, \alpha_2, \alpha_3)$ with real or complex entries, and put the usual inner product on $\mathbf{C}^3$: if $a = (\alpha_1, \alpha_2, \alpha_3)$ and $b = (\beta_1, \beta_2, \beta_3)$, then $\langle a, b \rangle = \alpha_1 \bar{\beta}_1 + \alpha_2 \bar{\beta}_2 + \alpha_3 \bar{\beta}_3$. Suppose that the matrix of the linear operator $T$ with respect to the standard basis $e_1 = (1, 0, 0)$,

$e_2 = (0, 1, 0)$, and $e_3 = (0, 0, 1)$ is

$$T = \begin{bmatrix} \tau_{11} & \tau_{12} & \tau_{13} \\ \tau_{21} & \tau_{22} & \tau_{23} \\ \tau_{31} & \tau_{32} & \tau_{33} \end{bmatrix}$$

Prove that $T$ is Hermitian if, and only if, the three diagonal entries $\tau_{11}$, $\tau_{22}$, $\tau_{33}$ are real and, for the off-diagonal entries, $\tau_{ji} = \overline{\tau_{ij}}$.

## 4.4 SOME GENERAL OPERATOR THEORY

Where do we stand? Section 4.1 introduced the notion of *compatibility*. I have put this first, for two reasons.

First, compatibility is indispensable for the study of differential operators in Hilbert space. Given a differential expression $l$, if we wish to define an operator $L$ in a Hilbert space $H$ whose law of action is given by $l$, then the largest possible domain for $L$ consists of those functions $f$ in $H$ for which $l(f)$ also belongs to $H$. Otherwise, for some $f$, $L(f)$ would not lie in $H$, and we would not have a Hilbert space operator. We might have a very interesting operator from some other point of view. But, if we want a Hilbert space operator, then we must keep its values in $H$, and that is precisely what the compatibility restriction accomplishes.

Second, compatibility is a new idea. It just doesn't come up in calculus, nor in finite-dimensional linear algebra either. Since it is new and unfamiliar, I have put it first for stress.

While compatibility is very important, it is nonetheless special for *differential* operators. In contrast, the concepts of eigenvalue and eigenfunction (Section 4.2) and the property "Hermitian" (Section 4.3) belong to the general theory of operators on Hilbert space. This section expounds some other aspects of this general theory. These general definitions and results are important in their own right; they also help us set in perspective the special topic of differential operators, which is our main interest.

We begin with the definition of linear operator, repeated from Section 4.1 but this time slightly generalized to an abstract Hilbert space (see Section 3.10).

**Definition 4.9.** A *linear operator* $L$ in a real or complex Hilbert space $H$ is a mapping, or assignment, that assigns to each element $f$ in a certain subspace $D$ of $H$, called the *domain* of the operator, another element $L(f)$ also in $H$ so that the following rules are satisfied:

$$L(f + g) = L(f) + L(g) \qquad \text{for all } f, g \text{ in } D$$

and

$$L(cf) = cL(f) \qquad \text{for all scalars } c \text{ and all } f \text{ in } D$$

Almost all operators encountered in practice have domains that are *dense* subspaces of their respective Hilbert spaces. And generally, in theoretical work, one almost always assumes that any operator under consideration is *densely defined*; that is, its domain is a dense subspace of its Hilbert space.

We discussed dense subspaces briefly in Section 3.10. The subspace $D$ of the Hilbert space $H$ is dense in $H$ if, given any element $h$ in $H$ and given any positive real number $\varepsilon$, however small, we can find an element $d$ in $D$ so that $\|h-d\| < \varepsilon$. Briefly, $D$ is dense in $H$ if there are elements of $D$ as close as we please to any element of $H$.

Dense proper subspaces don't exist in finite-dimensional linear algebra. If $H$ has finite dimension, say $n$, and $D$ is a subspace of $H$ of dimension $m$, say, then $m \leqslant n$. If $m < n$, then $D \subsetneq H$. In that case, we know from our linear algebra course that $H = D + D^{\perp}$, where $D^{\perp}$ is the subspace of $H$ consisting of all elements orthogonal to every element of $D$. The dimension of the subspace $D^{\perp}$ is $n - m$. Select $h \in D^{\perp}$ with $\|h\| = 1$. Then no element of $D$ can be any closer than 1 unit of distance from $h$ because $\langle h, d \rangle = 0$, whence $\|h-d\|^2 = \|h^2\| + \|d\|^2 \geqslant \|h\|^2 = 1$. Hence no proper subspace of a finite-dimensional Hilbert space can be dense. If $m = n$, then $D = H$, so $D$ is then certainly dense in $H$.

Only in infinite-dimensional Hilbert spaces do we meet proper dense subspaces. One example of a proper dense subspace is provided in Exercise 1 at the end of this section. Many other examples will occur along the way, because again and again we will construct orthogonal bases of infinite-dimensional $L^2$ spaces, and each such basis gives rise to a proper dense subspace, namely the subspace it spans (all finite linear combinations of the basis vectors)—refer to Lemma 3.23 in Chapter 3.

With regard to Definition 4.9, a question naturally arises—why not require by definition that every linear operator be defined on all of its Hilbert space $H$? Why bother considering operators that are defined only on proper subspaces of $H$, even though those proper subspaces are dense?

Answer: If you wish to study differential operators—and almost all the interesting applications involve differential operators—then your theory must be able to handle operators that are not everywhere defined, because differential operators on $L^2$ spaces *cannot* be everywhere defined.

Why not? is the natural follow-up question. The explanation is a simple and practical one: the domain of a second order linear differential expression $l$ is limited to the class of functions compatible with $l$. As the compatibility restriction is much more stringent than the square-integrability requirement, there will always be functions in a given $L^2$ space that fail to be compatible with $l$. For example, consider the differential expression $l(y) \equiv -y''$ in the space $L^2(a, b)$, $-\infty < a < b < \infty$. The *only* test that a function $f(x)$ must pass in order to be admitted into the space $L^2(a, b)$ is that the integral $\int_a^b |f|^2 \, dx$ be finite. *Every* bounded function, no matter how nondifferentiable or discontinuous, passes this test. (There is also the tacit requirement of Lebesgue measurability, but that is no restriction at all in practice.) But a function that is compatible with $l(y) \equiv -y''$ must at least be continuous and have a continuous first derivative at each $x$, $a < x < b$ (Theorem 4.2). Obviously, then, $L^2(a, b)$ will contain plenty of functions not compatible with $l$. So that is the explanation: the compatibility requirement inherent in the definition of a differential operator severely limits its domain.

Even though the differential expression itself can apply only to the compatible functions, one might wonder whether or not it would be possible to extend the definition of the operator to the whole Hilbert space by some other means. Of course, you would want any such extended operator to still carry the essential information contained in the differential expression $l$. There *are* such extension processes, but they won't extend a differential operator to the whole Hilbert space either. Proofs of these facts are heavily laden with pure mathematics and go far beyond our interest, which lies in the applications. They also are well beyond the scope of this course. For those who might at some future date be interested in pursuing the purely mathematical aspect of these questions, there is an excellent graduate-level book by J. Weidmann, *Linear Operators in Hilbert Spaces* (Springer, 1980).

We consider now an important classification of linear operators.

**Definition 4.10.** A linear operator $L$ on an inner product space $\{V, \langle \cdot, \cdot \rangle\}$ is said to be *bounded* or *continuous* if there exists a positive constant $M$ so that

$$\frac{\|L(f)\|}{\|f\|} \leqslant M \tag{4.4.1}$$

for all nonzero $f$ in the domain of $L$. If no such constant $M$ exists, then we call $L$ *unbounded* (sometimes *discontinuous*).

The norm in (4.4.1) is that derived from the inner product, $\|f\| = \langle f, f \rangle^{1/2}$. Hence, by squaring both sides of (4.4.1), we can rewrite it in the equivalent form $\langle L(f), L(f) \rangle / \langle f, f \rangle \leqslant M^2$.

Consider, as an example, the operator $L$ of Section 4.2 which has the differential expression $l(y) \equiv -y''$ and domain consisting of all functions $y$ in $L^2(0, 1)$ that are compatible with $l$ and satisfy $y(0) = y(1) = 0$. We proved that $L$ has the eigenvalues $n^2\pi^2$, with corresponding eigenfunctions $\sin n\pi x$, $n = 1, 2, \ldots$. Hence $L(\sin n\pi x) = n^2\pi^2 \sin n\pi x$. The domain of $L$ certainly contains all the functions $\sin n\pi x$ as these are eigenfunctions of $L$. Compute the ratio in (4.4.1) for these functions (use $\|\alpha f\| = |\alpha| \, \|f\|$):

$$\frac{\|L(\sin n\pi x)\|}{\|\sin n\pi x\|} = \frac{\|n^2\pi^2 \sin n\pi x\|}{\|\sin n\pi x\|} = \frac{n^2\pi^2 \|\sin n\pi x\|}{\|\sin n\pi x\|} = n^2\pi^2$$

There is certainly no constant $M$ that satisfies $n^2\pi^2 \leqslant M$ for all $n = 1, 2, \dots$. Hence our operator $L$ is unbounded.

This example is typical: all differential operators on $L^2$ spaces are unbounded. For our purpose, we shall need a rigorous proof only of this result: *a linear operator with an infinite sequence of eigenvalues* $\lambda_n$, $n = 1, 2, \dots$, *such that* $|\lambda_n| \to \infty$ *as* $n \to \infty$ *is unbounded* (Exercise 4), because all the differential operators that we shall study have such a sequence of eigenvalues.

Unbounded and bounded operators behave quite differently, especially as regards the extension of the linearity property to infinite sums. The essence of linearity can be expressed this way (refer to Definition 4.9): if $f_1$ and $f_2$ belong to the domain of $L$, and $c_1$ and $c_2$ are any scalars, then $L(c_1 f_1 + c_2 f_2) = c_1 L(f_1) + c_2 L(f_2)$. This formula is valid for any *finite* sum: if the $f_i$ belong to the domain of $L$ and the $c_i$ are any scalars, $1 \leqslant i \leqslant n$, then $L(c_1 f_1 + c_2 f_2 + \cdots + c_n f_n) = c_1 L(f_1) + c_2 L(f_2) + \cdots + c_n L(f_n)$. Proof: Do first for $n = 3$, writing $c_1 f_1 + c_2 f_2 + c_3 f_3 = (c_1 f_1 + c_2 f_2) + c_3 f_3$. By the basic axiom from Definition 4.9, $L((c_1 f_1 + c_2 f_2) + c_3 f_3) = L(c_1 f_1 + c_2 f_2) + L(c_3 f_3)$. Then apply the $n = 2$ case again. When $n = 4$, group as $3 + 1$ and apply the previous results, etc. (induction).

But there is nothing in the definition of linearity that allows us to apply $L$ this way to *infinite* sums, in general. Additional conditions are needed. I shall state and prove one theorem that spells out such conditions. This theorem applies to unbounded operators and is the result of primary value to us. There is another result for bounded operators whose statement and proof are left to the exercises.

**Theorem 4.11.** *Hypotheses: Let* $\{H, \langle \cdot, \cdot \rangle\}$ *be a real or complex Hilbert space, L a Hermitian operator whose domain is a subspace of H. Suppose L has an infinite sequence of eigenfunctions* $\phi_n$, $n = 1, 2, \dots$, *that form an orthogonal basis of H. Thus, for every f in H,*

$$f = \sum_{n=1}^{\infty} c_n \phi_n \qquad \text{where } c_n = \frac{\langle f, \phi_n \rangle}{\langle \phi_n, \phi_n \rangle} \tag{4.4.2}$$

*Let* $\mu_n$, $n = 1, 2, \dots$, *be the eigenvalue to which* $\phi_n$ *belongs, so* $L(\phi_n) = \mu_n \phi_n$, $n = 1, 2, \dots$.

*Conclusion: If f also belongs to the domain of L, then L may be applied term by term to (4.4.2):*

$$L(f) = \sum_{n=1}^{\infty} L(c_n \phi_n) = \sum_{n=1}^{\infty} c_n \mu_n \phi_n \tag{4.4.3}$$

Before the proof of Theorem 4.11, some comments.

Note the restrictive hypotheses in this theorem. First, $L$ has to be Hermitian and has to have an infinite sequence of eigenfunctions that form an orthogonal basis of the underlying Hilbert space $H$. Second, we are not applying $L$ to an arbitrary infinite sum but only to the general Fourier expansion with respect to that particular orthogonal basis.

Despite its restrictive hypotheses, Theorem 4.11 will be a very useful result for us, because each of the differential operators that we shall consider is Hermitian and does possess an infinite sequence of eigenfunctions that forms a basis. When we are able to find such an infinite sequence of eigenfunctions, we say that we have *diagonalized* the Hermitian operator $L$. The importance of this concept resides in the fact that we may then compute the action of $L$ by formula (4.4.3), independent of the differential expression from which $L$ arose. Additional advantages accrue from (4.4.3). Suppose we want to compute $L^2(f) = L(L(f))$. Formally, just apply $L$ term by term to (4.4.3):

$$L^2(f) = \sum_{n=1}^{\infty} L(c_n \mu_n \phi_n) = \sum_{n=1}^{\infty} c_n \mu_n^2 \phi_n$$

Of course, one has to be concerned about finding the proper domain for the operator $(L)^2$, but even that is easily handled. We have a general result, the heart of the "functional calculus" for operators:

**Corollary 4.12.** *Maintain the hypotheses of Theorem 4.11. Let $F(x)$ be a real- or complex-valued function such that all $F(\mu_n)$, $n = 1, 2, \ldots$, are defined (so, for example, if one or more $\mu_n = 0$, then $F(x) = 1/x$ is ruled out). Then the formula*

$$F(L)(f) = \sum_{n=1}^{\infty} c_n F(\mu_n) \phi_n \qquad (4.4.4)$$

*defines a Hilbert space operator with domain consisting of all $f = \Sigma c_n \phi_n$ in $H$ such that the numerical series*

$$\sum_{n=1}^{\infty} |c_n|^2 |F(\mu_n)|^2 \langle \phi_n, \phi_n \rangle \qquad (4.4.5)$$

*converges.*

There is very little to prove here (we have not yet proved Theorem 4.11— we'll get to that in a moment). Formula (4.4.4) defines a linear operator, i.e., $F(L)(af + bg) = aF(L)(f) + bF(L)(g)$, and the $f = \Sigma c_n \phi_n$ for which the series (4.4.5) converges constitute exactly the class of functions which make the series on the right side of (4.4.4) a proper element of $H$ (see Exercise 9 of Section 3.10).

The choice of functions $F(x) = e^{\pm ix}$ in Corollary 4.12 gives some intriguing results. Even though $L$ may be unbounded, the operators $e^{iL}$, $e^{-iL}$ are bounded and everywhere defined. They also have other interesting properties—see Exercise 6. This functional calculus is fascinating, but we won't have occasion to make much further use of it.

We come now, finally, to the proof of Theorem 4.11. Study this proof carefully.

*Proof of Theorem 4.11.* We are assuming that $f$ belongs to the domain of the Hermitian operator $L$. Hence, we can apply $L$ to $f$, and the resulting element $L(f)$ belongs to $H$. By hypothesis, the $\phi_n$, $n = 1, 2, \ldots$, form an orthogonal basis of $H$. This means that *every* element $g$ of $H$ has a general Fourier expansion with respect to the $\phi_n$ that converges to $g$ in the "mean-square" metric of $H$. In particular, this is true of $L(f)$, so

$$L(f) = \sum_{n=1}^{\infty} a_n \phi_n$$

where

$$a_n = \frac{\langle L(f), \phi_n \rangle}{\langle \phi_n, \phi_n \rangle}, \qquad n = 1, 2, \ldots$$

Now $L$ is Hermitian, so $\langle L(f), \phi_n \rangle = \langle f, L(\phi_n) \rangle$. And the $\phi_n$ are eigenfunctions of $L$, so $L(\phi_n) = \mu_n \phi_n$. Thus $\langle L(f), \phi_n \rangle = \langle f, L(\phi_n) \rangle = \langle f, \mu_n \phi_n \rangle = \bar{\mu}_n \langle f, \phi_n \rangle$. And $\bar{\mu}_n = \mu_n$ (why?), so

$$a_n = \frac{\langle L(f), \phi_n \rangle}{\langle \phi_n, \phi_n \rangle} = \mu_n \frac{\langle f, \phi_n \rangle}{\langle \phi_n, \phi_n \rangle} = \mu_n c_n$$

where $c_n = \langle f, \phi_n \rangle / \langle \phi_n, \phi_n \rangle$ are the general Fourier coefficients of $f$. Thus

$$L(f) = \sum_{n=1}^{\infty} \mu_n c_n \phi_n$$

But this is exactly what we would get if we applied $L$ term by term to (4.4.2). That completes the proof.

I shall close this section with an off-the-record explanation of what a selfadjoint operator is. This calls for the introduction of terms from the advanced theory, so I shall be quite vague and descriptive. For those who might want to look into this more deeply at some future date, there is the advanced graduate text *Linear Operators in Hilbert Spaces* by J. Weidmann that I mentioned before. I have included this discussion because the term "selfadjoint" is used frequently, and I wanted you to have an appreciation of the difference between Hermitian and selfadjoint.

Given any densely defined operator $T$ in a Hilbert space $H$, one can associate to it another operator $T^*$ called its *adjoint*. If $T$ is bounded and everywhere defined, then the adjoint is uniquely determined by the formula $\langle T(f), g \rangle = \langle f, T^*(g) \rangle$ for all $f, g$ in $H$ and is also bounded and everwhere defined. If $T$ is unbounded and densely defined, the definition of $T^*$ gets more involved. But, in any case, we always have this adjoint operator $T^*$. In terms of the adjoint operator, we can characterize Hermitian operators this way: $T$ is Hermitian if, and only if, the domain of $T$ is contained in that of $T^*$ and, on the smaller domain (that of $T$), $T$ and $T^*$ agree. We can write this symbolically: $T$ is Hermitian if, and only if, $T \subseteq T^*$. Now you can guess what a selfadjoint operator is: $T$ is *selfadjoint* when $T = T^*$. For bounded and everywhere defined operators—in particular for operators on finite-dimensional spaces—Hermitian and selfadjoint mean the same thing. Only for unbounded operators is there a difference. But, in all cases, every selfadjoint operator is Hermitian.

For us, Hermitian is the important concept. Memorize the definition: $T$ is *Hermitian* when it can be moved from one side of the inner product to the other without affecting the value of the inner product. In symbols, $T$ is Hermitian when $\langle Tf, g \rangle = \langle f, Tg \rangle$ holds for all $f, g$ in the domain of $T$. And, remember, operators that we call Hermitian, other authors often call "symmetric."

### Exercises

1. Consider the space $l^2(\mathbf{C})$ consisting of all $\infty$-tuples $(\alpha_1, \alpha_2, \ldots)$ of real or complex entries that are square summable: $\Sigma |\alpha_i|^2 < \infty$. We know that $l^2(\mathbf{C})$ is a Hilbert space under the inner product

   $$\langle a, b \rangle = \sum_{i=1}^{\infty} \alpha_i \bar{\beta}_i$$

   where $a = (\alpha_1, \alpha_2, \ldots)$ and $b = (\beta_1, \beta_2, \ldots)$—see Exercise 11 of Section 3.9. Let $D$ stand for the subset of $l^2(\mathbf{C})$ consisting of sequences that have only finitely many nonzero entries.
   (a) Prove that $D$ is a subspace of $l^2(\mathbf{C})$. That is, prove that $D$ is closed under sum and closed under scalar multiplication.
   (b) Does the sequence $v = (1, 1/2, 1/3, \ldots, 1/n, \ldots)$ belong to $l^2(\mathbf{C})$? Does it belong to $D$? Is $D$ a proper subspace of $l^2(\mathbf{C})$? Explain.
   (c) Prove that $D$ is dense in $l^2(\mathbf{C})$.
2. (Continuation)   Given $x = (\xi_1, \xi_2, \ldots, \xi_n \ldots) \in l^2(\mathbf{C})$, define $Tx = (\xi_1, \xi_2/2, \xi_3/3, \ldots, \xi_n/n, \ldots)$. Prove that $T$ is a linear operator. Prove that $\|Tx\|^2 \leqslant \|x\|^2$ for all $x \in l^2(\mathbf{C})$, and thus show that $T$ is bounded. Does the inequality $\|Tx\|/\|x\| \leqslant M$ hold for all nonzero $x$ in $l^2(\mathbf{C})$ for any constant $M$ *less* than 1? (You can justify an answer "no" by finding a specific $x$ with $\|Tx\| = \|x\|$.)

3. (Continuation)   Given $x = (\xi_1, \xi_2, \ldots, \xi_n \ldots) \in l^2(C)$, define $Sx = (\xi_1, 2\xi_2, 3\xi_3, \ldots, n\xi_n, \ldots)$. Prove that $S$ is a linear operator. Let $e_n = (0, 0, \ldots, 1(n\text{th place}), 0, \ldots)$. Show that each $e_n$ is an eigenvector for $S$ and find the corresponding eigenvalue. Is $S$ bounded? Prove your answer. Does the domain of $S$ contain the subspace $D$ of Exercise 1? Is the subspace of vectors compatible with $S$ dense in $l^2(C)$?

4. Prove: A linear Hilbert space operator with an infinite sequence of eigenvalues $\lambda_n$, $n = 1, 2, \ldots$, such that $|\lambda_n| \to \infty$ as $n \to \infty$ is unbounded.

5. Suppose $T$ is a bounded and everywhere defined linear operator on a Hilbert space $H$.
   (a) Prove that if $x_n \to 0$ in $H$, then $T(x_n) \to 0$.
   (b) Prove that if $x_n \to x$ in $H$, then $Tx_n \to Tx$.
   (c) Prove that if $x = \Sigma x_n$ in $H$, then $Tx = \Sigma Tx_n$.

6. Let $H$ be a Hilbert space with orthogonal basis $\phi_n$, $n = 1, 2, \ldots$, and let $L$ be a Hermitian operator that has the $\phi_n$ as its complete set of eigenvalues. Thus $L$ is diagonalized by the $\phi_n$. Let $\mu_n$, $n = 1, 2, \ldots$, be the corresponding eigenvalues: $L(\phi_n) = \mu_n \phi_n$. Construct the operators $U = e^{iL}$, $V = e^{-iL}$ by the functional calculus described in Corollary 4.12.
   (a) Show that $U$ and $V$ are everywhere defined.
   (b) Show that $\langle Uf, Ug \rangle = \langle f, g \rangle$ for all $f, g$ in $H$. Putting $f = g$, conclude that $U$ is bounded.
   (c) Show that $UVf = f = VUf$ for every $f \in H$, hence $V = U^{-1}$.
   (d) Show that $\langle Uf, g \rangle = \langle f, Vg \rangle$ for all $f, g$ in $H$. This means that $V$ is the *adjoint* of $U$; in symbols $V = U^*$ or $e^{-iL} = (e^{iL})^*$.
   (e) Show that the $\phi_n$ are eigenfunctions for $U$. What are the eigenvalues?

## 4.5   THE ONE-DIMENSIONAL LAPLACIAN

The term "one-dimensional Laplacian" refers to a class of Hilbert space operators, all based on the differential expression $l(y) \equiv -y''$. Two Hilbert spaces are involved: $L^2(a, b)$ for a finite interval $[a, b]$ and $L^2(C)$ where $C$ is the circle obtained by removing the interval $[a, b]$ from the real line, bending it around, and joining the point $x = a$ to the point $x = b$. I'll describe these two Hilbert spaces in more detail.

The first, $L^2(a, b)$, is built on the one-dimensional manifold $[a, b] = \{x : -\infty < a \leqslant x \leqslant b < \infty\}$ which has the two boundary points $x = a$ and $x = b$. The space $L^2(a, b)$ consists of all real- or complex-valued functions $f(x)$, $a \leqslant x \leqslant b$, that satisfy the square-integrability condition

$$\int_a^b |f(x)|^2 \, dx < \infty \tag{4.5.1}$$

The space $L^2(a, b)$ is very "democratic"; you need pass only the test (4.5.1) to get in. All bounded functions, no matter how nondifferentiable or discontinuous, pass this test. Certain unbounded functions do also (see Section 3.9). The inner product in $L^2(a, b)$ is

$$\langle f, g \rangle = \int_a^b f(x)\overline{g(x)}\, dx \qquad (4.5.2)$$

When we speak of functions being orthogonal in $L^2(a, b)$, we of course always mean orthogonal with respect to the inner product (4.5.2).

The second Hilbert space $L^2(C)$ is built on the circle $C$ that is constructed by removing the finite interval $[a, b]$ from the real line, bending it around, and joining the point $x = a$ to $x = b$. See Figure 4.1. This is precisely the same method as we used in Section 3.2 to construct the unit circle (radius 1) by bending the interval $[-\pi, \pi]$. There, we used the symbol $C$ to stand for that unit circle, and we shall let the same symbol $C$ serve also for this general circle of radius $(b - a)/2\pi$.

We shall also use the same generic descriptive phrase here as we did in Section 3.2: the circle $C$ is an instance of what is called a *one-dimensional manifold without boundary*. "One-dimensional" because one parameter serves to locate each point on $C$, and "without boundary" because $C$ has no boundary points. Put another way, every point on $C$ is an *interior point*: given any point $p$ on the circle $C$, you can construct an open interval (piece of arc) centered on $p$ which is entirely contained in $C$. (In fact, any interval centered at $p$ will lie entirely in $C$.) In contrast, no open interval centered at the point

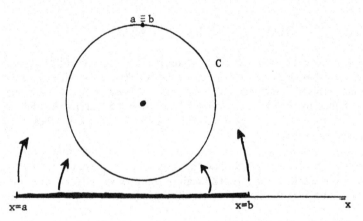

**Figure 4.1** Circle $C$ of radius $(b - a)/2\pi$ obtained by bending round the interval $[a, b]$ and identifying $x = a$ with $x = b$.

$x = a$ will lie entirely within the interval $[a, b]$. The left-hand half of that open interval will always be outside $[a, b]$. Thus the point $x = a$ is not an interior point of $[a, b]$; it is a boundary point. Similarly for the point $x = b$.

Here is a more homely metaphor to point up the difference between the interval $[a, b]$ and the circle $C$: you can walk round the circle without ever meeting an obstruction, but if you walk left on the interval you will hit the point $x = a$, and if you walk right, $x = b$.

The presence or absence of a boundary affects the definition of differential operators. Boundary conditions at $x = a$ and $x = b$ enter naturally into the definition of a differential operator in $L^2(a, b)$. In contrast, boundary conditions never apply in $L^2(C)$ because the underlying manifold has no boundary.

The space $L^2(C)$ consists of real- or complex-valued functions defined on the circle $C$ that are square integrable over $C$:

$$\int_C |f(x)|^2 \, dx < \infty \tag{4.5.3}$$

Now "function" always means single-valued function, so any function in $L^2(C)$ has to take only *one* value at the amalgamated point $a \equiv b$. The inner product in $L^2(C)$ is

$$\langle f, g \rangle = \int_C f(x)\overline{g(x)} \, dx \tag{4.5.4}$$

The integrals in (4.5.3) and (4.5.4) are taken counterclockwise once around the circle. In practice, we generally "transplant" functions defined on the interval $a \leqslant x \leqslant b$ to the circle $C$. For such transplanted functions, the integrals (4.5.3) and (4.5.4) can be computed simply as integrals from $x = a$ to $x = b$. This integration corresponds to integration once counterclockwise around the circle starting at the point $a \equiv b$ and returning to the same point. But we can start at any point (Exercise 1).

**Definition 4.13.** By the *one-dimensional Laplacian*, denoted $\Delta$, we shall mean any one of the following four classes of differential operators based on the differential expression $l(y) \equiv -y''$ and acting either in the space $L^2(C)$ or in $L^2(a, b)$:

1. In $L^2(C)$, domain$(\Delta)$ = all functions $y$ compatible with $l(y) \equiv -y''$.

   In $L^2(a, b)$, domain$(\Delta)$ = all functions $y$ compatible with $l(y) \equiv -y''$ and satisfying either the

2. Dirichlet boundary condition $y(a) = y(b) = 0$,
3. Neumann boundary condition $y'(a) = y'(b) = 0$, or
4. Mixed boundary condition $y(a) = y'(b) = 0$ or $y'(a) = y(b) = 0$.

Although we could include other boundary conditions, we shall use our term "one-dimensional Laplacian" only for the three common boundary conditions listed in the definition.

In this chapter we will be doing the one-dimensional case exclusively, so shall sometimes say simply "Laplacian," one-dimensional being understood.

### Theorem 4.14

(a) *The Laplacian $\Delta$ is Hermitian positive semidefinite. With the Dirichlet or the mixed boundary condition, it is also positive definite.*

(b) *The Laplacian has an infinite sequence of eigenvalues*

$$(0 = \lambda_0) < \lambda_1 < \lambda_2 < \cdots < \lambda_n < \lambda_{n+1} < \cdots$$

*such that $\lambda_n \to \infty$ as $n \to \infty$. $\lambda_0 = 0$ is a simple eigenvalue in $L^2(C)$, or in $L^2(a, b)$ with the Neumann boundary condition; in the Dirichlet and mixed cases 0 is not an eigenvalue. In $L^2(C)$, the $\lambda_n$, $n \geq 1$, have multiplicity 2; in all other cases, the eigenvalues are simple.*

(c) *In all cases, the orthogonal eigenfunctions of the Laplacian form an orthogonal basis of the underlying $L^2$ space.*

As for the proof of Theorem 4.14, I shall discuss (a) and (b) in general terms and leave most of the details for Exercises 1 through 4. I won't prove (c), but refer instead to J. R. Higgins' *Completeness and Basis Properties of Sets of Special Functions* (Cambridge, 1977) §§ 4.1 and 2.2. That book conveniently contains proofs that each of the orthogonal families of eigenfunctions that we shall study forms a basis of its respective $L^2$ space, so we shall refer to Higgins for all such proofs. Although we won't go into the proof of (c), we shall discuss in detail its meaning.

Look at (a) of Theorem 4.14. We need to prove that the Laplacian $\Delta$ is Hermitian, that is, $\langle \Delta f, g \rangle = \langle f, \Delta g \rangle$ for all $f, g$ in the domain of $\Delta$. This boils down to integration by parts, as does every proof that a differential operator is Hermitian. You learned integration by parts in your calculus course, and in this course you will get plenty of practice in it. Here are some hints that I find helpful. We have $\langle l(f), g \rangle = -\int f'' g \, dx$. In this integral, write $f'' \, dx = d(f')$, thus get $\int f'' g \, dx = \int g \, d(f')$, which makes it easy to apply the formula for integration by parts. When you integrate by parts, you will get an "integrated term" involving the values of the functions at the upper and lower limits of integration. The integrated term is handled differently in $L^2(C)$ than in $L^2(a, b)$.

First, $L^2(C)$. In this case the integration is over the circle $C$, once around counterclockwise. Although in principle there is no particular preferred starting (and ending) point for this integration, in practice the amalgamated point $a \equiv b$ is always used because of the way we constructed the circle, by bending round the interval $a \leqslant x \leqslant b$. So, for us, functions on the circle are functions that have been transplanted from the interval $[a, b]$, and to make sure that a transplanted function $y$ is compatible with $l(y) \equiv -y''$ we must make sure that at the joined point the necessary conditions of Theorem 4.2 are satisfied; that is, $y$ and $y'$ must be continuous at the single point $a \equiv b$. Thus we must have $y(a) = y(b)$ and $y'(a) = y'(b)$. Of course, $y$ and $y'$ must be continuous at *every* point of the circle. But, in practice, we have to focus our attention on the point $a \equiv b$ because our original function $y$ was defined separately at $x = a$ and $x = b$, and we must match its graph together smoothly when we join $x = a$ to $x = b$. The conditions $y(a) = y(b)$, $y'(a) = y'(b)$ are compatibility conditions. There are no boundary conditions; the circle $C$ has no boundary.

In $L^2(a, b)$ the integrated term is evaluated by using the boundary conditions at the end points of the interval.

In either $L^2(C)$ or $L^2(a, b)$, Theorem 4.8 (or Theorem 4.62) assures us that the integrated term is finite. In Exercise 2 you will supply the details of the proof of part (a) of Theorem 4.14.

As for part (b) of that same theorem, you will do the proof in Exercises 3 and 4 by finding the eigenvalues explicitly in all four cases.

That brings us to part (c) of Theorem 4.14, namely the assertion that, in all cases, the orthogonal family of eigenfunctions of the Laplacian forms an orthogonal basis of the underlying space. Our purpose here is to study the meaning of (c), to highlight some of the more surprising features of this assertion. To fix the ideas, let's focus on one particular example, the Laplacian in $L^2(-1, 1)$ with Dirichlet boundary conditions that we analyzed in Section 4.2 as an operator we called there $M$. There we found that the eigenvalues of $\Delta$ are all simple and are given by $\lambda_n = n^2\pi^2/4$, $n = 1, 2, \ldots$. The corresponding eigenfunctions are

$$\phi_n(x) = \begin{cases} \cos(n\pi x/2), & n = 1, 3, 5, \ldots \\ \sin n\pi x, & n = 2, 4, 6, \ldots \end{cases} \tag{4.5.5}$$

The assertions in Theorem 4.14(c) that the $\phi_n(x)$, $n = 1, 2, \ldots$, comprise an orthogonal basis of $L^2(-1, 1)$ mean that, for *every* $f$ in $L^2(-1, 1)$, we have

$$f(x) = \sum_{n=1}^{\infty} c_n \phi_n(x) \tag{4.5.6}$$

where

$$c_n = \frac{\langle f, \phi_n \rangle}{\langle \phi_n, \phi_n \rangle} = \frac{\int_{-1}^{1} f(x)\phi_n(x)\, dx}{\int_{-1}^{1} \phi_n(x)^2\, dx} \tag{4.5.7}$$

Look at (4.5.6). The series on the right contains the eigenfunctions of $\Delta$ given by (4.5.5); they are all simple trigonometric functions and all satisfy $\phi_n(-1) = \phi_n(1) = 0$, $n = 1, 2, \ldots$. The function on the left is subject *only* to the requirement

$$\int_{-1}^{1} |f(x)|^2 \, dx < \infty \qquad (4.5.8)$$

Such a function can be nondifferentiable, discontinuous, even unbounded as long as (4.5.8) is fulfilled. The values $f(-1)$, $f(1)$ need not be zero; they can be anything. How can we possibly have equality in (4.5.6)?

The answer lies in the meaning attached to the equality (4.5.6). The series on the right in (4.5.6) converges to $f(x)$ in the mean-square metric, which means that

$$\left\| f(x) - \sum_{n=1}^{N} c_n \phi_n(x) \right\|^2 = \int_{-1}^{1} \left| f(x) - \sum_{n=1}^{N} c_n \phi_n(x) \right|^2 dx \to 0 \qquad (4.5.9)$$

as $N \to \infty$. The value of the function at individual points does not affect the value of its integral, so (4.5.6) says nothing, on the face of it, about convergence at individual points. According to (4.5.9), equation (4.5.6) means simply this: if we take $N$ large enough, we can make the mean-square area between the graph of $f(x)$ and that of the $N$th partial sum of the series as small as we wish.

Another point arises in connection with this example. If we had chosen other boundary conditions, we would have obtained other orthogonal bases for $L^2(-1, 1)$. The natural question then arises: Which basis should I use? The answer often comes from the physical problem being modeled. The physics usually dictates the boundary conditions and thus selects the basis. When it is just a matter of expanding a function with respect to a basis, then choose the basis whose first term, or first few terms, most closely resembles the function being expanded. That way you get a better approximation in (4.5.9) with smaller $N$.

A final comment about Theorem 4.14(c). It says that we have "diagonalized" the Laplacian, in the language of Section 4.4. Many interesting applications unfold once you have diagonalized an operator, but we shall not detail these here for the Laplacian. We shall take them up later for Legendre's and Hermite's operator.

We have been dealing here with the one-dimensional Laplacian, our term for a class of Hilbert space operators based on the simplest possible second order linear differential expression, namely $l(y) \equiv -y''$. We close this section with some remarks about the general second order linear differential expression $l(y) \equiv A(x)y'' + B(x)y' + C(x)y$ and Hilbert space operators based on it.

First, it is best to put the general expression in Sturmian form (Exercises 10 and 11 of Section 2.2).

**Theorem 4.15.** *Suppose $A(x)$, $B(x)$, and $C(x)$ are analytic real-valued functions in the (finite or infinite) interval $a < x < b$ and $A(x) > 0$ in $a < x < b$. Then there are functions $p(x)$, $q(x)$, $r(x)$ likewise analytic and real valued in $a < x < b$ such that both $p(x) > 0$ and $r(x) > 0$ in $a < x < b$ and*

$$A(x)y'' + B(x)y' + C(x)y = \frac{1}{r(x)}(p(x)y')' + \frac{q(x)}{r(x)}y \tag{4.5.10}$$

*identically in* $y$.

*Proof.* If we carry out the differentiation on the right side of the displayed equation and equate like terms, we get $p/r = A$, $p'/r = B$, and $q/r = C$. The first two equations give $p'/p = B/A$, where division by $A$ is legitimate as $A > 0$. Solving, we get

$$p = \exp\left(\int \frac{B}{A}dx\right), \qquad r = \frac{1}{A}\exp\left(\int \frac{B}{A}dx\right), \qquad q = \frac{C}{A}\exp\left(\int \frac{B}{A}dx\right)$$

As $B$ and $A$ are real valued, we have $p > 0$, and as $A > 0$, also $r > 0$. As composites of analytic functions, $p$, $q$, and $r$ are also analytic. The implied constant of integration may be taken as zero, as its exponential will cancel in the ratios.

The expression on the right side of (4.5.10) is called the Sturmian form of the expression on the left. Having cast our expression in Sturmian form, we are able to answer immediately a frequently asked question: What $L^2$ space goes with a given differential expression? Answer: An $L^2$ space with weight function $r(x)$. The reason for this choice is direct enough: you want your operator to be Hermitian, and this choice of weight function makes integration by parts work. The choice of interval $[a, b]$ in $L^2(a, b; r(x))$ cannot be read off from the Sturmian form. The physics or engineering problem being modeled will usually determine the interval. In any case, the interval cannot contain a zero of $A(x)$ in its interior.

The Laplacian differential expression arises as the special case of (4.5.10) with $r(x) = 1$, $p(x) = 1$, and $q(x) = 0$. (Actually, we use the negative of (4.5.10), so that our resulting Hilbert space operator will be positive semidefinite.) Using that very special differential expression and assuming a finite interval $[a, b]$, we constructed an operator in the Hilbert space $L^2(a, b)$, the Laplacian, that had the useful and important properties listed in Theorem 4.14: Hermitian positive semidefinite, an increasing sequence of nonnegative eigenvalues tending to $\infty$, and an orthogonal basis of eigenfunctions. The same conclusions obtain under less stringent assumptions on $p(x)$ and $r(x)$, namely that both are strictly positive in the closed finite interval $a \leqslant x \leqslant b$. This can actually be shown to be equivalent to the result for the Laplacian by the Liouville transformation (Exercises 12, 13, and 14 of Section 2.2). Exercises 6 through 9 at the end of this section, which are independent of the earlier exercises from Chapter 2, lead you through a specific calculation of this type, for the operator $L$ with differential expression $l(y) \equiv -y'' - 2y'$ $= -e^{-2x}(e^{2x}y')'$ and domain consisting of all compatible $y$ in $L^2(-1, 1; e^{2x})$ that satisfy $y(-1) = e^2 y'(1)$, $y'(-1) = -e^2 y(1)$.

If either of the functions $p(x)$ or $r(x)$ goes to zero at the end points of the interval $[a, b]$, or if this interval is infinite, we have no general statement like Theorem 4.14. The Legendre, Hermite, Laguerre, and Bessel operators all fall into this second category, and we shall treat them on a case-by-case basis. In the earlier theory of boundary value problems, the former Laplacian-like examples were referred to as "regular" and other instances as "singular." This terminology seems less apt for the Hilbert space approach. The general procedure for defining a Hilbert space differential operator $L$ is first to deduce the restrictions on the domain of $L$ implied by the compatibility condition and then to add either the physically imposed boundary conditions or other necessary conditions to make the operator Hermitian.

Even though we don't have a general result like Theorem 4.14 in the singular case, there is a formal assertion which provides a helpful perspective on when an operator is Hermitian.

**Theorem 4.16.** *Let $a \leqslant x \leqslant b$ be a finite or infinite interval, and let $r(x)$ be a function analytic and positive on $a < x < b$. Consider the space $L^2(a, b; r)$ with inner product*

$$\langle f, g \rangle = \int_a^b f(x)\overline{g(x)}r(x)\, dx$$

*If $L$ is an operator in $L^2(a, b; r)$ with differential expression*

$$l(y) \equiv -\frac{1}{r(x)}\frac{d}{dx}\left(p(x)\frac{dy}{dx}\right) - \frac{q(x)}{r(x)}y$$

*then*

$$\langle L(f), g \rangle - \langle f, L(g) \rangle = p(b)W(b) - p(a)W(a) \qquad (4.5.11)$$

*where*

$$W(x) = W(f, \bar{g})(x) = f(x)\bar{g}'(x) - f'(x)\bar{g}(x) = \begin{vmatrix} f(x) & \bar{g}(x) \\ f'(x) & \bar{g}'(x) \end{vmatrix}$$

*Equation (4.5.11) holds for all functions f, g in $L^2(a, b, r)$ that are compatible with l and for which the indicated limits*

$$p(a)W(a) = \lim_{x \to a+} p(x)W(x) \quad and \quad p(b)W(b) = \lim_{x \to b-} p(x)W(x)$$

*exist.*

The proof employs the usual integration by parts and is omitted.

**Corollary 4.17.** *Same setup as Theorem 4.16. Let L be the Hilbert space operator in $L^2(a, b; r)$ defined by:*

*Differential expression $l(y) \equiv -(1/r)(py')' - (q/r)y$*
*Domain(L) a subspace D of $L^2(a, b; r)$ consisting of functions that are compatible with l and such that*

$$p(a)W(a) = p(b)W(b)$$

*for all f, g in D.*

*Then L is Hermitian.*

The corollary follows immediately from the theorem because

$$\langle L(f), g \rangle - \langle f, L(g) \rangle = p(b)W(b) - p(a)W(a) = 0$$

For the simple differential expression $l(y) \equiv -y''$ on a finite interval, compatibility implies a finite value at the end points (Theorem 4.8). In this case, Corollary 4.17 applies directly—see Exercises 10 through 12.

**Exercises**

1. A circle $C$ of radius $1/2\pi$ is constructed by bending round the interval $0 \leqslant x \leqslant 1$ and joining the point $x = 0$ to the point $x = 1$. Define a function $f(x)$ on $C$ by $f(x) = x$, $0 < x < 1$, and $f(0) = f(1) = \frac{1}{2}$. The inner product of this $f(x)$ with the constant function $g(x) = 1$ is

$$\langle f(x), 1 \rangle = \int_C f(x)\, dx \tag{1}$$

   (a) By integrating from $x = 0$ to $x = 1$, compute the integral (1).
   (b) Compute the integral (1) by starting at $x = \frac{1}{2}$, integrating once counterclockwise around $C$, and returning to $x = \frac{1}{2}$.
   (c) Explain why you get the same answer in (a) and (b).
2. (a) Show that, if $f$ and $g$ are compatible with $l(y) \equiv -y''$ on $L^2(a, b)$, then

$$\int_a^b l(f)\bar{g}\, dx = f'(a)\bar{g}(a) - f'(b)\bar{g}(b) + \int_a^b f'\overline{g'}\, dx \tag{1}$$

   where the values of $f$ and $f'$ in the integrated terms are finite.

(b) Using (1), show that if $f$ is compatible with $l$, then $\int_a^b |f'|^2\, dx$ is finite.

(c) Using (a) and (b), show that $\langle \Delta f, g \rangle = \langle f', g' \rangle$ in all four cases of Definition 4.13; then complete the proof of Theorem 4.14(a).

3. As in the text, construct a circle $C$ of radius $r = (b-a)/2\pi$ by bending round the interval $a \leqslant x \leqslant b$ and identifying the point $x = a$ with the point $x = b$. Consider the Laplacian $\Delta$ in $L^2(C)$.

(a) Show that $\lambda_0 = 0$ is a simple eigenvalue of $\Delta$, and $y = \frac{1}{2}$ spans the corresponding eigenspace.

(b) Show that $\lambda_n = n^2/r^2$, $n = 1, 2, \ldots$, are all the remaining eigenvalues. Show that each has multiplicity two, and find a pair of real eigenfunctions spanning each corresponding eigenspace.

[Hint: Follow the steps in Section 4.2. Compare your answer with a series from Section 3.4.2.]

4. Find the eigenvalues $\lambda_n$ of the Laplacian in $L^2(a, b)$ under

(a) Dirichlet boundary conditions. [Ans.: $\lambda_n = n^2\pi^2/(b-a)^2$, $n = 1, 2, \ldots$.]

(b) Neumann boundary conditions.

(c) Mixed boundary conditions.

Using the results you have obtained in Exercises 3 and 4, prove the first sentence of Theorem 4.14(b).

5. Expand the function $f(x) = 1 - |x|$ with respect to the orthogonal basis (4.5.5). Does this function belong to the domain of the Laplacian?

Exercises 6 through 9 below represent one project and must be done in sequence. This block of exercises deals with the operator $L$ in the space $L^2(-1, 1; e^{2x})$ with differential expression $l(y) \equiv -y'' - 2y'$ and boundary conditions $y(-1) = e^2 y'(1)$, $y'(-1) = -e^2 y(1)$. The purpose of this sequence of exercises is to show that this operator $L$ is equivalent to an operator $M$ in $L^2(-1, 1)$ with simple differential expression $m(z) \equiv -z''$ and, by using this equivalence, to show how we might find the eigenvalues and eigenfunctions of $L$. Although this equivalence is essentially the "Liouville transformation" of Exercises 12, 13, and 14 of Section 2.2, this set of exercises is self-contained and does not rely on those earlier ones.

6. Consider the Hilbert space $L^2(-1, 1; e^{2x})$ with its inner product $\langle \cdot, \cdot \rangle$. Likewise consider the Hilbert space $L^2(-1, 1)$ and denote its inner product $[\cdot, \cdot]$. Recall that

$$\langle f, g \rangle = \int_{-1}^{+1} f\bar{g} e^{2x}\, dx \qquad \text{and} \qquad [f, g] = \int_{-1}^{+1} f\bar{g}\, dx$$

(a) Let $z = e^x y$. Prove that $y$ belongs to $L^2(-1, 1; e^{2x})$ if and only if $z$ belongs to $L^2(-1, 1)$.

(b) Show that if $u = e^x f$ and $v = e^x g$ then

$$\langle f, g \rangle = [u, v]$$

7. Let $L$ be the operator on $L^2(-1, 1; e^{2x})$ described before Exercise 6, and let $M$ be the operator on $L^2(-1, 1)$ with differential expression $m(z) \equiv -z''$ and domain consisting of compatible functions $z$ such that $z(-1) = z'(1) - z(1)$ and $z'(-1) = z'(1) - 2z(1)$.

(a) Show that if $z = e^x y$, then $e^{-x} m(z) = l(y) - y$ and $e^x l(y) = m(z) + z$.

(b) Deduce from the formulas in (a) that $y$ is compatible with $l$ if and only if $z = e^x y$ is compatible with $m$. That is, prove both these implications:

    (i) if both $y$ and $l(y)$ belong to $L^2(-1, 1; e^{2x})$, then both $z = e^x y$ and $m(z)$ belong to $L^2(-1, 1)$.

    (ii) if both $z$ and $m(z)$ belong to $L^2(-1, 1)$, then both $y = e^{-x} z$ and $l(y)$ belong to $L^2(-1, 1; e^{2x})$.

(c) Show that if $y$ satisfies $y(-1) = e^2 y'(1)$, $y'(-1) = -e^2 y(1)$, then $z = e^x y$ satisfies $z(-1) = z'(1) - z(1)$, $z'(-1) = z'(1) - 2z(1)$. Show also the converse.

(d) From (b) and (c) deduce that $y$ belongs to the domain of $L$ if and only if $z = e^x y$ belongs to the domain of $M$.

8. If $u = e^x f$ and $v = e^x g$, where $f$ and $g$ belong to the domain of $L$, show that $[M(u), v] = \langle L(f), g \rangle - \langle f, g \rangle$ and that $[u, M(v)] = \langle f, L(g) \rangle - \langle f, g \rangle$. From these formulas, and your knowledge that $M$ is Hermitian, deduce that $L$ is Hermitian.

9. Prove that if $\lambda$ is an eigenvalue for $L$ with corresponding eigenfunction $y$, then $\lambda - 1$ is an eigenvalue for $M$ with corresponding eigenfunction $z = e^x y$.

Show conversely that if $\lambda$ is an eigenvalue for $M$ with corresponding eigenfunction $z$, then $\lambda + 1$ is an eigenvalue for $L$ with corresponding eigenfunction $y = e^{-x} z$.

Summarizing the results of Exercises 6 through 9: We have shown that the substitution $z = e^x y$ effects a complete equivalence between these two operators:

(I) The Hermitian operator $L$ acting in the Hilbert space $L^2(-1, 1; e^{2x})$ which has

Differential expression $l(y) \equiv -y'' - 2y'$
Domain consisting of functions $y$ that are compatible with $l$ and satisfy $y(-1) = e^2 y'(1)$, $y'(-1) = -e^2 y(1)$

(II) The Hermitian operator $M$ acting in the Hilbert space $L^2(-1, 1)$ which has

Differential expression $m(z) \equiv -z''$
Domain consisting of functions $z$ that are compatible with $m$ and satisfy
$\quad z(-1) = z'(1) - z(1)$, $z'(-1) = z'(1) - 2z(1)$

Thus we can find the eigenvalues and eigenfunctions of $L$ by first finding those of $M$ and then transforming back via Exercise 9. This would seem to provide a considerable advantage, as $M$ has a very simple differential expression and operates in an $L^2$ space that has weight function 1.

However, there is a catch! The somewhat involved boundary conditions for $M$ make its eigenvalue problem quite complicated. If you like, you may do the eigenvalue problem for $M$ as a project. You will find that $M$ has exactly one negative eigenvalue and an infinite sequence $\lambda_n$, $n = 1, 2, \ldots$, of positive eigenvalues such that $\lambda_n \approx n^2\pi^2/4$ for large $n$. All eigenspaces are one-dimensional.

Exercises 10 through 13 concern the operator $L$ in $L^2(a, b)$ with differential expression $l(y) \equiv -y''$ and domain $D$ consisting of functions $y$ that are compatible with $l$ and satisfy the boundary conditions listed. In each case, you are to apply the criterion of Corollary 4.17. The symbols $\alpha, \beta$, etc. represent constants.

10. Boundary conditions $y'(a) = \alpha y(a)$, $y'(b) = \beta y(b)$. Prove that if $\alpha$ and $\beta$ are real (i.e., $\bar{\alpha} = \alpha$ and $\bar{\beta} = \beta$), then $L$ is Hermitian.
11. Boundary conditions $y(a) = \alpha y'(a)$, $y(b) = \beta y'(b)$. Prove that if $\alpha$ and $\beta$ are real, then $L$ is Hermitian.
12. Boundary conditions $y(b) = \alpha y(a)$, $y'(b) = \beta y'(a)$. Prove that if $\alpha\bar{\beta} = 1$, then $L$ is Hermitian. (Note that if $\alpha\bar{\beta} = 1$, then $1 = \bar{1} = \overline{\alpha\bar{\beta}} = \bar{\alpha}\beta$.)
13. Boundary conditions $y'(a) = \alpha y(b)$, $y'(b) = \beta y(a)$. Prove that if $\alpha + \bar{\beta} = 0$, then $L$ is Hermitian.

## 4.6   LEGENDRE'S OPERATOR AND ITS EIGENFUNCTIONS, THE LEGENDRE POLYNOMIALS

**Definition 4.18.** Consider the Hilbert space $L^2(-1, 1)$ with inner product

$$\langle f, g \rangle = \int_{-1}^{+1} f(x)\overline{g(x)} \, dx$$

The following operator $L$, acting in $L^2(-1, 1)$, with differential expression $l(y) \equiv -(d/dx)((1 - x^2) \, dy/dx)$ and domain consisting of functions that are compatible with $l$ and satisfy the boundary conditions that $y(-1)$ and $y(1)$ are both finite, is called *Legendre's operator*.

The form of the differential expression for $L$ is a special case of that considered in Section 4.5. There we considered the general form

$$l(y) \equiv -\frac{1}{r(x)}\left[\frac{d}{dx}\left(p(x)\frac{dy}{dx}\right) + q(x)y\right]$$

This becomes Legendre's differential expression if we set $r(x)=1$, $p(x)=1-x^2$, and $q(x)=0$.

Consider next the specification in Definition 4.18 of the domain of $L$. Recall what it means for a function $y$ to be compatible with $l$. First, $y$ must itself belong to the Hilbert space $L^2(-1,1)$. That is, we must have

$$\int_{-1}^{+1} |y(x)|^2\, dx < \infty$$

Second, $l(y)$ must also belong to $L^2(-1,1)$. That is, we must have

$$\int_{-1}^{+1} \left|\frac{d}{dx}\left((1-x^2)\frac{dy}{dx}\right)\right|^2 dx < \infty$$

These two conditions constitute the compatibility requirement. The domain is further restricted by the boundary conditions that the values $y(1)$ and $y(-1)$ are finite. Remember that when we write $y(1)$ we mean the limit of $y(x)$ as $x$ tends to 1 from the left. Similarly, $y(-1)$ means the limit of $y(x)$ as $x$ tends to $-1$ from the right.

**Lemma 4.19.** *If $y(x)$ is compatible with Legendre's differential expression, then $(1-x^2)y'(x)$ has finite limits as $x \to \pm 1$. If, additionally, the values $y(1)$ and $y(-1)$ are both finite, then $\lim_{x\to\pm 1}(1-x^2)y'(x)=0$.*

The proof of Lemma 4.19, for which you are not responsible, is given in 4.65 of Appendix 4.A. However, you will need the facts stated in Lemma 4.19 to prove the vitally important result that Legendre's operator $L$ is Hermitian. The key to this proof is the following consequence of Lemma 4.19: If $f, g$ belong to the domain of Legendre's operator, then

$$\lim_{x\to\pm 1}(1-x^2)f'(x)\overline{g(x)} = \left[\lim_{x\to\pm 1}(1-x^2)f'(x)\right]\left[\lim_{x\to\pm 1}\overline{g(x)}\right] = 0 \times \text{finite} = 0$$

**Theorem 4.20.** *Legendre's operator is Hermitian.*

Theorem 4.20 means that $\langle L(f), g\rangle = \langle f, L(g)\rangle$ for all functions $f, g$ in the domain of Legendre's operator $L$. The inner product referred to here, as in this entire section, is that of the Hilbert space $L^2(-1, 1)$:

$$\langle f, g \rangle = \int_{-1}^{+1} f(x)\overline{g(x)}\, dx$$

Bear this in mind, as we shall not keep repeating it.

Exercise 3 outlines a direct proof that Legendre's operator is Hermitian. The direct proof also provides the additional information that Legendre's operator is *positive semidefinite*. From Theorem 4.20 and this latter fact, we deduce:

**Theorem 4.21.** *The eigenvalues of Legendre's operator are all $\geqslant 0$, and eigenfunctions belonging to different eigenvalues are orthogonal.*

This is an application of Theorems 4.6 and 4.7.

But note the crucial role played by the boundary conditions! If we had defined a different operator by keeping the same differential expression but broadening the domain to include all compatible functions, we could not be sure that the integration by parts implicit in the proof of Theorem 4.20 is valid and thus could not be sure that the resulting operator was Hermitian. Then there would be no justification for the important conclusion of Theorem 4.21, namely that eigenfunctions belonging to different eigenvalues are orthogonal. So we see once again that the specification of the domain is a crucial aspect of the definition of a differential operator in Hilbert space.

One point worth noting here is this: the domain of Legendre's operator contains all polynomials. First, any polynomial belongs to $L^2(-1, 1)$ because it is bounded on $-1 \leqslant x \leqslant 1$ and therefore certainly square integrable there. And, after we apply Legendre's expression $l(y) \equiv -((1-x^2)y')'$ to a polynomial $y$, the result is another polynomial which again belongs to $L^2(-1, 1)$. Hence any polynomial is compatible with Legendre's expression. The boundary conditions, namely that $y(-1)$ and $y(1)$ are both finite, are obviously satisfied by any polynomial $y$. Thus the domain of Legendre's operator contains all polynomials.

For a discussion of some more subtle aspects of the definition of the domain of Legendre's operator, see Appendix 4.A.

In Section 2.8 of Chapter 2 we used the method of undetermined coefficients to solve Legendre's equation

$$(1 - x^2)y'' - 2xy' + n(n + 1)y = 0 \tag{4.6.1}$$

where $n$ is a nonnegative integer. As we know, equation (4.6.1), like any second order differential equation, has two linearly independent solutions. In Chapter 2 we showed that, for each nonnegative integer $n$, one of the solutions of Legendre's equation was a polynomial $P_n(x)$ of degree $n$

containing only even powers of $x$ when $n$ is even and only odd powers when $n$ is odd (Theorem 2.13, Section 2.8, Chapter 2). The $P_n(x)$ are *Legendre's polynomials*. We list the first seven below.

*The first seven Legendre polynomials*

$P_0(x) = 1$        $P_4(x) = \frac{1}{8}(35x^4 - 30x^2 + 3)$
$P_1(x) = x$        $P_5(x) = \frac{1}{8}(63x^5 - 70x^3 + 15x)$
$P_2(x) = \frac{1}{2}(3x^2 - 1)$      $P_6(x) = \frac{1}{16}(231x^6 - 315x^4 + 105x^2 - 5)$
$P_3(x) = \frac{1}{2}(5x^3 - 3x)$

Legendre's polynomials are normalized so that the coefficient of the leading term in $P_n(x)$ is $(2n)!/2^n(n!)^2$. We shall see the reason for this choice in the next section.

Using the fact that $(1-x^2)y'' - 2xy' = ((1-x^2)y')'$, we can write Legendre's equation (4.6.1) as

$$l(y) = n(n+1)y \tag{4.6.2}$$

where $l(y) = -((1-x^2)y')'$ is Legendre's differential expression. Equation (4.6.2) is equivalent to equation (4.6.1) and is therefore also satisfied by the $n$th Legendre polynomial $y = P_n(x)$. Since $P_n(x)$ is contained in the domain of Legendre's operator, we can rewrite equation (4.6.2) as $L(P_n) = n(n+1)P_n$ where $L$ is Legendre's operator. Thus we have

**Theorem 4.22.** *Legendre's operator $L$ has the eigenvalues $n(n+1)$, $n = 0, 1, 2, \ldots$. The Legendre polynomial $P_n(x)$ is an eigenfunction belonging to the eigenvalue $\lambda = n(n+1)$; i.e.*

$$L(P_n) = n(n+1)P_n, \qquad n = 0, 1, 2, \ldots$$

We shall see shortly that the integers $n(n+1) = 0, 2, 6, 12, 20, \ldots$ are *all* the eigenvalues of Legendre's operator and that each of these eigenvalues is simple (Theorem 4.27).

As Legendre's operator is a Hermitian operator in $L^2(-1, 1)$, its eigenfunctions that belong to different eigenvalues will be orthogonal with respect to the inner product of $L^2(-1, 1)$. Hence:

**Corollary 4.23.** *The Legendre polynomials $P_n(x)$, $n = 0, 1, 2, \ldots$, form an orthogonal family in the Hilbert space $L^2(-1, 1)$, i.e.,*

$$\langle P_m, P_n \rangle = \int_{-1}^{+1} P_m(x)P_n(x)\,dx = 0 \qquad \text{when } m \neq n$$

In Section 4.8 we shall show that the norm of the Legendre polynomials $P_n(x)$ is

$$\langle P_n(x), P_n(x) \rangle = \int_{-1}^{+1} (P_n(x))^2 \, dx = \frac{2}{2n+1}, \qquad n = 0, 1, 2, \ldots \qquad (4.6.3)$$

a result that we shall take for granted now.

As the $P_n(x)$ form an orthogonal family with respect to the inner product $\langle f, g \rangle = \int_{-1}^{+1} f\bar{g} \, dx$, our basic approximation theorem (Theorem 3.10 et seq.) yields the following.

**Theorem 4.24.** *Given f in* $L^2(-1, 1)$*, the linear combination of the first* $N + 1$ *Legendre polynomials,*

$$c_0 P_0 + c_1 P_1 + \cdots + c_N P_N$$

*that offers the best mean-square approximation to f in the* $L^2(-1, 1)$ *metric is that obtained by choosing the* $c_n$ *to be the Fourier coefficients of f with respect to the* $P_n$:

$$c_n = \frac{\langle f, P_n \rangle}{\langle P_n, P_n \rangle} = \frac{2n+1}{2} \int_{-1}^{+1} f(x) P_n(x) \, dx, \qquad n = 0, 1, 2, \ldots \qquad (4.6.4)$$

By "best mean-square approximation in the $L^2(-1, 1)$ metric" we mean this: When the $c_n$ are chosen as the Fourier coefficients, then the quantity

$$\| f - (c_0 P_0 + c_1 P_1 + \cdots + c_n P_n) \|^2$$
$$= \int_{-1}^{+1} |f - (c_0 P_0 + c_1 P_1 + \cdots + c_n P_n)|^2 \, dx$$

is an absolute minimum. Any alteration of the $c_n$ will increase this mean-square error.

Our next result we state without proof (see Higgins's book cited earlier).

**Theorem 4.25.** *The Legendre polynomials* $P_n(x)$*,* $n = 0, 1, 2, \ldots$*, form an orthogonal basis of* $L^2(-1, 1)$.

Thus, given any function $f(x)$ in $L^2(-1, 1)$, we have

$$f(x) = \sum_{n=0}^{\infty} c_n P_n(x) \qquad (4.6.5)$$

where the coefficients $c_n$ are the Fourier coefficients of $f(x)$ with respect to the Legendre polynomials as given in equation (4.6.4).

The equality in equation (4.6.5) holds in the sense of mean-square convergence:

$$\left\| f(x) - \sum_{n=0}^{N} c_n P_n(x) \right\| = \left( \int_{-1}^{+1} \left| f(x) - \sum_{n=0}^{N} c_n P_n(x) \right|^2 dx \right)^{1/2} \to 0 \text{ as } N \to \infty$$

We refer to (4.6.5) as the *Legendre series*, or *Fourier-Legendre series*, of $f(x)$.

The coefficients in the Legendre series (4.6.5) are uniquely determined in the following sense. Given a function $f(x)$ in $L^2(-1, 1)$, suppose we secure somehow an expansion of $f(x)$ in terms of the Legendre polynomials, $f(x) = \sum a_n P_n(x)$, where equality holds in the sense of mean-square convergence. Then the coefficients $a_n$ in this expansion are necessarily the Fourier coefficients $c_n$ given by equation (4.6.4). Thus there is only one Legendre expansion. You are asked to prove this uniqueness in Exercise 4.

For example, consider the constant function $f(x) = 1$. Refer back to the listing of the Legendre polynomials given earlier in this section. We see that $P_0(x) = 1$, so $1 = P_0(x)$ is the Legendre series of $f(x) = 1$. We don't have to do the integration specified in equation (4.6.4). Likewise, $x = P_1(x)$ is the Legendre series for $f(x) = x$. To get the series for $f(x) = x^2$ note that $\frac{2}{3}P_2(x) = x^2 - \frac{1}{3}$, so $x^2 - \frac{2}{3}P_2(x) = \frac{1}{3} = \frac{1}{3}P_0$, so

$$x^2 = \tfrac{2}{3}P_2(x) + \tfrac{1}{3}P_0(x)$$

is the Legendre series for $f(x) = x^2$. Higher powers of $x$ are dealt with similarly. If $f(x)$ is a polynomial, then its Legendre series (4.6.5) has only finitely many terms. (How many?)

Consider next the Legendre series for $f(x) = \sin \pi x$. The coefficients are given by

$$c_n = \frac{2n + 1}{2} \int_{-1}^{+1} \sin \pi x P_n(x) \, dx$$

The Legendre polynomials $P_n(x)$ are even when $n$ is even, odd when $n$ is odd. The function $\sin \pi x$ is odd, so when $n$ is even, the integrand is (odd) × (even) = odd, thus the integral is zero. Hence $c_n = 0$ for $n = 0, 2, 4, \ldots$. When $n$ is odd the integrand is even, so

$$c_n = (2n + 1) \int_{0}^{1} \sin \pi x P_n(x) \, dx, \qquad n \text{ odd}$$

These integrals are evaluated in terms of the integrals $\int x^n \sin \pi x \, dx$. A double integration by parts yields

$$\int_{0}^{1} x^n \sin \pi x \, dx = \frac{1}{\pi} - \frac{n(n-1)}{\pi^2} \int_{0}^{1} x^{n-2} \sin \pi x \, dx \qquad (4.6.6)$$

Start the calculation with a straight integration by parts for $n = 1$:

$$\int_0^1 x \sin \pi x \, dx = \frac{1}{\pi^2} \sin \pi x - \frac{x}{\pi} \cos \pi x \Big|_0^1 = \frac{1}{\pi}$$

To evaluate the integrals when $n = 3$, use equation (4.6.6):

$$\int_0^1 x^3 \sin \pi x \, dx = \frac{1}{\pi} - \frac{(3)(3-1)}{\pi^2} \int_0^1 x \sin \pi x \, dx = \frac{1}{\pi} - \frac{6}{\pi^3}$$

Repeat for $n = 5$:

$$\int_0^1 x^5 \sin \pi x \, dx = \frac{1}{\pi} - \frac{(5)(5-1)}{\pi^2} \int_0^1 x^3 \sin \pi x \, dx = \frac{1}{\pi} - \frac{20}{\pi^3} + \frac{120}{\pi^5}$$

Hence

$$c_1 = (2 \cdot 1 + 1) \int_0^{+1} P_1(x) \sin \pi x \, dx = 3 \int_0^1 x \sin \pi x \, dx = \frac{3}{\pi}$$

$$c_3 = (2 \cdot 3 + 1) \int_0^1 P_3(x) \sin \pi x \, dx = 7 \int_0^1 (1/2)(5x^3 - 3x) \sin \pi x \, dx$$

$$= \frac{35}{2} \int_0^1 x^3 \sin \pi x \, dx - \frac{21}{2} \int_0^1 x \sin \pi x \, dx$$

$$= \frac{35}{2} \left( \frac{1}{\pi} - \frac{6}{\pi^3} \right) - \frac{21}{2} \left( \frac{1}{\pi} \right) = \frac{7}{\pi} - \frac{105}{\pi^3}$$

and

$$c_5 = (2 \cdot 5 + 1) \int_0^1 P_5(x) \sin \pi x \, dx = 11 \int_0^1 (1/8)(63x^5 - 70x^3 + 15x) \sin \pi x \, dx$$

$$= \frac{(11)(63)}{8} \int_0^1 x^5 \sin \pi x \, dx - \frac{(11)(70)}{8} \int_0^1 x^3 \sin \pi x \, dx + \frac{(11)(15)}{8} \int_0^1 \sin \pi x \, dx$$

$$= \frac{(11)(63)}{8} \left( \frac{1}{\pi} - \frac{20}{\pi^3} + \frac{120}{\pi^5} \right) - \frac{(11)(70)}{8} \left( \frac{1}{\pi} - \frac{6}{\pi^3} \right) + \frac{(11)(15)}{8} \left( \frac{1}{\pi} \right)$$

$$= \frac{11}{\pi} - \frac{1155}{\pi^3} + \frac{10,395}{\pi^5}$$

Therefore the Legendre series for $\sin \pi x$ starts out

$$\sin \pi x = \frac{3}{\pi} P_1(x) + \left( \frac{7}{\pi} - \frac{105}{\pi^3} \right) P_3(x) + \left( \frac{11}{\pi} - \frac{1155}{\pi^3} + \frac{10,395}{\pi^5} \right) P_5(x) + \cdots$$

$$(4.6.7)$$

**Table 4.1**  Tables of the $P_n(x)$, $2 \leqslant n \leqslant 5$

| $x$ | $P_2(x)$ | $P_3(x)$ | $P_4(x)$ | $P_5(x)$ |
|-----|----------|----------|----------|----------|
| 0   | −0.5     | 0        | 0.375    | 0            |
| 0.1 | −0.485   | −0.1475  | 0.3379375 | 0.178828750 |
| 0.2 | −0.44    | −0.28    | 0.232    | 0.30752      |
| 0.3 | −0.365   | −0.3825  | 0.0729375 | 0.345386250 |
| 0.4 | −0.26    | −0.44    | −0.113   | 0.27064      |
| 0.5 | −0.125   | −0.4375  | −0.2890625 | 0.089843750 |
| 0.6 | 0.04     | −0.36    | −0.408   | −0.15264     |
| 0.7 | 0.235    | −0.1925  | −0.4120625 | −0.365198750 |
| 0.8 | 0.46     | 0.08     | −0.233   | −0.39952     |
| 0.9 | 0.715    | 0.4725   | 0.2079375 | −0.04114125 |
| 1.0 | 1.0      | 1.0      | 1.0      | 1.0          |

On the HP-15C one can evaluate the Legendre polynomials via subroutines. The first two, $P_0 = 1$ and $P_1 = x$, are no problem. Programs to evaluate $P_2$ through $P_5$ are listed below. Tabulated values and graphs are given in Table 4.1 and Figure 4.2.

$P_2(x) = \frac{1}{2}(3x^2 - 1)$

```
g P/R
f LBL 2
g x²
3 × 1 −
2 ÷
g RTN
g P/R
```

$P_3(x) = \frac{1}{2}(5x^3 - 3x) = (x/2)(5x^2 - 3)$

```
g P/R
f LBL 3
ENTER
gx²
5 × 3 −
× 2 ÷
g RTN
g P/R
```

$P_4(x) = \frac{1}{8}(35x^4 - 30x^2 + 3)$
$= \frac{1}{8}(x^2(35x^2 - 30) + 3)$

```
g P/R
f LBL 4
g x²
ENTER, ENTER
35 × 30 −
× 3 +
8 ÷
g RTN
g P/R
```

$P_5(x) = \frac{1}{8}(63x^5 - 70x^3 + 15x)$
$= (x/8)(x^2(63x^2 - 70) + 15)$

```
g P/R
f LBL 5
ENTER
g x²
ENTER, ENTER
63 × 70 −
× 15 +
× 8 ÷
g RTN
g P/R
```

The following program computes the right side of equation (4.6.7).

```
g P/R
f LBL A
STO 0
RCL × 1
RCL 0
GSB 3          evaluates P₃(x)
RCL × 2
+
STO 4          the subroutine for P₅ empties the stack
RCL 0
GSB 5          evaluates P₅(x)
RCL × 3
RCL + 4
g RTN
g P/R
```

To run this program enter $3/\pi = 0.954929658$ in storage register 1, $7/\pi - 105/\pi^3$ in register 2, and the third coefficient in register 3. The tabulated values are given in Table 4.2. As the fit is within $\frac{1}{2}\%$ over most of the range, a graphical comparison would show one curve. We call the right side of equation (4.6.7) the six-term partial sum of the Legendre series, even though three of the terms are zero. We are working in $L^2(-1, 1)$, so $x$ is restricted to the range $-1 \leqslant x \leqslant 1$. Outside this range, equation (4.6.7) is meaningless. As the functions involved are odd, we need only compute in the range $0 \leqslant x \leqslant 1$.

We have also included in Table 4.2 the six-term power series approximation: $\sin \pi x \cong \pi x - (\pi x)^3/3! + (\pi x)^5/5!$ (omitting the three zero terms). Notice how the six-term partial sum of the power series fits very accurately for small $x$ but fits very poorly near $x = 1.0$. But the six-term partial sum of the Legendre series fits quite accurately over the entire range $0 \leqslant x \leqslant 1$. You see, the coefficients of the power series are determined from the value of $f(x)$ and its derivatives *at the single point* $x = 0$. Therefore one should expect it to be the more accurate, the smaller $x$. The Legendre polynomial approximation $S_6(x)$, on the other hand, makes an absolute minimum the *average over the whole interval* $-1 \leqslant x \leqslant 1$ of $|f(x) - S_6(x)|^2$. Thus we might say that the Legendre polynomials offer a *global* approximation to $f(x)$ on $-1 \leqslant x \leqslant 1$, while the power series gives a *local* approximation at $x = 0$. Both the power series and the Legendre series approximations are odd 5th degree polynomials of the form $ax + bx^3 + cx^5$.

Associated with the orthogonal expansion (4.6.5), as with any orthogonal expansion, there is a Parseval equality (refer to Section 3.8, Chapter 3). The

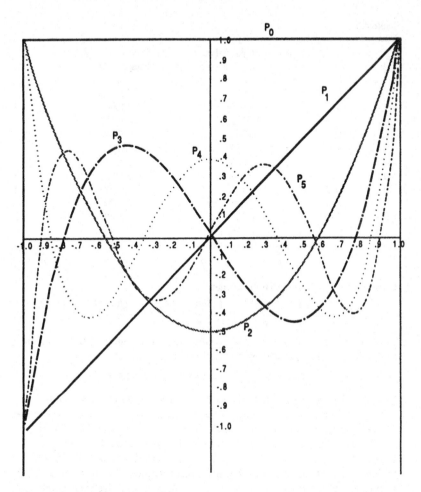

**Figure 4.2** Legendre polynomials.

Parseval equality corresponding to (4.6.5) is

$$\langle f, f \rangle = 2 \sum_{n=0}^{\infty} \frac{|c_n|^2}{2n + 1} \tag{4.6.8}$$

(see Exercise 8). For $f(x) = \sin \pi x$ we have

$$\langle \sin \pi x, \sin \pi x \rangle = \int_{-1}^{+1} (\sin \pi x)^2 \, dx = 2 \int_0^1 \sin^2 \pi x \, dx = 1$$

**Table 4.2**

| $x$ | $\sin \pi x$ | Six-term partial sum of the Legendre series | Six-term power series for $\sin \pi x$ |
|-----|--------------|---------------------------------------------|----------------------------------------|
| 0   | 0.0          | 0.0                                         | 0.0                                    |
| 0.1 | 0.3090       | 0.3055                                      | 0.3090                                 |
| 0.2 | 0.5878       | 0.5827                                      | 0.5878                                 |
| 0.3 | 0.8090       | 0.8052                                      | 0.8091                                 |
| 0.4 | 0.9510       | 0.9509                                      | 0.9520                                 |
| 0.5 | 1.0          | 1.0039                                      | 1.005                                  |
| 0.6 | 0.9510       | 0.9565                                      | 0.9670                                 |
| 0.7 | 0.8090       | 0.8113                                      | 0.8552                                 |
| 0.8 | 0.5878       | 0.5837                                      | 0.7030                                 |
| 0.9 | 0.3090       | 0.3032                                      | 0.5660                                 |
| 1.0 | 0.0          | 0.0160                                      | 0.5240                                 |

Combining (4.6.7) and (4.6.8), we get

$$\frac{1}{2} \cong \frac{1}{3}\left(\frac{3}{\pi}\right)^2 + \frac{1}{7}\left(\frac{7}{\pi} - \frac{105}{\pi^3}\right)^2 + \frac{1}{11}\left(\frac{11}{\pi} - \frac{1155}{\pi^3} + \frac{10,395}{\pi^5}\right)^2$$

for the (approximate) Parseval equality for $f(x) = \sin \pi x$. If we multiply through this last equation by $\pi^{10}$ and combine terms, we get

$$\pi^{10} - 42\pi^8 + 5040\pi^6 - 287,280\pi^4 + 4,365,900\pi^2 - 19,646,550 \cong 0$$

The actual value on the left side is 3.47. We have found a polynomial $f(x)$ with integer coefficients, leading coefficient 1, such that $f(\pi) \cong 0$. The number $\pi$ is transcendental and cannot be a root of any polynomial with integer coefficients. Roots of such polynomials are called algebraic numbers. Our polynomial $f(x) = x^{10} - 42x^8 + 5040x^6 -$ etc. has the root 3.141582448..., which is thus an algebraic number that approximates $\pi$ to within 0.0003%. By taking more and more terms in our series we can find algebraic numbers of higher and higher degree that approximate $\pi$ to greater and greater accuracy.

We next deduce a very important and useful consequence of fact that the Legendre polynomials form an orthogonal basis of $L^2(-1, 1)$ (Theorem 4.25).

**Theorem 4.26.** *If the function $f(x)$ belongs to $L^2(-1, 1)$ and is orthogonal to every $P_n(x)$, $n = 0, 1, 2, \ldots$, then $f = 0$.*

*Proof.* Since $f(x)$ belongs to $L^2(-1, 1)$, it has, by Theorem 4.25, a Legendre series expansion

$$f = \sum_{n=0}^{\infty} c_n P_n, \qquad c_n = \frac{\langle f, P_n \rangle}{\langle P_n, P_n \rangle} = \frac{2n+1}{2} \int_{-1}^{+1} f(x) P_n(x)\, dx$$

which converges in the $L^2(-1, 1)$ metric. By assumption, $\langle f, P_n \rangle = 0$ for all $n$, hence all the coefficients $c_n$ are zero, hence $f = 0$.

Using this result, we can complete Theorem 4.22.

**Theorem 4.27.** *Legendre's operator has no eigenvalues other than the integers* $n(n + 1)$, $n = 0, 1, 2, \ldots$, *and each of these eigenvalues is simple.*

*Proof.* Suppose first that there were an eigenvalue $\lambda$ different from all the $n(n + 1)$. Then there would exist a nonzero function $y$ in the domain of $L$ such that $L(y) = \lambda y$ (where $L$ denotes Legendre's operator). Now use the fact that, for a Hermitian operator, eigenfunctions belonging to different eigenvalues are orthogonal. The function $y$ is an eigenfunction belonging to $\lambda$, and the Legendre polynomials $P_n$ are eigenfunctions belonging to the $n(n + 1)$, $n = 0, 1, 2, \ldots$. Since $\lambda$ is different from all the $n(n + 1)$, $y$ is orthogonal to every $P_n$. By Theorem 4.26, this makes $y$ equal to zero, in contradiction to the assumption that $\lambda$ was an eigenvalue. Hence there is no other eigenvalue; the $n(n + 1)$, $n = 0, 1, 2, \ldots$, form a complete list.

A similar argument shows that each of the eigenvalues $n(n + 1)$ is simple. The details are left for Exercise 15.

In particular, none of the Legendre functions of the second kind (Section 2.8, Exercise 2) are eigenfunctions of Legendre's operator.

One final note on Theorem 4.25, which asserts that the Legendre polynomials form an orthogonal basis of $L^2(-1, 1)$. The Legendre polynomials are, in fact, polynomials and are therefore very nicely behaved as functions. They all lie in the domain of Legendre's operator, which is a proper subspace of $L^2(-1, 1)$; indeed, they are the eigenfunctions of Legendre's operator. It is, therefore, quite remarkable that we have a valid Fourier-Legendre expansion $f = \Sigma c_n P_n$, not only for the $f$ in the domain of Legendre's operator, but for *every* $f$ that satisfies

$$\int_{-1}^{+1} |f(x)|^2 \, dx < \infty$$

no matter how otherwise badly behaved that $f$ might be.

**Exercises**

1. Check that the function

$$f(x) = \ln \frac{1 + x}{1 - x}$$

satisfies $-((1 - x^2)f')' = 0$. Thus $f(x)$ is an eigenfunction for this operator belonging to the eigenvalue $\lambda = 0$. Check also that $g(x) = x$ satisfies

$-((1-x^2)g')' = 2g$. Thus $g(x)$ is an eigenfunction belonging to the eigenvalue $\lambda = 2$. As $0 \neq 2$, these eigenfunctions belong to different eigenvalues, so they must be orthogonal, i.e.,

$$\langle f(x), g(x) \rangle = \int_{-1}^{+1} x \ln \frac{1+x}{1-x} dx = 0$$

But

$$x \ln \frac{1+x}{1-x} \geqslant 0 \qquad \text{on } -1 \leqslant x \leqslant 1$$

so this integral cannot be 0! What is wrong?

2. Does the function $f(x) = \frac{2}{3}|x|^{3/2}$ belong to the domain of Legendre's operator? Give an explicit and careful argument to support your answer.

3. Show that for all functions $f, g$ in the domain of Legendre's operator $L$ we have

$$\langle L(f), g \rangle = \int_{-1}^{1} (1-x^2) f'(x) \bar{g}'(x) dx$$

$$= \langle (1-x^2)^{1/2} f'(x), (1-x^2)^{1/2} g'(x) \rangle$$

Deduce that Legendre's operator is Hermitian and positive semidefinite.

4. Prove that the coefficients $c_n$ in the Legendre series are uniquely determined and are given by equation (4.6.4) of the text. [Hint: Take the inner product of both sides with $P_m$; see Theorem 3.12.]

5. Find the Legendre series for $x^3$. For $x^4$. For $35x^4 - 30x^2 + 3$. Show that the Legendre series for a polynomial of degree $n$ has $n+1$ terms (some of which may be zero).

6. (Continuation) Prove that if $Q(x)$ is a polynomial of degree $\leqslant n-1$, then $\langle Q, P_n(x) \rangle = 0$.

7. Let $L$ be Legendre's operator.
   (a) If $f(x) = 5P_3(x)$, find $L(f)$. [Soln.: $L(f) = L(5P_3) = 5L(P_3) = (5)(3)(3+1)P_3 = 60P_3$.]
   (b) If $f = 2P_3 + 3P_5$, find $L(f)$.
   (c) Using your solution to Exercise 5, find $L(x^3)$ and $L(x^4)$.
   (d) Using the differential expression for $L$, find $L(x^3)$ and $L(x^4)$. Check that your answers agree with those in (c).

8. From the expansion $f = \sum_{n=0}^{\infty} c_n P_n$ derive the Parseval equality

$$\frac{1}{2} \langle f, f \rangle = \sum_{n=0}^{\infty} \frac{|c_n|^2}{2n+1}$$

*Calculating and Graphing Partial Sums of Legendre Series*

In Exercises 9 through 14 you are asked to compute partial sums of various Legendre series and compare them with their functions. The general pattern of these calculations is roughly the same as that used in Section 3.2 of Chapter 3 for trigonometric Fourier series, with one major difference—most scientific calculators don't have buttons for the Legendre polynomials. But the $P_n(x)$ can be calculated via subroutines, as shown in the section, if your calculator has this capability. The illustrative HP-15C program in the section is quite fast. If your calculator doesn't do subroutines, you will need tabular values for the Legendre polynomials, which are included in Table 4.1. Figure 4.2 shows the graphs of $P_0$ through $P_5$.

9. Let $f(x)=1$ for $0<x<1$, $f(x)=-1$ for $-1<x<0$, and $f(-1)=f(0)=f(1)=0$. Find the six-term partial sum of the Legendre series of $f(x)$ and graph it against the function.
10. Find the six-term partial sum of the Legendre series for $f(x)=|x|$, and graph it against the function.
11. Let $f(x)=2x$ for $-\frac{1}{2}\leqslant x\leqslant\frac{1}{2}$, $f(x)=2(1-x)$ for $\frac{1}{2}\leqslant x\leqslant 1$, and $f(x)=2(-1-x)$ for $-1\leqslant x\leqslant -\frac{1}{2}$. Find the six-term partial sum of the Legendre series for $f(x)$, and graph it against the function.
12. Find the six-term partial sum of the Legendre series for $f(x)=\cos\pi x$, and compare the tabulated values against the function. Also calculate the six-term partial sum of the power series for $\cos\pi x$. Compare all three on the same graph. (In the six-term partial sum of the Legendre series, which goes from $P_0$ through $P_5$, only $P_2$ and $P_4$ will have nonzero coefficients. In the six-term partial sum of the power series, which goes from $x^0$ through $x^5$, only three terms are nonzero.)
13. Find the six-term partial sum of the Legendre series for $f(x)=e^{-x^2/2}$, and compare the tabulated values against the function. [Hints: You will need the integral

$$\int_0^1 e^{-x^2/2}\,dx = 0.855624392$$

which can be evaluated using the hard-wired $\int_y^x$ numerical integration program in the HP-15C (optional). Use this, together with the recursion

$$\int_0^1 x^{2n}e^{-x^2/2}\,dx = e^{-1/2} + (2n-1)\int_0^1 x^{2(n-1)}e^{-x^2/2}\,dx$$

which you should derive using integration by parts.)

14. The function $f(x) = \ln((1+x)/(1-x))$ belongs to $L^2(-1, 1)$ (Exercise 5 of
    Section 3.9), so has a Legendre series expansion. Find the six-term partial
    sum of the Legendre series for $f(x) = \ln((1+x)/(1-x))$ and compare the
    tabulated values against the function. Also calculate the six-term partial
    sum of the power series for $\ln((1+x)/(1-x))$. Compare all three on the
    same graph.
    [Hints: First use integration by parts to derive the following antiderivate
    formulas:

$$\int x \ln\left(\frac{1+x}{1-x}\right) dx = \frac{1}{2}(x^2 - 1)\ln\left(\frac{1+x}{1-x}\right) + x$$

$$\int x^3 \ln\left(\frac{1+x}{1-x}\right) dx = \frac{1}{4}(x^4 - 1)\ln\left(\frac{1+x}{1-x}\right) + \frac{x^3}{6} + \frac{x}{2}$$

$$\int x^5 \ln\left(\frac{1+x}{1-x}\right) dx = \frac{1}{6}(x^6 - 1)\ln\left(\frac{1+x}{1-x}\right) + \frac{x^5}{15} + \frac{x^3}{9} + \frac{x}{3}$$

(neglecting additive constants). Next use these formulas to evaluate

$$\int_0^1 x^n \ln\left(\frac{1+x}{1-x}\right) dx$$

for $n = 1, 3, 5$. Then, remembering that $\ln((1+x)/(1-x))$ is odd, evaluate

$$c_n = \frac{\langle f, P_n \rangle}{\langle P_n, P_n \rangle} = (2n+1)\int_0^1 P_n(x)\ln\left(\frac{1+x}{1-x}\right) dx, \qquad n = 1, 3, 5$$

Derive the power series by integrating both series

$$\frac{1}{1-t} = 1 + t + t^2 + t^3 + \cdots$$

$$\frac{1}{1+t} = 1 - t + t^2 - t^3 + \cdots$$

term by term from 0 to $x$, then adding the results.]

The following exercises do not involve calculation.

15. Prove that all the eigenvalues of Legendre's operator $L$ are simple. [Hint:
    Suppose, to the contrary, that one eigenvalue, say $a(a+1)$, has multiplic-
    ity two where $a$ is some fixed nonnegative integer. Then there would exist
    a nonzero function $Q$ in the domain of $L$, linearly independent of $P_a$, such
    that $L(Q) = a(a+1)Q$.]
16. (a) Show that the kernel of Legendre's operator is one-dimensional, and
        find a convenient function that spans it.

    (b) Find $\ker(L-2I)$ and $\ker(L-6I)$, where $L$ is Legendre's operator and $I$ is the identity operator.

    (c) Show that $\ker(L-3I)=0$.

17. If $y$ is a function in the domain of Legendre's operator $L$ such that $L(y)=h$, prove that $\langle h, P_0 \rangle = 0$.

## 4.7  SOLVING OPERATOR EQUATIONS WITH LEGENDRE'S OPERATOR

In the previous lecture we have shown that Legendre's operator $L$ has a family of eigenfunctions, the Legendre polynomials, that form an orthogonal basis of the Hilbert space $L^2(-1,1)$. The existence of such an orthogonal basis for $L$ provides us with a new method for dealing with the operator $L$. The purpose of this section is to explain this new method.

For example, suppose we wish to find $L(x^4)$. We know (Exercise 5 of Section 4.6) that

$$x^4 = \tfrac{1}{5}P_0 + \tfrac{4}{7}P_2 + \tfrac{8}{35}P_4$$

Using the linearity of Legendre's operator $L$ and the fact that the Legendre polynomials $P_n$ are eigenfunctions for $L$, $L(P_n)=n(n+1)P_n$, we get

$$L(x^4) = \tfrac{1}{5}L(P_0) + \tfrac{4}{7}L(P_2) + \tfrac{8}{35}L(P_4)$$
$$= \tfrac{1}{5}(0)(0+1)P_0 + \tfrac{4}{7}(2)(2+1)P_2 + \tfrac{8}{35}(4)(4+1)P_4$$
$$= \tfrac{24}{7}P_2(x) + \tfrac{32}{7}P_4(x)$$

We can calculate the same quantity using the differential expression for $L$: $L(x^4) = -((1-x^2)(x^4)')' = 20x^4 - 12x^2$, which is equal to the last displayed formula.

This method obviously works for any *finite* linear combination of Legendre polynomials. It also works for an infinite series of Legendre polynomials when the sum function belongs to the domain of $L$.

**Theorem 4.28.** *Suppose $y(x)$ lies in the domain of Legendre's operator $L$. Then the equation obtained by applying $L$ term by term to the Legendre series expansion for $y(x)$ is valid. Put another way: If $y = \sum_{n=0}^{\infty} c_n P_n$, where $c_n = \langle y, P_n \rangle / \langle P_n, P_n \rangle$, and if $y \in \text{domain}(L)$ then*

$$L(y) = \sum_{n=0}^{\infty} c_n L(P_n) = \sum_{n=0}^{\infty} n(n+1)c_n P_n \qquad (4.7.1)$$

*Proof.* As $y$ belongs to the domain of $L$, $L(y)$ belongs to the Hilbert space $L^2(-1,1)$. This is the meaning of the compatibility requirement. *Any* function in $L^2(-1, 1)$ has a Legendre series expansion convergent in the mean-square metric, so

$$L(y) = \sum_{n=0}^{\infty} a_n P_n$$

when the uniquely determined coefficients $a_n$ are the Fourier-Legendre coefficients of the function $L(y)$ with respect to the Legendre polynomials:

$$a_n = \frac{\langle L(y), P_n \rangle}{\langle P_n, P_n \rangle} \qquad (4.7.2)$$

Now Legendre's operator $L$ is Hermitian, so we may move it from one side of the inner product to the other, thus: $\langle L(y), P_n \rangle = \langle y, L(P_n) \rangle$. And the $P_n$ are eigenfunctions for $L$: $L(P_n) = n(n+1)P_n$. Combining these two formulas and substituting the result into (4.7.2), we get

$$a_n = \frac{\langle y, n(n+1)P_n \rangle}{\langle P_n, P_n \rangle} = n(n+1)\frac{\langle y, P_n \rangle}{\langle P_n, P_n \rangle} = n(n+1)c_n$$

which completes the proof.

Note especially that in equation (4.7.1), it is imperative that the function $y$ belongs to the domain of $L$; otherwise the whole procedure makes no sense and will lead to nonsensical results. Accordingly, when applying Theorem 4.28, it is useful to have some quick procedures available to test whether or not a function $y$ belongs to the domain of Legendre's operator $L$. The requirements to belong to this domain are these:

1. $\displaystyle\int_{-1}^{+1} |y(x)|^2 \, dx < \infty$, i.e., $y \in L^2(-1, 1)$.

2. $\displaystyle\int_{-1}^{+1} \left| \frac{d}{dx}\left( (1-x^2)\frac{dy}{dx} \right) \right|^2 \, dx < \infty$, i.e., $l(y) \in L^2(-1, 1)$, the compatibility requirement.

3. $y(\pm 1)$ are finite.

Now generally we deal with functions that are analytic except at isolated points, like $1/x-1$, $1/\sin x$, etc. So a good practical rule of thumb is the following: *if the singularities of $y(x)$ lie strictly outside the interval $-1 \leqslant x \leqslant 1$ (or if it has no singularities at all), then $y(x)$ belongs to the domain of Legendre's operator.* For example, consider the function $y(x) = \cos x/\sqrt{5-4x}$. This function has one singularity at $x = 5/4$, but, as this point lies outside $[-1, 1]$, $y(x)$ belongs to the domain of Legendre's operator.

If $y(x)$ has singularities in $[-1, 1]$, then we must check the conditions one by one; there is no automatic decision either way.

In Section 4.6 we found the first three nonzero terms in the Legendre series expansion of sin $\pi x$ (equation (4.6.7)):

$$\sin \pi x = \frac{3}{\pi} P_1(x) + \frac{1}{\pi^3}(7\pi^2 - 105)P_3(x)$$

$$+ \frac{1}{\pi^5}(11\pi^4 - 1155\pi^2 + 10{,}395)P_5(x) + \cdots$$

The function sin $\pi x$ belongs to the domain of Legendre's operator $L$, so we may apply $L$ to both sides of this equation. We get

$$L(\sin \pi x) = \frac{6}{\pi} P_1(x) + \frac{12}{\pi^3}(7\pi^2 - 105)P_3(x)$$

$$+ \frac{30}{\pi^5}(11\pi^4 - 1155\pi^2 + 10{,}395)P_5(x) + \cdots$$

We can calculate the same quantity using the differential expression for $L$. $L(\sin \pi x) = -((1-x^2)(\sin \pi x)')' = \pi^2(1-x^2) \sin \pi x + 2\pi x \cos \pi x$. Combining this with the expression above, we get the first three nonzero terms of the Legendre expansion of the latter function.

So those calculations show one major advantage of the Hilbert space approach: we can calculate the action of Legendre's operator $L$ on a function $y$ (in the domain of $L$) by applying $L$ term by term to the Legendre series expansion of $y$.

A second feature of the new method arises in the solution of operator equations $L(y) = h$. The problem is this: we are given a function $h(x)$ in $L^2(-1, 1)$, and we seek a function $y(x)$ in the domain of $L$ such that $L(y) = h$. So $h$ is known, given beforehand, and $y$ is unknown.

There are three aspects to this problem. First, does a solution $y$ exist at all? Second, if a solution exists, to what extent is it unique? Third, when a solution exists, how can we find it effectively?

The following result addresses the first aspect (see Exercise 17 of Section 4.6).

**Theorem 4.29.** *If the equation $L(y) = h$, where $h \in L^2(-1, 1)$, has a solution $y$ in the domain of $L$, then $\langle h, P_0 \rangle = 0$.*

*Proof.* We have $\langle h, P_0 \rangle = \langle L(y), P_0 \rangle = \langle y, L(P_0) \rangle = \langle y, 0 \rangle = 0$, which is all there is to the proof. Where did we use the fact that Legendre's operator $L$ is Hermitian?

Now $P_0 = 1$, so the condition $\langle h, P_0 \rangle = 0$ means $\int_{-1}^{+1} h \, dx = 0$. We may accordingly rephrase Theorem 4.29 thus:

**Corollary 4.30.** *Consider Legendre's operator L acting in the Hilbert space $L^2(-1, 1)$, and consider the equation $L(y) = h$ where h is a given function in $L^2(-1, 1)$. If*

$$\int_{-1}^{+1} h(x)\, dx \neq 0$$

*then $L(y) = h$ has no solution y in the domain of L.*

So in the Hilbert space theory we have a restriction limiting the cases where $L(y) = h$ is solvable, a restriction that has no counterpart in the theory of Chapter 2. For example, $L(y) = 1$ has no solution, and neither does $L(y) = x^2$, because

$$\int_{-1}^{+1} 1\, dx = 2 \int_{0}^{1} dx = 2 \neq 0 \quad \text{and} \quad \int_{-1}^{+1} x^2\, dx = 2 \int_{0}^{1} x^2\, dx = \frac{2}{3} \neq 0$$

When $\langle h, P_0 \rangle = 0$, can we then solve $L(y) = h$? Here is a solution method that uses the orthogonal basis. Since h belongs to $L^2(-1, 1)$, it has a Fourier-Legendre expansion. Since $\langle h, P_0 \rangle = 0$, the first term in the expansion will vanish. Thus

$$h = \sum_{n=1}^{\infty} c_n P_n \quad \text{where } c_n = \frac{\langle h, P_n \rangle}{\langle P_n, P_n \rangle} = \frac{2n+1}{2} \int_{-1}^{+1} h P_n\, dx$$

As h is a known function, all the coefficients $c_n$ are known. We seek a function y that belongs to the domain of L and satisfies $L(y) = h$. Let us assume that such a y exists (which is plausible because we have met the restriction $\langle h, P_0 \rangle = 0$). Then y will have a Legendre expansion

$$y = \sum_{n=0}^{\infty} a_n P_n$$

where the coefficients $a_n$ are unknown. As y belongs to the domain of L we have (Theorem 4.28):

$$L(y) = \sum_{n=0}^{\infty} n(n+1) a_n P_n$$

and our equation $L(y) = h$ has taken the form

$$\sum_{n=0}^{\infty} n(n+1) a_n P_n = \sum_{n=1}^{\infty} c_n P_n$$

The $n = 0$ term on the left is $0(0+1)a_0 P_0 = 0$, which is consistent with the fact that there is no $n = 0$ term on the right. Thus, we see again where the restriction $\langle h, P_0 \rangle = 0$ comes into play. Now the coefficients of a Legendre expansion are unique, so we must have $n(n+1)a_n = c_n$ for $n \geqslant 1$, or

$$a_n = \frac{c_n}{n(n+1)}, \qquad n = 1, 2, \ldots$$

This determines all the expansion coefficients for the function $y(x)$ except $a_0$, which can be arbitrarily given. Thus our solution $y$ to $L(y) = h$ is

$$y = a_0 + \sum_{n=1}^{\infty} \frac{c_n}{n(n+1)} P_n \qquad \text{where } c_n = \frac{\langle h, P_n \rangle}{\langle P_n, P_n \rangle} = \frac{2n+1}{2} \int_{-1}^{+1} h P_n \, dx$$

and $a_0$ is an arbitrary constant.

For example, suppose we wished to solve $L(y) = x^3$. Now $x^3 = \frac{2}{5}P_3(x) + \frac{3}{5}P_1(x)$. Then, if $y = \Sigma a_n P_n$, our equation $L(y) = h$ becomes

$$\sum_{n=0}^{\infty} n(n+1)a_n P_n = \frac{2}{5}P_3 + \frac{3}{5}P_1$$

Hence $a_n = 0$ when $n \neq 1, 3$; $3(3+1)a_3 = \frac{2}{5}$; and $1(1+1)a_1 = \frac{3}{5}$. So our solution $y$ is

$$y(x) = a_0 + \tfrac{3}{10}P_1(x) + \tfrac{1}{30}P_3(x)$$

where $a_0$ is an arbitrary constant.

We can also use the differential calculus approach to the same problem. In terms of the differential expression, our equation $L(y) = x^3$ is

$$-((1 - x^2)y')' = x^3$$

Integrating once, we get

$$-\int_0^x ((1-t^2)y'(t))' \, dt = -(1-x^2)y'(x) + y'(0) = \int_0^x t^3 \, dt = \frac{x^4}{4}$$

In this equation, put $x = 1$, and remember that (refer to Lemma 4.19) $(1 - x^2)y'(x) = 0$ when $x = 1$. We get $y'(0) = 1/4$. Hence

$$-(1 - x^2)y'(x) = \frac{x^4}{4} - \frac{1}{4} = \frac{1}{4}(x^2 - 1)(x^2 + 1)$$

so

$$y'(x) = \tfrac{1}{4}(x^2 + 1)$$

which yields, after integration,

$$y(x) - y(0) = \tfrac{1}{4}\left(\frac{x^3}{3} + x\right)$$

or

$$y(x) = y(0) + \tfrac{1}{12}(x^3 + 3x)$$

Comparing this solution with that above, we get $y(0) = a_0$ and

$$\tfrac{1}{12}(x^3 + 3x) = \tfrac{3}{10}P_1(x) + \tfrac{1}{30}P_3(x)$$

which can be independently checked.

We shall take for granted that this method we have just illustrated always works. That is, when $\langle h, P_0 \rangle = 0$, then we can solve $L(y) = h$ by expanding the unknown function $y$ in its Legendre series and solving for the coefficients. The assumption that this method always works is equivalent to the assumption that the range of Legendre's operator is all of $P_0^\perp$, the subspace of $L^2(-1, 1)$ consisting of functions that are orthogonal to $P_0$.

As for uniqueness, the kernel of Legendre's operator is the one-dimensional subspace spanned by $P_0$. Hence, if we have a solution $y_p$ to $L(y) = h$ (a particular solution), then $y = y_p + a_0 P_0$ is also a solution for any constant $a_0$, and every solution has this form for some $a_0$ ($y_p + a_0 P_0$ is the general solution). We summarize our conclusions about the equation $L(y) = h$.

**Theorem 4.31.** *Consider Legendre's operator $L$ acting in the Hilbert space $L^2(-1, 1)$. Given $h$ in $L^2(-1, 1)$, the equation $L(y) = h$ has a solution $y$ in the domain of $L$ if and only if*

$$\langle h, P_0 \rangle = \int_{-1}^{+1} h(x)\, dx = 0$$

*In this case the general solution $y$ has the Legendre expansion*

$$y = a_0 + \sum_{n=1}^{\infty} \frac{c_n}{n(n+1)} P_n(x)$$

*where the $c_n$ are the Fourier-Legendre coefficients of the known function $h$,*

$$c_n = \frac{\langle h, P_n \rangle}{\langle P_n, P_n \rangle} = \frac{2n+1}{2} \int_{-1}^{+1} h(x) P_n(x)\, dx, \qquad n = 1, 2, \ldots$$

*and $a_0$ is an arbitrary constant.*

*Summary of this section.* The eigenfunctions of Legendre's operator, the Legendre polynomials, form an orthogonal basis of $L^2(-1, 1)$. In the terminology of Section 4.4, we have *diagonalized* Legendre's operator. Having diagonalized $L$, we can compute $L(f)$ and can solve $L(y) = h$ wholly in terms of the eigenfunctions. Having diagonalized the operator, we can almost forget about its original differential expression.

**Exercises**

In all the exercises below, $L$ stands for Legendre's operator.

1. (a) Using the equation $x^3 = \frac{2}{5}P_3 + \frac{3}{5}P_1$, compute $L(x^3)$ as a finite linear combination of Legendre polynomials.

   (b) Using the differential expression for $L$, compute $L(x^3)$ and check that your answer agrees with that in (a).

2. (a) Express $x^5$ as a finite linear combination of Legendre polynomials, and use this expression to compute $L(x^5)$ as a finite linear combination of Legendre polynomials.

   (b) Using the differential expression for $L$, compute $L(x^5)$, and check that your answer agrees with that in (a).

3. Suppose $f(x)$ is a finite linear combination of Legendre polynomials, $f(x) = c_0 P_0(x) + c_1 P_1(x) + \cdots + c_N P_N(x)$. Express $L(f)$ as a finite linear combination of Legendre polynomials.

4. (a) Check that $y(x) = \cos \pi x$ belongs to the domain of Legendre's operator.

   (b) In Exercise 12 of Section 4.6 we found the first two nonzero terms of the Legendre expansion of $\cos \pi x$:

   $$\cos \pi x = -\frac{15}{\pi^2}P_2(x) + \frac{45}{\pi^4}(21 - 2\pi^2)P_4(x) + \cdots$$

   Using this, find the beginning Legendre expansion of $L(\cos \pi x)$.

   (c) Using the differential expression for $L$, calculate $L(\cos \pi x)$.

5. (a) Check that $y(x) = (17 - 8x)^{-1/2}$ belongs to the domain of Legendre's operator.

   (b) In the next section we shall derive the Legendre series expansion

   $$4(17 - 8x)^{-1/2} = \sum_{n=0}^{\infty} 4^{-n}P_n(x)$$

   Using this, find the Legendre series expansion of $4L((17-8x)^{-1/2})$.

   (c) Using the differential expression for $L$, show that $4L((17-8x)^{-1/2}) = 32(17x - 6 - 2x^2)(17 - 8x)^{-5/2}$.

6. (Continuation) Combining (b) and (c) of Exercise 5, show that

$$32(17x - 6 - 2x^2)(17 - 8x)^{-5/2} = \sum_{n=0}^{\infty} n(n+1)4^{-n}P_n(x)$$

Compute the six-term partial sum of the series on the right (as we did in the exercises for Section 4.6), and compare the values with those of the function on the left in a table, $-1 \leqslant x \leqslant 1$, $(.1)$. (The six-term partial sum means $P_1$ through $P_5$, as the $P_0$ term is zero.)

7. In Exercise 14 of Section 4.6 we found the Legendre series expansion

$$\ln\left(\frac{1+x}{1-x}\right) = 3P_1(x) + \frac{7}{6}P_3(x) + \frac{11}{15}P_5(x) + \cdots$$

Show that, if you apply Legendre's operator $L$ to both sides of this equation, you get

$$0 = 6P_1(x) + 14P_3(x) + 22P_5(x) + \cdots$$

But this is nonsense, because the function $f(x)=0$ has the unique Legendre expansion

$$0 = 0P_0(x) + 0P_1(x) + 0P_2(x) + 0P_3(x) + 0P_4(x) + 0P_5(x) + \cdots$$

What is wrong?
8. Solve $L(y) = P_3(x)$ for $y$.
9. Does $L(y) = x^5$ have a solution? If so, find it as a linear combination of Legendre polynomials. Be sure to include all arbitrary coefficients.
10. Does $L(y) = 5x^4 - 3x^2$ have a solution? If so, find it.

## 4.8   MORE ON LEGENDRE'S POLYNOMIALS: RODRIGUES' FORMULA, THE RECURSION RELATION, AND THE GENERATING FUNCTION

Legendre's polynomials are a sequence of polynomials such that the $n$th polynomial in the sequence has degree exactly $n$. We begin with a useful general fact about any such sequence of polynomials.

**Lemma 4.32.** *Suppose we are given a sequence of polynomials* $Q_n(x)$, $n = 0, 1, 2, \ldots$, *such that, for each* $n$, $Q_n(x)$ *has degree* $n$. *Then, for each* $n = 0, 1, 2, \ldots$, $x^n$ *is a linear combination of* $Q_0, Q_1, \ldots, Q_n$. *Moreover, any polynomial of degree* $n$ *is a linear combination of* $Q_0, Q_1, \ldots, Q_n$.

*Proof.* Begin with $n = 0$. To say that $Q_0$ has degree 0 means that it is a nonzero constant. (The constant 0 is assigned degree $-1$.) Say $Q_0 = a_0 \neq 0$. Then $x^0 = (1/a_0)Q_0$, which proves our result for $n = 0$. To say that $Q_1$ has degree 1 means $Q_1 = b_1 x + b_0$ with $b_1 \neq 0$. Then $x^1 = (1/b_1)Q_1 - (b_0/a_0 b_1)Q_0$, which shows that $x^1$ is a linear combination of $Q_0$ and $Q_1$ as claimed. Likewise, if $Q_2 = c_2 x^2 + c_1 x + c_0$, $c_2 \neq 0$, then $x^2 = (1/c_2)Q_2 - (c_1/b_1 c_2)Q_1 + (b_0 c_1 - b_1 c_0)/a_0 b_1 c_2$, which is our assertion for $n = 2$.

Thus the first assertion of the lemma is true for $n = 0$, 1, and 2. We shall prove it now for all $n$ by an interesting mathematical technique known as *mathematical induction*. Let $S(n)$ stand for the statement "$x^n$ is a linear combination of $Q_0, Q_1, \ldots, Q_n$." Thus $S(0)$ is the statement "$x^0$ is a linear combination of $Q_0$," which we have just shown to be true. Likewise, $S(1)$ is the statement "$x^1$ is a linear combination of $Q_0$ and $Q_1$," also just proved. Same for $S(2)$, which is the statement "$x^2$ is a linear combination of $Q_0$, $Q_1$, and $Q_2$." Our objective is to show that $S(n)$ is true for all $n$. We do this by establishing what is called the

*Induction step: If $S(j)$ is true for $0 \leqslant j \leqslant n$, then $S(n+1)$ is true.*

Here is a proof of the induction step. We are assuming that our statement $S(j)$ is true for $0 \leqslant j \leqslant n$. Given that, we want to show that it must then be true for $n+1$, i.e., that $x^{n+1}$ is a linear combination of $Q_0$ through $Q_{n+1}$. Now $Q_{n+1}$ is a polynomial of degree $n+1$: $Q_{n+1} = a_{n+1}x^{n+1} + a_n x^n + \cdots + a_0$, $a_{n+1} \neq 0$. So $x^{n+1} - (1/a_{n+1})Q_{n+1}$ is a polynomial of degree $\leqslant n$, because the $(n+1)$st term cancels. Now by our assumption (which is called the *induction hypothesis*) that $S(j)$ is true for $0 \leqslant j \leqslant n$, each power $x^k$ in the polynomial $x^{n+1} - (1/a_{n+1})Q_{n+1}$ can be written as a linear combination of $Q_0$ through $Q_k$; hence the polynomial $x^{n+1} - (1/a_{n+1})Q_{n+1}$ itself can be written as a linear combination of $Q_0$ through $Q_n$ by combining terms. Hence

$$x^{n+1} - \frac{1}{a_{n+1}}Q_{n+1} = \text{linear combination of } Q_0 \text{ through } Q_n$$

so

$$x^{n+1} = \frac{1}{a_{n+1}}Q_{n+1} + \text{linear combination of } Q_0 \text{ through } Q_n$$

The last displayed formula shows that $x^{n+1}$ is a linear combination of $Q_0$ through $Q_{n+1}$. Thus the statement $S(n+1)$ is true, given the truth of $S(j)$ for $0 \leqslant j \leqslant n$. The induction step is proved.

Now back to the lemma. In the paragraph immediately following the statement of the lemma we showed that $S(j)$ is true for $j=0$, $j=1$, and $j=2$. Apply the induction step. It says that $S(3)$ has to be true. So now we know that $S(j)$ is true for $0 \leqslant j \leqslant 3$. Apply the induction step again. It says $S(4)$ must be true. So now we know that $S(j)$ is true for $0 \leqslant j \leqslant 4$. Apply the induction step again, etc.

A more careful finish to the argument, without using "etc.," goes as follows. Suppose there were some positive integer for which our statement were false. Then there would be a smallest such positive integer, call it $m$. Since $m$ is the smallest integer for which $S$ is false, $S(j)$ must be true for $0 \leqslant j \leqslant m-1$. But then the induction step says that $S(m)$ is true. That contradiction shows that there is no integer for which $S$ is false: $S(n)$ is true for all $n$.

The last sentence of Lemma 4.32, namely that any polynomial $p(x)$ of degree $n$ is a linear combination of $Q_0$ through $Q_n$, now follows easily. Just write each term $x^k$ as a linear combination of $Q_0$ through $Q_k$ and combine terms. Lemma 4.32 is proved.

Lemma 4.32 applies in particular to the Legendre polynomials. Thus any polynomial of degree $n$ is a linear combination of $P_0$ through $P_n$. Which is just another way of saying that in the Fourier-Legendre series of a polynomial of degree $n$, all coefficients $c_k$ for $k > n$ are zero.

**Lemma 4.33.** *If $p(x)$ is a polynomial of degree $m$, then $\langle p(x), P_n(x) \rangle = 0$ for $n > m$. In particular, $\langle x^m, P_n(x) \rangle = 0$ for $n > m$.*

We are using here, and throughout this section, the inner product of $L^2(-1, 1)$: $\langle f, g \rangle = \int_{-1}^{1} f\bar{g} \, dx$. To prove Lemma 4.33, simply note that by Lemma 4.32 $p(x) = c_0 P_0 + c_1 P_1 + \cdots + c_m P_m$, so

$$\langle p(x), P_n(x) \rangle = \sum_{k=1}^{m} c_k \langle P_k, P_n \rangle = 0$$

because for each term in the sum, $k \leqslant m < n$.

We turn now to a classic formula for the Legendre polynomials known as *Rodrigues' formula*.

**Theorem 4.34.** $P_n(x) = \dfrac{1}{2^n n!} \dfrac{d^n}{dx^n} (x^2 - 1)^n.$

*Proof.* Let $Q_n(x) = (1/2^n n!) \, d^n/dx^n (x^2 - 1)^n$. Our objective is to prove $Q_n = P_n$.

First observe the effect successive differentiations have on $(x^2 - 1)^n$:

$$\frac{d}{dx}(x^2 - 1)^n = (x^2 - 1)^{n-1}(2nx)$$

$$\frac{d^2}{dx^2}(x^2 - 1)^n = (x^2 - 1)^{n-2}(2n(2n-1)x^2 - 2n)$$

$$\frac{d^3}{dx^3}(x^2 - 1)^n = (x^2 - 1)^{n-3} \text{ (polynomial of degree 3)}$$

In general, for $k \leqslant n$,

$$\frac{d^k}{dx^k}(x^2 - 1)^n = (x^2 - 1)^{n-k} \text{ (polynomial of degree $k$)}$$

Hence, for $k < n$,

$$\frac{d^k}{dx^k}(x^2 - 1)^n \Big|_{-1}^{1} = 0$$

because the term $x^2 - 1$ vanishes at $x = -1$ and at $x = 1$. For $k = n$ we have $d^n/dx^n(x^2 - 1)^n$ = polynomial of degree $n$. Thus $Q_n(x)$ is a polynomial of degree $n$.

Take $m \leqslant n$ and integrate by parts:

$$\left\langle x^m, \frac{d^n}{dx^n}(x^2 - 1)^n \right\rangle = \int_{-1}^1 x^m \frac{d^n}{dx^n}(x^2 - 1)^n \, dx = \int_{-1}^1 x^m \frac{d}{dx}\left[ \frac{d^{n-1}}{dx^{n-1}}(x^2 - 1)^n \right] dx$$

$$= \int_{-1}^1 x^m d\left[ \frac{d^{n-1}}{dx^{n-1}}(x^2 - 1)^n \right]$$

$$= x^m \frac{d^{n-1}}{dx^{n-1}}(x^2 - 1)^n |_{-1}^1 - m \int_{-1}^1 x^{m-1} \frac{d^{n-1}}{dx^{n-1}}(x^2 - 1)^n \, dx$$

By the observation just made, the integrated term is zero. Hence we have

$$\left\langle x^m, \frac{d^n}{dx^n}(x^2 - 1)^n \right\rangle = -m \int_{-1}^1 x^{m-1} \frac{d^{n-1}}{dx^{n-1}}(x^2 - 1)^n \, dx$$

Integrate by parts $m$ times to get

$$\left\langle x^m, \frac{d^n}{dx^n}(x^2 - 1)^n \right\rangle = (-1)^m m! \int_{-1}^1 \frac{d^{n-m}}{dx^{n-m}}(x^2 - 1)^n \, dx \tag{4.8.1}$$

When $m < n$ the integrand is a derivative of a function that vanishes at $x = \pm 1$, hence is zero. Thus $\langle x^m, Q_n(x)\rangle = 0$ when $m < n$. Moreover, $\langle p(x), Q_n(x)\rangle = 0$ for any polynomial $p(x)$ of degree $< n$.

Now look at the Fourier-Legendre series of $Q_n(x)$. Since $Q_n(x)$ has degree $n$, Lemma 4.33 tells us that the series is finite, all coefficients $c_k$ being zero for $k > n$. Thus

$$Q_n = c_0 P_0 + c_1 P_1 + \cdots + c_n P_n$$

where $c_k = \langle Q_n, P_k\rangle / \langle P_k, P_k\rangle$. But $\langle Q_n, P_k\rangle = 0$ when $k < n$ because $P_k$ is a polynomial of degree $k$. Hence $Q_n = c_n P_n$. We shall prove $c_n = 1$ by showing that $P_n$ and $Q_n$ have the same leading coefficient. Our original convention in Chapter 2 put the leading coefficient of $P_n$ as $(2n)!/2^n(n!)^2$. The leading term in $Q_n$ comes from differentiating $x^{2n}$ $n$ times:

$$\frac{1}{2^n n!} \frac{d^n}{dx^n} x^{2n} = \frac{1}{2^n n!}(2n)(2n - 1) \cdots (2n - n + 1)x^n$$

$$= \frac{1}{2^n n!} \frac{(2n)!}{n!} = \frac{(2n)!}{2^n(n!)^2}$$

Thus $Q_n = P_n$ and Rodrigues' formula is proved.

Next we compute the norm of $P_n$, a result we have used frequently in the previous two sections.

**Theorem 4.35.**  $\langle P_n, P_n \rangle = 2/(2n + 1)$.

*Proof.*  Write $P_n = [(2n)!/2^n(n!)^2]x^n +$ terms of lower degree. Then

$$\langle P_n, P_n \rangle = \frac{(2n)!}{2^n(n!)^2} \langle x^n, P_n \rangle + \left\langle \begin{array}{c} \text{sum of terms} \\ \text{of degree } <n \end{array}, P_n \right\rangle$$

The last inner product is zero, by Lemma 4.33. In the first inner product on the right side, substitute Rodrigues' formula. We get

$$\langle P_n, P_n \rangle = \frac{(2n)!}{2^n(n!)^2} \frac{1}{2^n n!} \left\langle x^n, \frac{d^n}{dx^n} (x^2 - 1)^n \right\rangle$$

Then use equation (4.8.1) with $m = n$:

$$\langle P_n, P_n \rangle = \frac{(2n)!(-1)^n}{4^n(n!)^2} \int_{-1}^{1} (x^2 - 1)^n \, dx$$

So it boils down to the evaluation of the integral

$$(-1)^n \int_{-1}^{1} (x^2 - 1)^n \, dx = \int_{-1}^{1} (1 - x^2)^n \, dx = 2 \int_{0}^{1} (1 - x^2)^n \, dx = 2A_n$$

Integrate by parts with $u = (1 - x^2)^n$, $dv = dx$, $n \geqslant 1$:

$$A_n = \int_{0}^{1} (1 - x^2)^n \, dx = x(1 - x^2)^n |_0^1 - 2n \int_{0}^{1} (-x^2)(1 - x^2)^{n-1} \, dx$$

$$= -2n \int_{0}^{1} ((1 - x^2) - 1)(1 - x^2)^{n-1} \, dx$$

$$= -2nA_n + 2nA_{n-1}$$

Thus $A_n = (2n/2n+1)A_{n-1}$. As $A_0 = 1$, we have $A_1 = 2/3$, $A_2 = (4/5)A_1 = 4 \cdot 2/3 \cdot 5$, $A_3 = (6/7)A_2 = 6 \cdot 4 \cdot 2/3 \cdot 5 \cdot 7$, and in general

$$A_n = \frac{(2n)(2n-2)(2n-4)\cdots(2)}{(2n+1)(2n-1)(2n-3)\cdots(1)} = \frac{[(2n)(2n-2)(2n-4)\cdots(2)]^2}{(2n+1)(2n)(2n-1)(2n-2)\cdots(1)}$$

$$A_n = \frac{(2^n n!)^2}{(2n+1)!} = \frac{4^n(n!)^2}{(2n+1)!}$$

Hence

$$\langle P_n, P_n \rangle = \frac{(2n)!(2)4^n(n!)^2}{4^n(n!)^2(2n+1)!} = \frac{2}{2n + 1}$$

which proves Theorem 4.35.

Our next result, the three-term recursion for the Legendre polynomials, has both theoretical and practical importance.

**Theorem 4.36.** $nP_n = (2n-1)xP_{n-1} - (n-1)P_{n-2}$, $n \geqslant 2$.

The proof is outlined in Exercise 6.

Suppose the variable $x$ is real and lies in the interval $[-1, 1]$. The function $1 - 2xw + w^2$ will be zero if and only if $w = x \pm \sqrt{1-x^2}\, i$, which forces $|w| = 1$. Hence, for a fixed value of $w$ satisfying $|w| < 1$, the function $1 - 2xw + w^2$ is never zero for $-1 \leqslant x \leqslant 1$; thus $(1 - 2xw + w^2)^{-1/2}$ is bounded on $[-1, 1]$. It therefore belongs to $L^2(-1, 1)$, so has a Fourier-Legendre expansion. The coefficients in this expansion are remarkably simple.

**Theorem 4.37**

$$\frac{1}{\sqrt{1 - 2xw + w^2}} = \sum_{n=0}^{\infty} w^n P_n(x), \qquad |w| < 1, \qquad -1 \leqslant x \leqslant 1$$

*Proof.* Set

$$h(x, w) = \sum_{n=0}^{\infty} w^n P_n(x) = P_0(x) + wP_1(x) + w^2 P_2(x) + \cdots$$

Fix $x$ and regard $h$ as a function of $w$ alone. Differentiate the power series term by term to get

$$\frac{dh}{dw} = \sum_{n=0}^{\infty} nw^{n-1} P_n(x) = P_1(x) + \sum_{n=2}^{\infty} w^{n-1} nP_n(x)$$

Then use the recursion relation (Theorem 4.36), and substitute $P_1 = x$:

$$\frac{dh}{dw} = x + \sum_{n=2}^{\infty} w^{n-1}((2n-1)xP_{n-1}(x) - (n-1)P_{n-2}(x))$$

$$= x + \sum_{n=2}^{\infty} w^{n-1}((2(n-1)+1)xP_{n-1}(x) - ((n-2)+1)P_{n-2}(x))$$

$$= x + 2wx \sum_{n=2}^{\infty} (n-1)w^{n-2} P_{n-1}(x) + x \sum_{n=2}^{\infty} w^{n-1} P_{n-1}(x)$$

$$- w^2 \sum_{n=2}^{\infty} (n-2)w^{n-3} P_{n-2}(x) - w \sum_{n=2}^{\infty} w^{n-2} P_{n-2}(x)$$

Consider the four series on the right. We have

$$\sum_{n=2}^{\infty} (n-1)w^{n-2} P_{n-1}(x) = P_1(x) + 2wP_2(x) + 3w^2 P_3(x) + \cdots = \frac{dh}{dw}$$

and

$$\sum_{n=2}^{\infty} w^{n-1} P_{n-1}(x) = wP_1(x) + w^2 P_2(x) + \cdots = h - 1$$

Also

$$\sum_{n=2}^{\infty} (n-2)w^{n-3} P_{n-2}(x) = P_1(x) + 2wP_2(x) + 3w^2 P_3(x) = \frac{dh}{dw}$$

Finally,

$$\sum_{n=2}^{\infty} w^{n-2} P_{n-2}(x) = P_0(x) + wP_1(x) + w^2 P_2(x) + \cdots = h$$

Combine these results

$$\frac{dh}{dw} = x + 2wx\frac{dh}{dw} + x(h-1) - w^2\frac{dh}{dw} - wh$$

then rearrange terms

$$(1 - 2xw + w^2)\frac{dh}{dw} = (x - w)h$$

or

$$\frac{dh}{h} = \frac{(x-w)\,dw}{1 - 2xw + w^2}$$

Remembering $x$ is fixed, integrate both sides.

$$\ln h = \int \frac{(x-w)\,dw}{1 - 2xw + w^2} = -\frac{1}{2}\int \frac{dt}{t} \qquad (t = 1 - 2xw + w^2)$$

$$= -\frac{1}{2}\ln t + C = +\ln\frac{1}{\sqrt{t}} + C$$

so

$$h(x, w) = \frac{\text{const}}{\sqrt{1 - 2xw + w^2}}$$

As $h(x, 0) = 1$, the constant is 1, and Theorem 4.36 is proved. $(1 - 2xw + w^2)^{-1/2}$ is called the *generating function* for the Legendre polynomials.

The Legendre coefficients are uniquely determined: if

$$f = \sum_{n=0}^{\infty} c_n P_n, \qquad \text{then } c_n = \frac{\langle f, P_n \rangle}{\langle P_n, P_n \rangle} = \frac{2n+1}{2}\langle f, P_n \rangle$$

Hence from Theorem 4.37 we deduce

$$w^n = \frac{2n+1}{2} \int_{-1}^{1} \frac{P_n(x)}{\sqrt{1-2xw+w^2}} \, dx, \qquad |w| < 1, \; n = 0, 1, 2, \ldots$$

## Exercises

1. Prove that the Legendre polynomials are uniquely determined up to scalar multiples by their orthogonality properties. That is, prove the following result. If $R_n$, $n = 0, 1, 2, \ldots$, is a sequence of polynomials such that (1) $R_n$ is of degree $n$ for each $n = 0, 1, 2, \ldots$ and (2) $\langle R_m, R_n \rangle = 0$, $m \neq n$, then $R_n = \mu_n P_n$, $n = 0, 1, 2, \ldots$, where the $\mu_n$ are nonzero scalars.

2. Using Rodrigues' formula, compute $P_0$, $P_1$, $P_2$, and $P_3$.

3. (a) Show that $d^n/dx^n(x^k) = k(k-1) \cdots (k-n+1)x^{k-n} = (k!/(k-n)!)x^{k-n}$ when $n \leqslant k$ and that $d^n/dx^n(x^k) = 0$ when $n > k$.

   (b) Using Rodrigues' formula and the result in (a), show that

   $$P_n(x) = \frac{1}{2^n n!} \left[ \frac{(2n)!}{n!} x^n - n \frac{(2n-2)!}{(n-2)!} x^{n-2} + \frac{n(n-1)}{2} \frac{(2n-4)!}{(n-4)!} x^{n-4} \right.$$
   $$\left. - \frac{n(n-1)(n-2)}{6} \frac{(2n-6)!}{(n-6)!} x^{n-6} + \cdots \right]$$
   $$= \frac{1}{2^n n!} \sum_{k=0}^{p} (-1)^k \binom{n}{k} \frac{(2n-2k)!}{(n-2k)!} x^{n-2k}$$

   where $p$ is the largest integer $\leqslant n/2$ and $\binom{n}{k}$ is the binomial coefficient, i.e., the $k$th coefficient in the expansion of $(a+b)^n$:

   $$(a+b)^n = \sum_{k=0}^{n} \binom{n}{k} a^{n-k} b^k, \qquad \binom{n}{k} = \frac{n(n-1)(n-2)\cdots(n-k+1)}{k!}$$

   for $1 \leqslant k \leqslant n$, $\binom{n}{0} = 1$. The formula

   $$\binom{n}{k} = \frac{n!}{(n-k)! \, k!}$$

   works for $0 \leqslant k \leqslant n$.

   (c) Using the formula in (b), compute $P_0$, $P_1$, $P_2$, and $P_3$.

4. Using the result in 3(b), show again that when $n$ is even, $P_n(x)$ contains only even powers of $x$; when $n$ is odd, only odd powers. Thus $P_n(x)$, as a function, has the same parity as $n$. Using this, prove $P_n(-x) = (-1)^n P_n(x)$ and $\langle xP_m, P_n \rangle = 0$, $n = 0, 1, 2, \ldots$.

5. Find the constant $a_n$ such that $x^n = a_n P_n + \pi_{n-2}$ where $\pi_{n-2}$ is a polynomial of degree $\leqslant n-2$. Using this result, prove that

   $$\langle x^n, P_n \rangle = \frac{2^{n+1}(n!)^2}{(2n+1)!} \qquad \text{and} \qquad \langle xP_{n-1}, P_n \rangle = \frac{2n}{(2n+1)(2n-1)}$$

6. Show that $P_n - ((2n-1)/n)xP_{n-1}$ is a polynomial of degree $\leqslant n-2$, hence

$$P_n - \frac{2n-1}{n} xP_{n-1} = c_0 P_0 + c_1 P_1 + \cdots + c_{n-2} P_{n-2}$$

Show that $c_{n-2} = -(n-1)/n$, $c_j = 0$ otherwise. Thus derive the *recursion formula for Legendre polynomials* (Theorem 4.36):

$$nP_n(x) = (2n-1)xP_{n-1}(x) - (n-1)P_{n-2}(x), \qquad n \geqslant 2$$

7. Taking $P_0 = 1$ and $P_1 = x$ as known, use the recursion formula to compute $P_2$, $P_3$, $P_4$, and $P_5$.

8. Prove that if $f(x) = \Sigma_{n=0}^{\infty} c_n P_n(x)$, then

$$xf(x) = \sum_{n=0}^{\infty} \left( \frac{n+1}{2n+3} c_{n+1} + \frac{n}{2n-1} c_{n-1} \right) P_n(x)$$

9. Taking $P_0(1) = 1$ and $P_1(1) = 1$ as known, use the recursion formula and mathematical induction to prove $P_n(1) = 1$ for all $n$. Prove that $P_n(-1) = (-1)^n$, $n = 0, 1, 2, \ldots$.

10. Integrate by parts to show that $\langle P'_{n+1}, x^m \rangle = 1 + (-1)^{m+n}$, $m-1 < n+1$. Using this, show $\langle P'_{n+1} - xP'_n, x^m \rangle = 0$, $0 \leqslant m < n$, and thus conclude $P'_{n+1} - xP'_n = c_n P_n$. Find $c_n$.

11. Put $w = 1/2$ in the generating function for the Legendre polynomials (Theorem 4.37) to get the Fourier-Legendre expansion

$$\frac{2}{\sqrt{5-4x}} = \sum_{n=0}^{\infty} \frac{1}{2^n} P_n(x)$$

Compare in a table the six-term partial sum ($P_0$ through $P_5$) with the function $2/\sqrt{5-4x}$, $-1 \leqslant x \leqslant 1$ (.1).

12. At each step of the following sequence of calculations, you may use the results in the preceding steps.

(a) Suppose $w$ is a fixed real number, $-1 < w < 1$. Show that $1 - 2xw + w^2 > 0$ for $-1 \leqslant x \leqslant 1$.

(b) With $w$ as in (a) and $f(x) = (1 - 2xw + w^2)^{-1/2}$, show that $w \langle f, f \rangle = \ln(1 + w/1 - w)$.

(c) Using the Fourier-Legendre coefficients of $f(x)$ given in Theorem 4.37, compute the series half of Parseval's equality (equation (4.6.8)).

(d) Combining (b) with (c), show that Parseval's equality for $f(x)$ is the power series expansion for $\ln(1 + w/1 - w)$.

(e) (Optional) Using analytic continuation, show that the series expansion in (d) is valid for all complex $w$, $|w| < 1$.

13. Derive the following formulas, all valid for $|w| < 1$:

(a) $\displaystyle \int_{-1}^{1} \frac{dx}{\sqrt{1 - 2xw + w^2}} = 2$

(b) $\displaystyle\int_{-1}^{1} \frac{x\,dx}{\sqrt{1-2xw+w^2}} = \frac{2w}{3}$

(c) $\displaystyle\int_{-1}^{1} \frac{x^4\,dx}{\sqrt{1-2xw+w^2}} = \frac{1}{(5)(7)(9)}(16w^4 + 72w^2 + 126)$

## 4.9 HERMITE'S OPERATOR AND ITS EIGENFUNCTIONS, THE HERMITE POLYNOMIALS

**Definition 4.38.** Consider the Hilbert space $L^2(-\infty, \infty; e^{-x^2/2})$ with inner product

$$\langle f,g \rangle = \int_{-\infty}^{\infty} f(x)\overline{g(x)}e^{-x^2/2}\,dx$$

*Hermite's operator* $L$ has the differential expression $l(y) \equiv -e^{x^2/2}(e^{-x^2/2}y')'$; the domain $D$ of $L$ consists of all functions in $L^2(-\infty, \infty; e^{-x^2/2})$ that are compatible with $l$. The action of $L$ is defined on its domain $D$ by $L(y) = l(y)$, $y \in D$.

The differential expression for $L$ is a special case of that considered in Section 4.5. There we considered the general Sturmian form

$$l(y) = -\frac{1}{r(x)}\left[\frac{d}{dx}\left(p(x)\frac{dy}{dx}\right) + q(x)y\right]$$

This becomes Hermite's differential expression if we set $r(x)=p(x)=\exp(-x^2/2)$, $q(x)=0$. Note also that $l(y)=-y''+xy'$.

Consider next the specification in Definition 4.38 of the domain of $L$. Recall what it means for a function $y$ to be compatible with the differential expression $l$. First, of course, $y$ must belong to the Hilbert space $L^2(-\infty, \infty; e^{-x^2/2})$. That is, we must have

$$\int_{-\infty}^{\infty} |y(x)|^2 e^{-x^2/2}\,dx < \infty$$

Second, $l(y)$ must also belong to $L^2(-\infty, \infty; e^{-x^2/2})$. That is, we must have

$$\int_{-\infty}^{\infty} \left|e^{x^2/2}\frac{d}{dx}\left(e^{-x^2/2}\frac{dy}{dx}\right)\right|^2 e^{-x^2/2}\,dx < \infty$$

These two conditions constitute the compatibility requirement. The functions that satisfy them constitute a subspace $D$ of $L^2(-\infty, \infty; e^{-x^2/2})$. Then $L(y)$ is defined for those, *and only those*, functions $y$ in $D$ by

$$L(y) = l(y)$$

While the differential expression $l$ makes sense for a much broader class of functions—there is nothing in the differential expression itself that restricts it to the given Hilbert space—Hermite's operator $L$ by definition acts only on functions in $D$.

The domain $D$ contains all polynomials. To prove this, it is enough to prove that $D$ contains all simple power functions $x^n$ because a polynomial is a linear combination of such powers. The power $x^n$ belongs to $L^2(-\infty, \infty; e^{-x^2/2})$; i.e., the integral

$$\int_{-\infty}^{+\infty} (x^n)^2 e^{-x^2/2}\, dx$$

is finite (see Exercise 7 of Section 3.9). And $l(x^n) = (x^n)'' - x(x^n)' = n(n-1)x^{n-2} - nx^n$ also belongs to $L^2$. Thus $x^n$ is compatible with $l$, which is the requirement defining membership in $D$. The domain $D$ also contains functions of the form (polynomial) $\times e^x$ (Exercise 3).

Note that $D$ is defined by the compatibility requirement alone—there are no explicit boundary conditions. It is a remarkable fact that certain boundary behavior is forced by the compatibility requirement:

**Lemma 4.39.**  *For any function $f(x)$ in the domain $D$ of Hermite's operator, we have $f'(x) \in L^2(-\infty, \infty; e^{-x^2/2})$. Moreover, for any pair $f$, $g$ of functions in $D$,*

$$\lim_{x \to \pm\infty} e^{-x^2/2} f'(x)\bar{g}(x) = 0$$

Hence, unlike the situation with Legendre's operator, where the specification of the domain had to include independently stated boundary conditions, the boundary conditions for Hermite's operator are already implicit in the compatibility requirement and don't need to be independently stated. Although the *result* in Lemma 4.39 is vitally important for us, because it will permit us to prove that the operator involved is Hermitian and thus that its eigenfunctions form an orthogonal family, the *proof* of Lemma 4.39 uses mathematical techniques foreign to our applications-oriented presentation, so this optional proof is given in the appendix to this chapter.

**Theorem 4.40.**  *Hermite's operator is Hermitian and positive semidefinite.*

For the Hermitian character of the operator, one can use Corollary 4.17. A direct proof is outlined in Exercise 8.

Theorem 4.40 means that $\langle L(f), g \rangle = \langle f, L(g) \rangle$ for all functions $f$, $g$ in the domain $D$ of Hermite's operator $L$. The inner product referred to here, as in this entire section and the following two, is that of the Hilbert space

$L^2(-\infty, \infty; e^{-x^2/2})$:

$$\langle f, g \rangle = \int_{-\infty}^{\infty} f(x)\overline{g(x)}e^{-x^2/2}\,dx$$

Bear this in mind, as we shall not keep repeating it.

**Theorem 4.41.** *The eigenvalues of Hermite's operator are all $\geq 0$, and eigenfunctions belonging to different eigenvalues are orthogonal.*

This is an application of Theorems 4.6 and 4.7.

In connection with Theorems 4.40 and 4.41, it is very important to keep in mind that the definition of Hermite's operator consists of *two* equally important components: the differential expression $l(y) \equiv -e^{x^2/2}(e^{-x^2/2}y')'$, and the domain $D$. Hermite's operator $L$ is defined only on $D$ and there is given by $L(y) = l(y)$. The formula $\langle L(f), g \rangle = \langle f, L(g) \rangle$ of Theorem 4.40 holds only for $f$, $g$ in $D$. It makes no sense otherwise. And the eigenfunctions of $L$ referred to in Theorem 4.41 are by definition in $D$. The differential expression $l$ will apply outside $D$, but then Theorems 4.40 and 4.41 may not hold. See Exercise 7.

In Section 2.9 we used the method of undetermined coefficients to solve Hermite's equation

$$y'' - xy' + ny = 0 \tag{4.9.1}$$

where $n$ is a nonnegative integer. As we know, equation (4.9.1), like any second order differential equation, has two linearly independent solutions. In Chapter 2 we showed that for each nonnegative integer $n$, one of the solutions of Hermite's equation was a polynomial $H_n(x)$ of degree $n$ containing only even powers of $x$ when $n$ is even and only odd powers when $n$ is odd (Theorem 2.14 in Chapter 2). The $H_n(x)$ are *Hermite's polynomials*. We list the first seven below.

*The first seven Hermite polynomials*

| | |
|---|---|
| $H_0(x) = 1$ | $H_4(x) = x^4 - 6x^2 + 3$ |
| $H_1(x) = x$ | $H_5(x) = x^5 - 10x^3 + 15x$ |
| $H_2(x) = x^2 - 1$ | $H_6(x) = x^6 - 15x^4 + 45x^2 - 15$ |
| $H_3(x) = x^3 - 3x$ | |

Hermite's polynomials are normalized so that the coefficient of the leading term is 1.

Using the fact that $y'' - xy' = e^{x^2/2}(e^{-x^2/2}y')'$, we can write Hermite's equation (4.9.1) as

$$l(y) = ny \qquad\qquad (4.9.2)$$

where $l(y) \equiv -e^{x^2/2}(e^{-x^2/2}y')'$ is the differential expression associated with Hermite's operator. Equation (4.9.2) defines an eigenvalue problem where $y$ is the sought-for eigenfunction belonging to the eigenvalue $\lambda = n$. As we have already noted, $y = H_n(x)$ belongs to the domain $D$ of Hermite's operator and satisfies (4.9.2). Hence we have

**Theorem 4.42.** *Hermite's operator $L$ has the eigenvalues $n = 0, 1, 2, \ldots$. The Hermite polynomial $H_n(x)$ is an eigenfunction belonging to the eigenvalue $\lambda = n$; i.e.,*

$$L(H_n) = nH_n, \qquad n = 0, 1, 2, \ldots$$

We shall see shortly that the integers $n = 0, 1, 2 \ldots$ are *all* the eigenvalues of Hermite's operator and that each of these eigenvalues is simple.

From Theorems 4.41 and 4.42 we deduce

**Corollary 4.43.** *The Hermite polynomials $H_n(x)$, $n = 0, 1, 2, \ldots$, form an orthogonal family in the Hilbert space $L^2(-\infty, \infty; e^{-x^2/2})$, i.e.,*

$$\langle H_m, H_n \rangle = \int_{-\infty}^{\infty} H_m(x)H_n(x)e^{-x^2/2}\, dx = 0, \qquad m \neq n$$

In a subsequent section we shall show that the norm (squared) of the Hermite polynomials is given by

$$\langle H_n(x), H_n(x) \rangle = \int_{-\infty}^{\infty} (H_n(x))^2 e^{-x^2/2}\, dx = \sqrt{2\pi}\, n!, \qquad n = 0, 1, 2, \ldots \quad (4.9.3)$$

a result that we shall take for granted now.

Next, we have the very important

**Theorem 4.44.** *The Hermite polynomials $H_n(x)$, $n = 0, 1, 2, \ldots$, form an orthogonal basis of $L^2(-\infty, \infty; e^{-x^2/2})$.*

For the proof we refer to J. R. Higgins's book, §2.1.

In Section 3.8 we covered the theory of general Fourier expansions in inner product spaces. *General Fourier series* is the term used to describe a series expansion in terms of an orthogonal basis of an inner product space. Let us review this material as it applies to the Hermite polynomials.

Theorem 4.44, when expanded in full detail, makes the following four assertions:

First, given any function $f(x)$ in $L^2(-\infty, \infty; e^{-x^2/2})$, we have

$$f(x) = \sum_{n=0}^{\infty} c_n H_n(x) \tag{4.9.4}$$

where the $c_n$ are the Fourier coefficients of $f$ with respect to the $H_n$:

$$c_n = \frac{\langle f, H_n \rangle}{\langle H_n, H_n \rangle} = \frac{\int_{-\infty}^{\infty} f(x) H_n(x) e^{-x^2/2}\, dx}{\int_{-\infty}^{\infty} (H_n(x))^2 e^{-x^2/2}\, dx}$$

$$= \frac{1}{\sqrt{2\pi\, n!}} \int_{-\infty}^{\infty} f(x) H_n(x) e^{-x^2/2}\, dx \tag{4.9.5}$$

The series on the right of (4.9.4) converges to $f(x)$ in the mean-square metric:

$$\left\| f(x) - \sum_{n=0}^{N} c_n H_n(x) \right\|$$

$$= \left[ \int_{\infty}^{\infty} \left| f(x) - \sum_{n=0}^{N} c_n H_n(x) \right|^2 e^{-x^2/2}\, dx \right]^{1/2} \to 0 \quad \text{as } N \to \infty$$

The coefficients $c_n$ are uniquely determined. That is, if we have another expansion $f(x) = \Sigma\, a_n H_n(x)$ where the series converges to $f$ in the mean-square metric, then $a_n = c_n$, $n = 0, 1, 2, \ldots$. The expansion (4.9.4) is called the *Hermite series* or *Fourier-Hermite series* of $f(x)$, and the remark we have just made, which is based on the theory developed in Section 3.8, shows that there is only one Hermite series of a given $f$ in the Hilbert space $L^2(-\infty, \infty; e^{-x^2/2})$.

Second, an expansion (4.9.4) is valid for *every* $f(x)$ in $L^2(-\infty, \infty; e^{-x^2/2})$. The *only* requirement on a function $f(x)$ that it have a Hermite series (4.9.4) convergent to it in the mean-square metric is that

$$\int_{-\infty}^{\infty} |f(x)|^2 e^{-x^2/2}\, dx < \infty$$

The power of this sweeping statement is all the more remarkable when you consider that the Hermite polynomials are eigenfunctions of Hermite's operator, thus lie in the domain $D$ of that operator, which is a restricted proper subspace of the Hilbert space $L^2(-\infty, \infty; e^{-x^2/2})$. Yet the expansion (4.9.4) holds not only for functions in $D$ but also for *every* function in the given Hilbert space, no matter how ragged or badly behaved, just so long as it is square integrable against the weight function $e^{-x^2/2}$. If we were to restrict ourselves to *finite* linear combinations of Hermite polynomials, we would always stay in $D$ as $D$ is a subspace. But when we use infinite series, we fill out the whole Hilbert space.

Third, if we cut off the series (4.9.4) at some point, then the resulting partial sum makes the mean-square error

$$\left\| f(x) - \sum_{n=0}^{N} c_n H_n(x) \right\| = \left[ \int_{-\infty}^{\infty} \left| f(x) - \sum_{n=0}^{N} c_n H_n \right|^2 e^{-x^2/2} \, dx \right]^{1/2}$$

an absolute minimum. With the $c_n$ given by equation (4.9.5) we make the best possible mean-square approximation to $f(x)$. Any alteration of these coefficients will only worsen the approximation. In the exercises we will compare various specific functions with the six-term partial sum of their Hermite series, and we can expect to see the familiar way the Hermite series "wobbles" around the function as it seeks to minimize the mean-square area. However, note the role played by the weight function $e^{-x^2/2}$. This function goes very rapidly to zero as $x \to \infty$. For example, if $x = 10$, then $\exp(-(10)^2/2) = \exp(-50) \cong 1.93 \times 10^{-22}$. Hence the *actual* discrepancy between $f(x)$ and the partial sum of its Hermite series can become quite large as $x \to \infty$ because this growth is masked by the weight function. So, in such a comparison, we should stick to reasonably small values of $x$. Alternatively, we can compare

$$f(x)e^{-x^2/4} \qquad \text{with} \qquad \sum_{n=0}^{N} c_n H_n(x)e^{-x^2/4}$$

over a much larger range of values of $x$.

Fourth, and finally, as we proved in Section 3.8, we may take the inner product of both sides of the series (4.9.4) with any function $g$ in $L^2(-\infty, \infty; e^{-x^2/2})$ and we obtain a valid numerical series

$$\langle f, g \rangle = \sum_{n=0}^{\infty} c_n \langle H_n, g \rangle$$

In particular, with $f = g$, we get the *Parseval relation for the Hermite series*:

$$\langle f, f \rangle = \sum_{n=0}^{\infty} |c_n|^2 \langle H_n, H_n \rangle = \sqrt{2\pi} \sum_{n=0}^{\infty} n! |c_n|^2$$

Let us look at some examples of Hermite series. We know by Lemma 4.32 that any power $x^n$ is a linear combination of $H_0$ through $H_n$. By uniqueness, this is the Hermite series for $x^n$. For example, $x^0 = 1 = H_0$ and $x = H_1$ are the Hermite series for $x^0$ and $x^1$, respectively. Also

$$x^2 = H_0 + H_2$$
$$x^3 = 3H_1 + H_3$$
$$x^4 = 3H_0 + 6H_2 + H_4$$

etc.

Fourier-Hermite expansions for other functions are dealt with in the exercises. For these exercises, one will need means for evaluating the Hermite polynomials. Here are HP-15C subroutines for evaluating $H_n(x)$, $2 \leqslant n \leqslant 5$, Tabular values are given in Table 4.3.

*HP-15C subroutines for the Hermite polynomials*

| | |
|---|---|
| $H_2(x) = x^2 - 1$ | $H_3(x) = x^3 - 3x$ |
| | $\quad = x(x^2 - 3)$ |
| g P/R | g P/R |
| f LBL 2 | f LBL 3 |
| $gx^2$ | ENTER |
| $1 -$ | $gx^2$ |
| g RTN | $3 -$ |
| g P/R | $\times$ |
| | g RTN |
| | g P/R |
| $H_4(x) = x^4 - 6x^2 + 3$ | $H_5(x) = x^5 - 10x^3 + 15x$ |
| $\quad = x^2(x^2 - 6) + 3$ | $\quad = x(x^2(x^2 - 10) + 15)$ |
| g P/R | g P/R |
| f LBL 4 | f LBL 5 |
| $gx^2$ | ENTER |
| ENTER, ENTER | $gx^2$ |
| $6 -$ | ENTER, ENTER |
| $\times 3 +$ | $10 - \times$ |
| g RTN | $15 + \times$ |
| g P/R | g RTN |
| | g P/R |

**Theorem 4.45.** *If the function $f(x)$ belongs to $L^2(-\infty, \infty; e^{-x^2/2})$ and is orthogonal to every $H_n(x)$, $n = 0, 1, 2, \ldots$, then $f = 0$.*

The proof, which parallels the proof of Theorem 4.25, uses the fact that the Hermite polynomials form an orthogonal basis of $L^2(-\infty, \infty; e^{-x^2/2})$. You are asked to provide the details in Exercise 13.

Using this result, we can complete Theorem 4.42.

**Theorem 4.46.** *Hermite's operator has no eigenvalues other than the integers $0, 1, 2, \ldots$, and each of these eigenvalues is simple.*

The proof is left as Exercise 14.

**Table 4.3**   Tables of the Hermite Polynomials

| $x$ | $H_2(x)$ | $H_3(x)$ | $H_4(x)$ | $H_5(x)$ |
|-----|----------|----------|----------|----------|
| 0   | $-1$     | 0        | 3.0      | 0        |
| 0.1 | $-0.99$  | $-0.299$ | 2.94     | 1.49     |
| 0.2 | $-0.96$  | $-0.592$ | 2.7616   | 2.92032  |
| 0.3 | $-0.91$  | $-0.873$ | 2.4681   | 4.23243  |
| 0.4 | $-0.84$  | $-1.136$ | 2.0656   | 5.37024  |
| 0.5 | $-0.75$  | $-1.375$ | 1.5625   | 6.28125  |
| 0.6 | $-0.64$  | $-1.584$ | 0.9696   | 6.91776  |
| 0.7 | $-0.51$  | $-1.757$ | 0.3001   | 7.23807  |
| 0.8 | $-0.36$  | $-1.888$ | $-0.4304$| 7.20768  |
| 0.9 | $-0.19$  | $-1.971$ | $-1.2039$| 6.80049  |
| 1.0 | 0        | $-2.0$   | $-2.0$   | 6.0      |
| 1.1 | 0.21     | $-1.969$ | $-2.7959$| 4.80051  |
| 1.2 | 0.44     | $-1.872$ | $-3.5664$| 3.20832  |
| 1.3 | 0.69     | $-1.703$ | $-4.2839$| 1.24293  |
| 1.4 | 0.96     | $-1.456$ | $-4.9184$| $-1.06176$ |
| 1.5 | 1.25     | $\div 1.125$ | $-5.4375$ | $-3.65625$ |
| 1.6 | 1.56     | $-0.704$ | $-5.8064$| $-6.47424$ |
| 1.7 | 1.89     | $-0.187$ | $\div 5.9879$ | $-9.43143$ |
| 1.8 | 2.24     | 0.432    | $-5.9424$| $-12.42432$ |
| 1.9 | 2.61     | 1.159    | $-5.6279$| $-15.32901$ |
| 2.0 | 3.0      | 2.0      | $-5.0$   | $-18.0$  |
| 2.1 | 3.41     | 2.961    | $-4.0119$| $-20.26899$ |
| 2.2 | 3.84     | 4.048    | $-2.6144$| $-21.94368$ |
| 2.3 | 4.29     | 5.267    | $\div 0.7559$ | $-22.80657$ |
| 2.4 | 4.76     | 6.624    | 1.6176   | $-22.61376$ |
| 2.5 | 5.25     | 8.125    | 4.5625   | $-21.09375$ |
| 2.6 | 5.76     | 9.776    | 8.1376   | $-17.94624$ |
| 2.7 | 6.29     | 11.583   | 12.4041  | $-12.84093$ |
| 2.8 | 6.84     | 13.552   | 17.4256  | $-5.41632$ |
| 2.9 | 7.41     | 15.689   | 23.2881  | 4.72149  |
| 3.0 | 8.0      | 18.0     | 30.0     | 18.0     |

## Exercises

1. Repeat Exercise 7 of Section 3.9, where the formula

$$\|x^n\|^2 = \langle x^n, x^n \rangle = \int_{-\infty}^{\infty} x^{2n} e^{-x^2/2}\, dx$$

$$= \frac{(2n)!}{n!\, 2^n} \sqrt{2\pi}, \qquad n = 0, 1, 2, \ldots$$

was derived. Conclude that $L^2(-\infty, \infty; e^{-x^2/2})$ contains every $x^n$, $n = 0, 1, 2, \ldots$, thus every polynomial.

2. (Continuation) Prove the formula

$$\int_{-\infty}^{\infty} x^m e^{2\alpha x} e^{-x^2/2}\,dx = e^{2\alpha^2} \int_{-\infty}^{\infty} (x + 2\alpha)^m e^{-x^2/2}\,dx$$

and use it to conclude that $L^2(-\infty, \infty; e^{-x^2/2})$ contains $x^n e^{\alpha x}$ for every $n = 0, 1, 2, \ldots$ and every real $\alpha$. Show also that $x^n e^{\alpha x}$ belongs to the domain $D$ of Hermite's operator for $n = 0, 1, 2, \ldots$.

3. Show that $f(x) \in L^2(-\infty, \infty; e^{-x^2/2})$ if, and only if, $f(x)e^{-x^2/4} \in L^2(-\infty, \infty)$.

4. Define a function $h(x)$ as follows: if $x \geqslant 0$, set $h(x) = 1$ when $n \leqslant x \leqslant n + 1/(n+1)^2$, $n = 0, 1, 2, \ldots$, and set $h(x) = 0$ for other positive values of $x$. For $x$ negative, continue $h(x)$ as an even function: $h(-x) = h(x)$. Define $g(x) = h(x)e^{x^2/4}$.
   (a) Sketch a graph of $h(x)$. Show that $h(x)^2 = h(x)$.
   (b) Show that

$$\|g(x)\|^2 = \int_{-\infty}^{\infty} h(x)^2\,dx = 2 \sum_{n=1}^{\infty} \frac{1}{n^2} = \frac{\pi^2}{3}$$

   Conclude, in particular, that $g(x) \in L^2(-\infty, \infty; e^{-x^2/2})$.
   (c) Does $|g(x)|^2 e^{-x^2/2} \to 0$ as $x \to \infty$?

5. Let $f(x) = (x^2 + 1)^{-1/2} e^{x^2/4}$.
   (a) Show that

$$\|f(x)\|^2 = \int_{-\infty}^{\infty} \frac{dx}{1 + x^2} = \pi$$

   Conclude, in particular, that $f(x) \in L^2(-\infty, \infty; e^{-x^2/2})$.
   (b) Show that

$$f'(x) = \frac{1}{2}\left(\frac{x^2 - 1}{x^2 + 1}\right)\left(\frac{x}{(x^2 + 1)^{1/2}}\right)e^{x^2/4}$$

   hence that

$$|f'(x)|^2 e^{-x^2/2} = \frac{1}{4}\left(\frac{x^2 - 1}{x^2 + 1}\right)^2 \left(\frac{x^2}{x^2 + 1}\right)$$

   Using this last equation, show $\lim_{x \to \pm\infty} |f'(x)|^2 e^{-x^2/2} = 1/4$. Does $f'(x) \in L^2(-\infty, \infty; e^{-x^2/2})$? Does $f(x)$ belong to the domain of Hermite's operator?

6. Using the functions $g(x)$ and $f(x)$ from Exercises 4 and 5, respectively, show that

$$e^{-x^2/2} f'(x)\overline{g(x)} = \frac{1}{2}\left(\frac{x^2 - 1}{x^2 + 1}\right)\left(\frac{x}{(x^2 + 1)^{1/2}}\right)h(x)$$

Describe the behavior of $e^{-x^2/2} f'(x)\overline{g(x)}$ for large $x$. Show that $\lim_{x \to \infty} e^{-x^2/2} f'(x)\overline{g(x)}$ does not exist. Discuss in the light of Lemma 4.39.

7. Check that the function $f(x) = \int_0^x e^{t^2/2}\, dt$ is odd and satisfies $f'(x) = e^{x^2/2}$. Check also that $f(x)$ satisfies $-e^{x^2/2}(e^{-x^2/2}f'(x))' = 0$, thus is an eigenfunction for this operator belonging to the eigenvalue $\lambda = 0$. Check also that $g(x) = x$ satisfies $-e^{x^2/2}(e^{-x^2/2}g')' = g$. Thus $g(x)$ is an eigenfunction belonging to the eigenvalue $\lambda = 1$. As $0 \neq 1$, these eigenfunctions belong to different eigenvalues, so they must be orthogonal, i.e.;

$$\langle f(x), g(x) \rangle = \int_{-\infty}^{\infty} xf(x)e^{-x^2/2}\, dx = 0$$

But $xf(x)$ is even, and $> 0$ for $x > 0$, so this integral cannot be 0! What is wrong? (Use Exercise 10 of Section 3.9.)

8. Show that for all functions $f$, $g$ in the domain $D$ of Hermite's operator $L$ we have

$$\int_r^s L(f)\bar{g}e^{-x^2/2}\, dx = -e^{-s^2/2} f'(s)\bar{g}(s) + e^{-r^2/2} f'(r)\bar{g}(r)$$

$$+ \int_r^s f'(x)\bar{g}'(x)e^{-x^2/2}\, dx$$

Deduce, by passage to the limit, that

$$\langle L(f), g \rangle = \langle f', g' \rangle$$

and from this prove that Hermite's operator is Hermitian and positive semidefinite.

9. Suppose we have a Hermite expansion $f = \Sigma\, a_n H_n$ convergent in the mean-square metric. By taking the inner product of both sides with $H_m$, prove that $a_n$ are the Fourier-Hermite coefficients as given in equation (4.9.5).

10. Let $f(x) = 1$ for $x > 0$, $f(x) = -1$ for $x < 0$, and $f(0) = 0$. Find the six-term partial sum of the Hermite series for $f(x)$ ($H_0$ through $H_5$) and graph it against the function, $-3.6 \leqslant x \leqslant 3.6$. [Hint: Integrate by parts to derive the formula

$$\int_0^{\infty} x^{2m+1} e^{-x^2/2}\, dx = 2m \int_0^{\infty} x^{2m-1} e^{-x^2/2}\, dx$$

then use this to evaluate the integral on the left side for $m = 1, 2$. Use these numbers to compute the Fourier-Hermite coefficients of $f(x)$.]

11. Find the six-term partial sum of the Hermite series for $f(x) = |x|$ and graph it against the function, $-4 \leqslant x \leqslant 4$. (Use the hint for Exercise 10.)

12. Find the six-term partial sum of the Hermite series for $f(x) = e^x$ and the six-term partial sum of its power series. Graph all three functions on four-cycle semilog paper, $-2 \leqslant x \leqslant 5$.

13. Prove Theorem 4.45.

14. Prove Theorem 4.46.

## 4.10  SOLVING OPERATOR EQUATIONS WITH HERMITE'S OPERATOR

In the previous section, we showed that Hermite's operator $L$ has a family of eigenfunctions, the Hermite polynomials, that form an orthogonal basis of the Hilbert space $L^2(-\infty, \infty; e^{-x^2/2})$. The existence of such an orthogonal basis provides us with a new method for dealing with Hermite's operator.

For example, suppose we wish to find $L(x^4)$. We know that

$$x^4 = 3H_0 + 6H_2 + H_4$$

Using the linearity of Hermite's operator $L$ and the fact that the Hermite polynomials $H_n$ are eigenfunctions for $L$, $L(H_n) = nH_n$, $n = 0, 1, 2, \ldots$, we get

$$
\begin{aligned}
L(x^4) &= 3L(H_0) + 6L(H_2) + L(H_4) \\
&= 3(0)H_0 + 6(2)H_2 + 4H_4 \\
&= 12H_2 + 4H_4
\end{aligned}
$$

We can calculate the same quantity using the differential expression for $L$: $l(x^4) = -(x^4)'' + x(x^4)' = -12x^2 + 4x^4$, which is equal to the last displayed formula.

This method obviously works for any *finite* linear combination of Hermite polynomials. It also works for an infinite series of Hermite polynomials when the sum function belongs to the domain of Hermite's operator.

**Theorem 4.47.**  *Suppose $y(x)$ lies in the domain of Hermite's operator $L$. Then the equation obtained by applying $L$ term by term to the Hermite series for $y(x)$ is valid. Put another way: If $y = \sum_{n=0}^{\infty} c_n H_n$, where $c_n = \langle y, H_n \rangle / \langle H_n, H_n \rangle$, and if $y \in \text{domain}(L)$, then*

$$L(y) = \sum_{n=0}^{\infty} c_n L(H_n) = \sum_{n=0}^{\infty} nc_n H_n \qquad (4.10.1)$$

The proof of Theorem 4.47, which is similar to that of Theorem 4.28, is left as Exercise 6.

Note especially that, while any function $y$ in $L^2(-\infty, \infty; e^{-x^2/2})$ has a Hermite expansion as given by equation (4.9.4), in order to apply Hermite's operator $L$ to both sides of (4.9.4) it is essential that the function $y$ belongs to the domain of $L$. Otherwise, in equation (4.10.1), the left side is undefined. If we ignore the domain restriction and simply apply the differential expression $l(y) = -y'' + xy'$, then equation (4.10.1) will be nonsense when $y$ does not belong to the domain of $L$.

In Section 4.11 we shall derive the following Hermite expansion of $e^x$:

$$e^x = e^{1/2} \sum_{n=0}^{\infty} \frac{1}{n!} H_n(x) \tag{4.10.2}$$

(See also Exercise 12 of Section 4.9.) The function $y = e^x$ belongs to the domain of Hermite's operator (Exercise 2 of Section 4.9) so we may apply Hermite's operator $L$ term by term to (4.10.2):

$$L(e^x) = e^{1/2} \sum_{n=0}^{\infty} \frac{1}{n!} L(H_n) = e^{1/2} \sum_{n=0}^{\infty} \frac{1}{n!} n H_n(x) \tag{4.10.3}$$

and get thereby the Hermite expansion for $L(e^x)$. The function $L(e^x)$ can be explicitly calculated using the differential expression for $L$ (Exercise 4).

These calculations show one major advantage of the Hilbert space method: we can calculate the action of Hermite's operator $L$ on a function $y$ (in the domain of $L$) by applying $L$ term by term to the Hermite series expansion of $y$.

A second feature of the new method arises in the solution of operator equations $L(y) = h$. The problem is this: we are given a function $h(x)$ in $L^2(-\infty, \infty; e^{-x^2/2})$, and we seek a function $y(x)$ in the domain of $L$ such that $L(y) = h$. So $h$ is known, given beforehand, and $y$ is unknown.

There are three aspects to this problem. First, does a solution $y$ exist at all? Second, if a solution does exist, to what extent is it unique? Third, when a solution exists, how can we find it effectively?

The following result addresses the first aspect.

**Theorem 4.48.** *If the equation* $L(y) = h$, *where* $h$ *belongs to* $L^2(-\infty, \infty; e^{-x^2/2})$, *has a solution* $y$ *in the domain of* $L$, *then* $\langle h, H_0 \rangle = 0$.

The proof is left as Exercise 7.

Now $H_0 = 1$, so the condition $\langle h, H_0 \rangle = 0$ means

$$\int_{-\infty}^{\infty} h(x) e^{-x^2/2} \, dx = 0$$

We may accordingly rephrase Theorem 4.48 thus:

**Corollary 4.49.** *Consider Hermite's operator L acting in the Hilbert space* $L^2(-\infty, \infty; e^{-x^2/2})$, *and consider the equation* $L(y) = h$ *where* $h$ *is a given function in* $L^2(-\infty, \infty; e^{-x^2/2})$. *If*

$$\int_{-\infty}^{\infty} h(x)e^{-x^2/2} \, dx \neq 0$$

*then* $L(y) = h$ *has no solution* $y$ *in the domain of* $L$.

So in the Hilbert space theory we have a restriction limiting the cases where $L(y) = h$ is solvable, a restriction that has no counterpart in the theory of Chapter 2. For example, $L(y) = 1$ has no solution, and neither does $L(y) = x^2$.

When $\langle h, H_0 \rangle = 0$, can we then solve $L(y) = h$? Here is a solution method that uses the orthogonal basis. Since $h$ belongs to $L^2(-\infty, \infty; e^{-x^2/2})$, it has a Fourier-Hermite expansion. Since $\langle h, H_0 \rangle = 0$, the first term in the expansion will vanish. Thus

$$h = \sum_{n=1}^{\infty} c_n H_n \quad \text{where } c_n = \frac{\langle h, H_n \rangle}{\langle H_n, H_n \rangle} = \frac{1}{\sqrt{2\pi n!}} \int_{-\infty}^{\infty} h H_n e^{-x^2/2} \, dx$$

As $h$ is a known function, all the coefficients $c_n$ are known. We seek a function $y$ that belongs to the domain of $L$ and satisfies $L(y) = h$. Let us assume that such a $y$ exists (which is plausible because we have met the restriction $\langle h, H_0 \rangle = 0$). Then $L(y)$ will have a Hermite expansion

$$L(y) = \sum_{n=0}^{\infty} n a_n H_n$$

and our equation $L(y) = h$ takes the form

$$\sum_{n=0}^{\infty} n a_n H_n = \sum_{n=1}^{\infty} c_n H_n$$

The term on the left is $0 a_0 H_0 = 0$, which is consistent with the fact that there is no $n = 0$ term on the right. Thus, we see again where the restriction $\langle h, H_0 \rangle = 0$ comes into play. Now the coefficients of a Hermite expansion are unique, so we must have $n a_n = c_n$ for $n \geq 1$, or

$$a_n = \frac{c_n}{n}, \quad n = 1, 2, \ldots$$

This determines all the expansion coefficients for the function $y(x)$ except $a_0$, which can be arbitrarily specified. Thus our solution $y$ to $L(y) = h$ is

$$y = a_0 + \sum_{n=1}^{\infty} \frac{c_n}{n} H_n \quad \text{where } c_n = \frac{\langle h, H_n \rangle}{\langle H_n, H_n \rangle} = \frac{1}{\sqrt{2\pi n!}} \int_{-\infty}^{\infty} h H_n e^{-x^2/2} \, dx$$

and $a_0$ is an arbitrary constant.

For example, suppose we wished to solve $L(y) = x^3$. Now $x_3 = 3H_1 + H_3$. Then, if $y = \Sigma a_n H_n$, our equation $L(y) = h$ becomes

$$\sum_{n=0}^{\infty} n a_n H_n = 3H_1 + H_3$$

Hence $1a_1 = 3$, $3a_3 = 1$, and the other $a_n$ with $n \geqslant 1$ are zero. So our solution $y$ is

$$y = a_0 + 3H_1 + \tfrac{1}{3}H_3 \tag{4.10.4}$$

where $a_0$ is an arbitrary constant.

We can also use the differential calculus approach to the same problem. In terms of the differential expression, our equation $L(y) = x^3$ is

$$-e^{x^2/2}(e^{-x^2/2}y')' = x^3$$

Integrating once, we get

$$e^{-x^2/2}y' = -\int_0^x t^3 e^{-t^2/2}\,dt + y'(0)$$

Use parts on the integral:

$$\int_0^x t^3 e^{-t^2/2}\,dt = 2 - (2 + x^2)e^{-x^2/2}$$

which leads to $e^{-x^2/2}y' = (2+x^2)e^{-x^2/2} + y'(0) - 2$, or

$$y'(x) = 2 + x^2 + (y'(0) - 2)e^{x^2/2}$$

Integrating once more:

$$y(x) - y(0) = 2x + \tfrac{1}{3}x^3 + \int_0^x (y'(0) - 2)e^{t^2/2}\,dt \tag{4.10.5}$$

In order to have an $L^2$ solution, we must have $y'(0) = 2$ (Exercise 8). Hence

$$y(x) = y(0) + 2x + \tfrac{1}{3}x^3$$

which agrees with equation (4.10.4) when we set $y(0) = a_0$.

We shall take for granted that this method we have just illustrated always works. That is, when $\langle h, H_0 \rangle = 0$, we can solve $L(y) = h$ by expanding the unknown function $y$ in its Hermite series and solving for the coefficients. The assumption that this method always works is equivalent to the assumption that the range of Hermite's operator is all of $H_0^\perp$, the subspace of $L^2(-\infty, \infty; e^{-x^2/2})$ consisting of functions that are orthogonal to $H_0$.

As for uniqueness, the kernel of Hermite's operator is the one-dimensional subspace spanned by $H_0$ (Exercise 5). Hence, if we have a solution $y_p$ to

$L(y) = h$ (a particular solution), then $y = y_p + a_0 H_0$ is also a solution for any constant $a_0$, and every solution has this form for some $a_0$ ($y_p + a_0 H_0$ is the general solution). We summarize our conclusions about the equation $L(y) = h$.

**Theorem 4.50.** *Consider Hermite's operator $L$ acting in the Hilbert space $L^2(-\infty, \infty; e^{-x^2/2})$. Given $h$ in $L^2(-\infty, \infty; e^{-x^2/2})$, the equation $L(y) = h$ has a solution $y$ in the domain of $L$ if and only if*

$$\langle h, H_0 \rangle = \int_{-\infty}^{\infty} h(x) e^{-x^2/2} dx = 0$$

*In this case the general solution $y$ has the Hermite expansion*

$$y = a_0 + \sum_{n=1}^{\infty} \frac{c_n}{n} H_n(x)$$

*where the $c_n$ are the Fourier-Hermite coefficients of the known function $h$,*

$$c_n = \frac{\langle h, H_n \rangle}{\langle H_n, H_n \rangle} = \frac{1}{\sqrt{2\pi}\, n!} \int_{-\infty}^{\infty} h(x) H_n(x) e^{-x^2/2} dx, \qquad n = 1, 2, \ldots$$

*and $a_0$ is an arbitrary constant.*

### Exercises

In all the exercises below, $L$ stands for Hermite's operator.

1. Derive the equation $x^3 = 3H_1 + H_3$, and use it to compute $L(x^3)$ as a linear combination of Hermite polynomials. Check your answer by using the differential expression for $L$ to compute $L(x^3)$.
2. Express $x^5$ as a finite linear combination of Hermite polynomials, and use your expression to compute $L(x^5)$ as a finite linear combination of Hermite polynomials. Check, using the differential expression for $L$.
3. Suppose $f(x)$ is a finite linear combination of Hermite polynomials, $f(x) = c_0 H_0(x) + c_1 H_1(x) + \cdots + c_N H_N(x)$. Express $L(f)$ as a finite linear combination of Hermite polynomials.
4. Use the differential expression for $L$ to evaluate the term $L(e^x)$ on the left side of equation (4.10.3), and thus derive the following Hermite expansion:

$$(x - 1)e^x = e^{1/2} \sum_{n=1}^{\infty} \frac{1}{(n-1)!} H_n(x)$$

$$= e^{1/2}(H_1 + H_2 + (\tfrac{1}{2})H_3 + (\tfrac{1}{6})H_4 + (\tfrac{1}{24})H_5 + \cdots)$$

Compute both the six-term Hermite expansion ($H_0$ through $H_5$) and the six-term power series expansion ($x^0$ through $x^5$) and compare them with

$(x-1)e^x$ over the range $-4 \leqslant x \leqslant 4.5$. Compare all three on a semilog graph over the range $1.1 \leqslant x \leqslant 4.5$. (To get the power series for $(x-1)e^x$ multiply the power series for $e^x$ by $x$ and then subtract the power series for $e^x$.)

5. The kernel of Hermite's operator $L$ consists of all functions $y$ in the domain of $L$ such that $L(y)=0$.
   (a) Show that the kernel of Hermite's operator is one-dimensional, and find a convenient function that spans it.
   (b) Find $\ker(L-2I)$ and $\ker(L-5I)$, where $L$ is Hermite's operator and $I$ is the identity operator.

6. Prove Theorem 4.47.

7. Prove Theorem 4.48. (Compare Theorem 4.29; also Exercise 17 of Section 4.6.)

8. Explain why, in equation (4.10.5), we must have $y'(0)=2$. (See Exercise 10 of Section 3.9.)

9. Solve $L(y)=H_3(x)$ for $y$.

10. Does $L(y)=x^5$ have a solution? If so, find it as a linear combination of Hermite polynomials.

11. Does $L(y)=x^4-3x^2$ have a solution? If so, find it.

## 4.11 MORE ON HERMITE POLYNOMIALS: RODRIGUES' FORMULA, THE RECURSION RELATION, AND THE GENERATING FUNCTION

Hermite's polynomials are a sequence of polynomials $H_n(x)$, $n=0,1,2,\ldots$, such that, for each $n$, $H_n(x)$ has degree $n$. Hence Lemma 4.32 applies, and we have

**Lemma 4.51.** *Any polynomial of degree $n$ is a linear combination of the first $n+1$ Hermite polynomials $H_0, H_1, \ldots, H_n$. In particular, $x^n$ is a linear combination of $H_0, H_1, \ldots, H_n$.*

*Proof.* This result is a special case of Lemma 4.32, which we proved by mathematical induction. Since we have already proved it once, we don't need to prove it again. But it probably won't do any harm to refresh our memories by briefly recalling the *idea* of the proof.

Clearly, it is enough to write $x^n$ as a linear combination of $H_0, H_1, \ldots, H_n$, because then to represent a polynomial as a linear combination of Hermite polynomials we simply represent each term and then add the results. The Hermite polynomials are normalized so that the leading coefficient is 1, hence $H_n(x)=x^n+\cdots$ where the ellipses represent terms of degree less than $n$. Hence $x^n-H_n(x)$ has degree $<n$, say $x^n-H_n(x)=a_k x^k+\cdots$ where $a_k \neq 0$ and $k<n$.

Then $x^n - H_n(x) - a_k H_k(x)$ has degree $<k$. We then continue this process until we get 0. Then $x^n = H_n(x) + a_k H_k(x) + \cdots$. That is the idea of the proof.

**Lemma 4.52.** *If $p(x)$ is a polynomial of degree $m$, then $\langle p(x), H_n(x) \rangle = 0$ for $n > m$. In particular, $\langle x^m, H_n(x) \rangle = 0$ for $n > m$.*

*Proof.* Recall that we are using throughout Sections 9, 10, and 11 the inner product of $L^2(-\infty, \infty; e^{-x^2/2})$:

$$\langle f, g \rangle = \int_{-\infty}^{\infty} f(x)\overline{g(x)}e^{-x^2/2} \, dx$$

To prove Lemma 4.52 use Lemma 4.51 to write $p(x) = c_0 H_0(x) + c_1 H_1(x) + \cdots + c_m H_m(x)$; then

$$\langle p(x), H_n(x) \rangle = \sum_{k=0}^{m} c_k \langle H_k, H_n \rangle = 0$$

because the Hermite polynomials are an orthogonal family and, for each term in the sum, $k \leqslant m < n$.

We turn now to an explicit formula for the Hermite polynomials, known as Rodrigues' formula (same name as the formula for Legendre polynomials).

**Theorem 4.53.** $H_n(x) = (-1)^n e^{x^2/2} \, d^n/dx^n(e^{-x^2/2})$.

*Proof.* The proof here will follow the same pattern as the proof of Theorem 4.34. Set $Q_n(x) = (-1)^n e^{x^2/2} \, d^n/dx^n(e^{-x^2/2})$. We shall show that the $Q_n(x)$ are a sequence of polynomials $Q_n$ of degree $n$, and they form an orthogonal family with respect to the inner product of $L^2(-\infty, \infty; e^{-x^2/2})$.

A little experimentation quickly convinces one that $Q_n$ is a polynomial of degree $n$ with leading coefficient 1.

$$Q_0(x) = (-1)^0 e^{x^2/2} \frac{d^0}{dx^0}(e^{-x^2/2}) = 1$$

$$Q_1(x) = (-1)^1 e^{x^2/2} \frac{d}{dx}(e^{-x^2/2}) = -e^{x^2/2}(-xe^{-x^2/2}) = x$$

$$Q_2(x) = (-1)^2 e^{x^2/2} \frac{d}{dx}\left(\frac{d}{dx}(e^{-x^2/2})\right) = e^{x^2/2} \frac{d}{dx}(-xe^{-x^2/2})$$

$$= e^{x^2/2}(x^2 - 1)e^{-x^2/2} = x^2 - 1$$

etc.

A complete proof can be given by mathematical induction (Exercise 2). Hence $Q_n(x)$ is a polynomial of degree $n$ with leading coefficient 1.

Next we prove orthogonality: $\langle Q_m, Q_n \rangle = 0$ when $m \neq n$. It is enough to show that $\langle x^m, Q_n(x) \rangle = 0$ when $m < n$. The proof is by integration by parts.

$$\langle x^m, Q_n \rangle = \int_{-\infty}^{\infty} x^m (-1)^n e^{x^2/2} \frac{d^n}{dx^n} (e^{-x^2/2}) e^{-x^2/2} \, dx$$

$$= (-1)^n \int_{-\infty}^{\infty} x^m \, d\left[ \frac{d^{n-1}}{dx^{n-1}} (e^{-x^2/2}) \right]$$

$$= (-1)^n \left[ x^m \frac{d^{n-1}}{dx^{n-1}} (e^{-x^2/2}) \Big|_{-\infty}^{\infty} - m \int_{-\infty}^{\infty} x^{m-1} \frac{d^{n-1}}{dx^{n-1}} (e^{-x^2/2}) \, dx \right]$$

The integrated term vanishes, and we get

$$\langle x^m, Q_n \rangle = m \langle x^{m-1}, Q_{n-1} \rangle$$

Continuing, we find

$$\langle x^m, Q_n \rangle = m! \langle 1, Q_{n-m} \rangle = m! (-1)^{n-m} \int_{-\infty}^{\infty} \frac{d^{n-m}}{dx^{n-m}} (e^{-x^2/2}) \, dx \qquad (4.11.1)$$

If $m < n$, then $n - m > 0$, so the integrand on the right is a perfect differential. Thus its integral is 0. This proves orthogonality.

Now by Lemma 4.52, $\langle Q_m(x), H_n(x) \rangle = 0$ for $m < n$. By Lemma 4.32, we may write

$$H_n = a_0 Q_0 + a_1 Q_1 + \cdots + a_n Q_n$$

Taking the inner product of both sides with $Q_m$, $m < n$, we get $0 = a_m \langle Q_m, Q_m \rangle$, which shows $a_m = 0$, $m < n$. Thus $H_n = a_n Q_n$. As both polynomials have leading coefficient 1, we have $H_n = Q_n$. Theorem 4.53 is proved.

**Theorem 4.54.** $\langle H_n, H_n \rangle = \sqrt{2\pi} \, n!$.

*Proof.* Put $m = n$ in equation (4.11.1) to get

$$\langle x^n, H_n \rangle = n! \int_{-\infty}^{\infty} e^{-x^2/2} \, dx = \sqrt{2\pi} \, n!$$

the last step following from Exercise 1 of Section 4.9.

The Hermite polynomials, like the Legendre polynomials, have a three-term recursion relation.

**Theorem 4.55.** $H_n = x H_{n-1} - (n-1) H_{n-2}$, $n \geqslant 2$.

The proof is outlined in Exercise 6.

For any number $w$ the function $e^{wx}$ belongs to $L^2(-\infty, \infty; e^{-x^2/2})$ (Exercise 2 of Section 4.9), so has a Hermite expansion. It, multiplied by $e^{-w^2/2}$, is the generating function of the Hermite polynomials.

**Theorem 4.56**

$$e^{wx - w^2/2} = \sum_{n=0}^{\infty} \frac{w^n}{n!} H_n(x)$$

*Proof.* Let

$$h(x, w) = \sum_{n=0}^{\infty} \frac{w^n}{n!} H_n(x) = H_0(x) + wH_1(x) + \frac{w^2}{2!} H_2(x) + \cdots$$

Fix $x$ and regard $h$ as a function of $w$ alone. Differentiate the power series term by term to get

$$\frac{dh}{dw} = \sum_{n=0}^{\infty} \frac{nw^{n-1}}{n!} H_n(x) = H_1(x) + \sum_{n=2}^{\infty} \frac{w^{n-1}}{(n-1)!} H_n(x)$$

Then use the recursion relation (Theorem 4.55) and substitute $H_1(x) = x$ to get

$$\frac{dh}{dw} = x + \sum_{n=2}^{\infty} \frac{w^{n-1}}{(n-1)!} (xH_{n-1}(x) - (n-1)H_{n-2}(x))$$

$$= x + x \sum_{n=2}^{\infty} \frac{w^{n-1}}{(n-1)!} H_{n-1}(x) - w \sum_{n=2}^{\infty} \frac{w^{n-2}}{(n-2)!} H_{n-2}(x)$$

For the first series we have

$$\sum_{n=2}^{\infty} \frac{w^{n-1}}{(n-1)!} H_{n-1}(x) = \sum_{n=1}^{\infty} \frac{w^n}{n!} H_n(x) = h - 1$$

The second series is just $h$, so we have

$$\frac{dh}{dw} = x + x(h - 1) - wh = (x - w)h$$

whence

$$\frac{dh}{h} = (x - w)\, dw$$

Remember that $x$ is fixed, and integrate both sides.

$$\ln h = (xw - w^2/2) + c$$
$$h = (\text{const})e^{xw - w^2/2}$$

As $h(x, 0) = 1$, the constant is 1 and Theorem 4.56 is proved.

The Hermite coefficients are uniquely determined, so as a by-product of the series of the generating function we get

$$\frac{w^n}{n!} = \frac{\langle e^{wx - w^2/2}, H_n(x) \rangle}{\langle H_n(x), H_n(x) \rangle} = \frac{1}{\sqrt{2\pi\, n!}} e^{-w^2/2} \int_{-\infty}^{\infty} e^{wx} H_n(x) e^{-x^2/2}\, dx$$

or

$$\int_{-\infty}^{\infty} e^{wx} H_n(x) e^{-x^2/2} \, dx = \sqrt{2\pi} \, w^n e^{w^2/2}, \qquad n = 0, 1, 2, \ldots$$

## Exercises

1. Prove that the Hermite polynomials are uniquely determined up to scalar multiples by their orthogonality properties. That is, prove the following result. If $R_n$, $n = 0, 1, 2, \ldots$, is a sequence of polynomials such that (1) $R_n$ is of degree $n$ for each $n = 0, 1, 2, \ldots$ and (2) $\langle R_m, R_n \rangle = 0$, $m \neq n$, then $R_n = \mu_n H_n$, $n = 0, 1, 2, \ldots$, where the $\mu_n$ are nonzero scalars.

2. Prove by mathematical induction that

$$(-1)^n e^{x^2/2} \frac{d^n}{dx^n} e^{-x^2/2}$$

   is a polynomial of degree $n$ with leading coefficient 1.

3. Using Rodrigues' formula, compute $H_3$ and $H_4$.

4. Using Rodrigues' formula, prove that

$$\frac{d}{dx}(e^{-x^2/2} H_n) = -e^{-x^2/2} H_{n+1} \qquad \text{and} \qquad H_n' = x H_n - H_{n+1}$$

5. Using integration by parts and Exercise 4 show that

$$\langle H_n', H_m \rangle = \langle H_n, H_{m+1} \rangle$$

   thus

$$\langle H_n', H_m \rangle = \begin{cases} 0, & m \neq n - 1 \\ \sqrt{2\pi} \, n!, & m = n - 1 \end{cases}$$

6. $H_n'$ is a polynomial of degree $n - 1$; hence $H_n' = \sum_{k=0}^{n-1} c_k H_k$. Using the result of Exercise 5, evaluate the constants $c_k$ and thus prove $H_n' = n H_{n-1}$. Combine this with Exercise 4 to prove the recursion relation for Hermite's polynomials:

$$H_{n+1} = x H_n - n H_{n-1}, \qquad n \geqslant 1$$

7. Starting with $H_0 = 1$ and $H_1 = x$, use the recursion relation to compute $H_2$ through $H_6$.

8. Prove that if $f(x) = \sum_{n=0}^{\infty} c_n H_n$, then

$$xf(x) = c_1 H_0(x) + \sum_{n=1}^{\infty} (c_{n-1} + (n+1)c_{n+1}) H_n(x)$$

9. Derive the general formula for Hermite's polynomials

$$H_n(x) = x^n - 1 \cdot \binom{n}{2} x^{n-2} + 1 \cdot 3 \binom{n}{4} x^{n-4} - 1 \cdot 3 \cdot 5 \binom{n}{6} x^{n-6} + \cdots$$

(The binomial coefficients $\binom{n}{k}$ have been defined in Exercise 3 of Section 4.8.)

10. Determine the Parseval relation corresponding to the Hermite expansion of the generating function, and show that it is a familiar power series.

## APPENDIX 4.A MATHEMATICAL ASPECTS OF DIFFERENTIAL OPERATORS IN $L^2$ SPACES

A function is called *absolutely continuous* if it is the integral of its derivative. In symbols, the function $f(x)$ is absolutely continuous if

$$f(x) = \int_a^x f'(t)\,dt + f(a) \tag{4.A.1}$$

All functions that we have ever seen in basic calculus and all functions we shall meet in applications do have this property—each such function is the integral of its derivative. However, G. Cantor and H. Lebesgue, in their researches in pure mathematics, have shown the existence of functions that fail to satisfy this condition. The Cantor-Lebesgue function, so-called, is a continuous increasing function $f(x)$ defined on the closed interval $0 \leqslant x \leqslant 1$ such that $f(0) = 0$ and $f(1) = 1$, yet $f'(x) = 0$ almost everywhere (in the parlance of the trade). Hence, in Lebesgue's theory of integration, the integral on the right in equation (4.1.A) is identically zero, while the function $f(x)$ on the left increases continuously from 0 to 1 as $x$ goes from 0 to 1. So equation (4.A.1) fails dramatically for the Cantor-Lebesgue function.

Such not absolutely continuous functions as the Cantor-Lebesgue example are foreign to the spirit of this text. We wish to reach the major applications from a starting point of basic calculus, so we exclude such functions tacitly. We have taken and shall continue to take equation (4.A.1) for granted. For example, integration by parts may fail if we admit functions of the Cantor-Lebesgue type. We simply ignored this possibility in our frequent application of integration by parts to Hermitian operators. We shall continue to do all our calculations in that spirit.

But this optional section will constitute a brief diversion from that program. Here we shall study the implications for the theory of differential operators in $L^2$ spaces of the explicit assumption of absolute continuity. And, for simplicity, we shall take all spaces as real.

We begin by summarizing certain properties of the integral that we shall need. For a thorough development of the theory of the Lebesgue integral one may consult the book *Measure and Integral* by R. L. Wheeden and A. Zygmund (Marcel Dekker, Inc. 1977).

We shall be dealing with a finite or infinite interval $(a, b)$ and with the space $L^2(a, b)$ of functions that are square integrable on $(a, b)$, that is, with the space of functions $f(x)$ that satisfy

$$\int_a^b |f(x)|^2 \, dx < \infty$$

We shall also be dealing with functions that are integrable on $(a, b)$, meaning that

$$\int_a^b |f(x)| \, dx < \infty$$

As we have discussed in detail in Section 3.9 of Chapter 3, the convergence of this integral implies that of $\int_a^b f(x) \, dx$ where the absolute value signs have been removed.

**Theorem 4.57.**   (a) *If $f(x)$ and $g(x)$ belong to $L^2(a, b)$, then the product $f(x)g(x)$ is integrable on $(a, b)$.*

(b) *If $f(x)$ is square integrable on a finite interval, then it is integrable there.*

(c) *If $f(x)$ is integrable on $(a, b)$ then it is integrable on any subinterval $(c, d)$ of $(a, b)$, $a \leqslant c < d \leqslant b$.*

*Proof.*   Item (a) was proved in Section 3.9 of Chapter 3. We can also view it as a consequence of the Schwarz inequality:

$$\int_a^b |f(x)| \, |g(x)| \, dx \leqslant \left( \int_a^b |f(x)|^2 \, dx \right)^{1/2} \left( \int_a^b |g(x)|^2 \, dx \right)^{1/2}$$

(b) This item is proved by the following special case of the Schwarz inequality:

$$\int_a^b |f(x)| \, dx \leqslant (b - a)^{1/2} \left( \int_a^b |f(x)|^2 \, dx \right)^{1/2}$$

which was Exercise 4c, Section 3.6, Chapter 3.

(c) This statement follows from the inequality

$$\int_c^d |f(x)| \, dx \leqslant \int_a^b |f(x)| \, dx$$

valid when $a \leqslant c < d \leqslant b$.

We shall also want a precise statement of the conditions under which integration by parts is valid. The following result is Theorem 7.32 from the above-mentioned text by Wheeden and Zygmund.

**Theorem 4.58.** *If both $u(x)$ and $v(x)$ are absolutely continuous on the closed finite interval $a \leqslant x \leqslant b$, then*

$$\int_a^b u \, dv = [u(b)v(b) - u(a)v(a)] - \int_a^b v \, du$$

With this groundwork laid, we can begin our mathematically more rigorous study of second order linear differential operators in $L^2$ spaces. We shall first consider the simple differential expression $l(y) \equiv -y''$, then Legendre's expression $l(y) \equiv -((1 - x^2)y')'$, and finally Hermite's expression $l(y) \equiv -e^{x^2/2}(e^{-x^2/2}y')'$.

### 4.A.1 Operators with Differential Expression $l(y) \equiv -y''$

**Definition 4.59.** Let $(a, b)$ be a finite or infinite interval, and consider the inner product space $L^2(a, b)$. We call a function $y$ in $L^2(a, b)$ *compatible* with the differential expression $l(y) \equiv -y''$ if both:

1. For some fixed $x_0$, $a < x_0 < b$, we have

   $$y'(x) - y'(x_0) = \int_{x_0}^x y''(t) \, dt$$

   for all $x$ satisfying $a < x < b$.
2. $y'' \in L^2(a, b)$.

In item 1 of Definition 4.59 we are assuming the absolute continuity of the *first derivative* $y'(x)$ on the open interval $a < x < b$. In Section 4.1 we noted that the continuity of $y'(x)$ was a necessary condition in order to ensure that $y''$ belonged to $L^2(a, b)$. Every absolutely continuous function is continuous, so item 1 of Definition 4.59, which states $y'(x)$ is absolutely continuous on $a < x < b$, is a strengthening of the earlier condition that $y'$ be continuous on the same open interval.

If we are considering $L^2(C)$ where $C$ is the circle obtained by bending the interval $(a, b)$ and identifying the point $x = a$ with the point $x = b$, then the absolute continuity is assumed to hold at *every* point, since there are no longer any boundary points to exclude. This case has to be treated somewhat differently from that involving an ordinary interval, and we shall not go into the details of the matter.

Let us restate Definition 4.59 to specifically include the function $y(x)$ itself. To say that $y$ is compatible means that these four conditions are satisfied:

1. $y(x)$ is differentiable with finite derivative $y'(x)$ at each point $x$ in the open interval $a < x < b$.
2. $\int_a^b |y(x)|^2 \, dx < \infty$.
3. $y'(x)$ is absolutely continuous on the open interval $a < x < b$. Put another way, there exists an $x_0$, $a < x_0 < b$, such that

$$y'(x) - y'(x_0) = \int_{x_0}^x y''(t) \, dt$$

   for all $x$ satisfying $a < x < b$.
4. $\int_a^b |y''(x)|^2 \, dx < \infty$.

So, looked at under the mathematical microscope, those are the compatibility conditions on the domain of an operator $L$ with differential expression $l(y) \equiv -y''$ acting in $L^2(a, b)$. We shall now derive some consequences of these conditions.

The first conclusion we draw is this: $y(x)$ *is itself absolutely continuous on each interval* $[c, d]$, $a < c < d < b$. Hence, for some fixed $x_0$, $a < x_0 < b$, we have

$$y(x) - y(x_0) = \int_{x_0}^x y'(t) \, dt, \qquad a < x < b \qquad (4.\text{A}.2)$$

Every absolutely continuous function is continuous, so the formula displayed above is the fundamental theorem of calculus.

The second conclusion is this:

**Lemma 4.60.** *If the right-hand end point $b$ is finite, $b < \infty$, then both limits*

$$\lim_{x \to b-} y(x) \qquad and \qquad \lim_{x \to b-} y'(x)$$

*exist and are finite.*

And, for the left-hand end point $a$ we have the same conclusion:

**Lemma 4.61.** *If $a$ is finite, $a > -\infty$, then both limits*

$$\lim_{x \to a+} y(x) \qquad and \qquad \lim_{x \to a+} y'(x)$$

*exist and are finite.*

The proof is as follows. The function $y''(x)$ is square integrable on $(a, b)$ and therefore also square integrable on any subinterval of $(a, b)$, in particular on the finite interval $(x_0, b)$. As $y''$ is square integrable on that finite interval, it is integrable there. Hence

$$\lim_{x \to b-} y'(x) = \lim_{x \to b-} \left( y'(x_0) + \int_{x_0}^{x} y''(t)\, dt \right)$$

$$= y'(x_0) + \lim_{x \to b-} \int_{x_0}^{x} y''(t)\, dt$$

$$= y'(x_0) + \int_{x_0}^{b} y''(t)\, dt$$

The argument to show that $y(b)$ is finite proceeds in a similar way using formula (4.A.2) plus the fact that $y'(x)$ is continuous on the closed finite interval $[x_0, b]$, therefore bounded there (Theorem 1.15 of Wheeden and Zgymund).

The same argument establishes the existence and finiteness of $\lim y(x)$ and $\lim y'(x)$ as $x \to a+$.

In the earlier sections in this chapter we studied the expression $l(y) \equiv -y''$ on various finite intervals such as $(0,1)$, $(-1,1)$, and $(-\pi, \pi)$. We can now justify the assertion that $y$ and $y'$ were finite at the end points of these finite intervals.

**Lemma 4.62.** *Consider the expression $l(y) \equiv -y''$ in the space $L^2(a, b)$ for a finite interval $(a, b)$. If the domain D of the operator L consists of functions that are compatible with l (Definition 4.59), then for any function y in D all four limits*

$$y(a) = \lim_{x \to a+} y(x), \qquad y(b) = \lim_{x \to b-} y(x)$$

$$y'(a) = \lim_{x \to a+} y'(x), \qquad y'(b) = \lim_{x \to b-} y'(x)$$

*exist and are finite.*

So in those earlier examples the finiteness of these boundary values was assured. Only the boundary conditions, $y(0) = y(1)$, $y'(0) = y'(1)$, etc., needed to be specified.

Next we look into the case where one or both of the end points are infinite. Suppose first that $a$ is finite and $b$ infinite, so we are dealing with the space $L^2(a, \infty)$, where $a > -\infty$. For concreteness, let us suppose $a = 0$.

We know already from Lemma 4.61 that $y(0)$ and $y'(0)$ are both finite. What can be said about the behavior of $y$ and $y'$ at $+\infty$?

**Lemma 4.63.** *Consider the expression $l(y) \equiv -y''$ in the space $L^2(0, \infty)$. If the domain D of the operator L consists of functions that are compatible with l (Definition 4.59), then for any function y in D*

$$\lim_{x \to \infty} y(x) = 0 \qquad and \qquad \lim_{x \to \infty} y'(x) = 0?$$

*and* $y' \in L^2(0, \infty)$.

*Proof.* As $y$ and $y'$ are absolutely continuous, the following integration by parts is valid by Theorem 4.58:

$$\int_0^x y''(t)y(t)\,dt = y(x)y'(x) - y(0)y'(0) - \int_0^x (y'(t))^2\,dt \tag{4.A.3}$$

The integrand on the left, which is the product of two square-integrable functions, is integrable, so

$$\lim_{x \to \infty} \int_0^x y''y\,dt = \int_0^\infty y''y\,dt$$

exists and is finite. Hence the limit of the quantity on the right in equation (4.A.3) exists and is finite.

Because $(y'(t))^2 \geqslant 0$, the integral on the far right of equation (4.A.3) is an increasing function of $x$. Therefore, there are only two possibilities: either

$$\lim_{x \to \infty} \int_0^x (y'(t))^2\,dt = \infty$$

or

$$\lim_{x \to \infty} \int_0^x (y'(t))^2\,dt \qquad \text{exists and is finite}$$

We shall show that the first contingency is impossible. Suppose that were the case. Then, in order to maintain the overall finiteness of the right side of (4.A.3) as $x \to \infty$, we would have to have $y(x)y'(x) \to \infty$ as $x \to \infty$. But we have

$$\int_0^x y(t)y'(t)\,dt = \tfrac{1}{2}(y^2(x) - y^2(0)) \tag{4.A.4}$$

If it were true that $y(x)y'(x) \to \infty$ as $x \to \infty$, then the integral on the left in equation (4.A.4) would go to $\infty$ as $x \to \infty$. This would entail, by equation (4.A.4), that $y^2(x) \to \infty$ as $x \to \infty$. But, if $y^2(x) \to \infty$ as $x \to \infty$, then $y(x)$ could not be square integrable over $(0, \infty)$ as we have assumed. Thus the first contingency is untenable, and we must have

$$\lim_{x \to \infty} \int_0^x (y'(t))^2\,dt \text{ is finite}$$

which means $y' \in L^2(0, \infty)$.

Look at equation (4.A.4) again. Because both $y$ and $y'$ are square integrable, their product, $yy'$, in integrable. Thus the integral on the left in equation (4.A.4) tends to a finite limit as $x \to \infty$. Hence the right side of equation (4.A.4) must also tend to a finite limit as $x \to +\infty$, which is to say $y^2(x)$ tends to a finite limit as $x \to \infty$. But the integral of $y^2(x)$ over $(0, \infty)$ is finite by assumption, so this limit has to be 0. We conclude

$$\lim_{x \to \infty} y(x) = 0$$

Finally, as $y'$ is absolutely continuous, we have

$$\int_0^x y'(t)y''(t)\, dt = \tfrac{1}{2}(y'(x)^2 - y'(0)^2) \tag{4.A.5}$$

The integrand on the left in equation (4.A.5) is integrable because it is the product of two square-integrable functions. Hence, the integral on the left tends to a finite limit as $x \to \infty$, thus so does $(y'(x))^2$. But this limit must be 0 as $(y'(x))^2$ is integrable over $(0, \infty)$. We conclude

$$\lim_{x \to \infty} y'(x) = 0$$

That concludes the proof of Theorem 4.63.

Theorem 4.63 shows that, under the compatibility condition required by the $L^2$ theory, certain boundary conditions are forced. The limits

$$\lim_{x \to 0+} y(x) = y(0) \quad \text{and} \quad \lim_{x \to 0+} y'(x) = y'(0)$$

must exist and be finite and, at the infinite end point,

$$\lim_{x \to \infty} y(x) = 0, \quad \lim_{x \to \infty} y'(x) = 0$$

Similar conclusions hold for the interval $(-\infty, +\infty)$, but we shall carry this analysis no further.

## 4.A.2 Legendre's Operator

Legendre's operator $L$ is based on the differential expression $l(y) \equiv -((1 - x^2)y')'$ and acts in the Hilbert space $L^2(-1, 1)$. Our objective is to define a domain for $L$ that makes it Hermitian. We want this domain to be as large as possible so that our operator is applicable to the largest possible class of functions. But we also know, as detailed in Section 4.1, that the domain can be no larger than the subspace of functions compatible with $l$. So we first examine this subspace. For simplicity, we shall assume that all functions are real valued.

**Definition 4.63.** A function $y(x)$ is compatible with Legendre's expression $l(y) \equiv -((1 - x^2)y')'$ acting in the Hilbert space $L^2(-1, 1)$ if the following four conditions are fulfilled.

1. $y \in L^2(-1, 1)$; i.e., $\int_{-1}^{+1} y(x)^2 \, dx < \infty$.
2. $y$ is differentiable with finite derivative $y'(x)$ at each point $x$ in the open interval $-1 < x < 1$.
3. $(1-x^2)y'(x)$ is absolutely continuous in the open interval $-1 < x < 1$; i.e.,

$$(1-x^2)y'(x) = \int_0^x ((1-t^2)y'(t))' \, dt + y'(0), \quad -1 < x < 1$$

4. $l(y) \in L^2(-1, 1)$; i.e., $\int_{-1}^{+1} [((1-x^2)y')']^2 \, dx < \infty$.

Let us denote by $C$ the subspace of $L^2(-1, 1)$ consisting of functions compatible with Legendre's differential expression $l(y) \equiv -((1-x^2)y')'$. Thus $C$ is the subspace of functions that satisfy Definition 4.64. The differential operator defined by the differential expression $l$ and the domain $C$ is clearly the Hilbert space operator of largest possible domain that we can base on $l$. The question arises as to whether or not this "maximal" operator is Hermitian. That is, do we have $\langle l(f), g \rangle = \langle f, l(g) \rangle$ for all $f, g$ in $C$?

The answer is "no." Consider the function $Q_0(x) = (1/2) \ln((1+x)/(1-x))$. This function lies in $L^2(-1, 1)$ (Section 3.9, Exercise 5). Hence $Q_0(x)$ satisfies condition 1 of Definition 4.64. Also, at each point of the open interval $-1 < x < 1$, $Q_0'(x) = 1/(1-x^2)$, so $Q_0(x)$ also satisfies condition 2 of 4.64. And $(1-x^2)Q_0' = 1$, which is not only absolutely continuous on $-1 < x < 1$ but also absolutely continuous everywhere. Thus condition 3 of 4.64 is fulfilled. Finally $l(Q_0) = -((1-x^2)Q_0')' = -(1)' = 0$, so obviously $l(Q_0) \in L^2(-1, 1)$, and condition 4 is satisfied. Thus the function $f(x) = Q_0(x)$ belongs to the subspace $C$ of compatible functions. The function $g(x) = x$ also belongs to $C$; I'll leave that check to you. And $l(x) = -((1-x^2)x')' = -(1-x^2)' = 2x$. Now $\langle l(Q_0), x \rangle = \langle 0, x \rangle = 0$, but

$$\langle Q_0, l(x) \rangle = \int_{-1}^1 Q_0(x)(2x) \, dx = \int_{-1}^1 x \ln\left(\frac{1+x}{1-x}\right) dx$$

As the functions $x$ and $l((1+x)/(1-x))$ are both square integrable, their product is integrable (4.57(a)). Hence the integral on the right is finite. Moreover, the functions $x$ and $\ln((1+x)/(1-x))$ are both odd, so their product is even and, as is easily seen, positive. Thus

$$\langle Q_0, l(x) \rangle = \int_{-1}^1 x \ln\left(\frac{1+x}{1-x}\right) dx > 0$$

Hence $\langle l(Q_0), x \rangle = 0$ but $\langle Q_0, l(x) \rangle > 0$, so the operator that has differential expression $l$ and domain $C$ is *not* Hermitian.

We would like to narrow this domain by the imposition of boundary conditions so as to create a Hermitian operator. But we don't want to restrict the domain unduly so that we unnecessarily limit the applicability of our operator. We have a somewhat delicate task. How to proceed?

Let $r$ and $s$ be real numbers satisfying $-1 < r < s < 1$. If the functions $f$ and $g$ belong to $C$, then on the closed interval $[r, s]$ the functions $(1 - x^2)f'(x)$ and $g(x)$ are absolutely continuous, so by Theorem 4.58 the following integration by parts is valid.

$$\int_r^s [(1 - x^2)f']'g \, dx = \int_r^s g \, d[(1 - x^2)f']$$

$$= (1 - x^2)f'(x)g(x)|_r^s - \int_r^s (1 - x^2)f'(x)g'(x) \, dx$$

This leads to

$$\int_r^s [(1 - x^2)f']'g \, dx = (1 - s^2)f'(s)g(s) - (1 - r^2)f'(r)g(r)$$

$$- \int_r^s (1 - x^2)f'(x)g'(x) \, dx \qquad (4.A.6)$$

If we let $s \to +1$ and $r \to -1$, then the integral on the left becomes $-\langle l(f), g \rangle$. (Use conditions 1 and 4 of Definition 4.6.4 and 4.57(a).) We can make our operator Hermitian if we narrow the domain down to a class of functions such that $\lim_{s \to +1} (1 - s^2)f'(s)g(s) = \lim_{r \to -1} (1 - r^2)f'(r)g(r) = 0$. This way the integrated term in (4.A.6) drops out. The following result suggests a way out.

**Theorem 4.65.** *If $y(x)$ is compatible with Legendre's differential expression, then*

1. *$(1 - x^2)y'(x)$ has finite limits as $x \to \pm 1$.*
2. *If the values $y(1) = \lim_{x \to 1} y(x)$ and $y(-1) = \lim_{x \to -1} y(x)$ are both finite, then the limits in 1 are both 0.*

*Proof of Part 1.* Condition 4 of Definition 4.64 asserts that $((1 - x^2)y')'$ is square integrable on $[-1, 1]$; thus, by 4.57(b), is integrable there. Then, by condition 3 of 4.64,

$$\lim_{x \to 1} (1 - x^2)y'(x) = \int_0^1 ((1 - t^2)y'(t))' \, dt + y'(0)$$

The right side is finite because of the integrability of $((1 - x^2)y')'$; hence the limit on the left is also finite. A similar equation holds at $x = -1$. So part 1 is proved.

*Proof of Part 2.* Suppose that the boundary value $y(1) = \lim_{x \to 1^-} y(x)$ is finite. We shall prove that then $\lim_{x \to 1^-} (1 - x^2)y'(x) = 0$. We know from part 1 that this limit is finite. So it is either positive or negative. Suppose first that it were positive; $\lim_{x\to 1-}(1-x^2)y'(x) = a > 0$. Then there exists a $\delta > 0$, $\delta < 1$, so that

$$(1-t^2)y'(t) > \frac{a}{2} \qquad \text{for } \delta \leqslant t < 1$$

As $1-t^2>0$ also in this range, we have $y'(t)>a/2(1-t^2)$, $\delta\leqslant t<1$. Now $1-t^2=(1-t)(1+t)>1-t$ when $\delta\leqslant t<1$, so

$$y'(t) > \frac{a}{2(1-t)}, \qquad \delta \leqslant t < 1 \tag{4.A.7}$$

Items 2 and 3 of Definition 4.64 imply that $y(t)$ is absolutely continuous in $-1<t<1$, so there is the integral of its derivative. Hence from (4.A.7),

$$y(x) - y(\delta) = \int_\delta^x y'(t)\, dt > \frac{a}{2}\int_\delta^x \frac{dt}{1-t} = \frac{a}{2}\ln\left(\frac{1-\delta}{1-x}\right), \qquad \delta \leqslant x < 1$$

This implies that $y(x) \to \infty$ as $x \to 1-$, contradicting the assumption that $y(1)$ is finite. A similar argument shows that $y(x) \to -\infty$ as $x \to 1-$ if $\lim_{x\to 1-}(1-x^2)y'(x)$ is negative. Hence $\lim_{x\to 1-}(1-x^2)y'(x)=0$ must hold. Similarly for the limit as $x = -1$. Theorem 4.65 is proved.

Return now to equation (4.A.6). Let $D$ stand for the subspace of functions $y(x)$ in $L^2(-1,1)$ that are compatible with Legendre's differential expression *and* fulfill the boundary condition that the values $y(\pm 1)$ are finite. If, in equation (4.A.6), $f$ and $g$ belong to this subspace $D$, then, applying 4.65,

$$\lim_{s\to 1-} (1-s^2)f'(s)g(s) = 0 \times (\text{finite}) = 0$$

and likewise for the other term. So, if we define Legendre's operator $L$ by the differential expression $l$ and the domain $D$, then (4.A.6) leads to

$$\langle L(f),g\rangle = \int_{-1}^1 (1-x^2)f'(x)g'(x)\, dx, \qquad f,g \text{ in } D$$

for real-valued functions. For complex-valued functions we get

$$\langle L(f),g\rangle = \int_{-1}^1 (1-x^2)f'(x)\overline{g'(x)}\, dx, \qquad f,g \text{ in } D$$

From this, one can prove directly that $L$ is Hermitian and positive semidefinite (Exercise 3 of Section 4.6). As a by-product we deduce that $\sqrt{1-x^2}\,f'(x)\in L^2(-1,1)$ for all $f(x)$ in $D$, because $\langle L(f),f\rangle=\int_{-1}^1(1-x^2)|f'(x)|^2\,dx$.

### 4.A.3 Hermite's Operator

**Definition 4.66.** A function $y(x)$ is compatible with Hermite's expression $l(y) \equiv -e^{x^2/2}(e^{-x^2/2}y')'$ acting in the Hilbert space $L^2(-\infty, \infty; e^{-x^2/2})$ if the following four conditions are fulfilled:

1.  $y \in L^2(-\infty, \infty; e^{-x^2/2})$; i.e., $\int_{-\infty}^{\infty} |y(x)|^2 e^{-x^2/2} dx < \infty$.
2.  $y$ is differentiable with finite derivative $y'(x)$ at each real $x$, $-\infty < x < \infty$.
3.  $e^{-x^2/2}y'(x)$ is absolutely continuous for $-\infty < x < \infty$; i.e.,

$$e^{-x^2/2}y'(x) = \int_0^x (e^{-t^2/2}y'(t))' \, dt + y'(0), \qquad -\infty < x < \infty$$

4.  $l(y) \in L^2(-\infty, \infty, e^{-x^2/2})$; i.e.,

$$\int_{-\infty}^{\infty} |e^{x^2/2}(e^{-x^2/2}y')'|^2 e^{-x^2/2} dx < \infty$$

Note that condition 4 is equivalent to

$$\int_{-\infty}^{\infty} |(e^{-x^2/2}y')'|^2 e^{x^2/2} dx < \infty^* \qquad (\text{*See End Note on p. 266.})$$

We shall assume, for simplicity, that we are working here in real Hilbert space—i.e., we are dealing with real-valued functions over the real field. The functions that satisfy Definition 4.66 constitute a subspace of the Hilbert space $L^2(-\infty, \infty; e^{-x^2/2})$. Call that subspace $D$.

**Theorem 4.67.** *Let $f$, $g$ be any functions in $D$. Then*

1.  $f'(x) \in L^2(-\infty, \infty; e^{-x^2/2})$; i.e.,

$$\int_{-\infty}^{\infty} (f'(x))^2 e^{-x^2/2} dx < \infty$$

*and*

2.  $\lim_{x \to \pm\infty} e^{-x^2/2} f'(x)g(x) = 0$.

Thus the functions in the subspace $D$, which is the subspace of functions compatible with Hermite's differential expression, necessarily obey the "boundary condition" 2. It is automatic and does not have to be separately assumed.

We shall break up the proof of Theorem 4.67 in a series of steps:

**Lemma 4.68.** *In Theorem 4.67, condition 1 implies condition 2. That is,*

IF $f \in D \Rightarrow f' \in L^2(-\infty, \infty; e^{-x^2/2})$
THEN $f, g \in D \Rightarrow \lim_{x \to \pm \infty} e^{-x^2/2} f'(x) g(x) = 0$

*Proof.* As $g$ is differentiable with continuous derivative for all real $x$, it is absolutely continuous on every finite interval. And condition 3 of 4.66 says that $e^{-x^2/2} f'(x)$ is absolutely continuous. Hence by Theorem 4.58 the following integration by parts is valid:

$$\int_0^x g(t) \, d(e^{-t^2/2} f'(t)) = e^{-t^2/2} f'(t) g(t) \big|_0^x - \int_0^x f'(t) g'(t) e^{-t^2/2} \, dt$$

Next observe that

$$\begin{aligned}
g(t) \, d(e^{-t^2/2} f'(t)) &= g(t)(e^{-t^2/2} f'(t))' \, dt \\
&= e^{t^2/2}(e^{-t^2/2} f'(t))' g(t) e^{-t^2/2} \, dt \\
&= -l(f) g e^{-t^2/2} \, dt
\end{aligned}$$

So the previous integration by parts can be written

$$\int_0^x l(f) g e^{-t^2/2} \, dt = e^{-x^2/2} f'(x) g(x) - f'(0) g(0)$$
$$- \int_0^x f'(t) g'(t) e^{-t^2/2} \, dt$$

By conditions 1 and 4 of Definition 4.66, both $l(f)$ and $g$ belong to $L^2(-\infty, \infty; e^{-x^2/2})$, so the integral

$$\int_{-\infty}^{\infty} l(f) g e^{-t^2/2} \, dt$$

is finite (use the Schwarz inequality), and the integral

$$\int_0^{\infty} l(f) g e^{-t^2/2} \, dt$$

is also finite, by 4.57(c). Hence $\lim_{x \to \infty} \int_0^x l(f) g e^{-t^2/2} \, dt$ exists.

Likewise, $\lim_{x \to \infty} \int_0^x f'(t) g'(t) e^{-t^2/2} \, dt$ exists because we are assuming that all functions in $D$ have $L^2$ derivatives. It follows that

$$\lim_{x \to \infty} e^{-x^2/2} f'(x) g(x) \qquad \text{exists and is finite}$$

But

$$\int_0^{\infty} f'(x) g(x) e^{-x^2/2} \, dx < \infty$$

because both $f'$ and $g$ are $L^2$. Therefore $\lim_{x \to \infty} e^{-x^2/2} f'(x) g(x) = 0$. Use a symmetric argument at $x = -\infty$.

So, to prove Theorem 4.67, all we have to show is that $f \in D \Rightarrow f' \in L^2(-\infty, \infty; e^{-x^2/2})$.

Given $f \in D$, let $u(x) = e^{-x^2/4} f(x)$. Then $u(x)^2 = e^{-x^2/2}(f(x))^2$ so $u(x) \in L^2(-\infty, \infty)$. And $f(x) = u(x)e^{x^2/4}$,

$$f'(x) = u'(x)e^{x^2/4} + \frac{x}{2}u(x)e^{x^2/4} = \left(u' + \frac{x}{2}u\right)e^{x^2/4}$$

$$e^{-x^2/2}f'(x) = \left(u' + \frac{x}{2}u\right)e^{-x^2/4}$$

$$(e^{-x^2/2}f'(x))' = \left(u' + \frac{x}{2}u\right)'e^{-x^2/4} + \left(u' + \frac{x}{2}u\right)\left(-\frac{x}{2}\right)e^{-x^2/4}$$

$$= \left(u'' + \frac{1}{2}u - \frac{x^2}{4}u\right)e^{-x^2/4}$$

$$-l(f) = e^{x^2/2}(e^{-x^2/2}f')' = \left(u'' + \frac{1}{2}u - \frac{x^2}{4}u\right)e^{+x^2/4}$$

Now $l(f) \in L^2(-\infty, \infty; e^{-x^2/2})$ by 4.66(4), so

$$\int_{-\infty}^{\infty} l(f)^2 e^{-x^2/2}\, dx = \int_{-\infty}^{\infty} \left(u'' + \frac{1}{2}u - \frac{x^2}{4}u\right)^2 dx < \infty$$

so $u'' + (1/2)u - (x^2/4)u \in L^2(-\infty, \infty)$. As also $u \in L^2$, we have $u'' - (x^2/4)u \in L^2(-\infty, \infty)$. So the product $u(u'' - (x^2/4)u)$ is integrable. Then

$$\int_0^x u\left(u'' - \frac{t^2}{4}u\right) dt = \int_0^x uu''\, dt - \int_0^x \left(\frac{tu}{2}\right)^2 dt$$

$$= u(x)u'(x) - u(0)u'(0) - \int_0^x (u'(t))^2\, dt - \int_0^x \left(\frac{tu}{2}\right)^2 dt$$

$$\int_0^x u\left(u'' - \frac{t^2}{4}u\right) dt = u(x)u'(x) - u(0)u'(0) - \int_0^x \left[(u'(t))^2 + \left(\frac{tu}{2}\right)^2\right] dt$$

As $x \to \infty$, the limit on the left is finite. The integral on the far right increases with $x$ so either tends to a finite limit or to $+\infty$. If to $+\infty$, then $\lim_{x \to \infty} u(x)u'(x) = +\infty$. But then

$$\int_0^x u(t)u'(t)\, dt = \tfrac{1}{2}(u(x)^2 - u(0)^2) \to +\infty \qquad \text{contradicting } u \in L^2$$

Hence

$$\lim_{x \to \infty} \int_0^x \left[ (u'(t))^2 + \left( \frac{tu}{2} \right)^2 \right] dt$$

is finite, so both

$$u'(t) \quad \text{and} \quad \frac{tu(t)}{2}$$

belong to $L^2(-\infty, \infty)$, so $u'(x) + (x/2)u(x) = f'(x)e^{-x^2/4} \in L^2(-\infty, \infty)$, which means

$$\int_0^\infty (f'(x))^2 e^{-x^2/2} \, dx < \infty$$

Use a symmetric argument at $-\infty$. Theorem 4.67 is proved.


**End Note**
**Section 4.A.3, p. 263**


Condition 4 of Definition 4.66 is also equivalent to

$$\int_{-\infty}^\infty \left| y'' - xy' \right|^2 e^{-x^2/2} \, dx < \infty$$

# 5

# Schrödinger's Equations in One Dimension

## 5.1 THE WAVE EQUATION BY THE HILBERT SPACE METHOD

In Section 3.1 we derived the wave equation that describes the vertical displacement $u(x, t)$, at position $x$ and time $t$ of a tightly stretched wire.

$$\frac{\partial^2 u}{\partial x^2} = \frac{1}{w^2} \frac{\partial^2 u}{\partial t^2} \tag{5.1.1}$$

Here $w = (T/\rho)^{1/2}$, $T$ is the tension in the wire, and $\rho$ is its linear density. The quantity $w$ has the dimensions of velocity. In Exercises 1 and 3 of Section 3.1, we were able to interpret $w$ as the velocity of propagation of a pulse along the wire.

In Section 3.1, we solved equation (5.1.1) by separating out the time variable. We wrote $u(x, t) = y(x)z(t)$. Equation (5.1.1) then yielded two ordinary differential equations in $y$ and $z$:

$$\frac{d^2 y}{dx^2} = -\lambda y \tag{5.1.2}$$

$$\frac{d^2 z}{dt^2} = -\lambda w^2 z \tag{5.1.3}$$

where $\lambda$ is a yet-to-be-determined constant (which I've made negative for convenience). In the specific physical problem considered back then, the wire was clamped at both ends, so $u(0, t) = u(L, t) = 0$ ($x = 0, L$ were taken as the coordinates of the ends of the wire), which subjected equation (5.1.2) to the *boundary conditions* $y(0) = y(L) = 0$. Then equation (5.1.2) with these additional conditions turned out to have nonzero solutions only when $\lambda$ had certain discrete values, the *eigenvalues*, $\lambda = n^2\pi^2/L^2$. The corresponding solutions were $y_n = \sin(n\pi x/L)$, $n = 1, 2, \ldots$. We then solved (5.1.3) with these values of $\lambda$, and our final answer was expressed as an infinite series in the *eigenfunctions* $y_n$.

Both the eigenvalues and eigenfunctions have physical significance. The eigenfunctions represent certain simple stationary configurations of the vibrating wire, and the corresponding eigenvalues $\varepsilon_n = n^2\pi^2/L^2$ are related to the frequency $\nu_n$ and wavelength $\lambda_n$ of that configuration by $\nu_n = (w/2\pi)\varepsilon_n^{1/2}$ and $\lambda_n = 2\pi\varepsilon_n^{-1/2}$ (Section 3.5).

Using the theory of differential operators in $L^2$ spaces developed in Chapter 4, we shall recast our approach to the wave equation in the language of Hilbert space.

We begin with (5.1.1) as before and perform the same separations of variables to get the equation

$$-\frac{d^2y}{dx^2} = \lambda y \qquad (5.1.4)$$

with certain boundary conditions at $x = 0$ and $x = L$.

We then set up the Hilbert space $L^2(0, L)$, which consists of functions $f(x)$ that satisfy the single condition

$$\int_0^L |f(x)|^2 \, dx < \infty$$

The Hilbert space $L^2(0, L)$ carries the inner product

$$\langle f, g \rangle = \int_0^L f(x)\overline{g(x)} \, dx$$

We define a differential operator $\Delta$ in $L^2(0, L)$ as follows. The differential expression for this operator is

$$l(y) \equiv -\frac{d^2y}{dx^2}$$

The domain of $\Delta$ will be the subspace of $L^2(0, L)$ consisting of functions that are compatible with $l$ (remember that $y \in L^2(0, L)$ is compatible with $l$ if $l(y) \in L^2(0, L)$) and that satisfy certain boundary conditions at $x = 0$ and $x = L$.

Besides the clamped boundaries we have considered in Chapter 3, various other boundary conditions are possible. For example, a frictionless slot at $x = L$ would allow the wire to position itself freely at that end, so $y(L)$ would be unrestricted. But a frictionless slot can support only a normal force, so $y'(L) = 0$. Likewise, we could have $y'(0) = 0$, $y(L) = 0$. Other combinations are possible, but whatever they are they must determine a *subspace* of $L^2(0, L)$.

As we have said, the domain $D$ of our Hilbert space operator $\Delta$ consists of functions in $L^2(0, L)$ that are compatible with $l$ and that satisfy the specified boundary conditions. Then the action of $\Delta$ is given by

$$\Delta(y) = l(y) = -y'' \quad \text{when } y \in D$$

and equation (5.1.4) now becomes an eigenvalue problem for the Hilbert space operator $\Delta$:

For what values of $\lambda$ does the equation

$$\Delta(y) = \lambda y \tag{5.1.5}$$

have a nonzero solution $y$ in the domain of $\Delta$?

Now refer back to Definition 4.13 and Theorem 4.14 of Section 4.5, which defined and described the *one-dimensional Laplacian*. Or, think back: do you remember the definition and properties of that Hilbert space operator? Either way, referring back or thinking back, you will come up with this: the *one-dimensional Laplacian* $\Delta$ is a Hilbert space operator that acts in the Hilbert space $L^2(a, b)$, where $[a, b]$ is a finite interval. The operator $\Delta$ is based on the differential expression $l(y) \equiv -y''$, and its domain consists of functions in $L^2(a, b)$ that are compatible with $l$ and satisfy either the Dirichlet, Neumann, or mixed boundary conditions. (There is also the one-dimensional Laplacian on the circle $C$; you will deal with this case in Exercise 2.)

The Hilbert space operator $\Delta$ that we have just constructed from the wave equation by separating out the space variable is exactly the one-dimensional Laplacian (set $a = 0$, $b = L$). Theorem 4.14 of Section 4.5 answers our question (5.1.5) and provides a general solution to the wave equation (5.1.1).

The operator $\Delta$ is Hermitian positive semidefinite. The values of $\lambda$ for which the operator equation $\Delta(y) = \lambda y$ has nonzero solutions $y$ in the domain of $\Delta$ are exactly the eigenvalues of $\Delta$ which, according to Theorem 4.14, form an infinite sequence $\lambda = \varepsilon_n$,

$$(0 = \varepsilon_0) < \varepsilon_1 < \varepsilon_2 < \cdots < \varepsilon_n < \varepsilon_{n+1} < \cdots$$

such that $\varepsilon_n \to \infty$ as $n \to \infty$. All eigenvalues are simple. The eigenvalue $\varepsilon_0 = 0$ occurs only for the Neumann boundary condition. The eigenfunctions of $\Delta$, say $\phi_n$, $n = 1, 2, \ldots$, form an orthogonal basis of $L^2(0, L)$. (Under the Neumann boundary conditions include $\phi_0 = 1$.) Thus every function $y$ in

$L^2(0, L)$ has a general Fourier expansion

$$y(x) = \sum_{n=0}^{\infty} c_n \phi_n(x) \qquad \text{where } c_n = \frac{\langle y, \phi_n \rangle}{\langle \phi_n, \phi_n \rangle} = \frac{\int_0^L y \bar{\phi}_n \, dx}{\int_0^L |\phi_n|^2 \, dx}$$

which converges to $y$ in the mean-square metric. If $y$ also belongs to the domain of $\Delta$, then the action of $\Delta$ may be computed by applying $\Delta$ term by term to the expansion of $y$:

$$\Delta(y) = \sum_{n=0}^{\infty} \varepsilon_n c_n \phi_n$$

For each eigenvalue $\lambda = \varepsilon_n$, the equation (5.1.3) in time $t$, namely $z'' = -\lambda w^2 z$, has the solution

$$z_n = A_n \cos \sqrt{\varepsilon_n} \, wt + B_n \sin \sqrt{\varepsilon_n} \, wt$$

and the solution to the linear wave equation we then take as

$$u(x, t) = \sum_{n=0}^{\infty} z_n(t) \phi_n(x) = \sum_{n=1}^{\infty} (A_n \cos \sqrt{\varepsilon_n} \, wt + B_n \sin \sqrt{\varepsilon_n} \, wt) \phi_n(x) \qquad (5.1.6)$$

The constants $A_n$ and $B_n$ are obtained from the initial position $u(x, 0)$ and initial velocity $(\partial u / \partial t)(x, 0)$. Put $t = 0$ in (5.1.6) to get

$$u(x, 0) = \sum_{n=1}^{\infty} A_n \phi_n(x)$$

then

$$A_n = \frac{\langle u(x, 0), \phi_n(x) \rangle}{\|\phi_n\|^2} = \frac{1}{\|\phi_n\|^2} \int_0^L u(x, 0) \overline{\phi_n(x)} \, dx \qquad (5.1.7)$$

because the $A_n$ are the unique general Fourier coefficients of $u(x, 0)$ with respect to the orthogonal basis $\phi_n$.

Next, differentiate (5.1.6) with respect to $t$; then put $t = 0$

$$\frac{\partial u}{\partial t}(x, 0) = \sum_{n=1}^{\infty} \sqrt{\varepsilon_n} \, w B_n \phi_n(x)$$

so

$$B_n = \frac{\langle (\partial u / \partial t)(x, 0), \phi_n(x) \rangle}{\sqrt{\varepsilon_n} \, w \|\phi_n\|^2} = \frac{1}{\sqrt{\varepsilon_n} \, w \|\phi_n\|^2} \int_0^L \frac{\partial u}{\partial t}(x, 0) \overline{\phi_n(x)} \, dx \qquad (5.1.8)$$

where in both cases $\|\phi_n(x)\|^2 = \langle \phi_n(x), \phi_n(x) \rangle = \int_0^L |\phi_n(x)|^2 \, dx$

For example, consider the case where the wire is clamped at $x = 0$ and is held in a frictionless slot at $x = L$. Then the boundary conditions are $y(0) = 0$,

$y'(L) = 0$. The operator $\Delta$ is defined by its differential expression

$$l(y) \equiv -\frac{d^2y}{dx^2}$$

and domain($\Delta$) = compatible functions $y$ in $L^2(0, L)$ that satisfy

$$y(0) = 0, \qquad y'(L) = 0$$

Calculations similar to those we have done many times before show that the eigenvalues for this problem are $\varepsilon_n = ((2n+1)^2/4L^2)\pi^2$ with corresponding eigenfunctions $\phi_n(x) = \sin((2n+1)\pi x/2L)$, $n = 0, 1, 2, \ldots$. The corresponding solution is

$$u_n(x, t) = \sum_{n=0}^{\infty} \left[ A_n \cos(2n+1)\frac{\pi wt}{2L} + B_n \sin(2n+1)\frac{\pi wt}{2L} \right] \sin(2n+1)\frac{\pi x}{2L}$$

The fundamental modes are $\sin(\pi x/2L)$, $\sin(3\pi x/2L)$, $\sin(5\pi x/2L)$, .... The first three are pictured in Figure 5.1. The fundamental frequency is $w/2L$, and the higher harmonics are odd multiples of this.

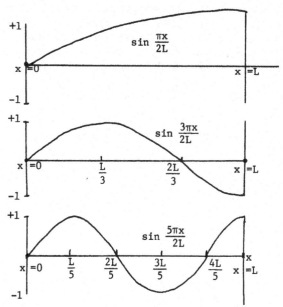

**Figure 5.1** Vibration of a wire with one free end.

**Exercises**

1. A wire of length $L$ is stretched between two frictionless slots at $x=0$ and $x=L$.
   (a) Determine its displacement $u(x, t)$ if the wire is given an initial displacement $u(x, 0)$ and an initial velocity $(\partial u/\partial t)(x, 0)$.

   $$[\text{Ans.:} \quad u(x, t) = A_0 + \sum_{n=1}^{\infty} \left( A_n \cos \frac{n\pi wt}{L} + B_n \sin \frac{n\pi wt}{L} \right) \cos \frac{n\pi x}{L}$$

   $$A_0 = \frac{1}{L} \int_0^L u(x, 0) \, dx, \qquad A_n = \frac{2}{L} \int_0^L u(x, 0) \cos \frac{n\pi x}{L} \, dx, \qquad n \geq 1$$

   $$B_n = \frac{2}{n\pi w} \int_0^L \frac{\partial u}{\partial t}(x, 0) \cos \frac{n\pi x}{L} \, dx, \qquad n \geq 1]$$

   (b) If $u(x, 0) = c$ and $(\partial u/\partial t)(x, 0) = 0$, find the coefficients of the expansion and show thus that $u(x, t) = c$. Explain on physical grounds.
   (c) If $u(x, 0) = \sin(\pi x/L)$ and $(\partial u/\partial t)(x, 0) = 0$, find $u(x, t)$.
   (d) If $u(x, 0) = \cos(\pi x/L)$ and $(\partial u/\partial t)(x, 0) = 0$, find $u(x, t)$. Show that in this case the series expansion collapses to one term. Describe the subsequent motion of the wire.

2. A wire of length $2\pi L$ is bent in a circle and maintained in tension. (This is physically difficult to do and we take it as a hypothetical case.) Take $x$ as measured along the wire, $0 \leq x \leq 2\pi L$, $0 \equiv 2\pi L$. Following the method developed in this section, define the appropriate operator $\Delta$, find its eigenvalues and eigenfunctions, and solve for the displacement $u(x, t)$ of the wire. (Note that there is *no* boundary in this configuration.)

   $$[\text{Ans.:} \quad u(x, t) = \frac{A_0}{2} + \sum_{n=1}^{\infty} \left( A_n \cos \frac{nwt}{L} + B_n \sin \frac{nwt}{L} \right) \cos \frac{nx}{L}$$

   $$+ \sum_{n=1}^{\infty} \left( C_n \cos \frac{nwt}{L} + D_n \sin \frac{nwt}{L} \right) \sin \frac{nx}{L}$$

   where

   $$A_n = \frac{1}{\pi L} \int_0^{2\pi L} u(x, 0) \cos \frac{nx}{L} \, dx, \qquad n \geq 0$$

   $$C_n = \frac{1}{\pi L} \int_0^{2\pi L} u(x, 0) \sin \frac{nx}{L} \, dx \qquad n \geq 1$$

   $$B_n = \frac{1}{nw\pi} \int_0^{2\pi L} \frac{\partial u}{\partial t}(x, 0) \cos \frac{nx}{L} \, dx \qquad n \geq 1$$

   $$D_n = \frac{1}{nw\pi} \int_0^{2\pi L} \frac{\partial u}{\partial t}(x, 0) \sin \frac{nx}{L} \, dx, \qquad n \geq 1]$$

## 5.2 THE HEAT EQUATION BY THE HILBERT SPACE METHOD

"Quantity of heat" is physically stored thermal energy. It is generally measured in calories. One (gram-) calorie, by definition the quantity of heat required to raise the temperature of 1 gram of water 1°C (degrees centigrade or celsius), equals 4.186 joules of energy.

The theory of heat is based on two principles. The first principle is that the rate of temperature increase in a fixed amount of material is proportional to the rate of increase of the quantity of heat. The proportionality constant is called the *specific heat* of the substance, and we shall denote it $\sigma$. It varies with the material and somewhat with the temperature, but we shall ignore the latter variation. Common units for $\sigma$ are cal/g-°C. Specific heats of some common materials are given in Table 5.1.

The second principle describes the flow of heat: the quantity of heat flowing across a unit area per second is proportional to the directional derivative of the temperature normal to that area, and the flow is in the direction of decreasing temperature. The proportionality constant here is called *thermal conductivity*, and we shall denote it $\kappa$. Common units are cal-cm/cm²-sec-°C. Thermal conductivities of some common materials are given in Table 5.1.

We deal with the one-dimensional situation: a rod of homogeneous material, uniform cross-sectional area $A$, and perfectly insulated along its sides. Heat can enter and leave the rod only through its ends. Coordinates are set up as in Figure 5.2. Distance along the rod is denoted by $x$, the ends of the rod being situated at $x=a$ and $x=b$. Let $u(x, t)$ denote the temperature of the rod at position $x$ and time $t$. According to the second principle of the theory of heat, the quantity of heat crossing the rod at position $x$ is

$$-\kappa A \frac{\partial u}{\partial x}(x, t)$$

**Table 5.1** Thermal Constants for Various Materials

| Material | Specific heat $\sigma$ (cal/g-°C) | Thermal conductivity $\kappa$ (cal-cm/cm²-sec-°C) | Density $\rho$ (g/cm³) |
|---|---|---|---|
| Aluminum | 0.217 | 0.49 | 2.7 |
| Copper | 0.093 | 0.92 | 8.9 |
| Steel | 0.107 | 0.12 | 7.8 |
| Glass | 0.199 | 0.002 | 2.6 |

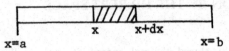

x=a                                                                          x=b

**Figure 5.2**  Coordinates for the heated rod.

and at position $x + dx$ is

$$-\kappa A \frac{\partial u}{\partial x}(x + dx, t)$$

so the net gain of heat in the small shaded section of the rod is

$$-\kappa A \frac{\partial u}{\partial x}(x, t) - \left(-\kappa A \frac{\partial u}{\partial x}(x + dx, t)\right) = \kappa A \frac{\partial^2 u}{\partial x^2} dx$$

in cal/sec. The first principle tells us that this small section suffers an increase in temperature at a rate, $\partial u / \partial t$, given by

$$\sigma(\rho A\, dx) \frac{\partial u}{\partial t} = \kappa A \frac{\partial^2 u}{\partial x^2} dx$$

where $\rho$ is the (uniform) density of the rod (g/cm$^3$). Canceling $A$ and $dx$ and rearranging, we get the *heat equation in one dimension*

$$\frac{\partial^2 u}{\partial x^2} = \frac{1}{a^2} \frac{\partial u}{\partial t} \qquad a^2 = \frac{\kappa}{\sigma\rho} \left(\frac{\text{cm}^2}{\text{sec}}\right) \qquad\qquad (5.2.1)$$

Note that the heat equation differs from the wave equation $\partial^2 u / \partial x^2 = (1/w^2) \partial^2 u / \partial t^2$ only in that it is first order in the time $t$. Like the wave equation, the heat equation is linear: if $u_1$ and $u_2$ are solutions, then so is $u = c_1 u_1 + c_2 u_2$ for any constants $c_1$ and $c_2$.

The method of separation of variables, which we used successfully in solving the wave equation, is worth trying for the heat equation. Writing $u(x, t) = y(x)z(t)$, we obtain as before two ordinary differential equations, the equation in the time variable now of first order.

$$-\frac{d^2 y}{dx^2} = \lambda y \qquad\qquad (5.2.2)$$

$$\frac{dz}{dt} = -\lambda a^2 z \qquad\qquad (5.2.3)$$

We solve (5.2.2) by the Hilbert space method. Set up the Hilbert space $L^2(a, b)$ of functions $f(x)$ that satisfy the single condition

$$\int_a^b |f(x)|^2 \, dx < \infty$$

The Hilbert space $L^2(a, b)$ carries the inner product

$$\langle f, g \rangle = \int_a^b f(x)\overline{g(x)} \, dx$$

We define a differential operator $\Delta$ in $L^2(a, b)$ as follows. The differential expression for this operator is

$$m(y) \equiv -\frac{d^2 y}{dx^2}$$

The domain $D$ of $\Delta$ will be the subspace of $L^2(a, b)$ consisting of functions that are compatible with $m$ (remember that $y$ is compatible with $m$ if $m(y) \in L^2(a, b)$) and that satisfy certain boundary conditions at $x = a$ and $x = b$.

As an example of the kind of boundary conditions one might encounter in heat conduction problems, the rod depicted in Figure 5.2 might have both ends in an ice bath. This would force the temperature $u$ at $x = a$ and $x = b$ to be 0°C, thus the boundary condition $u(a, t) = u(b, t) = 0$ or $y(a) = y(b) = 0$. Other boundary conditions on $y$ are of course possible, but whatever they are, they must define a *subspace* of $L^2(a, b)$. In heat conduction problems, some adjustments are sometimes necessary to achieve boundary conditions of this type. For example, if the ends of the rod were maintained at 10°C rather than 0°C, the boundary conditions would be $y(a) = y(b) = 10$. But these boundary conditions do not define a subspace of $L^2(a, b)$. In this case we recognize that $u = 10$ is a solution of the heat equation, the *steady-state* solution, so $v(x, t) = u(x, t) - 10$ is also a solution that satisfies the homogeneous boundary conditions $v(a, t) = v(b, t) = 0$. So we solve for $v$; then $u = v + 10$ is the final answer.

In any case, the domain $D$ of our Hilbert space operator $\Delta$ consists of functions in $L^2(a, b)$ that are compatible with $m$ and that satisfy the specified (adjusted) boundary conditions. Then $\Delta$ is defined by

$$\Delta(y) = m(y) = -y'' \qquad \text{when } y \in D$$

and equation (5.2.2) now becomes an eigenvalue problem for the Hilbert space operator $\Delta$:

For what values of $\lambda$ does the equation $\Delta(y) = \lambda y$          (5.2.4)
have a nonzero solution $y$ in the domain of $\Delta$?

As in Section 5.1, we are dealing here with the one-dimensional Laplacian $\Delta$. Once again, the facts regarding this Hilbert space operator: it is Hermitian positive semidefinite. As to question (5.2.4), the values of $\lambda$ for which the operator equation $\Delta(y) = \lambda y$ has nonzero solutions $y$ in the domain of $\Delta$ are exactly the eigenvalues of $\Delta$ which, according to Theorem 4.14, form an infinite sequence $\lambda = \varepsilon_n$,

$$(0 = \varepsilon_0) < \varepsilon_1 < \varepsilon_2 < \cdots < \varepsilon_n < \varepsilon_{n+1} \cdots$$

such that $\varepsilon_n \to \infty$ as $n \to \infty$. All eigenvalues are simple. The eigenvalue $\varepsilon_0 = 0$ occurs only for the Neumann boundary condition. The eigenfunctions of $\Delta$, say $\phi_n$, $n = 1, 2, \ldots$, form an orthogonal basis of $L^2(a, b)$. (Under the Neumann boundary conditions include $\phi_0 = 1$.) Thus every function $y$ in $L^2(a, b)$ has a general Fourier expansion

$$y(x) = \sum_{n=0}^{\infty} c_n \phi_n(x) \qquad \text{where } c_n = \frac{\langle y, \phi_n \rangle}{\langle \phi_n, \phi_n \rangle} = \frac{\int_a^b y \bar{\phi}_n \, dx}{\int_a^b |\phi_n|^2 \, dx}$$

which converges to $y$ in the mean-square metric. If $y$ also belongs to the domain of $\Delta$, then the action of $\Delta$ may be computed by applying $\Delta$ term by term to the expansion of $y$:

$$\Delta(y) = \sum_{n=0}^{\infty} \varepsilon_n c_n \phi_n$$

For each eigenvalue $\lambda = \varepsilon_n$, the equation (5.2.3) in time $t$, namely $z' = -\lambda a^2 z$, has the solution

$$z_n = A_n e^{-\varepsilon_n a^2 t} \qquad \text{where } A_n \text{ is a constant}$$

Then for each $n = 0, 1, 2, \ldots$, $u_n(x, t) = \phi_n(x) z_n(t) = A_n \phi_n(x) \exp(-\varepsilon_n a^2 t)$ is a solution of the heat equation (5.2.1), and we take as the general solution to (5.2.1).

$$u(x, t) = \sum_{n=0}^{\infty} \phi_n(x) z_n(t) = \sum_{n=0}^{\infty} A_n \phi_n(x) e^{-\varepsilon_n a^2 t} \tag{5.2.5}$$

The constants $A_n$ are determined from the initial temperature distribution in the rod, $u(x, 0)$. Put $t = 0$ in (5.2.5) to get

$$u(x, 0) = \sum_{n=0}^{\infty} A_n \phi_n(x) \tag{5.2.6}$$

Then

$$A_n = \frac{\langle u(x, 0), \phi_n(x) \rangle}{\langle \phi_n(x), \phi_n(x) \rangle} = \frac{1}{\|\phi_n\|^2} \int_a^b u(x, 0) \overline{\phi_n(x)} \, dx \tag{5.2.7}$$

because the $A_n$ are the unique general Fourier coefficients of $u(x, 0)$ with respect to the orthogonal basis $\phi_n$.

Consider the following example. An aluminum rod 100 cm long, of uniform cross-sectional area, is initially 0°C. At time $t = 0$, one end of the rod is abruptly raised to 100°C and held at that temperature, while the other end is maintained at 0°C. What is the temperature function $u(x, t)$?

If we take the ends of the rod at $x = 0$ and $x = L$ ($L = 100$ cm), then the boundary conditions are

$$u(0, t) = 0°C, \qquad u(L, t) = 100°C$$

We wish to translate these conditions into boundary conditions defining the domain of our operator. The trick is as follows. Instead of $u(x, t)$ we consider

$$v(x, t) = u(x, t) - x$$

where $x$ is multiplied by the constant $1 = 100°C/100$ cm, which converts $x$ in cm to °C to match $u$. Now $v$ also satisfies the heat equation because

$$\frac{\partial v}{\partial x} = \frac{\partial u}{\partial x} - 1, \qquad \frac{\partial^2 v}{\partial x^2} = \frac{\partial^2 u}{\partial x^2}, \quad \text{and} \quad \frac{\partial v}{\partial t} = \frac{\partial u}{\partial t}$$

so $\partial^2 v/\partial x^2 = (1/a^2)\, \partial v/\partial t$. The boundary conditions for $v$ are

$$v(0, t) = u(0, t) - 0 = 0°C$$

and

$$v(L, t) = u(L, t) - L = 100 - 100 = 0°C$$

Setting $v(x, t) = y(x)z(t)$, equation (5.2.2) now leads to the one-dimensional Laplacian $\Delta$:

Differential expression $m(y) \equiv -d^2 y/dx^2$

Domain($\Delta$) = compatible functions $y$ in $L^2(0, L)$ that satisfy $y(0) = y(L) = 0$

The next step is to find its eigenvalues $\varepsilon_n$ and eigenfunctions $\phi_n$. The usual methods yield

$$\varepsilon_n = \frac{n^2 \pi^2}{10^4}, \qquad \phi_n(x) = \sin \frac{n\pi x}{100}, \qquad n = 1, 2, \ldots$$

The constant $a^2$ for our aluminum rod is

$$a^2 = \frac{\kappa}{\sigma \rho} = \frac{0.49}{(0.217)(2.7)} = 0.836 \quad \text{cm}^2/\text{sec}$$

Thus our expansion is

$$v(x, t) = \sum_{n=1}^{\infty} A_n \sin \frac{n\pi x}{100} \exp\left( -\frac{n^2 \pi^2}{10^4}(0.836)t \right)$$

We have $v(x, 0) = u(x, 0) - x = 0 - x$, so

$$A_n = \frac{\langle -x, \sin(n\pi x/100)\rangle}{\|\sin(n\pi x/100)\|^2} = \frac{-1}{\|\sin(n\pi x/100)\|^2} \int_0^{100} x \sin \frac{n\pi x}{100} dx$$

Now

$$\left\|\sin \frac{n\pi x}{100}\right\|^2 = \int_0^{100} \left(\sin \frac{n\pi x}{100}\right)^2 dx = 50 \quad \text{and} \quad \int_0^{100} x \sin \frac{n\pi x}{100} dx = \frac{10^4}{n\pi}(-1)^{n+1}$$

so

$$A_n = \left(\frac{-1}{50}\right)\left(\frac{10^4}{n\pi}(-1)^{n+1}\right) = (-1)^n \frac{200}{n\pi}, \qquad n = 1, 2, \dots$$

Thus

$$v(x, t) = \frac{200}{\pi} \sum_{n=1}^{\infty} \frac{(-1)^n}{n} \sin \frac{n\pi x}{100} \exp\left(-\frac{n^2\pi^2(0.836)t}{10^4}\right) \tag{5.2.8}$$

The temperature distribution is then $u(x, t) = x + v(x, t)$.

In equation (5.2.8), $t$ is measured in seconds, $x$ in centimeters. For numerical calculations it is convenient to replace $x$ by $x/100$, so $0 \leqslant x \leqslant 1$, and to measure $t$ in hours, $t_{\text{hours}} = t_{\text{sec}}/3600$. Then

$$\frac{\pi^2(0.836)(3600)}{10^4} = 2.97$$

and our final answer for the temperature distribution $u(x, t)$ is

$$u(x, t) = 100x + \frac{200}{\pi} \sum_{n=1}^{\infty} \frac{(-1)^n}{n} \sin n\pi x \, e^{-2.97n^2 t} \tag{5.2.9}$$

or

$$u = \frac{200}{\pi}\left[\frac{\pi x}{2} - \sin \pi x \, e^{-2.97t} + \frac{1}{2}\sin 2\pi x \, e^{-(4)(2.97)t}\right.$$
$$\left. - \frac{1}{3}\sin 3\pi x \, e^{-(9)(2.97)t} + \frac{1}{4}\sin 4\pi x \, e^{-(16)(2.97)t} + \cdots\right]$$

$$\tag{5.2.10}$$

where $0 \leqslant x \leqslant 1$ and time $t$ is measured in hours.

When you calculate $u(x, t)$ from equation (5.2.10), fix a time $t$, say $t = 0.1$ hour, and then calculate, tabulate, and graph $u(x, 0.1)$ versus $x$, $0 \leqslant x \leqslant 1$. The graph of $u(x, 0.1)$ shows the temperature distribution in the rod at the fixed time $t = 0.1$ hour. Repeat the calculation for other times.

Early times are the most interesting. However, for small times one must take quite a few terms in the series (5.2.10) to get an accurate answer because the exponential term goes to zero slowly. If you program on a large machine, this is no problem. But you probably won't need to take more than 20 terms in any case.

On the HP-15C we can use the ISG command (Increment, and Skip if Greater Than) to control the number of terms we compute in the series (5.2.10). Refer to Section 10 of the HP-15C manual. Here is a sample program for the 15C.

| | | |
|---|---|---|
| g P/R | RCL × 0 | + |
| f LBL C | SIN | f ISG I |
| g π × | RCL .0 | GTO 2 |
| STO 0 | g x² | RCL 4 |
| 2 ÷ | RCL × 1 | STO I |
| f LBL 2 | eˣ × | ↓ |
| RCL I | RCL ÷ .0 | RCL × 3 |
| g INT | gπ RCL × .0 COS | g RTN |
| STO .0 | × | g P/R |

To run this program select your time $t$ and put $-a^2t$ in register 1. In our example we put $-2.97t$ in register 1. Put $200/\pi$ in register 3. In register 4 and register I put $N = 1.pqr01$, where $N = pqr$ is the number of terms you wish to sum. For smaller times, use larger $N$. Table 5.2 contains the results for the temperature $u(x, t)$ of our 100-cm aluminum rod which is initially at 0°C, then at time $t = 0$ the 100 cm end ($x = 1$) is brought to 100°C and held at that temperature. The graphs of $u(x, t)$ are shown in Figure 5.3.

### Exercises

1. A copper rod 100 cm long, of uniform cross section, is initially at 0°C. At time $t = 0$ both ends are raised to 100°C and maintained at that temperature. Find the subsequent temperature distribution $u(x, t)$ in the rod as a function of position $x$ and time $t$. Calculate and graph $u(x, t)$ versus $x$ for $t = 0.1, 2, 10,$ and 60 minutes.

2. A steel rod 100 cm long of uniform cross section is heated initially to a temperature distribution $u(x, 0) = 100 - (x/5)^2$°C, where $x$ is the distance in cm from the center of the rod. At time $t = 0$ the source of heat is removed, and all parts of the rod, including the ends, are insulated. Find the

**Table 5.2** $u(x, t) = 100x + (200/\pi) \Sigma_1^N ((-1)^n/n)(\sin n\pi x) \exp(-2.97n^2 t)$

|  | $x$ | $u(x, 0.01)$ | $u(x, 0.1)$ | $u(x, 1)$ |
|---|---|---|---|---|
|  | 0 | 0 | 0 | 0 |
|  | 0.1 | 0.0014 | 0.023 | 8.99 |
|  | 0.2 | 0.0007 |  | 18.1 |
|  | 0.3 |  |  | 27.4 |
|  | 0.4 |  | 1.45 | 36.9 |
|  | 0.5 | 0.0006 | 4.15 | 46.7 |
|  | 0.6 |  | 10.3 | 56.9 |
|  | 0.7 |  | 22.1 | 67.4 |
|  | 0.8 | 0.992 | 41.5 | 78.1 |
|  | 0.85 | 5.32 |  |  |
|  | 0.875 | 10.7 |  |  |
|  | 0.9 | 19.7 | 68.4 | 89.0 |
|  | 0.91 |  |  |  |
|  | 0.925 | 33.4 |  |  |
|  | 0.93 |  |  |  |
|  | 0.94 |  |  |  |
| experimenting | 0.95 | 51.9 (15 terms) |  |  |
| to check | 0.95 | 52.2 (10 terms) |  |  |
| accuracy | 0.95 | 56.2 (6 terms) |  |  |
|  | 0.96 |  |  |  |
|  | 0.975 | 74.7 |  |  |
|  | 0.98 |  |  |  |
|  | 0.99 |  |  |  |
|  | 1.0 | 100 | 100 | 100 |
|  |  | ↑ | ↑ | ↑ |
|  |  | $N = 15$ terms | $N = 5$ terms | $N = 3$ |
|  |  | except as shown | except as shown |  |

subsequent temperature distribution $u(x, t)$ in the rod. Calculate and graph $u(x, t)$ versus $x$ for $t = 0, 10, 30$, and 120 minutes. Show that as $t \to \infty$, $u(x, t)$ approaches the constant value

$$u(x, \infty) = \frac{1}{100} \int_{-50}^{50} u(x, 0) \, dx$$

and give a physical explanation of this fact.

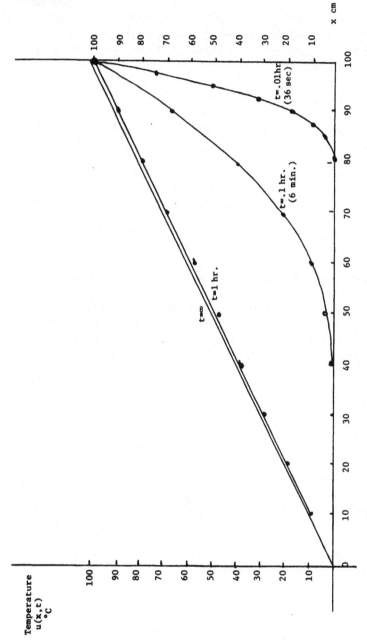

**Figure 5.3** Temperature distribution $u(x, t)$ in a 100-cm aluminum rod of cross-sectional area 1 cm$^2$. Initial condition $u(x, 0) = 0$. Boundary conditions $u(0, t) = 0$, $u(100, t) = 100$.

## 5.3   QUANTA AS EIGENVALUES—THE TIME-INDEPENDENT SCHRÖDINGER EQUATION IN ONE DIMENSION

In a familiar high school experiment, we sprinkle a sodium salt into a flame and watch the familiar yellow light emitted by the glowing particles. This yellow light is characteristic of the element sodium, and the experiment is an instance of a general fact: individual elements emit radiation at discrete colors or wavelengths.

Spectrographic studies of the emissions from pure elements began in the early 1800s. A typical (simplified) experiment might run like this. Produce an electrical discharge in a sealed tube containing pure hydrogen gas at low pressure, pass the emergent light through a narrow slit and then through a prism that separates the different colors or wavelengths, and allow the processed light to strike a strip of photographic film. One then finds on the film discrete lines separated by varying distances. The interpretation: the exited hydrogen gas emits its radiation only at discrete wavelengths, not with a continuous distribution of wavelengths.

Similar spectrographic studies done during the 1800s on many elements showed the same discrete spectra. These sequences of spectral lines proved characteristic of the individual elements and were used to identify elements. For hydrogen, Balmer found a simple formula that gave the wavelengths $\lambda$ of one sequence of lines:

$$\frac{1}{\lambda} = \text{constant} \times \left( \frac{1}{2^2} - \frac{1}{n^2} \right), \qquad n = 3, 4, 5, \ldots$$

Other empirical formulas emerged, but no explanation of the basic underlying phenomenon: Why emissions at discrete wavelengths rather than a continuous spectrum?

In the early 1900s Rutherford, on the basis of experiments on the scattering of $\alpha$-particles (bare helium nuclei), concluded that an individual atom consisted of a compact concentrated nucleus, which contained virtually all the mass, and electrons circulating around it at comparatively great distances. Rutherford's conclusion—that an atom in its internal structure resembled a miniature solar system—was based on measurements of the angular distribution of $\alpha$-particles scattered from thin targets. This distribution matched that calculated from a layer of point masses. Thus emerged the following picture of the atom: virtually all the mass is concentrated at the central point with a cloud of electrons rotating around it at relatively great distances.

Niels Bohr, writing in the *Philosophical Magazine*, vol. 26, July 1913, pp. 1–25, acknowledging Rutherford's "kind and encouraging interest," took

Rutherford's atomic model and, on the basis of a few simple but radical assumptions, derived the formula for the emission spectrum of hydrogen. In this paper, Bohr refers to earlier work by J. W. Nicholson published in a series of notes in the *Monthly Notices of the Royal Astronomical Society* (1912), pp. 49, 139, 677, 693, and 729.

Bohr assumed that the electrons revolved in a circular orbit around the nucleus of the atom. Suppose that the radius of an orbit is $r$ and the angular velocity of the electron is $\omega$ (radians/sec), so that the velocity $v$ of the electron is $v = \omega r$ and its angular momentum $A$ is

$$A = m\omega r^2 = mvr$$

where $m$ denotes the electron mass. Bohr introduced the hypothesis that the angular momentum $A$ can take only values that are integral multiples of $h/2\pi$, where $h =$ Planck's constant $= 6.624 \times 10^{-27}$ erg sec. Thus $A = nh/2\pi$, or $mvr = nh/2\pi, n = 1, 2, \ldots$ Now the electron must rotate at such a velocity that its centrifugal force balances the electric attraction pulling it toward the nucleus. Using appropriate units, the electrostatic force attracting the electron toward the nucleus is $e^2/r^2$, and the centrifugal force is $mv^2/r$. Thus $mv^2/r = e^2/r^2$ or $(mvr)v = e^2$.

The kinetic energy, $T$, of the electron is $T = \frac{1}{2}mv^2$. Its potential energy, $V = V(r)$, with reference $V(\infty) = 0$, is $V(r) = \int_r^\infty -(e^2/x^2)\,dx = e^2/x|_r^\infty = -e^2/r$. As $e^2/r = mv^2$, the electron's total energy, $E = T + V$, is then $E = \frac{1}{2}mv^2 - mv^2 = -\frac{1}{2}mv^2$. Substituting $v^2 = e^4/(mvr)^2 = 4\pi^2e^4/n^2h^2$, we get $E = -2\pi^2me^4/n^2h^2, n = 1, 2, \ldots$, as the possible values of the total energy $E$ of the electron in the possible "stable states" hypothesized by Bohr. Bohr assumed that the atom emits energy in transitions between its stable states, the amount of energy emitted equaling the difference in total energy between the initial and final states. Thus Bohr's theory predicts emission energies

$$E_{\text{initial}} - E_{\text{final}} = \frac{2\pi^2me^4}{h^2}\left(\frac{1}{k^2} - \frac{1}{n^2}\right), \qquad k < n$$

For $k = 2$ and $n = 3, 4, \ldots$, this gives exactly the Balmer series.

Bohr's spectacular accomplishment faded somewhat as it failed to apply easily to and give correct predictions for many-electron atoms. Also, Bohr did not *explain* the discrete character of the spectrum but deduced it from the assumption of discrete angular momenta.

The currently accepted explanation was given by Heisenberg in 1925 (matrix mechanics) and by Schrödinger in 1926 (wave mechanics). We discuss Schrödinger's version of these essentially equivalent theories.

Schrödinger's work was presented in a series of four papers in the *Annalen der Physik*: Parts I and II in vol. 79 (1926), Part III in vol. 80 (1926), and Part IV in vol. 81 (1926). These four papers and five related papers, all by

Schrödinger, have been translated into English and bound in a small book entitled *Wave Mechanics*, published by Chelsea in 1927. I have taken all quotations from this book, except that I have substituted the current terms "eigenvalue" and "eigenfunction" for the out-of-date English terms "proper value" and "proper function" used there.

Schrödinger's theory ties directly to the mathematics we have done in Chapter 4 because he explains the discreteness of the spectrum by deducing the discrete values as eigenvalues of a differential operator. His four papers all carry the same title: "Quantization as a problem of eigenvalues." He refers to the ad hoc quantum conditions of Bohr as "of such a completely strange and incomprehensible nature" (p. 28), but observes that his equation "carries within itself the quantum conditions" (as eigenvalues). In another place (p. 9) he says, "The essential thing seems to me to be, that the postulation of 'whole numbers' no longer enters into the quantum rules mysteriously, but that we have traced the matter a step further back, and found the 'integralness' to have its origin in the finiteness and single-valuedness of a certain space function." Thus the classical eigenvalue theory of differential operators, created in response to the problem of describing such classical physical problems as the vibrating wire, provided in the end an explanation for the discreteness of the emission and absorption atomic spectra studied so thoroughly in the century 1800–1900.

In his papers, Schrödinger mentions particularly the 1924 work of Louis de Broglie and work of Einstein. Einstein noted that much direct evidence suggested that light consisted of particles with energy $E$ related to the frequency $v$ of the light by the equation $E = hv$, $h = $ Planck's constant ($= 6.624 \times 10^{-27}$ erg sec). Einstein based this conclusion on his analysis of the photoelectric effect and the analysis of the equilibrium of a radiation field with gaseous matter that led to a new derivation of Planck's formula. So light sometimes acts like a particle and in other situations acts like a wave (diffraction). Likewise, a beam of electrons under suitable conditions exhibits a wavelike diffraction pattern. So electrons, under certain circumstances, behave like waves. On these general grounds, de Broglie associated to a point mass $m$ moving with a velocity $v$ a wavelength

$$\lambda = \frac{h}{mv} = \frac{h}{p}$$

where $p = mv$ is the momentum. Louis de Broglie's paper, entitled "Recherches sur la théorie des quanta," was published in the *Annales de Physique* (10), vol. 3 (1925), pp. 22–128. De Broglie's precise assertion was this (p. 111 freely translated from the French): "If the speed is weak enough to permit neglecting relativistic terms, the wavelength associated to the movement of a particle whose speed is $v$ will be $\lambda = \cdots = h/m_0 v$."

If the (nonrelativistic) point mass has total energy $E$ and potential energy $V$, then the kinetic energy $T = E - V$ is related to the momentum by $E - V = \frac{1}{2}mv^2 = p^2/2m$. So $p = \sqrt{2m(E - V)}$, and

$$\lambda = \frac{h}{\sqrt{2m(E - V)}} \tag{5.3.1}$$

is the wavelength associated to the moving point mass $m$ that has total energy $E$ and potential energy $V$.

And now Schrödinger set about the task of devising a "wave equation" to describe these matter waves of de Broglie. We shall quote mainly from the second of his series of four papers entitled "Quantization as a problem of eigenvalues (Part II)." Sections 1 and 2 of this paper, entitled respectively "The Hamiltonian analogy between mechanics and optics" and "'Geometrical' and 'undulatory' mechanics," provide the background for his derivation. We shall quote only a few comments from this extended discussion.

In Section 1 of the paper cited, Schrödinger suggests an analogy between geometrical optics and classical mechanics. He then points out that *"we know today, in fact, that our classical mechanics fails for very small dimensions of the path and for very great curvatures."* He concludes at the end of this section, "Then it becomes a question of searching for an undulatory mechanics. . . ."

He begins the next section, Section 2, by discussing the time variation and frequency of the matter waves. He assumes that the time variation is sinusoidal, $\sin(2\pi\nu t)$, and that the frequency $\nu$ is related to the *total* $E$ by

$$E = h\nu \tag{5.3.2}$$

the same formula as Einstein's for photons. Note that equation (5.3.2) relates the frequency of the matter wave to the total energy of the point mass, not to its kinetic energy, thus not in any obvious way to its velocity.

Since the $E$ in formula (5.3.2) is total energy, kinetic plus potential, one sees a puzzling indeterminacy in the application of (5.3.2) to material particles, because the potential energy in classical mechanics contains an arbitrary constant, which depends on the selection of a reference point. Thus the frequency $\nu$ is equally indeterminate. Schrödinger points this out, asserting (p. 19) that (5.3.2) was true only "if $E$ is absolute and not, as in classical mechanics, indefinite to the extent of an arbitrary constant." He does not say how to make $E$ absolute. He goes on to point out that formula (5.3.1) for the wavelength $\lambda$ is independent of the additive constant as it involves the kinetic energy $E - V = T$ only. Thus the wave velocity $\lambda\nu$ is then indeterminate to the same extent that $\nu$ is. These problems do not come up in the case of photons, which all travel with velocity $c$ and have zero mass. In that case (Einstein's original equation) $E$ is the absolute energy of the photon entity. But the indeterminacy in (5.3.2) does not affect the applications to material particles because adding a constant to $V$ merely shifts the energy eigenvalues by that amount, as we shall see.

Equation (5.3.2) had to wait until the second paper for its unequivocal statement. In Schrödinger's first paper, he noted that $E$ was an eigenvalue and thus should be proportional to the *square* of the frequency. (Refer to our discussion of the vibrating string in Section 5.1.) He felt this couldn't be correct, for it would not give the Bohr frequency conditions, so he argued thus: because of the indeterminacy in $E$, perhaps $C + E$ where $C$ is a very large constant (possibly $C = mc^2$, the rest energy of the point mass) is proportional to $v^2$. Then $v^2 = \text{const} \times (C + E)$ so

$$v = \text{const} \times \left(1 + \frac{E}{C}\right)^{1/2} = \text{const} \times \left(1 + \frac{1}{2}\frac{E}{C} + \cdots\right)$$

This will give the frequency differences approximately proportional to $E$ differences. That was Schrödinger's argument in his first paper. In the second paper, he stated equation (5.3.2), $E = hv$, unequivocally and referred to his first attempt "merely as an approximate equation, derived from pure speculation."

Now Schrödinger comes to his wave equation. We quote verbatim from Section 2 of Part II:

> In what way now shall we have to proceed to the undulatory representation of mechanics for those cases where it is necessary? We must start, not from the fundamental equations of mechanics, but from a wave equation for $q$-space and consider the manifold of processes possible *according to it.* The wave equation has not been explicitly used or even put forward in the communication. The only datum for its construction is the *wave velocity,* which is given by (6) or (6') as a function of the mechanical energy parameter or frequency respectively, and by this datum the wave equation is evidently not uniquely defined. It is not even decided that it must be definitely of the second order. Only the striving for simplicity leads us to try this to begin with. We will then say that for the wave function $\Psi$ we have
>
> $$\text{div grad } \Psi - \frac{1}{u^2}\ddot{\Psi} = 0$$
>
> valid for all processes which only depend on the time through a factor $e^{2\pi i v t}$.

Let us analyze Schrödinger's proposed equation.

The function $\Psi$ is the *amplitude* of the wave; in our discussion (Chapter 3) we denoted it $u$. Schrödinger's notation seems to have been universally adopted. I would guess that every subsequent book and paper on wave mechanics ever written refers to the "psi function" $\Psi$. Our original wave function $u$ stood for something quite concrete: the vertical displacement of the wire at position $x$ and time $t$. The meaning of the $\Psi$ function in wave mechanics is a whole story unto itself, whose telling we shall postpone for the moment.

The expression "div grad" from vector analysis is, in one dimension, simply $\partial^2\Psi/\partial x^2$. The double dot over $\Psi$ means the second time derivative, $\partial^2\Psi/\partial t^2$. So, in one dimension Schrödinger's wave equation, as he states it in his second paper, is

$$\frac{\partial^2\Psi}{\partial x^2} - \frac{1}{u^2}\frac{\partial^2\Psi}{\partial t^2} = 0 \tag{5.3.3}$$

which is exactly the wave equation we have studied in several previous lectures. Now this time-dependent equation turned out to be wrong, and Schrödinger later corrected it. However, he makes no direct use of equation (5.3.3) in this paper but uses it to derive his time-independent equation, as follows.

Schrödinger assumes an explicit time dependence $e^{2\pi i v t}$, so we have an automatic separation of variables

$$\Psi = \Psi e^{2\pi i v t}$$

in an easily understood convention where the $\Psi$ on the right depends on $x$ alone. Then

$$\frac{\partial^2\Psi}{\partial t^2} = -4\pi^2 v^2 \Psi e^{2\pi i v t}$$

Substituting in (5.3.3) and canceling $e^{2\pi i v t}$, we get

$$\frac{\partial^2\Psi}{\partial x^2} + \frac{4\pi^2 v^2}{u^2}\Psi = 0 \tag{5.3.4}$$

for the time-independent wave function $\Psi$.

To arrive at the final equation, we must decide on a value for the wave velocity $u$. (In our notation in Chapter 3, we used $w$ for the wave velocity.) Bear in mind that $u$ is the *wave* velocity of the matter wave associated to our point mass and does not have any obvious connection with its classical velocity of motion $v$. We shall take the wave velocity $u$ as the frequency of the wave times its wavelength: $u = \lambda v$. The wavelength $\lambda$ is given by equation (5.3.1) and the frequency $v$ by equation (5.3.2). Thus

$$u = \lambda v = \left(\frac{h}{\sqrt{2m(E-V)}}\right)\left(\frac{E}{h}\right) = \frac{E}{\sqrt{2m(E-V)}} \tag{5.3.5}$$

whence the coefficient of $\Psi$ in equation (5.3.4) becomes

$$\frac{4\pi^2 v^2}{u^2} = \frac{4\pi^2 v^2(2m(E-V))}{E^2} = \frac{8\pi^2 m(E-V)}{h^2}$$

where, in the last step, we have used $E^2/v^2 = h^2$. It is customary to write $\hbar = h/2\pi$. Making this substitution and noting that $\Psi$ depends now on $x$ alone so that $\partial^2\Psi/\partial x^2 = d^2\Psi/dx^2$, we can rewrite equation (5.3.4) as

$$\frac{d^2\Psi}{dx^2} + 2m/\hbar^2 (E - V)\Psi = 0$$

The potential energy function $V$ is assumed known and assumed to depend on $x$ alone. It is then natural to rearrange this equation as follows:

$$-\frac{\hbar^2}{2m}\frac{d^2\Psi}{dx^2} + V\Psi = E\Psi \tag{5.3.6}$$

which is *Schrödinger's time-independent wave equation in one dimension for a single particle of mass m with given potential energy $V = V(x)$.*

If we write the differential expression on the left of equation (5.3.6) as $H$ so that

$$H(\Psi) \equiv -\frac{\hbar^2}{2m}\frac{d^2\Psi}{dx^2} + V\Psi$$

then Schrödinger's equation becomes

$$H(\Psi) = E\Psi \tag{5.3.6'}$$

which obviously suggests an eigenvalue problem, where the values of the energy $E$ are the eigenvalues of the operator $H$ and the corresponding functions $\Psi$ are its eigenfunctions. Thus for good reason did Schrödinger entitle his series of papers "Quantization as a problem of eigenvalues" and assert that his equation (5.3.6) *"carries within itself the quantum conditions."*

To properly set (5.3.6) as an eigenvalue problem within our framework, we have to specify a domain for $H$ within a given $L^2$ space, and we shall do that in the next section.

### Exercises

Some basic physical constants

| | |
|---|---|
| Planck's constant | $h = 6.6254 \times 10^{-27}$ erg-sec |
| | $\hbar = h/2\pi = 1.0544652 \times 10^{-27}$ erg-sec |
| Mass of the electron | $m = 9.1086 \times 10^{-28}$ g |
| Charge of the electron | $e = 4.80294 \times 10^{-10}$ esu |
| Speed of light | $c = 2.997928 \times 10^{10}$ cm/sec |
| Useful conversion factor | $1$ eV $= 1.6021 \times 10^{-12}$ ergs |

1. Use Bohr's theory to answer the following.
   (a) Show that the maximum kinetic energy of the orbiting electron is $me^4/2\hbar^2 = 13.6$ eV.
   (b) Show that the radius $a$ of the innermost electronic orbit is $a = \hbar^2/me^2 = 5.2917 \times 10^{-9}$ cm (Bohr radius).
   (c) Show that the de Broglie wavelengths $\lambda_n$ of the orbiting electrons are given by $\lambda_n = (2\pi a)n$, $n = 1, 2, \ldots$, thus are simply integral multiples of the "Bohr circumference" $2\pi a$.
   (d) Show that the velocities $v_n$ of the orbiting electrons are $v_n = (1/n)(e^2/\hbar)$, $n = 1, 2, \ldots$. For $n = 1$ compute this in cm/sec and mph. [Ans.: 4,894,000 mph.]
   (e) The ratio of the velocity $e^2/\hbar$ of the innermost orbiting electron to the velocity of light $c$ is called the fine structure constant $\alpha = e^2/\hbar c$. A recent experimental value is $\alpha^{-1} = 137.036$. In a paper that appeared in the *Comptes Rendus*, vol. 269, p. 743 (1969), Armand Wyler claimed to prove that

$$\alpha^{-1} = \frac{16\pi^4}{9}\left(\frac{5!}{\pi^5}\right)^{1/4}$$

How close is this to the experimental value? (Wyler's derivation is not generally accepted.)

2. Show that the wave velocity of a nonrelativistic particle whose energy is purely kinetic is one half of its particle velocity.

3. (a) Solve Schrödinger's equation for a nonrelativistic particle of mass $m$ in a constant potential field.
   [Ans.: $\Psi = c_1 \cos(2\pi x/\lambda) + c_2 \sin(2\pi x/\lambda)$ where $\lambda = h/\sqrt{2m(E-V)}$ is the de Broglie wavelength.]
   (b) Compute the de Broglie wavelength $\lambda$ in cm for
   (i) A 1-eV electron
   (ii) A 1-g mass moving at 1 cm/sec

## 5.4 INTERPRETATION OF THE $\Psi$ FUNCTION. THE TIME-DEPENDENT SCHRÖDINGER EQUATION IN ONE DIMENSION

In the last section we derived Schrödinger's time-independent wave equation as follows. Given a particle of mass $m$ and momentum $p$, we associated to it a de Broglie "matter wave" of wavelength $\lambda$ given by $\lambda = h/p$, where $h$ is Planck's constant. We noted next that if our particle has total energy $E$ and potential energy $V$, then its kinetic energy $E - V$ is related to its momentum $p$ by $E - V = \frac{1}{2}mv^2 = p^2/2m$, so that $p = \sqrt{2m(E-V)}$. Thus the wavelength of the de Broglie matter wave is given by

$$\lambda = \frac{h}{\sqrt{2m(E - V)}}$$  (5.4.1)

Next, with Schrödinger, we associated to the de Broglie matter wave a frequency $v$ given by

$$v = \frac{E}{h}$$  (5.4.2)

where, as before, $E$ is the *total* energy and $h$ is Planck's constant. Having both the wavelength $\lambda$ and frequency $v$ of the matter wave, we could compute the *wave velocity* $w = \lambda v$:

$$w = \lambda v = \left(\frac{h}{\sqrt{2m(E - V)}}\right)\left(\frac{E}{h}\right) = \frac{E}{\sqrt{2m(E - V)}}$$  (5.4.3)

Substituting this wave velocity into the classical one-dimensional wave equation that we used to describe the vibrating wire (Section 5.1), assuming a time dependence of the form $e^{2\pi i v t}$, and canceling this time factor from the resulting equation, we obtained *Schrödinger's time-independent wave equation in one dimension for a single particle of mass m with given potential energy* $V = V(x)$ $(\hbar = h/2\pi)$:

$$-\frac{\hbar^2}{2m}\frac{d^2\Psi}{dx^2} + V\Psi = E\Psi$$  (5.4.4)

Thus far we have not ascribed any meaning to the "wave function" $\Psi$. In the theory of the vibrating wire, the corresponding wave amplitude function, there denoted $u(x, t)$, has a concrete physical interpretation: it is the vertical displacement of the wire from its equilibrium position at point $x$ and time $t$. The function $u(x, t)$ describes completely the state of the wire. The corresponding question for quantum mechanics is this: what is the physical interpretation of the wave function $\Psi$ in equation (5.4.4)?

We have today a widely accepted interpretation of the meaning of $\Psi$, but before I state that interpretation I should like to describe briefly how Schrödinger in his 1926 series of papers treated $\Psi$. I think this is particularly instructive, given the not-yet-settled controversy about the foundations of quantum physics.

In the first paper Schrödinger introduces the function $\Psi$ via a change in variables in the Hamilton-Jacobi differential equation from which he derives his wave equation by a variational procedure. The bulk of this first paper is then concerned with the eigenvalues (energies) of the hydrogen atom; the wave function $\Psi$ itself is ignored. Of course, Schrödinger is aiming to explain the discrete observed energies, so his concentration on the energy eigenvalues is perfectly understandable. In the closing section of this paper, §3, he does remark:

> §3. It is, of course, strongly suggested that we should try to connect the function $\Psi$ with some *vibration process* in the atom, which would more nearly approach reality than the electronic orbits, the real existence of which is being very much questioned today. I originally intended to found the new quantum conditions in this more intuitive manner, but finally gave them the above neutral mathematical form, because it brings more clearly to light what is really essential. The essential thing seems to me to be, that the postulation of "whole numbers" no longer enters into the quantum rules mysteriously, but that we have traced the matter a step further back, and found the "integralness" to have its origin in the finiteness and single-valuedness of a certain space function.

(I have included again, in context, the latter part of this paragraph which was quoted earlier.)

In a note added in the proof to this first paper, Schrödinger notes that the variational problem from which he derived his equation at the beginning of the paper could be formulated in a neater way, which he outlines. In this outline the $\Psi$ function is introduced simply as a mathematical entity that minimizes a certain "Hamiltonian integral" subject to the accessory condition $\int \Psi^2 \, d\tau = 1$ (integral over all space).

And that is all that was said on the $\Psi$ function in the first paper ever written on wave mechanics.

The second paper by Schrödinger in his series of four papers we have already quoted from extensively. This long paper consists of three sections. The first two sections give a discursive derivation—which we have just outlined—of equation (5.4.4), Schrödinger's wave equation. The third section of this paper, entitled "Applications to examples," goes on to apply equation (5.4.4) to specific physical configurations: the Planck oscillator, rotator with fixed axis, rigid rotator with free axis, and the nonrigid rotator (diatomic molecule). In this last example he does mention interpreting the $\Psi$ function.

> The direct interpretation of the wave function of *six* variables in *three*-dimensional space meets, at any rate initially, with difficulties of an abstract nature.

Beyond that, in this long paper, Part II in his series, where Schrödinger rederives his wave equation and then applies it to many problems of physical interest, he makes no attempt whatever to attach a meaning to the "wave amplitude" function $\Psi$.

Between the publication of Parts II and III, Schrödinger published another paper (also included in the Chelsea *Wave Mechanics* volume) in which he demonstrates the equivalence of Heisenberg's matrix mechanics and his wave mechanics. In this paper he does make an assumption about the physical significance of the $\Psi$ function. Noting that when time variation is included, the $\Psi$ function will contain factors $\exp(2\pi i v t)$ and thus be complex, he goes on to say

> We now make the *assumption* that the space density of electricity is given by the real part of
>
> $$\Psi \frac{\partial \bar{\Psi}}{\partial t}$$
>
> The bar is to denote the conjugate complex function.

So Schrödinger does not give physical significance to $\Psi$ directly, rather to $\Psi \, \partial\bar{\Psi}/\partial t$, but he does give the latter *direct* physical meaning: its real part is to be the "space density of electricity."

In Part III in the main series of papers (which followed the one just quoted), in a footnote added in the proof, he refers to the above expression $\Psi \, \partial\bar{\Psi}/\partial t$:

> Now I recognize this $\partial/\partial t$ to be incorrect, though I hoped it would make the later relativistic generalization easier. Statement (36), *loc. cit.* [the expression $\Psi \, \partial\bar{\Psi}/\partial t$], is to be replaced by $\Psi\bar{\Psi}$.

As $\Psi\bar{\Psi}$ is real, we must take it that $\Psi\bar{\Psi}$ is to be the space density of electricity. Nowhere else in this third paper does he discuss the physical significance of $\Psi$.

In the fourth, and last paper in this series he mentions the mistake in the earlier paper and reaffirms that $\Psi\bar{\Psi}$ should be used, and not $\Psi \, \partial\bar{\Psi}/\partial t$. He then goes on to say (Section 2)

> According to the *heuristic hypothesis* on the electrodynamical significance of the field scalar $\Psi$, the present quantity $[\Psi\bar{\Psi}]$—apart from a multiplicative constant— represents the electrical density as a function of the space coordinates and the time, *if x stands for only three space coordinates*, i.e. if we are dealing with the problem of *one* electron.

Finally, in Section 7 of this last paper, which section is entitled "On the physical significance of the field scalar," he addresses head-on for the first time the physical meaning of $\Psi$. In this section Schrödinger again advocates interpreting $\Psi\bar{\Psi}$ as the density of electricity. He then goes on to say

> This rule is now equivalent to the following conception, which allows the true meaning of $\Psi$ to stand out more clearly. $\Psi\bar{\Psi}$ is a kind of *weight-function* in the system's configuration space. The *wave-mechanical* configuration of the system is a

*superposition* of many, strictly speaking of *all*, point-mechanical configurations kinematically possible. Thus each point-mechanical configuration contributes to the true wave-mechanical configuration with a certain *weight*, which is given precisely by $\Psi\bar{\Psi}$. If we like paradoxes, we may say that the systems exists, as it were, simultaneously in all the positions kinematically imaginable, but not "equally strongly" in all. In macroscopic motions, the weight-function is practically concentrated in a small region of positions, which are practically indistinguishable. The centre of gravity of this region in configuration space travels over distances which are macroscopically perceptible. In problems of microscopic motions, we are in any case interested *also*, and in certain cases even *mainly*, in the varying *distribution* over the region.

This new interpretation may shock us at first glance, since we have often previously spoken in such an intuitive concrete way of the "$\Psi$-vibrations" as though of something quite real. But there is something tangibly real behind the present conception also, namely, the very real electrodynamically effective fluctuations of the electric space-density. The $\Psi$-function is to do no more and no less than permit of the totality of these fluctuations being mastered and surveyed mathematically by a single partial differential equation. We have repeatedly called attention to the fact that the $\Psi$-function itself cannot and may not be interpreted directly in terms of three-dimensional space—however much of the one-electron problem tends to mislead us on this point—because it is in general a function in configuration space, not real space.

He then goes on to argue that conservation of charge requires that "its integral [referring to $\Psi\bar{\Psi}$] over the whole configuration space ... remain constantly normalized to the same unchanging value, preferably to unity." That is, Schrödinger sets the condition that

$$\int \Psi\bar{\Psi}\, dv = \int |\Psi|^2\, dv = 1 \qquad \text{independent of time}$$

Schrödinger then goes on to show by calculation that if $\Psi$ is obtained by solving the time-dependent wave equation (which we shall get to in a moment), then $\int \Psi\bar{\Psi}\, dv$ remains constant in time, as is consistent with the above interpretation of $\Psi\bar{\Psi}$. And thus ends this series of four extraordinary papers that founded the subject known today as wave mechanics. All four papers were published in a six-month period in 1926.

The current interpretation of the significance of $\Psi\bar{\Psi}$ is different from that proposed by Schrödinger. Today the interpretation is that $\Psi\bar{\Psi} = |\Psi|^2$ is the *probability density* of finding the particle in a given differential volume element. Thus, referring to our one-dimensional equation (5.4.4), the integral

$$\int_c^d |\Psi(x)|^2\, dx$$

represents the probability of finding the electron between $x = c$ and $x = d$. Of course, for this to make sense we must have $\int_a^b |\Psi(x)|^2 \, dx = 1$; that is, the probability of finding the particle *somewhere* is 1. (Here $a \leqslant x \leqslant b$ is the interval of interest, and $a \leqslant c \leqslant d \leqslant b$.) Inasmuch as the wave equation (5.4.4) is linear, any solution $\Psi$ multiplied by a constant remains a solution. Thus we may secure the above normalization so long as we have

$$0 < \int_a^b |\Psi(x)|^2 \, dx < \infty \tag{5.4.5}$$

where $a \leqslant x \leqslant b$ is the interval of interest in any particular physical problem and may be finite, semi-infinite, or fully infinite. This is the same as the condition required by Schrödinger, even though his interpretation of $\Psi \bar{\Psi}$ was different. The important thing to note is that *condition (5.4.5) if we include $\Psi = 0$ defines an $L^2$ space, $L^2(a, b)$, which thus arises naturally in quantum mechanics.* This space carries an intrinsic inner product, so that the whole machinery of Hilbert space seems to be intrinsically linked with quantum mechanics. Schrödinger's operator $H$ now has its domain contained in $L^2(a, b)$ specified by various boundary conditions that arise in specific problems, the eigenfunctions of $H$ (which turns out to be Hermitian) form an orthogonal family in $L^2(a, b)$, etc.

The same mathematics would have resulted from Schrödinger's hypothesis, or any other hypothesis that required $\int \Psi \bar{\Psi} \, dx < \infty$. However, there is a basic philosophical difference between Schrödinger's interpretation of $\Psi \bar{\Psi}$ and the probabilistic interpretation. Schrödinger gave $\Psi \bar{\Psi}$ a real physical meaning, the "density of electricity," whereas under the modern-day interpretation $\Psi \bar{\Psi}$ is a probability density, not a physical quantity. This latter interpretation makes quantum mechanics a statistical theory rather than a deterministic one.

It is, in retrospect, a little puzzling that Schrödinger would have put forward his "charge density" hypothesis, because there is nothing in the derivation of his equation that requires the mass-particle to have any charge at all. The de Broglie waves apply to any particle, as does the entire theory, even though most of the original applications were to atomic electrons. For an uncharged particle, a "mass density" interpretation would presumably be the equivalent of Schrödinger's charge density hypothesis, but Schrödinger never explicitly deals with $\Psi \bar{\Psi}$ for uncharged particles.

The probabilistic interpretation was given in 1926, by Born. Schrödinger does refer to it in a later paper. He says (in connection with the "energy-momentum theorem"):

> The question whether the solution of the difficulty is really to be found only in the purely *statistical* interpretation of the field theory which has been proposed in

several quarters must for the present be left unsettled. Personally, I no longer regard this interpretation as a finally satisfactory one, even if it proves useful in practice. To me it seems to mean a renunciation, much too fundamental in principle, of all attempt to understand the individual process.

Be that as it may, the probabilistic interpretation is the one currently accepted and is the one we shall follow. It is nonetheless interesting that all interpretations refer to $\Psi\bar{\Psi} = |\Psi|^2$ and *not* to $\Psi$ itself. As to the (complex) wave amplitude $\Psi$, Heisenberg has this to say (*The Physical Principles of the Quantum Theory*, English edition, Dover, 1930, p. 51):

> It must be noted, however, that no experiment can ever measure the amplitude directly, as is evident from the fact that the de Broglie waves are complex.

and, a little further on, he says

> It is thus highly satisfactory that there is no experiment which will measure $\Psi$ at a given point at a given time.

Thus Schrödinger's equation (5.4.4) (and the time-dependent equation that we shall derive in a moment) has for solution a function $\Psi$ that itself has no direct physical interpretation, but $|\Psi|^2$ represents a probability density.

But the physical interpretations, or lack of them, do not affect the mathematics. We look at (5.4.4) thus: the differential expression

$$H(\Psi) = -\frac{\hbar^2}{2m}\frac{d^2\Psi}{dx^2} + V\Psi$$

when its domain is properly specified as a subspace of the $L^2$ space

$$L^2(a, b) = \left\{\Psi(x): \int_a^b |\Psi(x)|^2 < \infty\right\}$$

determines an operator in that Hilbert space, and the mathematical problem is then to determine the eigenvalues and eigenfunctions of that operator. The eigenvalues will be the possible discrete energies of our mass-particle.

Let us now turn finally to the derivation of Schrödinger's time-dependent equation. The classical wave equation is $\partial^2\Psi/\partial x^2 - (1/w^2)(\partial^2\Psi/\partial t^2) = 0$, where $w$ is wave velocity. This equation appeared in Schrödinger's second paper— refer to the previous section. If we substitute the expression (5.4.3) for the wave velocity $w$ into this equation, we get

$$\frac{\partial^2\Psi}{\partial x^2} - \frac{2m(E - V)}{E^2}\frac{\partial^2\Psi}{\partial t^2} = 0$$

This equation is *wrong*. The true time-dependent wave equation of quantum mechanics is of first order in the time $t$, not second order. Schrödinger comes to this point in the fourth (and last) paper of the series. We quote from the first section of that paper.

The wave equation (18) or (18″) of Part II., viz.

(1)     $$\nabla^2\Psi - \frac{2(E-V)}{E^2}\frac{\partial^2\Psi}{\partial t^2} = 0$$

or

(1′)     $$\nabla^2\Psi + \frac{8\pi^2}{h^2}(E-V)\Psi = 0$$

which forms the *basis* for the re-establishment of mechanics attempted in this series of papers, suffers from the disadvantage that it expresses the law of variation of the "mechanical field scalar" $\Psi$, neither *uniformly* nor *generally*. Equation (1) contains the energy- or frequency-parameter $E$, and is valid, as is expressly emphasized in Part II., with a *definite* $E$-value inserted, for processes which depend on the time exclusively through a *definite* periodic factor:

(2)     $$\Psi \sim \text{real part of}\left(\exp\left(\pm\frac{2\pi iEt}{h}\right)\right)$$

Equation (1) is thus not really any more general than equation (1′), which takes account of the circumstance just mentioned and does not contain the time at all.

Thus, when we designated equation (1) or (1′), on various occasions, as "the wave equation," we were really wrong and would have been more correct if we had called it a "vibration-" or an "amplitude-" equation.

Schrödinger then goes on to derive the *real* wave equation, as he calls it. We may get it simply by eliminating $E$ from equation (5.4.4). If $\Psi$ contains the time factor $\exp(-2\pi iEt/h) = \exp(-iEt/\hbar)$, then

$$\frac{\partial\Psi}{\partial t} = \frac{-iE}{\hbar}\Psi \qquad \text{so } E\cdot\Psi = -\frac{\hbar}{i}\frac{\partial\Psi}{\partial t} = i\hbar\frac{\partial\Psi}{\partial t}$$

Substituting on the right side of (5.4.4), we get

$$\boxed{-\frac{\hbar^2}{2m}\frac{\partial^2\Psi}{\partial x^2} + V\Psi = i\hbar\frac{\partial\Psi}{\partial t}} \tag{5.4.6}$$

or

$$H(\Psi) = i\hbar\frac{\partial\Psi}{\partial t} \tag{5.4.6′}$$

Equation (5.4.6) (or (5.4.6')) is Schrödinger's time-dependent wave equation in one dimension for a single particle of mass $m$. It is, as we mentioned, first order in the time $t$, similar to the heat equation, but is irretrievably complex, unlike any equation of classical mathematical physics. Note that if we solve equation (5.4.6) for $\Psi$, then we know the complex conjugate $\bar\Psi$, which also satisfies (5.4.6) with $i$ replaced by $-i$. Thus it does not matter whether we use $+i$ or $-i$ in (5.4.6).

**Exercises**

1. In the following manner, solve Schrödinger's time-independent equation for a particle of mass $m$ restrained to the interval $0 \leqslant x \leqslant L$ inside which $V = 0$.

   (a) Since the particle cannot appear outside the interval $0 \leqslant x \leqslant L$, we must have $|\Psi|^2 = 0$ for $x < 0$ and $x > L$. If we wish to have $\Psi$ continuous, which is a quite reasonable condition given $\Psi$'s meaning, we shall have to have $\Psi(0) = \Psi(L) = 0$. Regard $H$ as an operator in $L^2(0, L)$ with the differential expression $l(\Psi) \equiv -(\hbar^2/2m)(d^2\Psi/dx^2)$ and domain consisting of all compatible functions $\Psi$ in $L^2(0, L)$ that satisfy $\Psi(0) = \Psi(L) = 0$. Show that $H$ is Hermitian positive definite.

   (b) Find the eigenvalues and eigenfunctions of the Hilbert space operator $H$ from part (a), and show thus that Schrödinger's equation $H(\Psi) = E\Psi$ has nonzero solutions $\Psi$ only when $E = E_n = n^2\hbar^2\pi^2/2mL^2$, $n = 1, 2, \ldots$, and that the corresponding solutions $\Psi_n$, normalized so that $\|\Psi_n\|^2 = 1$, are $\Psi_n(x) = (2/L)^{1/2}\sin(n\pi x/L), n = 1, 2, \ldots$.

   (c) Discuss the solution you have found in (b) along the following lines: The solution (b) means that, in quantum mechanics, the total energy $E$ of a bound particle can have only certain discrete values; i.e., the total energy is *quantized* (and, since $V = 0$, also the kinetic energy). How are these discrete energies related to the operator $H$? Comment in this connection on Schrödinger's assertion that his equation "*carries within itself the quantum conditions.*"

   How are we assured in advance that these discrete energy values must be positive real numbers?

   How are we assured in advance that the stationary "eigenstates" $\Psi_n$ are orthogonal with respect to the inner product $\langle\Psi, \Phi\rangle = \int_0^L \Psi\Phi\,dx$?

   (d) Compute $E_1$ in eV for an electron with L = Bohr radius = $5.2917 \times 10^{-9}$ cm. Compute $E_1$ in eV for a 1-g mass bound in a 10-cm interval. Compare and discuss.

   (e) Plot $(L/2)|\Psi_1(x)|^2$, $0 \leqslant x \leqslant L$. Where is it most probable to find the particle? Same question for the eigenstate $n = 2$. In the case $n = 2$, at what interior point will you never find the particle?

2. (Continuation) In the following manner, solve Schrödinger's time-dependent wave equation for the same physical problem as in Exercise 1: a particle of mass $m$ restrained to the interval $0 \leqslant x \leqslant L$ inside which $V = 0$.

  (a) Separate variables by setting $\Psi(x, t) = X(x)Z(t)$, use Exercise 1, and thus derive

  $$\Psi(x, t) = \sqrt{\frac{2}{L}} \sum_{n=1}^{\infty} A_n \sin \frac{n\pi x}{L} \exp\left(-i\left(E_n/\hbar\right)t\right)$$

  (b) Show that the $A_n$ depend on the initial state $\Psi(x, 0)$ and find an explicit formula for them. If $\Psi(x, 0) = \sqrt{2/L} \sin(\pi x/L)$, find $\Psi(x, t)$.

  (c) Noting that $\int_0^L |\Psi(x, t)|^2 \, dx = \langle \Psi, \Psi \rangle$, find $\int_0^L |\Psi(x, t)|^2 \, dx$. Show that this quantity is independent of time $t$. Using your answer to (b), show that

  $$\int_0^L |\Psi(x, 0)|^2 \, dx = \int_0^L |\Psi(x, t)|^2 \, dx$$

  Thus if $\int_0^L |\Psi(x, 0)|^2 \, dx = 1$, then $\int_0^L |\Psi(x, t)|^2 \, dx = 1$ for all time $t$.

3. Solve Schrödinger's time-independent wave equation with $V = 0$ for a particle of mass $m$ restrained to a circle of radius $a$. (Follow the pattern of Exercise 1. You will need to set up the domain of $H$ properly. Note that there is no boundary.)

4. (Continuation) Solve Schrödinger's time-dependent wave equation for the same physical problem as in Exercise 3: a particle of mass $m$ restrained to a circle of radius $a$. Follow the pattern of Exercise 2.

## 5.5 THE QUANTUM LINEAR OSCILLATOR

In this section we shall solve Schrödinger's time-independent equation for the linear oscillator. This is the quantum mechanical version of the classical mass-on-a-spring problem that we solved in Section 2.5. We shall follow in detail Schrödinger's solution as given in Section 3 of his second paper.

We begin with Schrödinger's description of the problem. The following quote is taken from his paper "The continuous transition from micro- to macro-mechanics." This paper is included in the book *Wave Mechanics* referred to many times earlier. The quote is on p. 41, footnote 4: "a particle of mass $m$ which, moving in a straight line, is attracted towards a fixed point in it, with a force proportional to its displacement $q$ from this point."

We shall use $x$ instead of $q$ to denote the distance of the particle of mass from the fixed point. The setup is depicted in Figure 5.4.

We take $x$ positive to the right. According to assumption, the particle is attracted toward the fixed point (labeled 0 in Figure 5.4) with a force

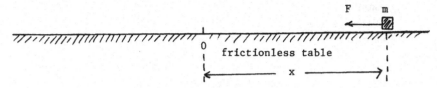

**Figure 5.4**  The quantum linear oscillator.

proportional to $x$. Let us call the proportionality constant $k$ so that $F = -kx$, the minus sign because the force is in the negative $x$ direction when $x$ is positive.

Schrödinger's equation is

$$-\frac{\hbar^2}{2m}\frac{d^2\Psi}{dx^2} + V\Psi = E\Psi$$

where $m$ is the mass of our particle, $V$ is its potential energy, and $E$ is its total energy, kinetic plus potential. For our mass particle, $V = \frac{1}{2}kx^2$. The natural frequency of our system, $\nu_0$, is given by $2\pi\nu_0 = (k/m)^{1/2}$. Solving for $k$, we get $k = 4\pi^2\nu_0^2 m$, and substituting this into the formula for the potential energy $V$ we get $V = 2\pi^2\nu_0^2 mx^2$. Now we substitute this expression for $V$ into Schrödinger's equation to get

$$-\frac{\hbar^2}{2m}\frac{d^2\Psi}{dx^2} + 2\pi^2\nu_0^2 mx^2\Psi = E\Psi \tag{5.5.1}$$

which is Schrödinger's time-independent equation for the linear oscillator with mass $m$ and natural frequency $\nu_0$. Our objective is to solve (5.5.1) for the "wave amplitude" $\Psi$.

The quantity $|\Psi|^2 = \Psi\bar{\Psi}$ represents the differential probability of finding our mass particle between $x$ and $x + dx$. Since the particle must be *somewhere*, we must have

$$\int_{-\infty}^{+\infty} |\Psi(x)|^2 \, dx = 1 \tag{5.5.2}$$

This sets the problem in the Hilbert space $L^2(-\infty, \infty)$, which consists of functions $f(x)$ that satisfy

$$\langle f, f \rangle = \int_{-\infty}^{+\infty} |f(x)|^2 \, dx < \infty$$

because any such $f(x)$ that is not zero can be normalized to secure (5.5.2). So we look on the left side of Schrödinger's equation (5.5.1) as the differential expression

$$s(\Psi) \equiv -\frac{\hbar^2}{2m}\frac{d^2\Psi}{dx^2} + 2\pi^2 v_0^{\,2} m x^2 \Psi$$

for a Hilbert space operator $H$ in $L^2(-\infty, \infty)$ whose domain consists of functions $\Psi$ that are compatible with $s$. Thus

domain($H$) = those functions $\Psi$ in $L^2(-\infty, \infty)$ such that $s(\Psi)$ also belongs to $L^2(-\infty, \infty)$

This compatibility condition is the only condition on the domain of $H$. There are no boundary conditions.

Thus the quantum linear oscillator has led to the following *Hilbert space eigenvalue problem*:

(5.5.3)  *Consider the Hilbert space $L^2(-\infty, \infty)$, and the following operator $H$ acting in this space: $H$ has the*

*Differential expression* $s(\Psi) \equiv -(\hbar^2/2m)\,d^2\Psi/dx^2 + 2\pi^2 v_0^{\,2} m x^2 \Psi$
*and*
*Domain($H$) consisting of those functions $\Psi$ in $L^2(-\infty, \infty)$ for which $s(\Psi)$ also belongs to $L^2(-\infty, \infty)$*

$$(5.5.3)$$

*For what values of the energy parameter $E$ does the equation*

$$H(\Psi) = E\Psi$$

*have nonzero solutions $\Psi$ in the domain of $H$?*

We shall solve this eigenvalue problem by transforming it to the eigenvalue problem for Hermite's operator. Following Schrödinger, we introduce the following quantities:

$$a = \frac{2mE}{\hbar^2}, \qquad b = \frac{16\pi^2 v_0^{\,2} m^2}{\hbar^2}$$

Then $E = \hbar^2 a/2m$ and $2\pi^2 v_0^{\,2} m x^2 = b\hbar^2 x^2/8m$. Making these substitutions in Schrödinger's equation (5.5.1), then canceling a term $\hbar^2/2m$, we find

$$-\frac{d^2\Psi}{dx^2} + \frac{b}{4}x^2\Psi = a\Psi$$

The quantity $b$ has units cm$^{-4}$, so $b^{1/4}$ has units cm$^{-1}$. Thus $z = xb^{1/4}$ is dimensionless. Substitute $x = z/b^{1/4}$ into the equation just derived to get

$$-\frac{d^2\Psi}{dz^2} + \frac{z^2}{4}\Psi = \frac{a}{\sqrt{b}}\Psi$$

The quantity $a/\sqrt{b} = E/h\nu_0$ is also dimensionless. Finally, we make the substitution

$$\Psi = e^{-z^2/4}y$$

Using the product rule for differentiation, we get

$$\frac{d\Psi}{dz} = \left(-\frac{z}{2}y + \frac{dy}{dz}\right)e^{-z^2/4}$$

and

$$\frac{d^2\Psi}{dz^2} = \left(\left(\frac{z^2}{4} - \frac{1}{2}\right)y - z\frac{dy}{dz} + \frac{d^2y}{dz^2}\right)e^{-z^2/4}$$

This final substitution leads to the equation

$$-\frac{d^2y}{dz^2} + z\frac{dy}{dz} = \lambda y \qquad \text{where } \lambda = \frac{E}{h\nu_0} - \frac{1}{2} \tag{5.5.4}$$

The differential expression on the left of (5.5.4), $l(y) \equiv -y'' + zy'$, is that defining Hermite's operator (Section 4.9). We can perhaps recognize this differential expression more easily if we put it in Sturmian form:

$$l(y) \equiv -y'' + zy' = -e^{z^2/2}(e^{-z^2/2}y')'$$

As $\Psi = e^{-z^2/4}y$, we have

$$\int_{-\infty}^{\infty} |\Psi(z)|^2 \, dz = \int_{-\infty}^{\infty} |y(z)|^2 e^{-z^2/2} \, dz$$

Hence $\Psi(z)$ belongs to the Hilbert space $L^2(-\infty, \infty)$ if and only if $y(z)$ belongs to the Hilbert space $L^2(-\infty, \infty; e^{-z^2/2})$. Thus the eigenvalue problem (5.5.3), which governs the behavior of the quantum linear oscillator, is equivalent to the eigenvalue problem for Hermite's operator:

(5.5.5) *Consider the Hilbert space* $L^2(-\infty, \infty; e^{-z^2/2})$ *and Hermite's operator L acting in this space: L has the*

*Differential expression* $l(y) = -y'' + zy'$
*and*
*Domain(L) consisting of those functions y in* $L^2(-\infty, \infty; e^{-z^2/2})$ *for which* $l(y)$
    *also belongs to* $L^2(-\infty, \infty; e^{-z^2/2})$.

$$\tag{5.5.5}$$

*For what values of the parameter* $\lambda$ *does the equation*

$$L(y) = \lambda y$$

*have nonzero solutions y in the domain of L?*

In Section 4.9 we solved the eigenvalue problem for Hermite's operator. The eigenvalues of Hermite's operator are the nonnegative integers $\lambda = n$, $n = 0, 1, 2, \ldots$, each eigenvalue is simple, and the $n$th Hermite polynomial $H_n(z)$ is an eigenfunction belonging to the eigenvalue $\lambda = n$.

The eigenvalue parameter $\lambda$ in (5.5.5) is related to the total-energy parameter $E$ of the linear oscillator by $\lambda = E/hv_0 - 1/2$ (see (5.5.4)). *Hence the total energy $E$ of the linear oscillator is quantized*, i.e., can have only discrete values

$$E_n = (n + \tfrac{1}{2})hv_0, \qquad n = 0, 1, 2, \ldots \tag{5.5.6}$$

If we set $E_0 = (1/2)hv_0$, the *zero-point energy*, then (5.5.6) becomes

$$E_n = (2n + 1)E_0, \qquad n = 0, 1, 2, \ldots \tag{5.5.7}$$

so the possible total energies of the quantum linear oscillator are all odd multiples of the zero-point energy $E_0$.

Nothing in classical physics hints at this quantization of energy. For the classical linear oscillator that we studied in Section 2.5 we could stretch the spring to an initial distance, $x_{max}$, which was totally arbitrary. In this process we would have imparted to the system an initial potential energy $V = 2\pi^2 v_0^2 m(x_{max})^2$. Assuming the initial velocity was zero, the total energy of the system would remain constant at this value. Upon release, the mass will oscillate sinusoidally between $x_{max}$ and $-x_{max}$ with an energy exchange between potential and kinetic, but the total energy $E$ would remain constant at the initial value $2\pi^2 v_0^2 m(x_{max})^2$, which was arbitrary.

The quantum oscillator behaves quite differently. The total energy $E$ *cannot* be arbitrarily specified. There is associated to any particular oscillator a zero-point energy, $E_0 = \tfrac{1}{2}hv_0$, and the possible total energies of this oscillator are the odd integral multiples of $E_0$. Total energy comes in lumps.

Refer back now to our original eigenvalue problem (5.5.3). We have shown this: *The Hilbert-space operator $H$ in $L^2(-\infty, \infty)$ defined as follows:*

*differential expression:* $s(\Psi) \equiv -(\hbar^2/2m)\, d^2\Psi/dx^2 + 2\pi^2 v_0^2 mx^2\Psi$
*domain: those $\Psi \in L^2(-\infty, \infty)$ such that also $s(\Psi) \in L^2(-\infty, \infty)$*

*has for its eigenvalues—i.e., the $E$ for which $H(\Psi) = E\Psi$ has a nonzero solution $\Psi$ in domain($H$)—exactly the possible energies of the quantum oscillator, namely $E_n = (2n+1)E_0$, $n = 0, 1, 2, \ldots$, where $E_0 = \tfrac{1}{2}hv_0$.*

Thus it is that Schrödinger explained the discreteness of nature, the fact that energy seemed to come in lumps rather than vary continuously. His time-independent equation for the system, when interpreted as an equation in Hilbert space, has solutions only for discrete values of the energy, the eigenvalues. His equation "*carries within itself the quantum conditions.*" In his lectures at the Royal Institution, London, in March 1928, Schrödinger emphasized the point again:

> It shows that, whatever the waves may mean physically, the theory furnishes a method of quantization which is absolutely free from arbitrary postulates that this or that quantity must be an integer.

Note that there are no boundary conditions that determine the domain of our operator. There is only the manifestly reasonable compatibility condition that $s(\Psi)$ belong to $L^2(-\infty, \infty)$ when $\Psi$ does. The fact that there are no boundary conditions bothered Schrödinger initially because he thought that you needed boundary conditions to get eigenvalues. In the same 1928 lectures to the Royal Institution, he says:

> A simplification in the problem of the "mechanical" waves (as compared with the fluid problem) consists in the absence of boundary conditions.
> I thought the latter simplification fatal when I first attacked these questions. Being insufficiently versed in mathematics, I could not imagine how proper vibration frequencies [eigenfrequencies] could appear *without* boundary conditions.

We know now that the compatibility condition forces certain boundary behavior: if $f$ and $g$ belong to the domain of Hermite's operator, then $\lim e^{-x^2/2} f'(x)\overline{g(x)} = 0$ at $x \pm \infty$. See Lemma 4.39 and the appendix to Chapter 4. This fact allowed us to prove that Hermite's operator is Hermitian, whence it follows that Schrödinger's operator $H$ is a Hermitian operator in $L^2(-\infty, \infty)$. No separate boundary conditions are required.

Let us now return to the detailed analysis of the quantum linear oscillator.

Suppose the two physical parameters of our oscillator are fixed: the mass $m$ of the mass particle and the natural frequency $\nu_0$ of the system, where $\nu_0$ is given by $2\pi\nu_0 = (k/m)^{1/2}$, $k$ being the spring constant. The total energy of this oscillator can have only the discrete values $E_n = (2n+1)E_0$, $n = 0, 1, 2, \ldots$, where $E_0 = \frac{1}{2}h\nu_0$. The values $E_n$ are the eigenvalues of Schrödinger's operator $H$ (see (5.5.3)). We have established the equivalence of Schrödinger's eigenvalue problem with that for Hermite's operator (see (5.5.5)), so the eigenfunction $\Psi_n$ belonging to the eigenvalue $E_n$ can be taken as any nonzero multiple of $H_n(z)e^{-z^2/4}$:

$$\Psi_n(z) = c_n H_n(z) e^{-z^2/4}$$

(See the discussion preceding equation (5.5.4)). We want to choose the constant $c_n$ so that $\Psi_n(z)$ is normalized with integral 1. Hence

$$1 = \int_{-\infty}^{\infty} |\Psi_n(z)|^2 \, dz = c_n^2 \int_{-\infty}^{\infty} H_n(z)^2 e^{-z^2/2} \, dz = c_n^2 \sqrt{2\pi}\, n!$$

Hence $c_n = (2\pi)^{-1/4}(n!)^{-1/2}$, and our normalized eigenfunctions are

$$\Psi_n(z) = \frac{1}{(2\pi)^{1/4}\sqrt{n!}} H_n(z)e^{-z^2/4} \qquad (5.5.8)$$

where the $H_n(z)$ are Hermite's polynomials. These eigenfunctions are normalized so that $\langle \Psi_n, \Psi_n \rangle = 1$ (in $L^2(-\infty, \infty)$) and form an orthogonal basis of $L^2(-\infty, \infty)$. The quantity

$$|\Psi_n(z)|^2 = \frac{1}{\sqrt{2\pi n!}} H_n^2(z)e^{-z^2/2}$$

represents the probability that the mass-particle lies between $z$ and $z + dz$ (when the oscillator is in the "eigenstate" corresponding to energy $E_n$).

In our analysis of the classical linear oscillator, we saw that the probability of finding the mass-particle between $u$, $u + du$, where $u = x/x_{max}$, was

$$p(u) = \begin{cases} \dfrac{1}{\pi\sqrt{1 - u^2}}, & |u| < 1 \\ 0, & |u| > 1 \end{cases}$$

so that there was zero probability of finding the particle at distances $x$ greater in absolute value than $x_{max}$, where

$$x_{max} = \frac{1}{\sqrt{2}\pi v_0} \sqrt{\frac{E}{m}}$$

(for all this, see Section 2.5). For the quantum oscillator, the probability $|\Psi_n(z)|^2$ is nonzero all the way to infinity, so there is a nonzero probability of finding the particle at arbitrarily large distances from the origin.

Our dimensionless variable $z$ is defined as

$$z = b^{1/4}x \qquad \text{where } b = \frac{16\pi^2 v_0^2 m^2}{\hbar^2}$$

Simple algebra leads to

$$x_{max} = \left(n + \frac{1}{2}\right)^{1/2} \frac{\hbar}{\sqrt{mE_0}} \qquad \text{and} \qquad b^{1/4} = \left(n + \frac{1}{2}\right)^{1/2} \frac{2}{x_{max}}$$

so $z = \sqrt{2}\sqrt{2n+1}\, u$ where $u = x/x_{max}$. Therefore the classical range $|u| \leqslant 1$ corresponds to

$$-\sqrt{2}\sqrt{2n + 1} \leqslant z \leqslant \sqrt{2}\sqrt{2n + 1}$$

To conveniently compare the various $\Psi_n(z)$ with each other and with the classical probability, it is best to express all functions in terms of the variable $u = x/x_{max}$ for which the classical range is $-1 \leqslant u \leqslant 1$. The equation

$$|\Psi_n(u)|^2 = \sqrt{2}\sqrt{2n+1}\,|\Psi_n(z)|^2 = \frac{1}{n!}\left(\frac{2n+1}{\pi}\right)^{1/2} H_n^2(z)e^{-z^2/2},$$

where

$$z = \sqrt{2}\sqrt{2n+1}\,u$$

*defines* the function $|\Psi_n(u)|^2$ which has the normalization

$$\int_{-\infty}^{\infty} |\Psi_n(u)|^2\,du = 1$$

For $n = 0$ we have

$$|\Psi_0(u)|^2 = \frac{1}{\sqrt{\pi}}\,e^{-z^2/2}, \quad z = \sqrt{2}\,u$$

The function is graphed in Figure 5.5. As you can see, this Gaussian is peaked at $u = 0$, where the particle in the $n = 0$ state is most likely to be found. We can compute the probability that the particle is in the classical range $-1 \leqslant u \leqslant 1$:

$$\int_{-1}^{1} |\Psi_0(u)|^2\,du = \frac{1}{\sqrt{\pi}}\int_{-1}^{1} e^{-u^2}\,du = \frac{2}{\sqrt{\pi}}\int_{0}^{1} e^{-u^2}\,du = 0.84270$$

Thus, in the $n = 0$ state, there is a 15.7% chance of finding the particle outside the classical range (where the kinetic energy is negative).

For $n = 3$ we have

$$|\Psi_3(u)|^2 = \frac{1}{3!}\left(\frac{7}{\pi}\right)^{1/2} H_3^2(z)e^{-z^2/2}, \quad z = \sqrt{14}\,u$$

The following HP-15C program can be used to compute $|\Psi_3(u)|^2$. Place $\sqrt{14} = 3.741657387$ in storage register 0, and place $(1/3!)(7/\pi)^{1/2} = 0.248784222$ in register 2. The subroutine GSB 3 evaluates $H_3(z)$. Enter the value of $u$. The program supplies $|\Psi_3(u)|^2$. Values are graphed in Figure 5.6 along with $|\Psi_{classical}(u)|^2 = (1/\pi)(1 - u^2)^{-1/2} = p(u)$.

| f LBL A | RCL 1 | RCL × 2 |
|---------|-------|---------|
| RCL × 0 | g $x^2$ | g RTN |
| STO 1 | 2 ÷ CHS | g P/R. |
| GSB 3 | $e^x$ | |
| g $x^2$ | × | |

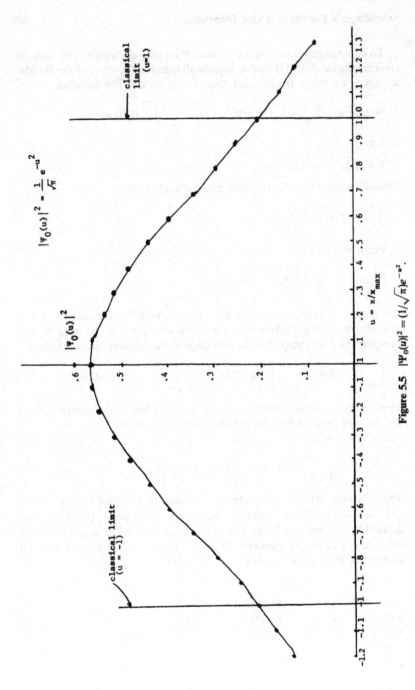

**Figure 5.5** $|\Psi_0(u)|^2 = (1/\sqrt{\pi})e^{-u^2}$.

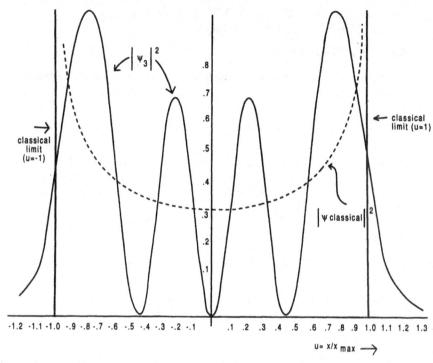

**Figure 5.6** $|\Psi_3(u)|^2$ versus $|\Psi_{classical}(u)|^2$.

## Exercises

1. Consider the electron as the mass-particle of a linear oscillator. For the zero-point energy $E_0$ of this linear oscillator, take the kinetic energy of the innermost orbiting electron in Bohr's model, to wit $E_0 = me^4/2\hbar^2$ (see Exercise 1a of Section 5.3). Show then that the $x_{max}$ for the $n = 0$ state is

$$x_{max} = \frac{\hbar^2}{me^2}$$

which is the Bohr radius (Exercise 1b of Section 5.3).

2. Compute $|\Psi_5(u)|^2$ and graph it along with $|\Psi_{classical}(u)|^2$ as in Figure 5.6.

## 5.6 SOLUTION OF THE TIME-DEPENDENT SCHRÖDINGER EQUATION

### 5.6.1 The General Case

The time-dependent Schrödinger equation for a wave function $\Psi(x, t)$ which depends on time $t$ is

$$-\frac{\hbar^2}{2m}\frac{\partial^2 \Psi}{\partial x^2} + V\Psi = i\hbar\frac{\partial \Psi}{\partial t} \tag{5.6.1}$$

We shall assume that the potential function $V$ is a function of $x$ alone, $V = V(x)$ independent of $t$, and shall show that under this assumption we can solve the time-dependent Schrödinger equation in exactly the same manner as the heat equation (Section 5.2).

We separate the variables in equation (5.6.1) in the usual way, writing $\Psi(x, t) = Y(x)Z(t)$. Substituting in (5.6.1), we get

$$-\frac{\hbar^2}{2m}\frac{d^2 Y}{dx^2}Z + VYZ = i\hbar Y\frac{\partial Z}{\partial t}$$

Now divide by $YZ$ to get

$$-\frac{\hbar^2}{2m}\frac{1}{Y}\frac{d^2 Y}{dx^2} + V = i\hbar\frac{1}{Z}\frac{\partial Z}{\partial t} \tag{5.6.2}$$

Under our assumption that $V$ depends on $x$ alone, the left side of (5.6.2) depends on $x$ alone, the right side only on $t$. Thus the left side is, in fact, independent of $x$ (because the right side is) and must be a constant, say $E$. So we get two equations:

$$-\frac{\hbar^2}{2m}\frac{d^2 Y}{dx^2} + VY = EY \tag{5.6.3}$$

$$i\hbar\frac{dZ}{dt} = EZ \tag{5.6.4}$$

Now equation (5.6.3) is just the time-independent Schrödinger eigenvalue equation $H(Y) = EY$, for which we expect nonzero solutions only for discrete values of $E$, say $E_n$, $n = 1, 2, \ldots$, with corresponding eigenfunctions $Y_n$, $n = 1, 2, \ldots$. The operator $H$ acts in some appropriate $L^2$ space $L^2(a, b)$, and we may normalize the eigenfunctions $Y_n$ so that

$$\int_a^b |Y_n(x)|^2\, dx = 1, \qquad n = 1, 2, \ldots$$

Equation (5.6.4) is easily integrated. We have $dZ/Z = -(iE_n/\hbar)\,dt$, which has the solution

$$Z_n = A_n \exp\left(-\frac{iE_n}{\hbar}t\right), \qquad n = 1, 2, \ldots$$

where $A_n$ is an arbitrary constant of integration. Thus we have a family of solutions of (5.6.1)

$$\Psi_n(x, t) = A_n Y_n(x) \exp\left(-\frac{iE_n}{\hbar}t\right)$$

and we sum these solutions, as we have before, to get the general solution

$$\Psi(x, t) = \sum_{n=1}^{\infty} A_n Y_n(x) \exp\left(-\frac{iE_n}{\hbar}t\right) \tag{5.6.5}$$

Just as with the heat equation, the constants $A_n$ are determined by the initial conditions. In the heat equation, they are determined by the initial temperature distribution $u(x, 0)$. In (5.6.5) the constants are determined by the wave function $\Psi(x, 0)$, which must be specified. Setting $t = 0$, taking the inner product of both sides of (5.6.5) with $Y_n$, and remembering $\langle Y_n, Y_n \rangle = \| Y_n \|^2 = 1$, we get

$$A_n = \langle \Psi(x, 0), Y_n \rangle = \int_a^b \Psi(x, 0)\overline{Y_n(x)}\,dx \tag{5.6.6}$$

With the constants $A_n$ determined, the solution (5.6.5) to the time-dependent equation is completely determined.

We need to check that the normalization is maintained. Since $|\Psi(x, t)|^2 = \Psi\overline{\Psi}$ is a probability density, we should have

$$\int_a^b |\Psi(x, t)|^2\,dx = 1$$

independent of the time $t$. Taking the inner product of each side of (5.6.5) with itself and remembering the orthonormality of the $Y_n$, we get

$$\langle \Psi(x, t), \Psi(x, t) \rangle = \sum_{n=0}^{\infty} |A_n|^2$$

As the right side is $\| \Psi(x, 0) \|^2$ (Parseval's equality), we see that if the initially given wave function is normalized (as it must be), then the wave function $\Psi(x, t)$ remains automatically normalized for all subsequent times.

### 5.6.2  Application to the Linear Oscillator

Let us apply this method of solution of (5.6.1) to derive the time-dependent behavior of the linear oscillator. There, our normalized wave functions are

$$Y_n(z) = \frac{1}{(2\pi)^{1/4}\sqrt{n!}} H_n(z)e^{-z^2/4}, \qquad \text{where } z = b^{1/4}x \qquad (5.6.7)$$

Then

$$\Psi(z,t) = \frac{e^{-z^2/4}}{(2\pi)^{1/4}} \sum_{n=0}^{\infty} \frac{A_n}{\sqrt{n!}} H_n(z)e^{-iE_n t/\hbar} = \sum_{n=0}^{\infty} A_n Y_n(z)e^{-iE_n t/\hbar} \qquad (5.6.8)$$

where $E_n = (n + \frac{1}{2})h\nu_0$, and $\Sigma_{n=0}^{\infty}|A_n|^2 = 1$. Equation (5.6.8) gives us the general time-dependent solution to the quantum oscillator. The $A_n$ are restricted by the condition $\Sigma_0^{\infty}|A_n|^2 = 1$. Specifying all the $A_n$, subject to this condition, is equivalent to giving the wave function $\Psi(z,0)$ at time $t = 0$

$$\Psi(z,0) = \frac{e^{-z^2/4}}{(2\pi)^{1/4}} \sum_{n=0}^{\infty} \frac{A_n}{\sqrt{n!}} H_n(z) = \sum_{n=0}^{\infty} A_n Y_n(z)$$

$$A_n = \langle \Psi(z,0), Y_n(z) \rangle = \int_{-\infty}^{+\infty} \Psi(z,0)\overline{Y_n(z)}\, dz \qquad (5.6.9)$$

$$= \frac{1}{(2\pi)^{1/4}\sqrt{n!}} \int_{-\infty}^{+\infty} \Psi(z,0)H_n(z)e^{-z^2/4}\, dz$$

Let us examine some aspects of the general solution (5.6.8). First, if we begin in an eigenstate $Y_n$, i.e., $\Psi(z,0) = Y_n(z)$, then by (5.6.9) $A_k = 0$ when $k \neq n$, and $A_n = 1$, so

$$\Psi(z,t) = \frac{1}{(2\pi)^{1/4}\sqrt{n!}} H_n(z)e^{-z^2/4}e^{-iE_n t/\hbar} = Y_n(z)e^{-iE_n t/\hbar} \qquad (5.6.10)$$

So, if we start at time $t = 0$ in the $n$th eigenstate, i.e., $\Psi_n(z,0) = Y_n(z)$, then the oscillator will remain in that eigenstate for all subsequent time. Exactly the same kind of behavior occurred with the vibrating wire (Section 3.5). In that case, if we put the wire initially in the form of one of its harmonics, $u(x,0) = \sin(n\pi x/L)$, then the subsequent motion of the wire would simply be that spatial configuration multiplied by the time variation: $u(x,t) = \sin(n\pi x/L)\cos(n\pi wt/L)$. So the eigenfunctions $Y_n(z)$ correspond to "standing waves." If the quantum oscillator is vibrating in one of these eigenstates, it will not make the transition to another state unless perturbed from outside the system.

One would hope that in the limit of large quantum numbers ($n \to \infty$) we could recapture from quantum mechanics the classical picture of a point mass, localized in time and in space, oscillating sinusoidally between $-x_{max}$ and $x_{max}$. If we are to recapture the classical description, we must do so from equation (5.6.8), which gives the time-dependent $\Psi$ function, $\Psi(z, t)$. If we are to have a correspondence with the classical linear oscillator, then the probability density $|\Psi(z, t)|^2$ should describe a particle following the law of motion $x = x_{max} \cos 2\pi\nu_0 t$.

Unfortunately, the simple expedient of using an eigenstate $\Psi(z, t) = Y_n(z)e^{-iE_nt/\hbar}$ for large $n$ does not work, because in this case $|\Psi(z, t)|^2 = |Y_n(z)|^2$ independent of time. That certainly does not describe the classical oscillating particle.

Schrödinger devised a very clever procedure for passing to the classical limit. He published his method in a paper entitled "The continuous transition from micro- to macro-mechanics." This paper is found in the same Chelsea volume that contains his other papers.

Schrödinger's trick is this: he substitutes into the series (5.6.8) for $\Psi(x, t)$ the coefficients

$$A_n = e^{-A^2/8} \left(\frac{A}{2}\right)^n \frac{1}{\sqrt{n!}} \qquad (5.6.11)$$

where $A$ is a fixed positive constant, much bigger than 1 ($A \gg 1$). The rationale behind the selection of these particular coefficients is as follows. Fix a positive real number $x$, and consider the sequence $c_n = x^n/n!$ Compute the ratio

$$\frac{c_{n+1}}{c_n} = \frac{x^{n+1}}{(n+1)!} \frac{n!}{x^n} = \frac{x}{n+1} .$$

When $c_{n+1} > c_n$, the sequence is increasing; when $c_{n+1} < c_n$, it is decreasing. The formula for the ratio $c_{n+1}/c_n$ shows that $c_n$ increases as long as $n + 1 < x$ and decreases when $n + 1 > x$. Thus a maximum occurs at approximately $x = n + 1 \cong n$, the latter approximation good when $n$ is large. Thus the sequence

$$\left(\frac{A}{2}\right)^n \frac{1}{\sqrt{n!}} = \left(\frac{(A^2/4)^n}{n!}\right)^{1/2}$$

will have a maximum at approximately $n \cong A^2/4$, or $A^2 \cong 4n$. Recall from the previous section that the classical limit for the variable $z$ is $z_{max}^2 = 4n + 2 \cong 4n \cong A^2$. When $A$ is large, the coefficients $A_n$ given by (5.6.11) have a maximum for large $n$, so the corresponding wave function $\Psi(x, t)$ would be a blend of eigenstates for large $n$. We might hope then that this wave function would correspond roughly to the classical motion $z = z_{max} \cos 2\pi\nu_0 t$. The weight factor $e^{-A^2/8}$ in (5.6.11) forces the $A_n$ to have the right normalization:

$$\sum_{n=0}^{\infty} |A_n|^2 = e^{-A^2/4} \sum_{n=0}^{\infty} \frac{(A^2/4)^n}{n!} = e^{-A^2/4} e^{A^2/4} = 1$$

Substituting (5.6.11) into (5.6.8), we get

$$\Psi(z,t) = \frac{e^{-z^2/4 - A^2/8}}{(2\pi)^{1/4}} \sum_{n=0}^{\infty} \frac{(A/2)^n e^{-iE_n t/\hbar}}{n!} H_n(z)$$

Now use $E_n = (n+\frac{1}{2})h\nu_0$ and $h/\hbar = 2\pi$:

$$\Psi(z,t) = \frac{e^{-z^2/4 - A^2/8 - i\pi t \nu_0}}{(2\pi)^{1/4}} \sum_{n=0}^{\infty} \frac{((A/2)e^{-2\pi i \nu_0 t})^n}{n!} H_n(z)$$

The series on the right corresponds to the generating function for the Hermite polynomials (Theorem 4.56 of Chapter 4):

$$e^{wx - w^2/2} = \sum_{n=0}^{\infty} \frac{w^n}{n!} H_n(x)$$

Put $w = (A/2)e^{-2\pi i \nu_0 t} = (A/2)(\cos 2\pi\nu_0 t - i \sin 2\pi\nu_0 t)$, and note that $w^2 = (A/2)^2 e^{-4\pi i \nu_0 t}$:

$$\Psi(z,t) = (2\pi)^{-1/4} \exp\left( -\frac{z^2}{4} - \frac{A^2}{8} - i\pi t\nu_0 + \frac{Az}{2}(\cos 2\pi\nu_0 t - i \sin 2\pi\nu_0 t) \right.$$
$$\left. -\frac{1}{2}\left(\frac{A}{2}\right)^2 (\cos 4\pi\nu_0 t - i \sin 4\pi\nu_0 t) \right)$$

Now compute $\Psi\bar{\Psi}$, remembering that if $x$ is real, then $(e^{ix})(\overline{e^{ix}}) = e^{ix}e^{-ix} = 1$:

$$|\Psi(z,t)|^2 = \Psi\bar{\Psi} = (2\pi)^{-1/2} \exp\left( -\frac{z^2}{2} - \frac{A^2}{4} + Az\cos 2\pi\nu_0 t - \left(\frac{A}{2}\right)^2 \cos 4\pi\nu_0 t \right)$$

Now $\cos 2\theta = 2\cos^2\theta - 1$, so $\cos 4\pi\nu_0 t = 2\cos^2 2\pi\nu_0 t - 1$. Thus:

$$|\Psi|^2 = (2\pi)^{-1/2} \exp\left( -\frac{z^2}{2} - \frac{A^2}{4} + Az\cos 2\pi\nu_0 t - \frac{A^2}{2}\cos^2 2\pi\nu_0 t + \frac{A^2}{4} \right)$$

$$|\Psi(z,t)|^2 = \frac{1}{\sqrt{2\pi}} \exp\left( -\frac{A^2}{2}\left(\frac{z}{A} - \cos 2\pi\nu_0 t\right)^2 \right) \qquad (5.6.12)$$

Since $A$ is very large, the exponential in equation (5.6.12) will be negligibly small except when

$$\frac{z}{A} - \cos 2\pi\nu_0 t \cong 0 \qquad \text{or} \qquad z \cong A\cos 2\pi\nu_0 t$$

Thus the probability density $|\Psi(z, t)|^2$ is virtually zero except when $z \cong A \cos 2\pi v_0 t$. Thus, with probability almost 1, we will find the particle oscillating in the classical mode $z = A \cos 2\pi v_0 t$.

This concordance with classical theory is most gratifying and provides added evidence for the correctness of Schrödinger's ideas. His theory of the linear oscillator predicts the discrete energy levels in the quantum region and agrees with the classical motion for high quantum numbers.

**Exercises**

1. On one piece of graph paper draw nine horizontal lines, evenly spaced from top to bottom. Mark each horizontal line as a $u$ axis, $-1.2 \leqslant u \leqslant 1.2$. Let $A = 20$, and express $|\Psi(z, t)|^2$ from equation (5.6.12) in terms of $u = z/A$. Using the topmost horizontal line, graph $|\Psi(z, t)|^2$ versus $u$ for $2\pi v_0 t = 0$. Using the second line from the top, repeat with $2\pi v_0 t = \pi/3$. Continue successively with $2\pi v_0 t = \pi/2$, $2\pi/3$, $\pi$, $4\pi/3$, $3\pi/2$, $5\pi/3$, and $2\pi$. Discuss, with reference to the classical case of an oscillating point mass. What effect would you expect to see if you used a larger value of $A$?

## 5.7 A BRIEF HISTORY OF MATRIX MECHANICS

Werner Heisenberg's paper "Quantum-theoretical re-interpretation of kinematic and mechanical relations," received by the journal *Zeitschrift für Physik* on 29 July 1925 (Schrödinger's first paper was received on 27 January 1926), has been bound in a volume titled *Sources of Quantum Mechanics*, edited by B. L. van der Waerden and published by Dover (1967). This book contains many other related papers, including those we refer to below, and has a detailed interesting historical introduction by van der Waerden. It is kind of a companion volume to the Chelsea book of the collected papers of Schrödinger (see Section 5.3 through 5.6).

Heisenberg, in the paper cited above, criticizes Bohr's theory of the atom (Section 5.3) on the grounds that it contains as basic elements quantities, such as the position and period of the electron, that are apparently unobservable. He sets out to try to establish a new theoretical quantum mechanics in which only observable quantities, such as energy, occur. (Incidentally, Max Born seems to have been the first to use the term "quantum mechanics" (in a paper published in 1924).) Heisenberg notes that in classical kinematics, which is the study of pure motion without reference to mass or force, calculations dealing with a classical quantity $x(t)$ often begin with a Fourier expansion in the time variable $t$. He cautions that in quantum mechanics one ought not to interpret $x(t)$ as position in space, because position is not observable. But frequency is observable. With that interpretation, he writes the classical expansion as

$$x(n, t) = \sum_{\alpha = -\infty}^{\alpha = \infty} A_\alpha(n) e^{i\omega(n)\alpha t} \qquad \text{(classical)}$$

where $\omega(n)$ is the angular frequency and the fixed integer $n$ presumably refers to some atomic level. Heisenberg then argues that the quantum mechanical version should have two integer arguments, presumably to account for transitions between levels, and the corresponding expansion in the quantum case would take the form

$$\sum_{\alpha = -\infty}^{\alpha = \infty} A(n, n - \alpha) e^{i\omega(n, n-\alpha)t} \qquad \text{(quantum)}$$

At that point he raises the question: What would the coefficients of $x^2$, or more generally $xy$, be in the quantum case? If $y$ has coefficients $B(n, n-\alpha)$, he puts down the coefficients $C(n, n-\beta)$ of the product $xy$ as

$$C(n, n - \beta) = \sum_{\alpha = -\infty}^{\alpha = \infty} A(n, n - \alpha) B(n - \alpha, n - \beta) \qquad (5.7.1)$$

because "this type of combination is an almost necessary consequence of the frequency combination rules." He points out that with this rule of multiplication $xy$ need not equal $yx$. So Heisenberg associates observable quantities with peculiar double-index symbols with multiplication prescribed by (5.7.1). In the final section of his paper, he turns to the dynamical problem, where forces are taken into account, and uses his calculus to analyze the anharmonic oscillator and the rotator.

Born and Jordan, in a paper received on 27 September 1925, two months after Heisenberg's paper, note that Heisenberg's symbolic multiplication (5.7.1) is matrix multiplication. They review the laws of matrix calculation and then describe the dynamical system by representing the spatial coordinate $q$ and momentum $p$ by infinite Hermitian matrices (thereby ignoring Heisenberg's admonition to avoid use of the position coordinate of the electron because it is unobservable). They formulate equations of motion, assert the "stronger quantum condition"

$$qp - pq = \hbar i I \qquad (5.7.2)$$

where $I$ is the identity matrix, and use it to demonstrate the energy conservation law. The paper closes with an investigation of the harmonic and anharmonic oscillators.

Born, Heisenberg, and Jordan, in a paper received on 16 November 1925, about two months later, give a systematic treatment of matrix mechanics and note that "the transformation of matrices can most easily be grasped if one regards them as a system of coefficients for linear transformations or bilinear forms." The term "linear transformation" means the same as linear operator. They make no further reference to linear transformations, but they describe in detail the connection with Hermitian forms and do mention eigenvalues in connection with the diagonalization of Hermitian forms. They close with some physical applications including the quantization of angular momentum.

Dirac (paper received 22 January 1926) and Pauli (paper received 17 January 1926) independently applied the new quantum mechanics to the hydrogen atom. The hydrogen atom was also analyzed by Schrödinger in his first paper (received 27 January 1926) using his method based on the eigenvalues of his wave equation (see Sections 5.3 through 5.6), a method that seemed to have nothing in common with Heisenberg's matrix mechanics.

During 1925–1926, Heisenberg was Born's assistant at Göttingen, Germany. Jordan was Born's pupil. Pauli, at Hamburg, Germany, was in close contact with the Göttingen group. Heisenberg gave advance copies of his paper (referred to above) to Pauli and Born, and it was Born who forwarded Heisenberg's paper to the *Zeitschrift für Physik* for publication. Meanwhile, Dirac, in Cambridge, England, was given proof sheets of Heisenberg's paper by R. H. Fowler (in September 1925); then began his otherwise largely independent work on quantum mechanics. (Heisenberg and Dirac were about 24–25 years old at this time.) Schrödinger, an Austrian, was at Zurich, Switzerland. He knew of Heisenberg's work but followed his own star, formulating and applying *his* equation to the problems of atomic physics. (Our Sections 5.3 through 5.6 are based on this work of Schrödinger.) Thus there were three separate groups involved in this intense and heroic period of physics: Heisenberg, Born, and others in Göttingen, Germany; Dirac in Cambridge, England; and Schrödinger in Zurich, Switzerland. Heisenberg and Born were vigorously applying Heisenberg's noncommutative matrix calculus to the problems of atomic physics; Dirac was doing the same, but pretty much independently of the Göttingen group; while Schrödinger, working alone, was getting all the same answers by his eigenvalue method.

Then, in a paper received on 18 March 1926, Schrödinger showed that matrix mechanics and his theory were equivalent. This paper is included in the Chelsea volume of the collected papers of Schrödinger that I have quoted from in Sections 5.3 through 5.6. (I shall give the idea of this proof of equivalence later on.) Thus it turned out that there was *one* theory of quantum mechanics. Heisenberg and Schrödinger had started at opposite ends and had met in the middle.

Dirac's book, *The Principles of Quantum Mechanics*, first published in 1930, presented a unified, amalgamated theory. Dirac's enormously influential book has gone through four editions, is now available in paperback, and has been reprinted six times. Most modern texts on quantum mechanics still follow the general formulation of quantum mechanics laid down by Dirac. We shall follow it too in Sections 5.8 and 5.9 but shall use our up-to-date mathematics developed in Chapters 3 and 4.

Dirac and Schrödinger received the Nobel Prize jointly in 1933; Heisenberg had been awarded the Nobel Prize one year earlier.

I shall close this historical essay with an anecdote taken from the book *Erwin Schrödinger, an Introduction to His Writings* by William T. Scott (University of Massachusetts Press, 1967, p. 8). Hans Thirring, who was to become a lifelong friend of Schrödinger, recalls the first time their paths crossed. They were both students at the University of Vienna in Austria. Thirring was sitting in the library, when a blond young man came into the room. Thirring's neighbor poked him, saying, "Das ist der *Schrödinger.*"

## 5.8   A GENERAL FORMULATION OF QUANTUM MECHANICS: STATES

The historical material in the preceding section is intended to provide some background to afford a better appreciation of the modern-day theory of quantum mechanics. In this section, and the next, I shall describe the modern theory, following in the main Dirac's exposition translated into our current mathematical idiom. I shall describe the theory as it is often applied—in a series of steps. Here are the steps:

1. Define the physical system.
2. Form the configuration space.
3. Identify the states and the state space.
4. Match observables to Hermitian operators.

Steps 1, 2, and 3 are explained in this section; step 4 in the next.

*Step 1. Define the Physical System*

The system might be the linear oscillator, for example. We defined that physical system in Section 5.5: "a particle of mass $m$ which, moving in a straight line, is attracted toward a fixed point in it, with a force proportional to its displacement from this point." The physical system might be a particle restricted to an interval—the idealized case of a zero force in an interval $a \leqslant x \leqslant b$ with an infinite restraining force at the end points (see Exercises 1 and 2 of Section 5.4). Those systems are one-dimensional, in keeping with our particular focus of attention in this chapter. We shall deal with problems in higher dimensions in the following chapters, treating such physical systems as the hydrogen atom, idealized as a particle moving in three-dimensional space under the influence of a force directed toward the origin, of magnitude inversely proportional to the square of the particle's distance from the origin. The physical system may consist of many particles. To define such a system, one defines the potential $V$ as a function of the positions of the particles, and one specifies the constraints, if any.

Quantum mechanical systems are generally defined in classical terms. Especially common—almost universal—is the use of the term "particle." While all books on quantum mechanics repeat and emphasize the wave-particle duality (which motivated Schrödinger's original discovery of wave mechanics), and all of them discuss the Heisenberg uncertainty principle (forbidding precise simultaneous values of certain pairs of observables like position and momentum), nonetheless they all use the term "particle" and, when defining their physical systems, stay true to the mental image the word "particle" conveys—that of a classical mass-point localized in space and time. This mental picture, which appears to be a tacit part of the definition of almost all physical systems, has the following justification: whenever actually observed in an experiment, whether the scintillation of a phosphor, a track in a photographic emulsion, a track in a Wilson cloud chamber, or the click of a Geiger-Müller counter, the fundamental entities of microphysics, like the photon, electron, positron, proton, and neutron, leave a record that is precisely located at a point in space, at an instant of time.

Nonetheless, even though defined classically, the *behavior* of a physical system, either as predicted by quantum mechanics or as observed by experiment, is completely at odds with the implications that the classical mental picture provides. We have seen this already in our study of the quantum linear oscillator (Section 5.5).

The term "particle" is a basic undefined term in quantum theory. While quantum mechanics does take due account of some individual properties of particles, such as the spin of the electron, it provides no explanation for the existence of particles such as electrons or for the nature of, or the quantization of, electric charge.

*Step 2. Form the Configuration Space*

By the *configuration space* of a physical system we mean the Cartesian space $\mathbf{R}^n$, of $n$-tuples of real numbers, of a dimension $n$ sufficient to specify the positions of the particles of the physical system under study, taking into account their degrees of freedom. Time does not enter, as we are dealing with the stationary case. The configuration space for the quantum linear oscillator, for example, is $\mathbf{R}^1 = \{x: -\infty < x < \infty\}$, as the oscillator consists of a single particle constrained to move in a straight line. The configuration space of the idealized hydrogen atom (immobile nucleus) is $\mathbf{R}^3 = \{(x, y, z): -\infty < x, y, z$

$< \infty$}. The configuration space of a system of $k$ unconstrained particles is $\mathbf{R}^{3k}$. Thus, for a complicated physical system, its configuration space can have a very high dimension. Coordinate systems other than Cartesian are often convenient, but the change of coordinates is usually made at a later stage.

*Step 3. Identify the States and the State Space*

The *state space* of a physical system whose configuration space is $\mathbf{R}^n$ consists of all (real- or) complex-valued functions $\Psi(x_1, x_2, \ldots, x_n)$ defined on the configuration space $\mathbf{R}^n$ and satisfying

$$\int_{\mathbf{R}^n} |\Psi(x_1, x_2, \ldots, x_n)|^2 \, dx_1 \, dx_2 \cdots dx_n < \infty \tag{5.8.1}$$

The functions that satisfy (5.8.1) form a vector space, which is to say: if $\Psi_1$ and $\Psi_2$ satisfy (5.8.1), then so does $\Psi_1 + \Psi_2$. The proof of this is the same as that given for the one-variable case in Section 3.9, Theorem 3.16.

Using exactly the same arguments that established Theorem 3.16, we can prove that the formula

$$\langle \Phi, \Psi \rangle = \int_{\mathbf{R}^n} \Phi(x_1, x_2, \ldots, x_n) \overline{\Psi(x_1, x_2, \ldots, x_n)} \, dx_1 \, dx_2 \cdots dx_n \tag{5.8.2}$$

defines an inner product on the state space. That is, the expression defined by equation (5.8.2) is conjugate bilinear, Hermitian symmetric, and positive definite (see Section 3.6). Thus the state space is a complex inner product space. In fact, it is a Hilbert space, as it is complete in the metric derived from its inner product. Although we won't make any use of the completeness, we shall nonetheless refer to the state space as a Hilbert space and shall use the notation $L^2(\mathbf{R}^n)$ for it. This is an obvious extension of the notation introduced in Section 3.9.

The only new feature of the Hilbert space $L^2(\mathbf{R}^n)$ as compared to the space $L^2(a, b)$ that we studied in Section 3.9 is the technical difficulty of integrating over several variables rather than one. We shall look into these technical matters in some detail in the next chapter, but for now we shall work in a general way with the space $L^2(\mathbf{R}^n)$. Note that, since the configuration space for a physical system may have a very high dimension, we must be prepared to deal with Hilbert spaces involving integrals of many variables.

For a function $\Psi$ in the state space, we define its norm $\|\Psi\|$ in terms of the inner product (5.8.2) in the usual way: $\|\Psi\| = \langle \Psi, \Psi \rangle^{1/2}$ (Definition 3.3 of Section 3.7).

So that is the definition of the state *space*—it is a Hilbert space. What is a state?

A *state* of a physical system, though somewhat elusive to define, is the fundamental concept of quantum mechanics. The dictionary definition of the word:

*state*: any of the various forms in which a thing is found to exist

seems to convey the essential meaning. Dirac puts it this way (*Principles*, p. 11):

A state of a system may be defined as an undisturbed motion that is restricted by as many conditions or data as are theoretically possible without mutual interference or contradiction.

The connection between this physical concept of state and the mathematical construct, the state space, is provided by a basic postulate of quantum mechanics: *to each state of the physical system corresponds a function* $\Psi$ *in its state space (excluding the essentially zero function,* $\|\Psi\| = 0$*) which completely characterizes the physical state.*

As time is not involved here, we are dealing with what are called stationary states. Equivalently, we could just work at a fixed instant of time. We shall clarify this aspect in a moment.

According to Born's interpretation, if $\Psi$ is a normalized state function ($\|\Psi\| = 1$), then $|\Psi|^2$ is the *probability density* of the system being in a given configuration (this idea was introduced in Section 5.4). For example, for a two-particle system, each particle having three degrees of freedom, the quantity

$$|\Psi(x_1, y_1, z_1, x_2, y_2, z_2)|^2 \, dx_1 \, dy_1 \, dz_1 \, dx_2 \, dy_2 \, dz_2$$

is the joint probability that particle 1 is in the volume element $dx_1 \, dy_1 \, dz_1$ and particle 2 in the element $dx_2 \, dy_2 \, dz_2$.

If $\Psi$ is any state function, then $\Psi/\|\Psi\|$ is normalized, so Born's probability interpretation applies to it. This fact leads to the surmise that if $\Psi_1$ and $\Psi_2$ are two state functions such that $\Psi_1/\|\Psi_1\| = \Psi_2/\|\Psi_2\|$, then $\Psi_1$ and $\Psi_2$ correspond to the same state of the physical system. It is customary to go further than this and assume that if $\Psi_1 = a\Psi_2$ where $a$ is any nonzero complex number, then $\Psi_1$ and $\Psi_2$ correspond to the same physical state. For this assumption to hold up, it must be the case that if $\Psi_1 = a\Psi_2$, then $\Psi_1$ and $\Psi_2$ provide exactly the same data about the physical system. Under the latter assumption, then, the actual correspondence sends each state of the physical system to a one-dimensional subspace $\{a\Psi: a \in \mathbf{C}\}$ (excluding zero) of the state space.

The concept of quantum state has some problematic features. To better explain this concept of state, to clarify its meaning, and to illustrate how one computes the time evolution of a system, I shall first discuss the traditional concept of state for the classical vibrating wire (Sections 3.1, 3.5, and 5.1) and then, by comparison, highlight the unique features of quantum states and explain their time evolution.

The vibrating wire is a physical system in the classical sense. We defined this classical physical system in Section 3.1: a tightly stretched wire of constant linear density $\rho$, under tension $T$, and clamped at both ends. Assuming the ends of the wire have coordinates $x=0$ and $x=L$, the configuration space of the wire is the interval $\{x: 0 \leqslant x \leqslant L\}$. Its state space is the Hilbert space $L^2(0, L)$. The state of the wire is completely specified by giving the vertical displacement $u(x, t)$ as a function of position $x$ and time $t$; for each fixed $t$, the function $u(x, t)$, considered as a function of $x$ alone, belongs to the state space $L^2(0, L)$. The function $u(x, t)$ gives complete information about the vibrating wire.

We showed in Section 5.1 that the initial deflection and initial velocity, namely $u(x, 0)$ and $(\partial u/\partial t)(x, 0)$, uniquely determine the state $u(x, t)$ for all $t > 0$. The wave equation, $\Delta u + (1/w^2)\, \partial^2 u/\partial t^2 = 0$, where $\Delta$ is the one-dimensional Laplacian, effects this determination. Our proof of that fact, given in Section 5.1, ran, in outline, as follows: the eigenfunctions of the one-dimensional Laplacian $\Delta$, say $\phi_n(x)$, $n = 1, 2, \ldots$, form an orthogonal basis of the state space $L^2(0, L)$. Therefore the time-dependent state function $u(x, t)$, for fixed $t$, has an expansion

$$u(x, t) = \sum_{n=1}^{\infty} c_n(t)\phi_n(x) \qquad (5.8.3)$$

convergent in the mean-square metric.

Equation (5.8.3) holds for any function $u(x, t)$ which, for each fixed $t$, satisfies $\int_0^L |u(x, t)|^2\, dx < \infty$. This is what we mean when we say that the $\phi_n(x)$ are an orthogonal basis of $L^2(0, L)$—*every* function in this Hilbert space has a convergent general Fourier expansion in the $\phi_n(x)$, where convergence is taken in the sense of the intrinsic mean-square metric. To utilize the wave equation, we must now apply the unbounded Hilbert space operator $\Delta$ to both sides of (5.8.3). This requires that, for each fixed $t$, $u(x, t)$ belongs to the domain of the Laplacian $\Delta$ (Theorem 4.11 of Section 4.4). Thus, at this point, the mathematics imposes a restriction on the physics: the state function $u(x, t)$ of the vibrating wire must belong to the domain of the Laplacian $\Delta$ in order to be "physically admissible." So not all functions in $L^2(0, L)$ describe states of the vibrating wire; only those in the domain of $\Delta$ do. In order that $u(x, t)$ belong to the domain of $\Delta$ it must be compatible with the differential expression $\partial^2/\partial x^2$, which requires at least that both $u$ and $\partial u/\partial x$ be continuous functions of $x$ ($t$ fixed) (Theorem 4.2, Section 4.1). As we are dealing with a wire, these restrictions are certainly reasonable. But note that they are dictated by the mathematics, namely the requirement that our operators act in the Hilbert space $L^2(0, L)$. Once we have postulated the Hilbert space setting for the description of our problem, these requirements fall naturally into place. They don't have to be added as independent physical assumptions. We shall see in a minute how the same thing happens in quantum theory.

Restricting our attention in equation (5.8.3) to "physically admissible" state functions $u(x, t)$, namely those which, for each fixed $t$, belong to the domain of the unbounded Hilbert space operator $\Delta$, we may apply $\Delta$ term by term to (5.8.3). Remembering that $\Delta$ acts only on the variable $x$, so that $\Delta(c_n(t)\phi_n(x)) = c_n(t)\Delta\phi_n(x)$, and that $\Delta\phi_n = \varepsilon_n\phi_n$, where $0 < \varepsilon_1 < \varepsilon_2 < \cdots$ are the eigenvalues of $\Delta$ (Theorem 4.14, Section 4.5), we get

$$\Delta u = \sum_{n=1}^{\infty} \varepsilon_n c_n(t)\phi_n(x) \tag{5.8.4}$$

Combining (5.8.4) with wave equation $\Delta u = -(1/w^2)\partial^2 u/\partial t^2$ and using the uniqueness of the general Fourier series, we find that the $c_n(t)$ must satisfy the constant-coefficient second order ordinary differential equation $c_n'' + w^2\varepsilon_n c_n = 0$ which has the general solution

$$c_n(t) = A_n \cos\sqrt{\varepsilon_n}\,wt + B_n \sin\sqrt{\varepsilon_n}\,wt, \qquad n = 1, 2, \ldots \tag{5.8.5}$$

where $A_n = c_n(0)$ and $B_n = (1/\sqrt{\varepsilon_n}\,w)c_n'(0)$. According to (5.8.3), $c_n(0) = \langle u(x, 0), \phi_n(x)\rangle/\langle\phi_n, \phi_n\rangle$ and $c_n'(0) = \langle(\partial u/\partial t)(x, 0), \phi_n(x)\rangle/\langle\phi_n, \phi_n\rangle$. Thus the initial deflection $u(x, 0)$ and initial velocity $\partial u(x, 0)/\partial t$ determine the $c_n(0)$ and $c_n'(0)$, which in turn determine $c_n(t)$ by (5.8.5), and thus yield the time-dependent state function $u(x, t)$ via (5.8.3).

That describes the situation for the classical vibrating wire. Quantum theory has a parallel setup, and comparing and contrasting the two is very instructive.

Suppose that we have defined a quantum mechanical physical system (Step 1), formed its configuration space (Step 2), and constructed its state space (Step 3). Then, as we have already said, a major tenet of quantum mechanics asserts that each time-independent state of the system is completely characterized by a function $\Psi$ in the state space (excluding the essentially zero function, $\|\Psi\| = 0$). Note the contrast with the classical case. The initial (time-fixed) state of the vibrating wire requires two real-valued functions, $u(x, 0)$ and $(\partial u/\partial t)(x, 0)$, to characterize it. In quantum mechanics, only one function, $\Psi$, is required (which cannot be the essentially zero function), but it is complex valued.

Another contrast between the classical and quantum case. for the classical vibrating wire, the functions $u(x, 0)$ and $(\partial u/\partial t)(x, 0)$ have a concrete physical meaning; $u(x, 0)$ is the initial deflection and $(\partial u/\partial t)(x, 0)$ the initial velocity. In quantum theory, $|\Psi|^2$ is a *probability density*, not a physical quantity.

Despite these contrasts between the classical and quantum theoretical constructs, the computation of the time evolution of the quantum system is closely parallel to that of the classical vibrating wire. For clarity, I shall assume that our quantum physical system consists of a single particle moving in one dimension.

The stationary state $\Psi(x)$, taken as the state of the system at time $t = 0$, uniquely determines its state $\Psi(x, t)$ for all $t > 0$. Schrödinger's time-dependent equation, $H(\Psi) = i\hbar(\partial\Psi/\partial t)$, where $H$ is Schrödinger's energy operator, effects this determination. Our proof of this fact, given in Section 5.6, ran, in outline, as follows: assume that the eigenfunctions of the Hilbert space operator $H$, say $Y_n(x)$, $n = 1, 2, \ldots$, form an orthogonal basis of the state space, $L^2(\mathbf{R})$. Therefore the time-dependent state function $\Psi(x, t)$, for fixed $t$, has an expansion

$$\Psi(x, t) = \sum_{n=1}^{\infty} C_n(t) Y_n(x) \tag{5.8.6}$$

convergent in the mean-square metric.

Equation (5.8.6) holds for any function $\Psi(x, t)$ which, for each fixed $t$, satisfies $\int_{-\infty}^{\infty} |\Psi(x, t)|^2 \, dx < \infty$. This is what we mean when we say that the $Y_n(x)$ are an orthogonal basis of $L^2(\mathbf{R})$—*every* function in this Hilbert space has a convergent general Fourier expansion with $Y_n(x)$, where convergence is taken in the sense of the intrinsic mean-square metric. To utilize Schrödinger's equation, we must now apply the unbounded Hilbert space operator $H$ to both sides of (5.8.6). This requires that, for each fixed $t$, $\Psi(x, t)$ belongs to the domain of the energy operator $H$ (Theorem 4.11, Section 4.4). Thus, at this point, the mathematics imposes a restriction on the physics: the state function $\Psi(x, t)$ of the quantum system must belong to the domain of the operator $H$ in order to be "physically admissible." So not all functions in $L^2(\mathbf{R})$ describe states of the quantum system; only those in the domain of $H$ do. In order that $\Psi(x, t)$ belong to the domain of $H$ it must be compatible with the differential expression $-(\hbar^2/2m)(\partial^2/\partial x^2)$, which requires at least that both $\Psi$ and $\partial\Psi/\partial x$ be continuous functions of $x$ ($t$ fixed) (Theorem 4.2, Section 4.1). While these restrictions may be physically reasonable, note that they are dictated by the mathematics, namely the requirement that our operators act in the Hilbert space $L^2(\mathbf{R}^n)$. Once we have postulated the Hilbert space setting for the description of our problem, these requirements fall naturally into place. They don't have to be added as independent physical assumptions.

Restricting our attention in equation (5.8.6) to "physically admissible" state functions $\Psi(x, t)$, namely those which, for each fixed $t$, belong to the domain of the unbounded Hilbert space operator $H$, we may apply $H$ term by term to (5.8.6). Remembering that $H$ acts only on the variable $x$, so that $H(C_n(t)Y_n(x)) = C_n(t)H(Y_n(x))$, and that $H(Y_n) = E_n Y_n$, where $0 < E_1 < E_2 < \cdots$

are the energy eigenvalues of $H$, we get

$$H(\Psi) = \sum_{n=1}^{\infty} E_n C_n(t) Y_n(x) \qquad (5.8.7)$$

Combining (5.8.7) with Schrödinger's equation $H(\Psi) = i\hbar(\partial\Psi/\partial t)$ and using the uniqueness of the general Fourier series, we find that the $C_n(t)$ must satisfy the constant-coefficient first order ordinary differential equation $i\hbar C'_n = E_n C_n$, which has the general solution

$$C_n(t) = A_n e^{-i(E_n/\hbar)t} \qquad (5.8.8)$$

where $A_n = C_n(0)$. According to (5.8.6), $C_n(0) = \langle \Psi(x,0), Y_n(x) \rangle / \langle Y_n, Y_n \rangle$. Thus the stationary state $\Psi(x,0) = \Psi(x)$ determines the $C_n(0)$, which in turn determine the $C_n(t)$ by (5.8.8), thus yielding the time-dependent state function via (5.8.6).

That describes the time evolution of the quantum system.

The paradoxical aspects of the concept of quantum state show up in the *principle of superposition of states*, which is this: *The functions that correspond to physical states of a quantum system form a subspace of the state space* (excluding the essentially zero function). This means, in particular, that if the functions $\Psi_1$ and $\Psi_2$ both correspond to physical states of the system, then so does $\Psi = c_1\Psi_1 + c_2\Psi_2$ for any (real or) complex constants $c_1$ and $c_2$ (excluding the case $\|\Psi\| = 0$).

We have already postulated that a function in the state space that is to correspond to an actual physical state of the quantum system must be in the domain of Schrödinger's energy operator $H$. While the domain of $H$ is a subspace of the state space (which is a Hilbert space), it cannot be the entire space, as we have repeatedly stressed in Chapter 4. (For a function $\Psi$ to be in the domain of $H$, $\Psi$ must be compatible with the differential expression defining $H$ and must satisfy the associated boundary conditions, if any. In order to satisfy these requirements, it is necessary that $\Psi$ be continuous and have a continuous first derivative. As the *only* requirement to belong to the state space is the square-integrability condition $\int |\Psi|^2 \, dv < \infty$, there will certainly be functions in the state space not in the domain of $H$. See Sections 4.1 and 4.4.) Accordingly, it seems a reasonable extension of the principle of superposition to postulate that *the subspace of functions that correspond to physical states coincides with the domain of Schrödinger's energy operator $H$* (excluding the zero function). This postulate allows for the formation of infinite linear combinations of state functions as long as the sum function belongs to the domain of Schrödinger's energy operator $H$. Of course, in testing for convergence of infinite series of state functions, we always use the intrinsic mean-square metric of our Hilbert space. With this identification of the physical states with the domain of $H$ comes the corollary that there will generally be nonzero functions in the state space that correspond to no physical state of the system.

To bring out the paradoxical consequences of the principle of super-position of states, we don't need infinite linear combinations of states; two is enough. Consider the quantum linear oscillator that we analyzed in Section 5.5. Its state space is the complex Hilbert space $L^2(\mathbf{R})$. Our analysis in Section 5.5 showed that the total energy $E$ of the oscillator is quantized. The total energy of the quantum oscillator in a stationary time-independent state is an odd integer multiple of the zero-point energy $E_0 = \frac{1}{2}h\nu_0$; $E_n = (2n+1)E_0$, $n = 0, 1, 2, \ldots$. The functions

$$\Psi_n(z) = \frac{1}{(2\pi)^{1/4}\sqrt{n!}} H_n(z)e^{-z^2/4} \qquad n = 0, 1, 2, \ldots$$

are normalized representatives of the corresponding eigenstates, where $H_n(z)$ are the Hermite polynomials. Thus, if an oscillator is in the eigenstate $\Psi_0$ and you measure its energy, you will get $E_0$ exactly, every time. Likewise, if an oscillator is in the eigenstate $\Psi_1$ and you measure its energy, you will get $3E_0$ exactly, every time.

Now consider the function $\Psi = \frac{3}{5}\Psi_0 + \frac{4}{5}\Psi_1$. Remembering that $\Psi_0$ and $\Psi_1$ are orthogonal, we see easily that $\|\Psi\| = 1$. According to the principle of superposition, this normalized function $\Psi$ corresponds to a physical state of the (single) quantum linear oscillator. What is the energy of the oscillator in this state? The answer provided by quantum mechanics is this: if you measure the energy of the oscillator in the state $\Psi = \frac{3}{5}\Psi_0 + \frac{4}{5}\Psi_1$ you will get *either* $E_0$ or $3E_0$, *never* anything else, *never* any in-between value.

You might counter: if I found the energy value $E_0$ then the oscillator must have been originally in the state $\Psi_0$, and if I found the energy value $E_1$ then the oscillator must have been originally in the state $\Psi_1$. That's logic! That's common sense!

Well, logic has a lot of trouble with quantum mechanics. And so does common sense. The quantum mechanical explanation is this. Prior to your measurement, the oscillator was in the state $\Psi = \frac{3}{5}\Psi_0 + \frac{4}{5}\Psi_1$. If your measurement yielded the energy value $E_0$, then after that measurement, the oscillator would be in the state $\Psi_0$. Your measurement caused the "collapse of the wave packet," so called. A subsequent measurement would again yield $E_0$. Similar considerations if you got $E_1$.

In reply to the question, what was the energy before measurement?, quantum mechanics can only say this: it was $E_0$ with probability $(3/5)^2$ and $E_1$ with probability $(4/5)^2$. Thus, even though the total energy of the oscillator is quantized, quantum theory can make no precise prediction, in general, of its energy, but can only give the probability that the energy is this or that allowed value. Quantum theory does yield an *average* value, as we shall see in the next section.

One final postscript to this (long) section on states. For us, a state function must be square integrable (at least). But there are occasions where it is convenient to step outside this boundary. For example, the physical system consisting of a free particle $(V = 0)$ moving without restriction in one dimension has the configuration space $\mathbf{R} = \{x: -\infty < x < \infty\}$, thus the state space $L^2(\mathbf{R}) = L^2(-\infty, \infty)$. Schrödinger's equation for this case is $-(\hbar^2/2m)\Psi'' = E\Psi$. Disregarding for the moment operator-theoretic considerations, the two linearly independent solutions to this equation are

$$\Psi = Ae^{\pm ipx/\hbar} \qquad (5.8.9)$$

where $p = \sqrt{2mE}$ is the momentum (see Exercise 3a of Section 5.3) and $A$ is an arbitrary complex constant. It is customary in quantum mechanics to refer to the function (5.8.9) as representing the *translational state* of the free particle ($+$ for motion in the positive $x$ direction, $-$ for motion in the negative $x$ direction). Although it is not strictly correct to speak of (5.8.9) as representing a state, as the function (5.8.9) is not square integrable and thus does not belong to the state space (unless $A = 0$, which corresponds to no state), it is a convenient and frequently used fiction. The function (5.8.9) is often called the "plane wave"; it and Dirac's delta function are essentially Fourier transforms of each other (see Chapter 8).

The strict Hilbert space analysis of the free particle goes as follows. The differential expression $-d^2/dx^2$, restricted to the domain of functions in $L^2(-\infty, \infty)$ that are compatible with it, defines a Hilbert space operator in $L^2(-\infty, \infty)$. It is natural to call it also the one-dimensional Laplacian $\Delta$, thus extending Definition 4.13. This Laplacian shares the same differential expression with the Laplacians of Definition 4.13 but operates in the Hilbert space $L^2(-\infty, \infty)$ (rather than $L^2(a, b)$ for a finite interval $[a, b]$ or $L^2(C)$ for a circle $C$). It has in common with the Laplacian defined in $L^2(C)$ that no boundary values enter into its definition. We shall study this Laplacian in Section 8.5 and shall prove there (Theorem 8.7) that it is Hermitian positive definite but has *no* eigenvalues. It has what is called a "continuous spectrum," but we won't get into that in this book.

Schrödinger's equation for the free particle, $-(\hbar^2/2m)\Psi'' = E\Psi$, properly set as an eigenvalue problem in Hilbert space, becomes

$$H(\Psi) = E\Psi \qquad (5.8.10)$$

where $H = (\hbar^2/2m)\Delta$ is Schrödinger's energy operator, $\Delta$ being the one-dimensional Laplacian just discussed. Equation (5.8.10) asks for the eigenvalues of $H$. But, as we shall prove in Theorem 8.7, $H$ has *no* eigenvalues. The conclusion to be drawn from this is that *the total energy $E$ of the quantum free* *particle is not quantized.* That is to say, the total energy $E$ of the free particle does not come in lumps, in contrast to the total energy of the quantum linear oscillator (Section 5.5). The free particle does, of course, *have* a total energy, and quantum mechanics can make probabilistic assertions about that energy, these assertions making use of the *spectral resolution* of the operator $H$. The spectral resolution, along with the continuous spectrum, is another topic we are not going to touch here. But we can, on the basis of the theory developed in this book, compute the *average* value of that energy, and we will show how in the next section.

That is how one analyzes the free particle if one sticks strictly to the Hilbert space method. The plane wave offers another perspective. If we let the differential expression $(-\hbar^2/2m)(d^2/dx^2)$ act on the plane wave (5.8.9), $\Psi = Ae^{\pm ipx/\hbar}$, we find $-(\hbar^2/2m)(d^2/dx^2)\Psi = E\Psi$, where $E = p^2/2m$ is the total energy of the particle (which equals its kinetic energy because $V = 0$). So, by stepping outside the Hilbert space, we get a sort of ersatz solution to (5.8.10). According to this point of view, (5.8.10) has the "eigenvalues" $E = p^2/2m$ with corresponding "eigenstates" $\Psi = Ae^{\pm ipx/\hbar}$. That perspective can be convenient in some circumstances. When it is convenient, use it. But keep in mind that the eigenvalue problem (5.8.10) has in fact no solution and that the energy of the free particle is in fact not quantized.

## 5.9  A GENERAL FORMULATION OF QUANTUM MECHANICS: OBSERVABLES

The *observables* of a physical system are its measurable physical quantities (or dynamical variables) such as energy, momentum, position, velocity, and angular momentum. A basic assumption of quantum mechanics is this: *to every observable of a dynamical system corresponds a Hermitian operator on its state space (which is a Hilbert space).*

Right off I have to say this: where I say *Hermitian* operator I should really say *selfadjoint* operator. But we are blurring that distinction in this book, because "selfadjoint" gets to be such a fussy concept for unbounded operators (see Section 4.4). So we'll go with Hermitian.

"Which operator goes with which observable?" is the natural question that follows on the above assumption. To provide an answer to that question, and to provide as well some insight into the reasoning underlying the assumption itself, we'll return to Schrödinger's time-independent equation.

Write Schrödinger's equation as $H(\Psi) = E\Psi$, where the Hilbert space operator $H$ has the differential expression (for a one-particle system whose motion is restricted to the $x$ axis)

$$l \equiv -\frac{\hbar^2}{2m}\frac{d^2}{dx^2} + V(x) \tag{5.9.1}$$

where $V(x)$ stands for the potential energy and, in equation (5.9.1), the term $V(x)$ means the operation of multiplication by the function $V(x)$. The domain of $H$ consists of all functions in the state space that are compatible with the differential expression (5.9.1) and satisfy the appropriate boundary conditions, if any. In Schrödinger's equation, $H(\Psi) = E\Psi$, the eigenvalue parameter $E$ stands for the total energy of the system. It is therefore natural to associate to the total energy of the system (which is an observable) the Hilbert space operator $H$. Thus we have our first connection between observables and Hermitian operators.

Look again at the differential expression (5.9.1), which we have associated to the total energy of the system. As $V(x)$ is potential energy, we would look on the term $-(\hbar^2/2m)\,d^2/dx^2$ in (5.9.1) as somehow corresponding to kinetic energy, also an observable. The classical formula for the kinetic energy $T$ is $T = \frac{1}{2}mv^2$, where $m$ is the mass of our single particle and $v$ its velocity. Momentum is $p = mv$, so $T = p^2/2m$. If we associate to momentum $p$ the differential expression $i\hbar\,d/dx$, then $T = p^2/2m \rightarrow -(\hbar^2/2m)(d^2/dx^2)$, which is the first term in the differential expression (5.9.1) of Schrödinger's energy operator $H$. Thus we are led to associate the differential expression $-i\hbar\,d/dx$ to the linear momentum observable $p$, the differential expression $-(\hbar^2/2m)(d^2/dx^2)$ to the kinetic energy $T$, and, as $v = p/m$, the differential expression $-i(\hbar/m)\,d/dx$ to the velocity $v$. All of these associations become quite plausible once we make the differential expression (5.9.1) correspond to the total energy $E$ of the system.

There is one correspondence not covered by the above: the position $x$ of the particle (which is an observable) corresponds to the operation of multiplication by the independent variable $x$. The law of action of this operator is given by $l(\Psi) = x\Psi$, and its domain consists of all $\Psi$ in the state space for which $x\Psi$ also belongs to the state space.

Table 5.3 summarizes these correspondences.

To implement the actual correspondence, which is between observables and Hermitian *operators*, we must define for each differential expression on the right side of Table 5.3 its associated Hermitian operator. The procedure for doing this occupied our attention through Chapter 4, where the method was explained in fine detail, and then illustrated in the worked examples of the one-dimensional Laplacian, Legendre's operator, and Hermite's operator. I'm sure that's all retained vividly in your memory. Nonetheless, let me run over it again for the record.

Given an $L^2$ space and a differential expression $l$, we begin by restricting $l$ to the subspace of functions $f$ that are compatible with it, i.e., the $f$ in the $L^2$ space for which $l(f)$ also belongs to the same space. The differential expression, restricted to that domain, defines a Hilbert space operator, the "maximal" operator associated with that expression. We call it the maximal operator for that differential expression because it is the operator of largest

**Table 5.3** Correspondence Between Certain Observables of a One-Particle, One-Dimensional Physical System and Their Differential Expressions

| Observable | Differential expression |
|---|---|
| Position $x$ | Multiplication by $x$ |
| Momentum $p$ | $-i\hbar \dfrac{d}{dx}$ |
| Velocity $v$ | $-i\dfrac{\hbar}{m}\dfrac{d}{dx}$ |
| Kinetic energy $T = \dfrac{p^2}{2m}$ | $-\dfrac{\hbar^2}{2m}\dfrac{d^2}{dx^2}$ |
| Potential energy $V(x)$ | Multiplication by $V(x)$ |
| Total energy $E$ | $-\dfrac{\hbar^2}{2m}\dfrac{d^2}{dx^2} + V(x)$ |

possible domain with that expression as its law of action. It may happen that that maximal operator is Hermitian, as was the case with Hermite's operator (Section 4.9). If so, fine. If not, then we shrink the domain further by the imposition of boundary conditions until we get a Hermitian operator. For the operators of physics, these boundary conditions often come from the physical problems. To get the final correspondence observable → operator we must carry out this procedure for each of the differential expressions on the right side of Table 5.1. How this is to be done will vary from case to case, depending on the particular $L^2$ space and the potential energy function $V(x)$.

The postulate that observables correspond to Hermitian operators has to be supplemented by a set of rules concerning the measured values the observables may have. These rules provide the predictions of quantum mechanics that are to be compared with experiment. For us, the two rules can be stated thus (we'll use the same symbol for an observable and its operator):

1. *If the observable L has a discrete set of eigenvalues $\lambda_1, \lambda_2, \ldots$ such that the corresponding eigenfunctions form an orthogonal basis of the state space, then the only possible values for L are the real numbers $\lambda_1, \lambda_2, \ldots$.*
2. *The average value of the observable L for a system in the normalized state $\Psi$ is $\langle L(\Psi), \Psi \rangle$.*

The first rule tells us when the values of a given observable are *quantized*, i.e., come in lumps. This is a very important physical fact—Schrödinger stressed this fact in his series of papers. But not all observables are quantized. They still have values nevertheless, and rule 2 tells us how to predict the average value of an observable in a given state. Note that the state $\Psi$ must be in the domain of the observable $L$ in order for rule 2 to make sense.

When computing $\langle L(\Psi), \Psi \rangle$ as required by rule 2 bear in mind that $L$ is the Hermitian *operator* associated to the given observable. Avoid using the differential expression whenever possible. It is always possible to avoid the differential expression when the Hermitian operator $L$ has been *diagonalized*; i.e. when we have secured an orthogonal basis of the underlying Hilbert space consisting of eigenfunctions of $L$. The method of computing $L(\Psi)$ using the diagonalization of $L$ was explained in Theorem 4.11 and the discussion that follows that theorem. Refer also to the specific applications in Theorems 4.28 and 4.47. Use this procedure when computing $\langle L(\Psi), \Psi \rangle$.

Finally, a word about the idea behind the proof that Heisenberg's matrix mechanics and Schrödinger's wave mechanics are equivalent theories. Heisenberg associates to each observable a Hermitian matrix. But to each Hermitian operator on a Hilbert space there is also associated a Hermitian matrix, once an orthogonal basis is chosen. Let $T$ be the operator, and suppose $e_1, e_2, \ldots$ is an orthogonal basis for the space, such that every $e_i$ belongs to the domain of $T$. Then the numbers $a_{ij} = \langle Te_i, e_j \rangle$, $i, j = 1, 2, \ldots$, make up the matrix of $T$ with respect to the $e_i$. This connection between matrices and operators provides the link that enabled Schrödinger to prove the equivalence of the two theories.

### Exercises

1. As in Table 5.3, let $Q$ denote the position observable in one dimension, $Q(\Psi) = x\Psi$, and let $P$ denote the momentum observable, $P(\Psi) = -i\hbar\, d\Psi/dx$. Show that for any $\Psi$ that lies in the domain of both $QP$ and $PQ$:
   (a) $QP(\Psi) = -i\hbar x\, d\Psi/dx$
   (b) $PQ(\Psi) = -i\hbar\Psi - i\hbar x\, d\Psi/dx$
   (c) $(QP - PQ)(\Psi) = i\hbar\Psi$
   Thus conclude that this $Q$ and $P$ satisfy the "stronger quantum condition" of Born and Jordan (equation (5.7.2)).

2. The quantum linear oscillator (Section 5.5) has the configuration space $L^2(-\infty, \infty)$, has energy eigenvalues $E_n = (2n+1)E_0$, $n = 0, 1, \ldots$, where $E_0 = \frac{1}{2}h\nu_0$ is the zero-point energy (equation (5.5.7)), and has corresponding normalized eigenfunctions $\Psi_n(z)$ which form an orthogonal basis of $L^2(-\infty, \infty)$. Suppose that the initial state $t = 0$ of the oscillator is $\Psi = \frac{3}{5}\Psi_0 + \frac{4}{5}\Psi_1$.
   (a) Utilizing the discussion in Section 5.8, find the state $\Psi(z, t)$ of the oscillator at time $t > 0$. [Ans.: $\Psi(z,t) = \frac{3}{5}e^{-i(E_0/h)t}\Psi_0(z) + \frac{4}{5}e^{-i(E_1/h)t}\Psi_1(z)$.]
   (b) Find the average energy of the oscillator at time $t$ and show that it is independent of $t$. What values could result from measurement of the oscillator's energy at time $t$?

(c) Using equation (5.5.8), show that

$$|\Psi(z,t)|^2 = \frac{e^{-z^2/2}}{25\sqrt{2\pi}}(16z^2 + 24z \cos 2\pi v_0 t + 9)$$

On one piece of graph paper, plot $\sqrt{2\pi}|\Psi(z,t)|^2$ versus $z$, $-3 \leqslant z \leqslant 3$, for $t = 0$, $t = 1/4v_0$, and $t = 1/2v_0$.

3. (Continuation—maintain the notation of Exercise 2.)

  (a) Let the time-dependent state of the quantum linear oscillator be

$$\Psi(z,t) = \sum_{n=0}^{\infty} A_n \Psi_n(z) e^{-i(E_n/\hbar)t}$$

  where $\sum_{n=0}^{\infty}|A_n|^2 = 1$. Assuming that $\Psi(z,t)$ belongs to the domain of Schrödinger's energy operator $H$, show that the average energy of the quantum oscillator in state $\Psi(z,t)$ is

$$E_{\text{quantum ave.}} = \sum_{n=0}^{\infty} E_n|A_n|^2 = \frac{hv_0}{2}\sum_{n=0}^{\infty}(2n+1)|A_n|^2$$

  independent of time.

  (b) Sum the series for $E_{\text{quantum ave.}}$ in (a) in closed form for Schrödinger's coefficients given in equation (5.6.11): $A_n = \exp(-A^2/8)(A/2)^n(n!)^{-1/2}$.

  (c) The quantum oscillator in the state described in (b), with $A \gg 1$, acts like a classical oscillator $z = A\cos 2\pi v_0 t$; see Section 5.6, equation (5.6.12) ff. and Exercise 1. The total energy of a classical oscillator is $E_{\text{classical}} = \frac{1}{2}k(x_{\text{max}})^2$ (see Section 2.5), thus $E_{\text{classical}} = \frac{1}{2}(k/\sqrt{b})(z_{\text{max}})^2 = \frac{1}{2}(k/\sqrt{b})A^2$. Using the value of $b$ from Section 5.5, show that

$$E_{\text{classical}} = \frac{hv_0 A^2}{4}$$

  (d) Combining your results from (b) and (c), show that

$$\frac{E_{\text{quantum ave.}}}{E_{\text{classical}}} = 1 + \frac{2}{A^2}$$

  Discuss.

4. (a) Compute the matrix of Legendre's operator $L$ with respect to the Legendre polynomials $P_n$, $n = 0, 1, 2, \ldots$ (see Section 4.6).

  (b) Compute the matrix of Hermite's operator $L$ with respect to the Hermite polynomials $H_n$, $n = 0, 1, 2, \ldots$ (see Section 4.9).

# 6

# Bessel's Operator and Bessel Functions

## 6.1 THE WAVE EQUATION AND OTHER EQUATIONS IN HIGHER DIMENSIONS; POLAR COORDINATES

We consider a thin elastic membrane stretched and clamped along some rigid boundary that lies in a plane. The drumhead is a typical example. This is a two-dimensional configuration corresponding to the one-dimensional vibrating wire studied in Chapter 3. As in that case, our objective here is to analyze and predict the motion of the membrane.

We will concern ourselves with very small vertical motions, although in Figure 6.1 we have exaggerated the displacement to make the figure clearer. We specify position in the equilibrium plane of the membrane by Cartesian coordinates $(x, y)$ and denote by $u(x, y, t)$ the vertical displacement of the membrane at position $(x, y)$ and time $t$. The vertical displacement is measured from the equilibrium position given by $u = 0$.

The tension $T$ in our membrane is measured in dynes/cm. We interpret $T$ as the force necessary to keep closed a slit in the membrane 1 cm long. We shall assume that $T$ is independent of direction. The symbol $\rho$ represents the areal density of the membrane (grams/cm$^2$), which we can measure by dividing the weight of the stretched portion by its area.

In Figure 6.1 we have pictured an isolated infinitesimal piece of our membrane. The symbols $X$ and $Y$ represent the forces acting on the ends of this displaced piece. Let us deal first with the forces $Y_1$ and $Y_2$ acting along the

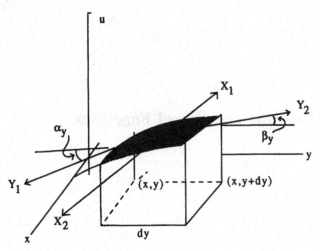

**Figure 6.1**   Section of the vibrating membrane.

$y$ direction. Under the reasonable assumption that the horizontal component of force is constant, we have $T\,dx = Y_2\cos(\beta_y) = Y_1\cos(\alpha_y)$, where the angles $\alpha_y$ and $\beta_y$ are shown in Figure 6.1 and $T$ is the (constant) tension in the membrane. The net vertical force $dF_y$ resulting from this pair of forces is

$$dF_y = Y_2\sin(\beta_y) - Y_1\sin(\alpha_y)$$

Taking into account the relation just derived for $Y_1$ and $Y_2$, we get $dF_y = T(\tan(\beta_y) - \tan(\alpha_y))\,dx$. The expression $\tan(\beta_y) - \tan(\alpha_y)$ represents the change in slope as we go from $y$ to $y + dy$ with $x$ constant. Since $\partial u/\partial y$ equals the slope, we have

$$\tan(\beta_y) - \tan(\alpha_y) = \frac{\partial}{\partial y}\left(\frac{\partial u}{\partial y}\right)dy = \frac{\partial^2 u}{\partial y^2}\,dy$$

Hence the net vertical force resulting from the pair $Y_1$, $Y_2$ is

$$dF_y = T\frac{\partial^2 u}{\partial y^2}\,dx\,dy$$

If we perform a similar analysis for the forces $X_1$, $X_2$ acting in the $x$ direction, we find

$$dF_x = T\frac{\partial^2 u}{\partial x^2}\,dx\,dy$$

Adding, we get the net vertical (upward) force on our piece of membrane

$$dF = T\left(\frac{\partial^2 u}{\partial x^2} + \frac{\partial^2 u}{\partial y^2}\right) dx\, dy$$

We equate this net force to mass times acceleration. The mass of our piece of membrane before displacement is $dm = \rho\, dx\, dy$, and this mass remains the same after displacement, even though it is distributed over a slightly larger stretched piece. The acceleration is $\partial^2 u/\partial t^2$. So Newton's law, force = mass × acceleration, yields

$$T\left(\frac{\partial^2 u}{\partial x^2} + \frac{\partial^2 u}{\partial y^2}\right) dx\, dy = (\rho\, dx\, dy)\frac{\partial^2 u}{\partial t^2}$$

Canceling $dx\, dy$, dividing by $T$, and setting $w = (T/\rho)^{1/2}$, which has units of velocity (Exercise 1), we get the *wave equation in two dimensions*:

$$\frac{\partial^2 u}{\partial x^2} + \frac{\partial^2 u}{\partial y^2} = \frac{1}{w^2}\frac{\partial^2 u}{\partial t^2} \tag{6.1.1}$$

This equation differs from the one-dimensional wave equation in that the displacement $u$ now depends on two space variables instead of one, and on the left side of the equation $\partial^2 u/\partial x^2 + \partial^2 u/\partial y^2$ replaces $\partial^2 u/\partial x^2$. On this basis, one may by reasonable extrapolation take it that the wave equation in three dimensions will involve a displacement $u$ that depends on three space variables $x$, $y$, and $z$ and will have the form

$$\frac{\partial^2 u}{\partial x^2} + \frac{\partial^2 u}{\partial y^2} + \frac{\partial^2 u}{\partial z^2} = \frac{1}{w^2}\frac{\partial^2 u}{\partial t^2}$$

where $w$ again is the wave velocity (cm/sec).

The differential expression that stands on the left side of the wave equation is called the Laplacian differential expression. Any Hilbert space operator determined by such an expression (actually its negative), whatever the dimension, we shall call a *Laplacian operator* and denote by $\Delta$. We studied the one-dimensional Laplacian in Section 4.5 and shall take up the two-dimensional Laplacian operator in the next chapter. In terms of the differential expression alone, disregarding the domain for the moment, we have

$$-\Delta = \begin{cases} \dfrac{d^2}{dx^2} & \text{in one dimension} \\[2ex] \dfrac{\partial^2}{\partial x^2} + \dfrac{\partial^2}{\partial y^2} & \text{in two dimensions} \\[2ex] \dfrac{\partial^2}{\partial x^2} + \dfrac{\partial^2}{\partial y^2} + \dfrac{\partial^2}{\partial z^2} & \text{in three dimensions} \end{cases}$$

Using this symbol, we may write

$$\Delta u + \frac{1}{w^2} \frac{\partial^2 u}{\partial t^2} = 0 \qquad \text{wave equation} \qquad\qquad (6.1.2)$$

independent of spatial dimension.

We may do the same for the heat equation and Schrödinger's equations. The heat equation is:

$$\Delta u + \frac{1}{\alpha^2} \frac{\partial u}{\partial t} = 0 \qquad \text{heat equation} \qquad\qquad (6.1.3)$$

where $u$ is temperature as a function of position and time, and

$$\alpha^2 = \frac{\kappa}{\sigma \rho}$$

where $\kappa$ is the thermal conductivity, $\sigma$ is the specific heat, and $\rho$ is the density.

Schrödinger's equations are

$$\frac{\hbar^2}{2m} \Delta \Psi + V\Psi = E\Psi \qquad \begin{array}{l}\text{Schrödinger's time–} \\ \text{independent equation}\end{array} \qquad (6.1.4)$$

and

$$\frac{\hbar^2}{2m} \Delta \Psi + V\Psi = i\hbar \frac{\partial \Psi}{\partial t} \qquad \begin{array}{l}\text{Schrödinger's time–} \\ \text{dependent equation}\end{array} \qquad (6.1.5)$$

where the symbols have the same meaning as before except that $\Psi$ and $V$ will now depend on one, two, or three space variables as the case may be. Also, $\Psi$ must satisfy the requirement

$$\int_S |\Psi|^2 \, dV = 1$$

where $S$ is the one-, two-, or three-dimensional region of interest.

For example, consider Schrödinger's time-independent equation describing a particle restrained to a rectangular region $S = \{(x, y) : 0 \leqslant x \leqslant a,$

$0 \le y \le b$. Since the particle cannot appear outside $S$, we must have $|\Psi(x, y)|^2 = 0$ for $(x, y)$ not in $S$. Thus the normalization condition becomes

$$\int_S |\Psi(x, y)|^2 \, dx \, dy = \int_0^a dx \int_0^b dy \, |\Psi(x, y)|^2 = 1$$

which leads us to the space $L^2(S)$ consisting of functions $f(x, y)$ which are square integrable over $S$:

$$L^2(S) = \left\{ f(x, y): \int_0^a dx \int_0^b dy \, |f(x, y)|^2 < \infty \right\}$$

(because any nonzero function in $L^2(S)$ may be made to satisfy the normalization condition simply by multiplying it by a constant). Applying the technique that has served us so well in the past, namely separation of variables, we write $\Psi(x, y) = X(x)Y(y)$, then observe that

$$\int_S |\Psi(x, y)|^2 \, dx \, dy = \left( \int_0^a |X(x)|^2 \, dx \right) \left( \int_0^b |Y(y)|^2 \, dy \right)$$

so that the "separated" function $\Psi(x, y) = X(x)Y(y)$ belongs to $L^2(S)$ if and only if both $X(x) \in L^2(0, a)$ and $Y(y) \in L^2(0, b)$.

Continuing our analysis of this example, we shall solve Schrödinger's time-independent equation for $\Psi$ under the assumption of zero potential energy: $V = 0$. Remember that $\Psi(x, y) = 0$ for $(x, y)$ outside the rectangle $S$, because the particle cannot appear outside $S$. If we wish to have $\Psi$ continuous, which is a quite reasonable condition given the interpretation of $|\Psi|^2$ as a probability density, we shall have to have $\Psi(0, y) = \Psi(a, y) = \Psi(x, 0) = \Psi(x, b) = 0$. For the components of the separated function, $\Psi(x, y) = X(x)Y(y)$, this forces $X(0) = X(a) = 0$ and $Y(0) = Y(b) = 0$.

Now substitute $\Psi(x, y) = X(x)Y(y)$ into Schrödinger's equation and then divide by the product $XY$ to get

$$-\frac{\hbar^2}{2m} \frac{1}{X} \frac{d^2 X}{dx^2} = \frac{\hbar^2}{2m} \frac{1}{Y} \frac{d^2 Y}{dy^2} + E$$

The left side of this equation depends on $x$ alone, the right side on $y$ alone; hence both must equal the same constant, say $\lambda$. We arrive at two equations:

$$-\frac{\hbar^2}{2m} \frac{d^2 X}{dx^2} = \lambda X, \qquad \int_0^b |X(x)|^2 \, dx < \infty, \qquad X(0) = X(a) = 0 \qquad (6.1.6)$$

and

$$-\frac{\hbar^2}{2m} \frac{d^2 Y}{dy^2} = (E - \lambda) Y, \qquad \int_0^b |Y(y)|^2 \, dy < \infty, \qquad Y(0) = Y(b) = 0 \qquad (6.1.7)$$

These equations translate to eigenvalue problems for the one-dimensional Laplacian $\Delta$ with Dirichlet boundary conditions (Section 4.5, Definition 4.13):

$$\frac{\hbar^2}{2m}\Delta(X) = \lambda X \qquad \text{in } L^2(0,a) \qquad\qquad (6.1.6')$$

$$\frac{\hbar^2}{2m}\Delta(Y) = (E - \lambda)Y \qquad \text{in } L^2(0,b) \qquad\qquad (6.1.7')$$

The eigenvalues and eigenfunctions, whose existence is guaranteed by Theorem 4.14 of Section 4.5, are easily computed in this case. For (6.1.6') we have eigenvalues $\lambda = p^2\hbar^2\pi^2/2ma^2$, $p = 1, 2, \ldots$, with corresponding eigenfunctions $X_p(x) = \sin(p\pi x/a)$. Likewise, for (6.1.7'), the eigenvalues are $E - \lambda = q^2\hbar^2\pi^2/2mb^2$, $q = 1, 2, \ldots$, with corresponding eigenfunctions $Y_q(y) = \sin(q\pi y/b)$. Thus the possible values for the total energy $E$ of this particle are given by $E = (p^2/a^2 + q^2/b^2)(\hbar^2\pi^2/2m)$, $p, q = 1, 2, \ldots$, with corresponding (nonnormalized) wave functions $\Psi_{p,q}(x, y) = \sin(p\pi x/a)\sin(q\pi y/b)$. The general solution $\Psi$ is a doubly infinite sum of the $\Psi_{p,q}$.

The exercises deal with similar solutions of the wave and heat equations in two dimensions using rectangular coordinates.

However, in many two-dimensional physical applications, polar coordinates are more convenient than rectangular coordinates. This is the case, for example, with the circular drumhead and the heated circular disk. To apply our equations to such problems, we need to express the two-dimensional Laplacian

$$-\Delta u = \frac{\partial^2 u}{\partial x^2} + \frac{\partial^2 u}{\partial y^2}$$

in polar coordinates. To do this set $x = r\cos\theta$, $y = r\sin\theta$ and apply the chain rule for partial differentiation:

$$\frac{\partial u}{\partial x} = \frac{\partial u}{\partial r}\frac{\partial r}{\partial x} + \frac{\partial u}{\partial \theta}\frac{\partial \theta}{\partial x}, \qquad \frac{\partial u}{\partial y} = \frac{\partial u}{\partial r}\frac{\partial r}{\partial y} + \frac{\partial u}{\partial \theta}\frac{\partial \theta}{\partial y} \qquad (6.1.8)$$

As $r^2 = x^2 + y^2$, we have $2r\,\partial r/\partial x = 2x$ and $2r\,\partial r/\partial y = 2y$, whence $\partial r/\partial x = x/r = \cos\theta$ and $\partial r/\partial y = y/r = \sin\theta$. To find $\partial\theta/\partial x$ we differentiate the equation $y = r\sin\theta$ with respect to $x$, holding $y$ constant:

$$0 = \frac{\partial}{\partial x}(y) = \frac{\partial}{\partial x}(r\sin\theta) = \frac{\partial r}{\partial x}\sin\theta + r\cos\theta\frac{\partial\theta}{\partial x}$$

Using the equation $\partial r/\partial x = \cos\theta$ just derived, we get

$$0 = \cos\theta\sin\theta + r\cos\theta\frac{\partial\theta}{\partial x}$$

whence $\partial\theta/\partial x = -\sin\theta/r$. Differentiating $x = r\cos\theta$ with respect to $y$ holding $x$ constant, we get $\partial\theta/\partial y = \cos\theta/r$. In summary:

$$\frac{\partial r}{\partial x} = \cos\theta, \qquad \frac{\partial r}{\partial y} = \sin\theta$$

$$\frac{\partial\theta}{\partial x} = -\frac{\sin\theta}{r}, \qquad \frac{\partial\theta}{\partial y} = \frac{\cos\theta}{r}$$

Substituting these into (6.1.8), we get

$$\frac{\partial u}{\partial x} = \cos\theta\,\frac{\partial u}{\partial r} - \frac{\sin\theta}{r}\frac{\partial u}{\partial\theta} \qquad \text{and} \qquad \frac{\partial u}{\partial y} = \sin\theta\,\frac{\partial u}{\partial r} + \frac{\cos\theta}{r}\frac{\partial u}{\partial\theta} \qquad (6.1.9)$$

Next we must repeat this procedure:

$$\frac{\partial^2 u}{\partial x^2} = \frac{\partial}{\partial x}\left(\frac{\partial u}{\partial x}\right) = \frac{\partial}{\partial r}\left(\frac{\partial u}{\partial x}\right)\frac{\partial r}{\partial x} + \frac{\partial}{\partial\theta}\left(\frac{\partial u}{\partial x}\right)\frac{\partial\theta}{\partial x}$$

$$= \cos\theta\,\frac{\partial}{\partial r}\left(\frac{\partial u}{\partial x}\right) - \frac{\sin\theta}{r}\frac{\partial}{\partial\theta}\left(\frac{\partial u}{\partial x}\right)$$

The partial derivatives in this last equation are to be computed from equation (6.1.9). I shall ask the reader to start at this point and complete the calculation (Exercise 6) to get the form of *the Laplacian in polar coordinates*

$$-\Delta u = \frac{\partial^2 u}{\partial r^2} + \frac{1}{r}\frac{\partial u}{\partial r} + \frac{1}{r^2}\frac{\partial^2 u}{\partial\theta^2}$$

Thus, in two dimensions, the wave and heat equations expressed in polar coordinates are

$$\frac{\partial^2 u}{\partial r^2} + \frac{1}{r}\frac{\partial u}{\partial r} + \frac{1}{r^2}\frac{\partial^2 u}{\partial\theta^2} = \begin{cases} \dfrac{1}{w^2}\dfrac{\partial^2 u}{\partial t^2} & \text{wave equation} \qquad (6.1.10) \\[2mm] \dfrac{1}{\alpha^2}\dfrac{\partial u}{\partial t} & \text{heat equation} \qquad (6.1.11) \end{cases}$$

and the Schrödinger equations are

$$-\frac{h^2}{2m}\left(\frac{\partial^2\Psi}{\partial r^2} + \frac{1}{r}\frac{\partial\Psi}{r} + \frac{1}{r^2}\frac{\partial^2\Psi}{\partial\theta^2}\right) + V\Psi$$

$$= \begin{cases} E\Psi & \text{time independent} \qquad (6.1.12) \\[2mm] i\hbar\,\dfrac{\partial\Psi}{\partial t} & \text{time dependent} \qquad (6.1.13) \end{cases}$$

In the special case where there is no $\theta$ dependence ($\partial/\partial\theta = 0$), the Laplacian takes the form

$$-\Delta = \frac{\partial^2}{\partial r^2} + \frac{1}{r}\frac{\partial}{\partial r}$$

and the various equations above are modified accordingly.

The solution of these equations in polar coordinates will lead us to the study of Bessel functions, the subject of this chapter.

### Exercises

1. Show that $w = (T/\rho)^{1/2}$ has the dimensions of velocity, where $T$ is the tension in our membrane and $\rho$ its planar density.

2. Show that the Laplacian is linear; i.e., $\Delta(\alpha u_1 + \beta u_2) = \alpha \Delta u_1 + \beta \Delta u_2$ for any constants $\alpha$ and $\beta$ and any functions $u_1$ and $u_2$. Show that the wave, heat, and the two Schrödinger equations are linear: if $u_1(\Psi_1)$ and $u_2(\Psi_2)$ are solutions, so is $\alpha u_1 + \beta u_2 (\alpha \Psi_1 + \beta \Psi_2)$.

3. (a) Show that if $f(\cdot)$ is a function of one variable and if $\alpha$ and $\beta$ are real constants that satisfy $\alpha^2 + \beta^2 = 1$, then $u(x, y, t) = f(\alpha x + \beta y - wt)$ satisfies the two-dimensional wave equation. Same for $f(\alpha x + \beta y + wt)$.

   (b) In part (a) consider $f(z) = 1$ for $0 \leqslant z \leqslant 1$, $f(z) = 0$ elsewhere (the "unit pulse"). Suppose $w = 2$ and $\alpha = \beta = 1/\sqrt{2}$. Sketch the boundaries of the traveling wave $f((x + y)/\sqrt{2} - 2t)$ for $t = 0, 2, 4$ (in the plane).

4. In the following steps, solve the wave equation for a rectangular drumhead $S = \{(x, y): 0 \leqslant x \leqslant a, 0 \leqslant y \leqslant b\}$ clamped along its boundaries.

   (a) Show that a separation of variables $u(x, y, t) = p(x, y)T(t)$ results in a pair of equations

   $$\Delta p = \sigma p, \qquad T'' = -w^2 \sigma T$$

   where $\sigma$ is a constant.

   (b) Show how a further separation of variables $p(x, y) = X(x)Y(y)$ leads to a pair of eigenvalue problems
   
   (i) $-X'' = \lambda X$, $X(0) = X(a) = 0$
   
   (ii) $-Y'' = (\sigma - \lambda)Y$, $Y(0) = Y(b) = 0$
   
   where $\sigma$ is the constant from part (a).

   (c) By adding the physically reasonable condition $\int_0^a |X(x)|^2 dx < \infty$, interpret (i) as a Hilbert space problem involving the one-dimensional Laplacian operator $\Delta$ with Dirichlet boundary conditions (Section 4.5). Show that $\Delta$ has eigenvalues $m^2\pi^2/a^2$, $m = 1, 2, \ldots$, and corresponding eigenfunctions $\sin(m\pi x/a)$.

   (d) Repeat part (c) for the eigenvalue problem (ii). Thus conclude that the constant $\sigma$ introduced in (a) has values $\sigma_{m,n} = \pi^2(m^2/a^2 + n^2/b^2)$, $m, n = 1, 2, \ldots$.

(e) Show that the equation $T'' = -w^2 \sigma_{m,n} T$ in part (a) has solutions

$$T_{m,n} = A_{m,n} \cos w\pi \left(\frac{m^2}{a^2} + \frac{n^2}{b^2}\right)^{1/2} t + B_{m,n} \sin w\pi \left(\frac{m^2}{a^2} + \frac{n^2}{b^2}\right)^{1/2} t$$

where $A_{m,n}$ and $B_{m,n}$ are arbitrary constants and $m,n = 1, 2, \ldots$. Thus arrive at a formal general solution $u(x, y, t)$ to the wave equation for the rectangular drumhead:

$$u(x, y, t) = \sum_{m=1}^{\infty} \sum_{n=1}^{\infty} (A_{m,n} \cos w\pi\sigma_{m,n} t$$

$$+ B_{m,n} \sin w\pi\sigma_{m,n} t) \sin \frac{m\pi x}{a} \sin \frac{n\pi y}{b}$$

5. Repeating the steps in Exercise 4, solve the heat equation $\Delta u + (1/\alpha^2)\, \partial u/\partial t = 0$ for a rectangular plate $S = \{(x, y): 0 \leqslant x \leqslant a, 0 \leqslant y \leqslant b\}$ whose edges are maintained at temperature $u = 0$. Thus arrive at a formal general solution $u(x, y, z)$ to the heat equation for a rectangular plate

$$u(x, y, t) = \sum_{m=1}^{\infty} \sum_{n=1}^{\infty} A_{m,n} \sin \frac{m\pi x}{a} \sin \frac{n\pi y}{b} e^{-\alpha^2 \sigma_{m,n} t}$$

where

$$\sigma_{m,n} = \pi^2 \left(\frac{m^2}{a^2} + \frac{n^2}{b^2}\right), \qquad m, n = 0, 1, 2, \ldots$$

6. (a) Starting from equation (6.1.9), express $\partial^2 u/\partial x^2$ in polar coordinates. (Combine terms, using the fact that $\partial^2 u/\partial r\, \partial\theta = \partial^2 u/\partial\theta\, \partial r$.)
   (b) Same as (a) for $\partial^2 u/\partial y^2$.
   (c) Adding your answers from (a) and (b), derive the form of the Laplacian in polar coordinates:

$$-\Delta u = \frac{\partial^2 u}{\partial r^2} + \frac{1}{r}\frac{\partial u}{\partial r} + \frac{1}{r^2}\frac{\partial^2 u}{\partial \theta^2}$$

   (d) What is the form of the Laplacian if $u$ does not vary with $\theta$?
7. (Continuation) If you do *not* assume that $\partial^2 u/\partial r\, \partial\theta = \partial^2 u/\partial\theta\, \partial r$, what is the resulting expression for the Laplacian in polar coordinates?

## 6.2 BESSEL'S EQUATION AND BESSEL'S OPERATOR OF ORDER ZERO

If the displacement $u$ depends only on the radius $r$, then the wave equation has the form

$$\frac{\partial^2 u}{\partial r^2} + \frac{1}{r}\frac{\partial u}{\partial r} = \frac{1}{w^2}\frac{\partial^2 u}{\partial t^2} \qquad\qquad (6.2.1)$$

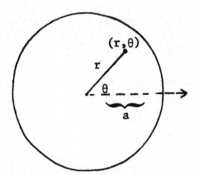

**Figure 6.2**   The drumhead.

We deal with a circular drumhead of radius $a$, the membrane clamped at the rim (Figure 6.2), so that $u(a, t) \equiv 0$ for all $t$.

We shall use again the method of separation of variables, setting $u = R(r)T(t)$. Substituting this into (6.2.1) and then dividing by $u = RT$ we get

$$\frac{R''}{R} + \frac{1}{r}\frac{R'}{R} = \frac{1}{w^2}\frac{T''}{T}$$

As the left side depends only on $r$ and the right side only on $t$, they are each equal to a constant, say $-\lambda$. We thus get two equations, one in the space variable $r$ and one in the time variable,

$$-R'' - \frac{1}{r}R' = \lambda R \qquad \text{and} \qquad T'' = -w^2\lambda T$$

As our drumhead is clamped at the rim $r = a$, we have $u(a, t) = R(a)T(t) = 0$ for all time $t$. This requires $R(a) = 0$, so our equation for the function $R(r)$ includes a boundary condition:

$$-R'' - \frac{1}{r}R' = \lambda R, \qquad R(a) = 0$$

This is an eigenvalue problem, with eigenvalue parameter $\lambda$. We seek to set this up as an eigenvalue problem for a Hilbert space operator. We thus require, first, an appropriate $L^2$ space of functions on $0 \leqslant r \leqslant a$. Our operator will have differential expression $-R'' - (1/r)R'$, and we then must settle on a domain, which will of course include the boundary condition $R(a) = 0$.

To get the appropriate $L^2$ space, we write our differential expression in Sturmian form and then read off the weight function from this form. This procedure was explained in Section 4.5 (after Theorem 4.15), to which the reader might want to refer. In Sturmian form, our differential expression looks thus:

$$-R'' - \frac{1}{r}R' = -\frac{1}{r}\frac{d}{dr}\left(r\frac{dR}{dr}\right)$$

so $f(r) = r$ is dictated as the weight function for our $L^2$ space. So we shall be working in the Hilbert space

$$L^2(0, a; r)$$

which consists of functions defined on the interval $0 \leqslant r \leqslant a$ that are square integrable against the weight function $w(r) = r$:

$$\int_0^a |f(r)|^2 r\, dr < \infty$$

The inner product in our Hilbert space is

$$\langle f, g \rangle = \int_0^a f(r)\overline{g(r)} r\, dr \qquad \text{(see Section 3.9)}$$

Having now secured the appropriate Hilbert space, we will complete the specification of "Bessel's operator" as a Hilbert space operator by defining its domain as an appropriate subspace of $L^2(0, a; r)$.

**Definition 6.1.** *Bessel's operator* $B$, which acts in the Hilbert space $L^2(0, a; r)$, has the *differential expression*

$$b(f) \equiv -\frac{d^2 f}{dr^2} - \frac{1}{r}\frac{df}{dr} = -\frac{1}{r}\frac{d}{dr}\left(r\frac{df}{dr}\right)$$

and *domain(B)* consisting of functions $f(r)$ that satisfy the following three conditions:

1. $f$ is compatible with $b$; i.e., both $f$ and $b(f) \in L^2(0, a; r)$.
2. $f(a) = 0$.
3. $\lim_{r \to 0+} f(r)$ exists and is finite.

For the functions $f$ in the domain of $B$, and *only* for those functions, Bessel's operator is defined by

$$B(f) = b(f) = -\frac{1}{r}\frac{d}{dr}\left(r\frac{df}{dr}\right), \qquad f \in \text{domain}(B)$$

The three conditions that define the domain of Bessel's operator have, of course, been chosen with the object of making the operator Hermitian. Let us look in detail at each of these three conditions.

The first condition, compatibility, restricts the domain to functions $f \in L^2(0, a; r)$ such that $b(f) = -(1/r)(rf')'$ also belongs to $L^2(0, a; r)$. This condition must be fulfilled if the resulting operator is to act *in* our $L^2$ space; i.e., at least this much of a restriction must be put on the domain in order to be within the scope of the Hilbert space theory at all. Of course one can, and frequently does, study differential equations involving the expression $b(f) = -(1/r)(rf')'$ without any $L^2$ restrictions on the functions $f$ involved. But to do Hilbert space theory, one needs the compatibility.

In prior cases we have seen that compatibility can sometimes bring with it certain restrictions on the boundary behavior. Such is the case again here.

**Theorem 6.2.** *Suppose that $f(r)$ satisfies the compatibility condition 1 to belong to the domain of Bessel's operator (refer to Definition 6.1). Then necessarily $f(a)$ and $f'(a)$ are finite, and $\lim_{r \to 0+} rf'(r)$ exists and is finite.*

You are not responsible for the proof of this mathematical result, which is given in the appendix to this chapter, but you should understand the statement.

**Theorem 6.3.** *If $f(r)$ satisfies both conditions 1 and 3 to belong to the domain of Bessel's operator, then $\lim_{r \to 0+} rf'(r) = 0$.*

So, from compatibility alone, we get $\lim_{r \to 0+} rf'(r)$ to be finite. If condition 3 is also satisfied, then this limit must be zero. The proof of Theorem 6.3 is also given in Appendix 6.A. You will need Theorems 6.2 and 6.3 to prove the next very important result.

**Theorem 6.4.** *Bessel's operator is Hermitian and positive definite, i.e.,*

$$\langle B(f), g \rangle = \langle f, B(g) \rangle \qquad \text{for all } f, g \text{ in domain } (B)$$

*and*

$$\langle B(f), f \rangle > 0 \qquad \text{whenever } 0 \neq f \in \text{domain } (B)$$

You are asked to supply the proof as Exercise 4.

Recall these definitions: An eigenfunction $f$ of $B$ by definition *belongs to domain(B)*, is not zero, and satisfies $B(f) = \lambda f$ for some constant $\lambda$. The $\lambda$'s for which this happens are the eigenvalues. The nonzero functions $f$ in the domain of $B$ that satisfy $B(f) = \lambda f$ are called the eigenfunctions belonging to the eigenvalue $\lambda$. The entire collection of such functions, with the zero function thrown in, is the eigenspace belonging to the eigenvalue $\lambda$.

**Corollary 6.5.** *All eigenvalues of B are real and positive, and eigenfunctions belonging to different eigenvalues are orthogonal.*

You are asked to give the proof in Exercise 5.

Corollary 6.5 tells us that, if $B$ has any eigenvalues at all, then they must be real and positive. We now take up the task of finding these eigenvalues and their corresponding eigenfunctions. The procedure here is the same as in previous cases. We first find the general solution to the homogeneous differential equation $y'' + (1/r)y' + \lambda y = 0$. We know the solution space is two-dimensional, so the general solution has the form $y = c_1 f_1 + c_2 f_2$, where $c_1$ and $c_2$ are arbitrary constants and $f_1$ and $f_2$ linearly independent solutions. Then we determine for what values of the constants $c_1, c_2$, and for what values of the parameter $\lambda$, the function $y$ will be in the domain of $B$.

Our differential equation is

$$\frac{d^2 y}{dr^2} + \frac{1}{r}\frac{dy}{dr} + \lambda y = 0$$

We make the change of variable $x = \sqrt{\lambda}\, r$, which keeps us in the real numbers as we know $\lambda > 0$. Then

$$\frac{dy}{dr} = \frac{dy}{dx}\frac{dx}{dr} = \sqrt{\lambda}\frac{dy}{dx} \quad \text{and} \quad \frac{d^2 y}{dr^2} = \lambda\frac{d^2 y}{dx^2}$$

Substituting these expressions into our equation, canceling $\lambda$, and multiplying through by $x$, we get

$$x\frac{d^2 y}{dx^2} + \frac{dy}{dx} + xy = 0 \tag{6.2.2}$$

This is called *Bessel's equation of order zero*. It is the equation we must solve to find the eigenvalues and eigenfunctions of Bessel's differential operator $B$.

**Exercises**

1. Let $A$ be a constant.
   (a) For what value of $A$ is the function $f(r) = Ar$ compatible with Bessel's differential expression?
   (b) Show that the function $f(r) = Ar^2$ satisfies conditions 1 and 3 of Definition 6.1 but that the boundary value $f(a)$ may be arbitrarily specified by choosing the constant $A$.
2. (a) Show that the function $f(r) = \ln(r/a)$, $0 < r \leqslant a$, $f(0) = 0$, satisfies the first two conditions of Definition 6.1 but fails to satisfy the third.
   (b) Show that the function $g(r) = a^2 - r^2$ satisfies all three conditions of Definition 6.1.

3. (Continuation) Define a Hilbert space operator $C$ in $L^2(0, a; r)$ as follows. The differential expression for $C$ is the same as Bessel's differential expression $b(y) = -(1/r)(ry')'$. The domain of $C$ is the subspace of functions that satisfy conditions 1 and 2 of Definition 6.1. Thus the operator $C$ acts in the same way that Bessel's operator acts, i.e., $C(f) = -(1/r)(rf')'$, but has a larger domain.

(a) Show that the functions $f(r) = \ln(r/a)$ and $g(r) = a^2 - r^2$ from Exercise 2 lie in the domain of $C$.

(b) For the functions $f, g$ given in (a), compute $\langle C(f), g \rangle$ and $\langle f, C(g) \rangle$. [Ans.: $\langle f, C(g) \rangle = -a^2$.]

(c) Is the operator $C$ Hermitian? Noting that the operator $C$ has the same differential expression as Bessel's operator, how do you reconcile this result with Theorem 6.4?

4. Suppose $f$ and $g$ belong to the domain of Bessel's operator.

(a) Let $\varepsilon$ be a small positive number, and derive the formula

$$\int_\varepsilon^a (rf')'\bar{g} \, dr = af'(a)\bar{g}(a) - \varepsilon f'(\varepsilon)\bar{g}(\varepsilon) - \int_\varepsilon^a f'(r)\bar{g}'(r)r \, dr$$

(b) Refer to Theorems 6.2 and 6.3, and pass to the limit as $\varepsilon \to 0+$ to derive the formula

$$\langle B(f), g \rangle = \langle f', g' \rangle$$

for all $f, g \in$ domain($B$).

(c) In your result in (b), interchange $f$ and $g$, take the complex conjugate of the result, and show therefrom that $B$ is Hermitian.

(d) From your result in (b) deduce that $B$ is positive definite.

5. (a) From the fact that $B$ is Hermitian, prove that its eigenvalues must be real.

(b) From the fact that $B$ is Hermitian, prove that eigenfunctions belonging to different eigenvalues are orthogonal.

(c) From the fact that $B$ is positive definite, prove that its eigenvalues are $> 0$.

6. Show that for any differential operator $A$ with differential expression the same as Bessel's and domain restricted by conditions 1 and 3 of Definition 6.1 we have $\langle A(f), g \rangle - \langle f, A(g) \rangle = a(f(a)\overline{g'(a)} - f'(a)\overline{g(a)})$. Show that if the domain is further restricted by the condition $\alpha y(a) + \beta y'(a) = 0$, where $\alpha$ and $\beta$ are real and not both zero, then the resulting operator $A$ is Hermitian.

The following four exercises deal with Bessel's differential equation of order zero, $xy'' + y' + xy = 0$, and with certain properties that any solution of this equation must possess. Thus we are deducing properties of a solution without actually solving the equation—we are getting our information directly from the equation. Do these exercises in sequence.

7. Show that if $y = y(x)$ is any function that solves Bessel's equation $xy'' + y' + xy = 0$, then $y$ satisfies the antiderivative equation

$$\int xy\, dx = -xy' + C \qquad \text{[Hint: } xy'' + y' = (xy')'.\text{]}$$

Show, conversely, if $y = y(x)$ satisfies the above antiderivative equation, then $y$ satisfies Bessel's equation.

8. Show that if $y$ is any solution to Bessel's equation, then

$$\int (1 + x^2)y\, dx = xy - x^2y' + C$$

9. Show that if $y$ is any solution to Bessel's equation, then

$$2\int xy^2\, dx = x^2(y')^2 + x^2y^2 + C$$

10. Show that if $y$ is any solution to Bessel's equation, then

$$\int x^3y\, dx = (4x - x^3)y' + 2x^2y + C$$

## 6.3 $J_0(x)$: THE BESSEL FUNCTION OF THE FIRST KIND OF ORDER ZERO

In the previous two sections we expressed the wave equation in polar coordinates so as to apply it more conveniently to the vibrating drumhead; then, assuming independence of $\theta$, we separated variables. The equation in the space variable, $-y'' - (1/r)y' = \lambda y$, $y(a) = 0$, contained a free parameter $\lambda$. We interpreted this as an eigenvalue problem for the Hermitian operator $B$ which operates in the space $L^2(0, a; r)$, has differential expression $b(y) = -y'' - (1/r)y'$, and has domain consisting of functions $f$ in $L^2(0, a; r)$ that (1) are compatible with $b$, (2) satisfy $f(a) = 0$, and (3) for which $\lim_{r \to 0+} f(r)$ exists and is finite. The problem of solving the space-variable half of our wave equation is equivalent to finding the eigenvalues of $B$ (allowable values of the parameter $\lambda$) and the corresponding eigenfunctions.

To find the eigenvalues $\lambda$ of the operator $B$ we must solve the differential equation

$$\frac{d^2y}{dr^2} + \frac{1}{r}\frac{dy}{dr} + \lambda y = 0$$

and then determine when such solutions lie in the domain of our operator. As in the previous section, we make the change of variable $x = \sqrt{\lambda} r$ (knowing $\lambda > 0$), and *Bessel's differential equation of order zero* results:

$$x\frac{d^2y}{dx^2} + \frac{dy}{dx} + xy = 0 \tag{6.3.1}$$

The parameter $\lambda$ has disappeared. Our job is now to solve equation (6.3.1).

Equation (6.3.1) has nonconstant coefficients, so we try a power series solution, a method that served us well in the past. (Refer to Sections 2.8 and 2.9.) We therefore begin by assuming a power series expansion for $y$:

$$y = a_0 + a_1 x + a_2 x^2 + \cdots + a_n x^n + \cdots$$

where the coefficients $a_i$ are to be determined so that $y$ satisfies (6.3.1). We set up the calculation this way:

$$
\begin{aligned}
xy = & \quad a_0 x + a_1 x^2 + a_2 x^3 + \cdots + a_{n-1}x^n + \cdots \\
y' = & \; a_1 + 2a_2 x + 3a_3 x^2 + 4a_4 x^3 + \cdots + (n+1)a_{n+1}x^n + \cdots \\
xy'' = & \quad 2a_2 x + 3{\cdot}2a_3 x^2 + 4{\cdot}3a_4 x^3 + \cdots + (n+1)(n)a_{n+1}x^n + \cdots
\end{aligned}
$$

Adding and making use of the identity $(n+1)n + (n+1) = (n+1)^2$, we get

$$xy'' + y' + xy = a_1 + (2^2 a_2 + a_0)x + (3^2 a_3 + a_1)x^2 + (4^2 a_4 + a_2)x^3 + \cdots$$
$$+ ((n+1)^2 a_{n+1} + a_{n-1})x^n + \cdots$$

If $y$ is to satisfy the differential equation (6.3.1), the power series on the right must vanish identically, so all its coefficients must be zero. Look first at the coefficients of $x^0, x^2, x^4, x^6, \ldots$. These are $a_1$, $3a_3^2 + a_1$, $5a_5^2 + a_3$, etc. So we must have $a_1 = 0$. Then as $3a_3^2 + a_1 = 0$, also $a_3 = 0$. Since $5a_5^2 + a_3 = 0$, also $a_5 = 0$, etc. Hence $a_n = 0$ for $n$ odd. Next examine the coefficients of $x$, $x^3$, $x^5, \ldots$. Setting these equal to zero, we get

$$2^2 a_2 + a_0 = 0$$
$$4^2 a_4 + a_2 = 0$$
$$6^2 a_6 + a_4 = 0$$
$$\vdots$$
$$(2n)^2 a_{2n} + a_{2n-2} = 0$$

Set $a_0 = 1$. Then

$$a_2 = \frac{-1}{2^2}$$

$$a_4 = -\frac{a_2}{4^2} = \frac{+1}{4^2 \cdot 2^2}$$

$$a_6 = -\frac{a_4}{6^2} = \frac{-1}{6^2 \cdot 4^2 \cdot 2^2}$$

$$\vdots$$

$$a_{2n} = \frac{(-1)^n}{(2n)^2(2n-2)^2 \cdots (2)^2}$$

Now $(2n)^2(2n-2)^2 \cdots (2)^2 = 2^{2n}(n!)^2$. Hence we have determined a power series solution of (6.3.1) which is called the *Bessel function of the first kind of order zero* and denoted $J_0(x)$

$$J_0(x) = \sum_{n=0}^{\infty} \frac{(-1)^n}{(n!)^2} \left(\frac{x}{2}\right)^{2n} = 1 - \left(\frac{x}{2}\right)^2 + \frac{1}{(2!)^2}\left(\frac{x}{2}\right)^4 - \frac{1}{(3!)^2}\left(\frac{x}{2}\right)^6 + \cdots$$

$$(6.3.2)$$

(Refer to Section 2.9, Exercise 5.)

Let us apply the ratio test to determine the region of convergence of the series (6.3.2). Recall the procedure: we compute for the series (6.3.2) the ratio $|a_{n+1}/a_n|$ and pass to the limit $L$ as $n \to \infty$. The series then converges in the disk determined by the condition $L < 1$. We have

$$\left|\frac{a_{n+1}}{a_n}\right| = \left|\frac{(-1)^{n+1}}{[(n+1)!]^2}\left(\frac{x}{2}\right)^{2(n+1)} \frac{(n!)^2}{(-1)^n}\left(\frac{2}{x}\right)^{2n}\right| = \frac{x^2}{4(n+1)^2} \to 0 \quad \text{as } n \to \infty$$

As $0 < 1$, we have convergence for all $x$. Thus the series (6.3.2) represents an entire function—one that is analytic at every point of the finite plane. It shares this property with the exponential and trigonometric functions. It is mildly surprising that $J_0$ is analytic at $x=0$, because from the fact that the coefficient of the leading term of Bessel's equation vanishes at $x=0$ we might have expected a singularity there. But $J_0(x)$ is analytic at $x=0$, and $J_0(0)=1$. The singularity does show up, however, in the second linearly independent solution.

We may obtain a second linearly independent solution to Bessel's equation by the methods of Chapter 2: find the Wronskian $W$ of the equation by Abel's formula; then solve the first order equation $y'J_0 - yJ_0' = W$ for the second linearly independent solution $y$. Abel's formula: $W = c\exp(-\int (a_1/a_0)\,dx)$. For Bessel's equation, $a_0$ (the leading coefficient) $= x$

and $a_1 = 1$. *So*

$$W = c \exp\left(-\int \frac{dx}{x}\right) = c \exp(-\ln x) = c \exp\left(\ln \frac{1}{x}\right) = \frac{c}{x}$$

We set $c = 1$ for convenience and solve $y'J_0 - yJ_0' = 1/x$. We have

$$y'J_0 - yJ_0' = J_0^2 \frac{d}{dx}\left(\frac{y}{J_0}\right)$$

so our equation becomes

$$J_0^2 \frac{d}{dx}\left(\frac{y}{J_0}\right) = \frac{1}{x}$$

which has the solution $y = J_0 \int dx/x J_0^2(x)$, a second linearly independent solution to Bessel's equation.

It is convenient for our purpose to recast this solution in a somewhat different form. $J_0$ is analytic everywhere, and $J_0(0) = 1$; the same is true of $J_0^2$. Hence, in a neighborhood of $x = 0$ we may write

$$\frac{1}{J_0^2(x)} = 1 + c_1 x + c_2 x^2 + \cdots$$

It follows from this that

$$\frac{1}{x J_0^2(x)} = \frac{1}{x} + \text{power series in } x$$

so

$$y = J_0 \int \frac{dx}{x J_0^2(x)} = J_0(x) \ln x + \text{power series in } x$$

As $J_0(0) = 1$, we conclude that *the second linearly independent solution to Bessel's equation has a logarithmic singularity at $x = 0$.*

If $a$ is a nonzero constant and $c$ any constant, then $ay + cJ_0$ is again a linearly independent solution to Bessel's equation, so we have a rather wide latitude in our normalization of this second solution. The following is a common choice.

$$Y_0(x) = \frac{2}{\pi}\left(\ln\left(\frac{x}{2}\right) + \gamma\right)J_0(x) + \frac{2}{\pi}\sum_{n=1}^{\infty}\frac{(-1)^{n+1}}{(n!)^2}\left(1 + \frac{1}{2} + \cdots + \frac{1}{n}\right)\left(\frac{x}{2}\right)^{2n}$$

$$(6.3.3)$$

where

$$\gamma = \lim_{p \to \infty}\left(\sum_{n=1}^{p}\frac{1}{n} - \ln p\right) = 0.577215\ldots$$

This is known as *the Bessel function of the second kind of order zero*. The general solution to Bessel's equation is $c_1 J_0(x) + c_2 Y_0(x)$.

What is important for us is the requirement that the solution belong to the domain of Bessel's operator. This requirement includes the condition that the solution be continuous at $r = 0$, which means also at $x = 0$ as $x = \sqrt{\lambda} r$. The general solution $c_1 J_0(x) + c_2 Y_0(x)$ is continuous at $x = 0$ if and only if $c_2 = 0$, because $Y_0(x)$ contains the term $\ln x$ which goes to $-\infty$ as $x$ tends to 0. Hence $J_0(x)$ is the only solution of Bessel's equation that lies in the domain of the operator (up to constant multiples).

Before returning to the eigenvalue problem, we shall look more closely at properties of $J_0(x)$.

### Exercises

The following five exercises should be done in sequence.

1. Let $c_n = \int_0^\pi \sin^{2n} \theta \, d\theta$. Show that $c_0 = \pi$. Show that for $n \geq 1$, $c_n = ((2n-1)/2n)c_{n-1}$. Using this formula show that

$$c_n = \frac{(2n)!}{2^{2n}(n!)^2} \pi$$

2. Using the power series for $\cos u$,

$$\cos u = \sum_{n=0}^{\infty} \frac{(-1)^n}{(2n)!} u^{2n}$$

and the formula for $\int_0^\pi \sin^{2n} \theta \, d\theta$ derived in Exercise 1, prove the following integral representation for $J_0(x)$:

$$J_0(x) = \frac{1}{\pi} \int_0^\pi \cos(x \sin \theta) \, d\theta$$

3. Let $p$ be a constant $> 0$. Using the representation for $J_0(x)$ derived in Exercise 2, show that

$$\int_0^\infty e^{-px} J_0(x) \, dx = \frac{p}{\pi} \int_0^\pi \frac{d\theta}{p^2 + \sin^2 \theta}$$

4. The integral $\int_0^\pi d\theta/(p^2 + \sin^2 \theta)$ can be evaluated several ways:
   (i) Look up in tables (no fun).
   (ii) Substitute $u = \cot \theta$; then use the formula

$$\int \frac{du}{a^2 + u^2} = \frac{1}{a} \arctan \frac{u}{a}$$

(iii) Write

$$\int_0^\pi \frac{d\theta}{p^2 + \sin^2\theta} = \frac{1}{2}\int_{-\pi}^{+\pi} \frac{d\theta}{p^2 + \sin^2\theta}$$

convert this latter integral to a complex line integral

$$\oint_{|z|=1} f(z)\, dz$$

by substituting $z = e^{i\theta}$, and then use the calculus of residues.
Using one of these methods, show that

$$\int_0^\pi \frac{d\theta}{p^2 + \sin^2\theta} = \frac{\pi}{p\sqrt{1+p^2}}, \qquad p > 0$$

5. Combining the results of Exercises 3 and 4, show that

$$\int_0^\infty e^{-px} J_0(x)\, dx = \frac{1}{\sqrt{1+p^2}}, \qquad p > 0$$

(This is the Laplace transform of $J_0(x)$.)

6. Starting from $J_0(x) = (1/\pi)\int_0^\pi \cos(x\sin\theta)\, d\theta$ (refer to Exercise 2), show that

$$J_0(x) = \frac{2}{\pi}\int_0^{\pi/2} \cos(x\cos\theta)\, d\theta = \frac{2}{\pi}\int_0^1 \frac{\cos xt}{\sqrt{1-t^2}}\, dt$$

$$= \frac{1}{\pi}\int_{-1}^1 \frac{\cos xt}{\sqrt{1-t^2}}\, dt = \frac{1}{\pi}\int_{-1}^1 \frac{e^{ixt}}{\sqrt{1-t^2}}\, dt$$

7. Do you think that $Y_0(x)$ belongs to the Hilbert space $L^2(0,1;x)$? Explain the reasons for your answer.

## 6.4 $J_0(x)$: CALCULATING ITS VALUES AND FINDING ITS ZEROS

The power series for $J_0(x)$,

$$J_0(x) = \sum_{n=0}^\infty \frac{(-1)^n}{(n!)^2}\left(\frac{x}{2}\right)^{2n}$$

$$= 1 - \left(\frac{x}{2}\right)^2 + \frac{1}{(2!)^2}\left(\frac{x}{2}\right)^4 - \frac{1}{(3!)^2}\left(\frac{x}{2}\right)^6 + \frac{1}{(4!)^2}\left(\frac{x}{2}\right)^8 - \cdots \qquad (6.4.1)$$

is a rapidly convergent alternating series. Because it is an alternating series, if we use a finite number of terms of it to compute $J_0(x)$ we make an error which is in absolute value less than the last term omitted. Hence, if we keep seven terms ($n=0$ through $n=6$), we will be in error by less than

$$\frac{1}{(7!)^2}\left(\frac{x}{2}\right)^{14} = \frac{1}{24,401,600}\left(\frac{x}{2}\right)^{14}$$

For $|x| \leqslant 4$ this error is less than 0.0007, so we may truncate (6.4.1) at seven terms and use the resulting polynomial to compute $J_0(x)$ for $|x| \leqslant 4$. Note that $J_0(x)$ is an even function, $J_0(-x)=J_0(x)$, so that we need only compute it for $0 \leqslant x \leqslant 4$.

To compute with (6.4.1) introduce the variable $u = -(x/2)^2$, whence (6.4.1) becomes

$$J_0(x) = \sum_{n=0}^{\infty}\frac{u^n}{(n!)^2} \cong 1 + u + \frac{u^2}{(2!)^2} + \frac{u^3}{(3!)^2} + \frac{u^4}{(4!)^2} + \frac{u^5}{(5!)^2} + \frac{u^6}{(6!)^2}$$

or

$$J_0(x) \cong 1 + u\left(1 + u\left(\frac{1}{(2!)^2} + u\left(\frac{1}{(3!)^2} + u\left(\frac{1}{(4!)^2} + u\left(\frac{1}{(5!)^2} + \frac{u}{(6!)^2}\right)\right)\right)\right)\right)$$

$$(6.4.2)$$

The approximation (6.4.2) is easily programmed. An HP-15C program looks like this:

| | |
|---|---|
| f LBL A | RCL × 0 |
| ENTER | .25 + |
| 2 ÷ gx² CHS STO 0 | RCL × 0 |
| 6 fx! gx² ÷ | 1 + |
| 5 fx! gx² 1/x + | RCL × 0 |
| RCL × 0 | 1 + |
| 4 fx! gx² 1/x + | g RTN |
| RCL × 0 | g P/R |
| 3 fx! gx² 1/x + | |

Calculations with this program take about 5 seconds per value. For $x=4$ the program gives $J_0(4) = -0.39654$ while the actual value is $-0.39715$ (see Table 6.3 at the end of this section), so the approximate value is in error by about 0.0006 as predicted. At $x=3$ our approximation gives $-0.260040855$ while the exact value is $-0.260051955$, an accuracy of 0.004%.

The power series (6.4.1) becomes awkward to use for large $x$, so we need another representation if we are to secure information about $J_0(x)$, and its zeros, for large values of the argument $x$. The following formula:

$$J_0(x) = \frac{1}{\pi}\int_{-1}^{1}\frac{e^{ixt}}{\sqrt{1-t^2}}\,dt \tag{6.4.3}$$

is useful here. This formula was derived in Exercise 6 of Section 6.3. That derivation went as follows: substitute $u = x \sin \theta$ in the power series for $\cos u$, integrate term by term to get $J_0(x) = (1/\pi) \int_0^\pi \cos(x \sin \theta) \, d\theta$, then make a few manipulations to derive (6.4.3).

In formula (6.4.3) the contour of integration is the real axis between $-1$ and $+1$. We deform that contour into the complex plane, integrating first from $-1$ to $-1 + iR$ ($R > 0$), then along the horizontal line from $-1 + iR$, to $+1 + iR$, then vertically down from $+1 + iR$ to $+1$. (See Figure 6.3.) As we are deforming through a region of analyticity of the integrand, which has its singularities at $t = \pm 1$, the value of the integral does not change. We have $J_0 = I + II + III$, where

$$I = \frac{1}{\pi} \int_{-1}^{-1+iR} \frac{e^{ixt}}{\sqrt{1 - t^2}} \, dt, \qquad II = \frac{1}{\pi} \int_{-1+iR}^{1+iR} \frac{e^{ixt}}{\sqrt{1 - t^2}} \, dt,$$

$$III = \frac{1}{\pi} \int_{1+iR}^{1} \frac{e^{ixt}}{\sqrt{1 - t^2}} \, dt$$

We shall deal with integrals $I$ and $III$ first. In integral $I$ we write $t = -1 + is$, $0 \leqslant s \leqslant R$, $dt = i \, ds$, $1 - t^2 = s^2 + 2is = s(s + 2i)$, and get

$$I = \frac{i}{\pi} \int_0^R \frac{e^{-sx} e^{-ix}}{\sqrt{s(s + 2i)}} \, ds$$

Operating similarly with integral $III$, we get

$$III = \frac{-i}{\pi} \int_0^R \frac{e^{-sx} e^{ix}}{\sqrt{s(s - 2i)}} \, ds$$

then adding

$$I + III = \frac{i}{\pi} \int_0^R \left( \frac{e^{-ix}}{\sqrt{s + 2i}} - \frac{e^{ix}}{\sqrt{s - 2i}} \right) \frac{e^{-sx}}{\sqrt{s}} \, ds$$

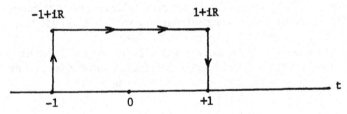

**Figure 6.3**  Integration contour for equation (6.4.3).

In integral $II$ we make the substitution $t = s + iR$, $-1 \leqslant s \leqslant 1$, $dt = ds$ to get

$$II = \frac{e^{-xR}}{\pi} \int_{-1}^{+1} \frac{e^{isx}\, ds}{\sqrt{1 - s^2 + R^2 - 2isR}}$$

This integral will be in absolute value less than the maximum of the absolute value of the integrand on the path of integration times the length of path. Using $|e^{ixs}| = 1$ and $|1 - s^2 + R^2 - 2isR| \geqslant R^2$, we get

$$|II| \leqslant \frac{2}{\pi} \frac{e^{-xR}}{R}$$

which tends rapidly to 0 as $R \to \infty$ (remembering $x > 0$). Letting $R \to \infty$, we then get

$$J_0(x) = \frac{i}{\pi} \int_0^\infty \left( \frac{e^{-ix}}{\sqrt{s + 2i}} - \frac{e^{ix}}{\sqrt{s - 2i}} \right) \frac{e^{-sx}}{\sqrt{s}}\, ds, \qquad x > 0 \tag{6.4.4}$$

When $x$ is large (which is the region of interest for us), the exponential factor $e^{-sx}$ tends to zero very rapidly. Therefore, taking into account the behavior of the remaining part of the integrand, we see that the major contribution to the integral comes from the neighborhood of $s = 0$. We shall take advantage of this to obtain our desired approximation.

Near $s = 0$ we have

$$\sqrt{s + 2i} \cong \sqrt{2i} = \sqrt{2}\sqrt{i} = \sqrt{2}(e^{i\pi/2})^{1/2} = \sqrt{2}\, e^{i\pi/4}$$

and

$$\sqrt{s - 2i} \cong \sqrt{-2i} = \sqrt{2}\sqrt{-i} = \sqrt{2}(e^{-i\pi/2})^{1/2} = \sqrt{2}\, e^{-i\pi/4}$$

so

$$J_0(x) \cong \frac{i}{\pi\sqrt{2}} \int_0^\infty [e^{-i(x+\pi/4)} - e^{i(x+\pi/4)}] \frac{1}{\sqrt{s}} e^{-sx}\, ds$$

$$= \frac{+\sqrt{2}}{\pi} \left( \frac{e^{i(x+\pi/4)} - e^{-i(x+\pi/4)}}{2i} \right) \int_0^\infty \frac{1}{\sqrt{s}} e^{-sx}\, ds$$

$$= \frac{\sqrt{2}}{\pi} \sin\left( x + \frac{\pi}{4} \right) \int_0^\infty \frac{1}{\sqrt{s}} e^{-sx}\, ds$$

In this last integral we change variables, $t^2 = xs$, $2t\, dt = x\, ds$, so

$$\int_0^\infty \frac{1}{\sqrt{s}} e^{-sx}\, dx = \int_0^\infty \frac{\sqrt{x}}{t} e^{-t^2} \frac{2t}{x}\, dt = \frac{2}{\sqrt{x}} \int_0^\infty e^{-t^2}\, dt = \frac{2}{\sqrt{x}} \frac{\sqrt{\pi}}{2} = \frac{\sqrt{\pi}}{\sqrt{x}}$$

where we have used the common trick to evaluate the Gaussian integral:

$$\left(\int_0^\infty e^{-t^2}\,dt\right)^2 = \left(\int_0^\infty e^{-t^2}\,dt\right)\left(\int_0^\infty e^{-s^2}\,ds\right)$$

$$= \int_0^\infty \int_0^\infty e^{-(s^2+t^2)}\,ds\,dt, \qquad s = r\cos\theta,\; t = r\sin\theta$$

$$= \int_0^{\pi/4} d\theta \int_0^\infty e^{-r^2} r\,dr = \frac{\pi}{4} \qquad \text{so} \int_0^\infty e^{-t^2}\,dt = \frac{\sqrt{\pi}}{2}$$

Thus

$$J_0(x) \cong \frac{\sqrt{2}}{\pi}\sin\left(x+\frac{\pi}{4}\right)\frac{\sqrt{\pi}}{\sqrt{x}} = \sqrt{\frac{2}{\pi x}}\sin\left(x+\frac{\pi}{4}\right)$$

As $\sin(x+\pi/4) = \cos(x-\pi/4)$ we have our final result:

$$J_0(x) \cong \sqrt{\frac{2}{\pi x}}\cos\left(x-\frac{\pi}{4}\right), \qquad x \gg 0 \tag{6.4.5}$$

**Table 6.1**  Asymptotic Approximation to $J_0(x)$

| $x$ | $J_0(x)$ | $(2/\pi x)^{1/2}\cos(x-\pi/4)$ |
|------|----------|-------------------------------|
| 4.0  | −0.3971  | −0.3979 |
| 4.1  | −0.3887  | −0.3882 |
| 4.2  | −0.3766  | −0.3749 |
| 4.3  | −0.3610  | −0.3583 |
| 4.4  | −0.3423  | −0.3386 |
| 4.5  | −0.3205  | −0.3160 |
| 4.6  | −0.2961  | −0.2909 |
| 4.7  | −0.2693  | −0.2634 |
| 4.8  | −0.2404  | −0.2340 |
| 4.9  | −0.2097  | −0.2029 |
| 5.0  | −0.1776  | −0.1704 |
| 6    | 0.1506   | 0.1568 |
| 7    | 0.3001   | 0.3009 |
| 8    | 0.1717   | 0.1683 |
| 9    | −0.0903  | −0.0938 |
| 10   | −0.2459  | −0.2468 |
| 15   | −0.0142  | −0.0159 |
| 16   | −0.1749  | −0.1757 |
| 17   | −0.1699  | −0.1692 |

The formula on the right side of equation (6.4.5) is easily programmed (be sure your calculator is in radian mode). For the HP-15C:

| | |
|---|---|
| g P/R | RCL .0 |
| f LBL B | g$\pi$ 4 ÷ − |
| STO .0 | COS × |
| g$\pi$ × | g RTN |
| 1/x 2 × | g P/R |
| $\sqrt{x}$ | |

Formula (6.4.5) turns out to be reasonably accurate for $x \geqslant 4$; see Table 6.1.
    The truncated power series (6.4.2) together with the asymptotic formula (6.4.5) gives us a reasonably effective means of calculating the values of $J_0(x)$. The following HP-15C program, labeled "C", combines the two approximations, branching automatically to "A" or "B" according as $x \leqslant 4$ or $x > 4$. (Use only positive $x$ as $J_0$ is even.) So this one program will compute $J_0(x)$

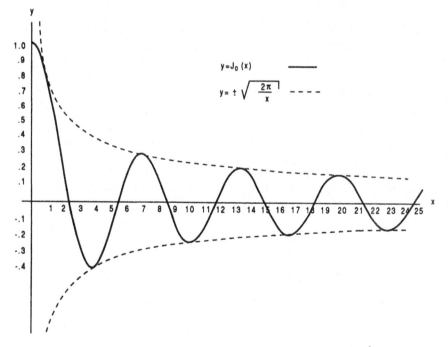

**Figure 6.4**  The Bessel function $J_0(x)$ and its envelope.

**Table 6.2**  Zeros of $J_0(x)$

| $n$ | $j_{0,n}$ | $n\pi - \pi/4$ |
|-----|-----------|----------------|
| 1   | 2.405     | 2.356          |
| 2   | 5.520     | 5.498          |
| 3   | 8.654     | 8.639          |
| 4   | 11.792    | 11.781         |
| 5   | 14.931    | 14.923         |
| 6   | 18.071    | 18.064         |
| 7   | 21.212    | 21.206         |
| 8   | 24.352    | 24.347         |
| 9   | 27.493    | 27.489         |
| 10  | 30.635    | 30.631         |

with reasonable accuracy for all $x$.

| | |
|----------|--------|
| gP/R     | g x≤y  |
| f LBL C  | GTO A  |
| ENTER    | GTO B  |
| 4        | g RTN  |
| x≤y      | g P/R  |

In Figure 6.4 I have plotted $J_0(x)$ versus $x$ for $0 \leqslant x \leqslant 25$. I also show on the same graph the envelope $(2/\pi x)^{1/2}$. The graph shows that $J_0(x)$ has infinitely many zeros. Let $j_{0,n}$ denote the $n$th zero of $J_0(x)$, $n = 1, 2, \ldots$. Calculations with the power series show that $2.4 < j_{0,1} < 2.5$. Equation (6.4.5) suggests that for large $n$, $j_{0,n}$ should equal approximately the $n$th zero of $\cos(x - \pi/4)$, so

$$j_{0,n} \cong n\pi - \frac{\pi}{4}, \qquad n = 1, 2, \ldots \tag{6.4.6}$$

In Table 6.2 we compare the first 10 zeros of $J_0(x)$ to the values given by (6.4.6). Even for $j_{0,1}$, formula (6.4.6) is accurate to 2%, and it gives the 10th zero to 0.01% (also the 20th zero to 0.003%).

Table 6.3 gives accurate values of $J_0(x)$, $0 \leqslant x \leqslant 15$ (0.1).

**Table 6.3** The Bessel Function $J_0(x)$

| $x$ | $J_0(x)$ | $x$ | $J_0(x)$ | $x$ | $J_0(x)$ |
|---|---|---|---|---|---|
| 0.0 | 1.00000 00000 | 5.0 | -0.17759 67713 | 10.0 | -0.24593 57644 |
| 0.1 | 0.99750 15620 | 5.1 | -0.14433 47470 | 10.1 | -0.24902 96505 |
| 0.2 | 0.99002 49722 | 5.2 | -0.11029 04397 | 10.2 | -0.24961 70698 |
| 0.3 | 0.97762 62465 | 5.3 | -0.07580 31115 | 10.3 | -0.24771 68134 |
| 0.4 | 0.96039 82266 | 5.4 | -0.04121 01012 | 10.4 | -0.24337 17507 |
| 0.5 | 0.93846 98072 | 5.5 | -0.00684 38694 | 10.5 | -0.23664 81944 |
| 0.6 | 0.91200 48634 | 5.6 | +0.02697 08846 | 10.6 | -0.22763 50476 |
| 0.7 | 0.88120 08886 | 5.7 | 0.05992 00097 | 10.7 | -0.21644 27399 |
| 0.8 | 0.84628 73527 | 5.8 | 0.09170 25675 | 10.8 | -0.20320 19671 |
| 0.9 | 0.80752 37981 | 5.9 | 0.12203 33545 | 10.9 | -0.18806 22459 |
| 1.0 | 0.76519 76865 | 6.0 | 0.15064 52572 | 11.0 | -0.17119 03004 |
| 1.1 | 0.71962 20185 | 6.1 | 0.17729 14222 | 11.1 | -0.15276 82954 |
| 1.2 | 0.67113 27442 | 6.2 | 0.20174 72229 | 11.2 | -0.13299 19368 |
| 1.3 | 0.62008 59895 | 6.3 | 0.22381 20061 | 11.3 | -0.11206 84561 |
| 1.4 | 0.56685 51203 | 6.4 | 0.24331 06048 | 11.4 | -0.09021 45002 |
| 1.5 | 0.51182 76717 | 6.5 | 0.26009 46055 | 11.5 | -0.06765 39481 |
| 1.6 | 0.45540 21676 | 6.6 | 0.27404 33606 | 11.6 | -0.04461 56740 |
| 1.7 | 0.39798 48594 | 6.7 | 0.28506 47377 | 11.7 | -0.02133 12813 |
| 1.8 | 0.33998 64110 | 6.8 | 0.29309 56031 | 11.8 | +0.00196 71733 |
| 1.9 | 0.28181 85593 | 6.9 | 0.29810 20354 | 11.9 | 0.02504 94416 |
| 2.0 | 0.22389 07791 | 7.0 | 0.30007 92705 | 12.0 | 0.04768 93107 |
| 2.1 | 0.16660 69803 | 7.1 | 0.29905 13805 | 12.1 | 0.06966 67736 |
| 2.2 | 0.11036 22669 | 7.2 | 0.29507 06914 | 12.2 | 0.09077 01231 |
| 2.3 | 0.05553 97844 | 7.3 | 0.28821 69476 | 12.3 | 0.11079 79503 |
| 2.4 | +0.00250 76832 | 7.4 | 0.27859 62326 | 12.4 | 0.12956 10265 |

**Table 6.3** The Bessel Function $J_0(x)$

| x | $J_0(x)$ | x | $J_0(x)$ | x | $J_0(x)$ |
|---|---|---|---|---|---|
| 2.5 | −0.04838 37764 | 7.5 | 0.26633 96578 | 12.5 | 0.14688 40547 |
| 2.6 | −0.09680 49543 | 7.6 | 0.25160 18338 | 12.6 | 0.16260 72717 |
| 2.7 | −0.14244 93700 | 7.7 | 0.23455 91395 | 12.7 | 0.17658 78885 |
| 2.8 | −0.18503 60333 | 7.8 | 0.21540 78077 | 12.8 | 0.18870 13547 |
| 2.9 | −0.22431 15457 | 7.9 | 0.19436 18448 | 12.9 | 0.19884 24371 |
| 3.0 | −0.26005 19549 | 8.0 | 0.17165 08071 | 13.0 | 0.20692 61023 |
| 3.1 | −0.29206 43476 | 8.1 | 0.14751 74540 | 13.1 | 0.21288 81975 |
| 3.2 | −0.32018 81696 | 8.2 | 0.12221 53017 | 13.2 | 0.21668 59222 |
| 3.3 | −0.34429 62603 | 8.3 | 0.09600 61008 | 13.3 | 0.21829 80903 |
| 3.4 | −0.36429 55967 | 8.4 | 0.06915 72616 | 13.4 | 0.21772 51787 |
| 3.5 | −0.38012 77399 | 8.5 | 0.04193 92518 | 13.5 | 0.21498 91658 |
| 3.6 | −0.39176 89837 | 8.6 | +0.01462 29912 | 13.6 | 0.21013 31613 |
| 3.7 | −0.39923 02033 | 8.7 | −0.01252 27324 | 13.7 | 0.20322 08326 |
| 3.8 | −0.40255 64101 | 8.8 | −0.03923 38031 | 13.8 | 0.19433 56352 |
| 3.9 | −0.40182 60148 | 8.9 | −0.06525 32468 | 13.9 | 0.18357 98554 |
| 4.0 | −0.39714 98098 | 9.0 | −0.09033 36111 | 14.0 | 0.17107 34761 |
| 4.1 | −0.38866 96798 | 9.1 | −0.11423 92326 | 14.1 | 0.15695 28770 |
| 4.2 | −0.37655 70543 | 9.2 | −0.13674 83707 | 14.2 | 0.14136 93846 |
| 4.3 | −0.36101 11172 | 9.3 | −0.15765 51899 | 14.3 | 0.12448 76852 |
| 4.4 | −0.34225 67900 | 9.4 | −0.17677 15727 | 14.4 | 0.10648 41184 |
| 4.5 | −0.32054 25089 | 9.5 | −0.19392 87476 | 14.5 | 0.08754 48680 |
| 4.6 | −0.29613 78165 | 9.6 | −0.20897 87183 | 14.6 | 0.06786 40683 |
| 4.7 | −0.26933 07894 | 9.7 | −0.22179 54820 | 14.7 | 0.04764 18459 |
| 4.8 | −0.24042 53272 | 9.8 | −0.23227 60275 | 14.8 | 0.02708 23145 |
| 4.9 | −0.20973 83275 | 9.9 | −0.24034 11055 | 14.9 | +0.00639 15448 |
| 5.0 | −0.17759 67713 | 10.0 | −0.24593 57644 | 15.0 | −0.01422 44728 |

*Source:* M. Abramowitz and I. A. Stegun (Editors), *Handbook of Mathematical Functions*, Applied Math. Series 55, N.B.S., 1965.

## Exercises

1. Let $z = e^{-ix}/\sqrt{s+2i}$. Show that formula (6.4.4) can be written

$$J_0(x) = \frac{1}{\pi} \int_0^\infty i(z - \bar{z}) \frac{e^{-sx}}{\sqrt{s}} ds = \frac{1}{\pi} \int_0^\infty [-2Im(z)] \frac{e^{-sx}}{\sqrt{s}} ds$$

2. (Continuation) Show that

$$z = \frac{1}{(s^2 + 4)^{1/4}} e^{-i(x+\theta/2)} \qquad \text{where } \tan\theta = \frac{2}{s}$$

and thus show that

$$J_0(x) = \frac{2}{\pi} \int_0^\infty \frac{\sin(x+\theta/2)}{\sqrt{s}(s^2+4)^{1/4}} e^{-sx} ds$$

3. (Continuation) Derive the identity $\arctan(2/s) + \arctan(s/2) = \pi/2$, $s \geqslant 0$.

4. Let $\phi = \arctan(s/2)$. Show that

$$\sin\left(x + \frac{\theta}{2}\right) = \sin\left(x + \frac{\pi}{4}\right)\cos\frac{\phi}{2} - \cos\left(x + \frac{\pi}{4}\right)\sin\frac{\phi}{2}$$

Show that

$$\cos\frac{\phi}{2} = \frac{\sqrt{u+2}}{\sqrt{2u}}, \qquad \sin\frac{\phi}{2} = \frac{\sqrt{u-2}}{\sqrt{2u}} \qquad \text{where } u = \sqrt{s^2+4}$$

5. (Continuation) Show that

$$J_0(x) = \frac{\sqrt{2}}{\pi} \cos\left(x - \frac{\pi}{4}\right) \int_0^\infty \left[\frac{(s^2+4)^{1/2}+2}{s^2+4}\right]^{1/2} \frac{1}{\sqrt{s}} e^{-xs} ds$$

$$- \frac{\sqrt{2}}{\pi} \cos\left(x + \frac{\pi}{4}\right) \int_0^\infty \left[\frac{(s^2+4)^{1/2}-2}{s^2+4}\right]^{1/2} \frac{1}{\sqrt{s}} e^{-xs} ds$$

By replacing the terms in square brackets by their values at $s = 0$, rederive formula (6.4.5) of the text. Discuss how you might improve on the accuracy of formula (6.4.5).

## 6.5 THE EIGENVALUES AND EIGENFUNCTIONS OF BESSEL'S OPERATOR OF ORDER ZERO

We are in a position now to determine the eigenvalues and eigenfunctions of Bessel's operator $B(y) = -y'' - (1/r)y' = -(1/r)(ry')'$ whose domain consists of compatible functions $y$ in $L^2(0, a; r)$ such that $y(0)$ is finite and $y(a) = 0$.

Any eigenfunction of $B$ must satisfy the differential equation $ry'' + y' + \lambda r y = 0$, which becomes Bessel's equation $xy'' + y' + xy = 0$ after the substitution $x = \sqrt{\lambda} r$. The general solution of Bessel's equation is $c_0 J_0(x) + c_1 Y_0(x)$ where $J_0$ and $Y_0$ are the Bessel functions of order zero of the first and second kind, respectively. Replacing $x$ by $\sqrt{\lambda} r$, we find that the general solution to $ry'' + y' + \lambda r y = 0$ (which is equivalent to the equation $b(y) = \lambda y$) is $c_0 J_0(\sqrt{\lambda} r) + c_1 Y_0(\sqrt{\lambda} r)$. And the question before us now is: *for what values of the constants $c_0$, $c_1$ and for what values of the parameter $\lambda$ does the function $f(r) = c_0 J_0(\sqrt{\lambda} r) + c_1 Y_0(\sqrt{\lambda} r)$ lie in the domain of $B$?* To get into the domain of $B$, the function $f(r)$ must pass through various gates. We list them one by one.

$f(0)$ *is finite*: This forces $c_1 = 0$ as $Y_0(\sqrt{\lambda} r)$ has a logarithmic singularity at $r = 0$. So $f(r) = c_0 J_0(\sqrt{\lambda} r)$.

*Both $f(r)$ and $b(f(r))$ must belong to $L^2(0, a; r)$*: As $f(r)$ satisfies the equation $b(f) = \lambda f$, it is enough to show that $f$ belongs to our $L^2$ space. But from the graph of $J_0$ it is clear that $|J_0(x)| \leqslant 1$ for all $x$, so $|f(r)| \leqslant |c_0|$, whence

$$\int_0^a |f(r)|^2 r \, dr \leqslant |c_0|^2 \int_0^a r \, dr = \frac{|c_0|^2 a^2}{2} < \infty$$

Thus $f(r)$ belongs to $L^2(0, a; r)$ and is compatible with $b$.

$f(a) = 0$: We must have $c_0 \neq 0$ lest $f$ equal the trivial zero solution, so $f(a) = 0$ means $J_0(\sqrt{\lambda} a) = 0$, which requires

$$\sqrt{\lambda} a = j_{0,n}$$

for some zero $j_{0,n}$, $n = 1, 2, \ldots$, of $J_0$. Thus

$$\lambda_n = \frac{j_{0,n}^2}{a^2}, \qquad n = 1, 2, \ldots \tag{6.5.1}$$

are eigenvalues of Bessel's operator. To each of these eigenvalues belongs the eigenfunction

$$\phi_n(r) = J_0\left(j_{0,n} \frac{r}{a}\right) \tag{6.5.2}$$

where we have used the convenient normalization $c_0 = 1$. Thus we have

$$B(\phi_n) = \lambda_n \phi_n \quad \text{or} \quad B\left(J_0\left(j_{0,n} \frac{r}{a}\right)\right) = \frac{j_{0,n}^2}{a^2} J_0\left(j_{0,n} \frac{r}{a}\right) \tag{6.5.2'}$$

$n = 1, 2, \ldots$, where $B$ is Bessel's operator.

We summarize the results proved so far:

**Theorem 6.6.** *Bessel's operator (refer to Definition 6.1) has eigenvalues*

$$\lambda_n = \frac{j_{0,n}^2}{a^2}, \qquad n = 1, 2, \ldots$$

*where $j_{0,n}$ is the nth zero of $J_0(x)$, the Bessel function of the first kind of order zero. To each of these eigenvalues belongs the eigenfunction*

$$\phi_n(r) = J_0\left(j_{0,n}\frac{r}{a}\right), \qquad n = 1, 2, \ldots$$

*These eigenfunctions form an orthogonal family in $L^2(0, a; r)$; i.e.,*

$$\langle \phi_m(r), \phi_n(r) \rangle = \int_0^a \phi_m(r)\phi_n(r)r\,dr = \int_0^a J_0\left(j_{0,m}\frac{r}{a}\right) J_0\left(j_{0,n}\frac{r}{a}\right) r\,dr = 0$$

*when $m \neq n$.*

The orthogonality comes from the fact that the $\phi_n$ are eigenfunctions of a Hermitian operator (Exercise 4c of Section 6.2) and the fact that, for a Hermitian operator, eigenfunctions belonging to different eigenvalues are orthogonal (Exercise 5b of Section 6.2).

If we make the substitution $x = r/a$, then the orthogonality relation becomes

$$\int_0^1 J_0(j_{0,m}x)J_0(j_{0,n}x)x\,dx = 0 \qquad \text{when } m \neq n$$

a fact that is difficult to verify directly. But in our approach it is an immediate consequence of the principle that, for a Hermitian operator in Hilbert space, eigenfunctions belonging to different eigenvalues are orthogonal.

Next, we shall determine the norm (squared) of the eigenfunctions $\phi_n$. This means evaluating the integrals

$$\|\phi_n(r)\|^2 = \langle \phi_n(r), \phi_n(r) \rangle = \int_0^a J_0(j_{0,n}r/a)^2 r\,dr$$

Make the substitution $x = j_{0,n}r/a$ to get

$$\|\phi_n\|^2 = \left(\frac{a}{j_{0,n}}\right)^2 \int_0^{j_{0,n}} J_0(x)^2 x\,dx$$

In Exercise 9 of Section 6.2 we have shown that, for any solution $y$ of Bessel's equation $xy'' + y' + xy = 0$, we have

$$2\int xy^2\,dx = x^2(y')^2 + x^2y^2 + C \tag{6.5.3}$$

The derivation of this result goes as follows. Integrate by parts with $u = y^2$, $dv = x\,dx$ to get

$$2 \int xy^2\,dx = x^2 y^2 - 2 \int x^2 yy'\,dx \qquad (6.5.4)$$

Using the fact $xy'' + y' = (xy')'$, Bessel's equation can be written $xy = -(xy')'$. Multiplying both sides by $xy'$ we get $x^2 yy' = -(xy')(xy')'$. Thus

$$\int x^2 yy'\,dx = -\int (xy')(xy')'\,dx = -\int (xy')\,d(xy') = -\frac{(xy')^2}{2} + C$$

Substituting this into (6.5.4), we get (6.5.3).

As $y = J_0(x)$ satisfies Bessel's equation, we can use (6.5.3) to evaluate our integral:

$$\int_0^{j_{0,n}} x J_0(x)^2\,dx = \frac{1}{2}[x^2(J_0'(x))^2 + x^2(J_0(x))^2]_0^{j_{0,n}}$$

$$= \frac{1}{2}(j_{0,n})^2 [J_0'(j_{0,n})]^2$$

where we have used the fact that $J_0(j_{0,n}) = 0$. Thus

$$\|\phi_n\|^2 = \left(\frac{a}{j_{0,n}}\right)^2 \frac{1}{2}(j_{0,n})^2 [J_0'(j_{0,n})]^2 = \frac{a^2}{2}(J_0'(j_{0,n}))^2$$

Hence, to know these norms, one needs to know the value of the derivative of $J_0$ at its zeros. These are given in Table 6.4. You will need these values to do Exercises 1 and 2.

**Theorem 6.7.** *The eigenfunctions* $\phi_n(r) = J_0(j_{0,n}r/a)$, $n = 1, 2, \ldots$, *form an orthogonal basis of* $L^2(0, a; r)$.

We shall not prove this result in this book. The book *Completeness and Basis Properties of Sets of Special Functions* by J. R. Higgins (Cambridge University Press, 1977) contains a proof of this theorem as well as the other orthogonal basis theorems that we have stated.

What does Theorem 6.7 mean? The general theory of orthogonal bases in inner product spaces was covered in Section 3.8 of Chapter 3. Let us review this material as it applies to the functions $\phi_n(r) = J_0(j_{0,n}r/a)$ in the Hilbert space $L^2(0, a; r)$ and in that way explain in detail the meaning of Theorem 6.7.

Theorem 6.7 makes the following four assertions.

First, given any function $f(r)$ in $L^2(0, a; r)$, we have

$$f(r) = \sum_{n=1}^{\infty} c_n \phi_n(r) = \sum_{n=1}^{\infty} c_n J_0\left(j_{0,n}\frac{r}{a}\right) \qquad (6.5.5)$$

**Table 6.4** Zeros of $J_0(x)$ and Associated Values of $J_0'(x)$

| $n$ | $j_{0,n}$ | $J_0'(j_{0,n})$ |
|---|---|---|
| 1 | 2.40483 | −0.519147 |
| 2 | 5.52008 | +0.340265 |
| 3 | 8.65373 | −0.271452 |
| 4 | 11.79153 | +0.232460 |
| 5 | 14.93092 | −0.206546 |
| 6 | 18.07106 | +0.187729 |
| 7 | 21.21164 | −0.173266 |
| 8 | 24.35247 | +0.161702 |
| 9 | 27.49348 | −0.152181 |
| 10 | 30.63461 | +0.144166 |

*Source*: M. Abramowitz and I. A. Stegun (Editors), *Handbook of Mathematical Functions*, Applied Math. Series 66, N.B.S., 1966.

where the $c_n$ are the "Fourier-Bessel" coefficients of $f$ with respect to the $\phi_n$:

$$c_n = \frac{\langle f, \phi_n \rangle}{\langle \phi_n, \phi_n \rangle} = \frac{\int_0^a f(r) J_0(j_{0,n} r/a) r \, dr}{\int_0^a (J_0(j_{0,n} r/a))^2 r \, dr}$$

$$= \frac{2}{a^2 (J_0'(j_{0,n}))^2} \int_0^a f(r) J_0 \left( j_{0,n} \frac{r}{a} \right) r \, dr \qquad (6.5.6)$$

The series on the right of (6.5.5) converges to $f(r)$ in the mean-square metric:

$$\left\| f(r) - \sum_{n=1}^N c_n \phi_n(r) \right\|^2 = \int_0^a \left| f(r) - \sum_{n=1}^N c_n \phi_n(r) \right|^2 r \, dr \to 0 \qquad \text{as } N \to \infty$$

$$(6.5.7)$$

The coefficients $c_n$ are uniquely determined. That is, if we have another expansion $f(r) = \Sigma a_n \phi_n(r)$ where the series converges to $f$ in the mean-square metric, then $a_n = c_n$, $n = 1, 2, \ldots$. The expansion (6.5.5) is called the *Fourier-Bessel series* of $f(r)$.

Second, if we cut off the series (6.5.5) at some point, the resulting partial sum makes the mean-square error

$$\left\| f(r) - \sum_{n=1}^N c_n \phi_n(r) \right\| = \left[ \int_0^a \left| f(r) - \sum_{n=1}^N c_n \phi_n(r) \right|^2 r \, dr \right]^{1/2}$$

an absolute minimum. With the $c_n$ given by equation (6.5.6), the partial sum $\sum_{n=1}^{N} c_n\phi_n(r)$ is the best possible mean-square approximation to $f(x)$. Any alteration of these coefficients will only worsen the approximation.

Third, the expansion (6.5.5) is valid in the mean-square sense for *every* $f(r)$ in $L^2(0, a; r)$. The *only* requirement on a function $f(r)$ that it have a Fourier-Bessel series (6.5.5) convergent to it in the mean-square metric is that

$$\int_0^a |f(r)|^2 r\, dr < \infty \tag{6.5.8}$$

Note that all the $\phi_n(r)$ vanish at $r = a$: $\phi_n(a) = 0$, $n = 1, 2, \ldots$, as does, of course, every function in the domain of Bessel's operator $B$. So in the series $f(r) = \sum c_n\phi_n(r)$, if we put $r = a$, the right side is zero. But $f(a)$ need not be zero; the only requirement on $f$ is the square-integrability requirement (6.5.8). So when we substitute $r = a$ into the equality $f(r) = \sum c_n\phi_n(r)$ we seem to get an incorrect result with 0 on the right side and a nonzero value on the left (generally). What is the explanation of this paradox? The explanation lies in the meaning attached to the equality (6.5.5). When we write $f(r) = \sum_{n=1}^{\infty} c_n\phi_n(r)$ we mean that the series on the right converges to $f(r)$ in the mean-square metric, which is exactly the meaning of the assertion (6.5.7). The series need not converge pointwise at each individual value of $r$; the mean-square metric is a measure of the area between the graphs. So functions can differ at finitely many points and yet have the mean-square distance between them zero.

The fourth and final consequence of Theorem 6.7 is this: we may take the inner product of both sides of the Fourier-Bessel series (6.5.5) with any function $g$ in $L^2(0, a; r)$ and get a valid numerical series

$$\langle f, g \rangle = \sum_{n=1}^{\infty} c_n \langle \phi_n, g \rangle$$

In particular, we may put $f = g$ and use the fact that $\langle \phi_n, f \rangle = \overline{\langle f, \phi_n \rangle} = \overline{c_n} \langle \phi_n, \phi_n \rangle$ to get the *Parseval relation for the Fourier-Bessel series*

$$\langle f, f \rangle = \sum_{n=1}^{\infty} |c_n|^2 \langle \phi_n, \phi_n \rangle = \frac{a^2}{2} \sum_{n=1}^{\infty} (J_0'(j_{0,n}))^2 |c_n|^2$$

or

$$\int_0^a |f(r)|^2 r\, dr = \frac{2}{a^2} \sum_{n=1}^{\infty} \frac{2}{(J_0'(j_{0,n}))^2} \left| \int_0^a f(r) J_0\left(j_{0,n}\frac{r}{a}\right) r\, dr \right|^2 \tag{6.5.9}$$

Here are some suggestions for doing practical calculations with the Fourier-Bessel expansion (6.5.5). In the exercises we will use four-term approximations, which will therefore take the form

$$f(x) \cong c_1 J_0(\mu_1 x) + c_2 J_0(\mu_2 x) + c_3 J_0(\mu_3 x) + c_4 J_0(\mu_4 x) \tag{6.5.10}$$

where the constants $c_i$, $\mu_i$ are known. If you use the HP-15C and have inserted into your machine the program labeled "C" given in Section 6.4 to calculate the values of $J_0(x)$, then the following program will calculate the right side of (6.5.10). Note that I avoid using the memory registers 0 and .0 because they are used in the subroutines A and B.

| | | |
|---|---|---|
| f LBL D | | |
| STO .5 | STO+5 | RCL×.4 |
| RCL×1 | RCL .5 | STO+5 |
| GSB C | RCL×3 | RCL 5 |
| RCL×.1 | GSB C | g RTN |
| STO 5 | RCL×.3 | g P/R |
| RCL .5 | STO+5 | |
| RCL×2 | RCL .5 | |
| GSB C | RCL×4 | |
| RCL×.2 | GSB C | |

Put the coefficients $c_i$, $\mu_i$ in the following storage registers:

$$\mu_1 \to \text{Reg 1} \qquad \mu_3 \to \text{Reg 3}$$
$$c_1 \to \text{Reg .1} \qquad c_3 \to \text{Reg .3}$$
$$\mu_2 \to \text{Reg 2} \qquad \mu_4 \to \text{Reg 4}$$
$$c_2 \to \text{Reg .2} \qquad c_4 \to \text{Reg .4}$$

The program "D" is a little slow, mainly because of the time involved in the subroutine "A" to calculate $J_0(x)$.

**Theorem 6.8.** *If the function $f(r)$ belongs to $L^2(0, a; r)$ and is orthogonal to every $\phi_n(r) = J_0(j_{0,n}r/a)$, $n = 1, 2, \ldots$, then $f = 0$.*

The proof uses the fact that the $\phi_n$ are an orthogonal basis of $L^2(0, a; r)$. You are asked to give the proof in Exercise 4.

**Theorem 6.9.** *Bessel's operator has no eigenvalues other than $\lambda_n = j_{0,n}{}^2/a^2$, $n = 1, 2, \ldots$.*

An easy proof can be given based on Theorem 6.8 and the principle that, for a Hermitian operator, eigenfunctions belonging to different eigenvalues are orthogonal. You are asked to do this in Exercise 5.

**Theorem 6.10.** *Each of the eigenvalues of Bessel's operator is simple.*

Assigned as Exercise 6.

**Theorem 6.11.** *Suppose $f(r)$ lies in the Hilbert space $L^2(0, a; r)$, so has a Fourier-Bessel expansion*

$$f = \sum_{n=1}^{\infty} c_n \phi_n \qquad \text{where } c_n = \frac{\langle f, \phi_n \rangle}{\langle \phi_n, \phi_n \rangle} \tag{6.5.11}$$

*If $f(r)$ also belongs to the domain of Bessel's operator $B$, then the equation obtained by applying $B$ term by term to (6.5.11) is valid:*

$$B(f) = \frac{1}{a^2} \sum_{n=1}^{\infty} j_{0,n}^2 c_n \phi_n \tag{6.5.12}$$

The proof is assigned as Exercise 7. The left side of (6.5.12) may be computed using the differential expression for $f$: $B(f) = -(1/r)(rf')'$.

#### Exercises

1. (a) Use the formulas (Exercises 7 and 10, Section 6.2)

$$\int xy \, dx = -xy' \qquad \text{and} \qquad \int x^3 y \, dx = (4x - x^3)y' + 2x^2 y$$

valid for any solution $y$ of Bessel's equation $xy'' + y' + xy = 0$, to derive the formulas

$$\int_0^1 x J_0(j_{0,n} x) \, dx = -\frac{1}{j_{0,n}} J_0'(j_{0,n})$$

and

$$\int_0^1 x^3 J_0(j_{0,n} x) \, dx = \frac{1}{j_{0,n}^3} (4 - j_{0,n}^2) J_0'(j_{0,n})$$

(b) Using the results in (a), derive the expansion

$$1 - x^2 = 8 \sum_{n=1}^{\infty} \frac{1}{j_{0,n}^3 (-J_0'(j_{0,n}))} J_0(j_{0,n} x) \qquad \text{in } L^2(0, 1; x)$$

In a table compare the function $1 - x^2$ and its four-term Bessel approximation, $0 \leqslant x \leqslant 1$ (.1).

2. (a) Verify that the function $f(x) = 1 - x^2$ belongs to the domain of Bessel's operator with $a = 1$.

(b) By applying Theorem 6.11 to the expansion in Exercise 1b, derive the expansion

$$1 = 2 \sum_{n=1}^{\infty} \frac{1}{j_{0,n}(-J_0'(j_{0,n}))} J_0(j_{0,n}x) \quad \text{in } L^2(0,1;x)$$

Compare graphically the function $f(x) = 1$ and its four-term Bessel approximation, $0 \leqslant x \leqslant 1$. Your graph should show the familiar "wobble around" effect.

3. Check that the function $f(r) = -\ln r$ satisfies

$$-\frac{1}{r}(rf')' = 0$$

thus $f$ is an eigenfunction belonging to the eigenvalue $\lambda = 0$. From Equation (6.5.2') with $a = 1$ and $g(r) = J_0(j_{0,1}r)$ we have

$$-\frac{1}{r}(rg')' = j_{0,1}^2 g$$

thus $g$ is an eigenfunction belonging to the eigenvalue $\lambda_1 = j_{0,1}^2 = (2.405)^2$. Since these eigenvalues are different, the corresponding eigenfunctions must be orthogonal, i.e.,

$$\langle f, g \rangle = \int_0^1 (-\ln r)J_0(j_{0,1}r)r \, dr = 0$$

But $-\ln r > 0$ for $0 < r < 1$, and $J(j_{0,1}r) > 0$ for $0 < r < 1$ (check the graph). Hence this integral cannot be 0. What is wrong?

4. Prove Theorem 6.8.
5. Prove Theorem 6.9.
6. Prove Theorem 6.10.
7. Prove Theorem 6.11.
8. What is the kernel of Bessel's operator? Explain.
9. Suppose $h$ is a given function in $L^2(0, a; r)$ and thus has a known Fourier-Bessel expansion $h = \sum_{n=1}^{\infty} c_n \phi_n$, $c_n = \langle h, \phi_n \rangle / \langle \phi_n, \phi_n \rangle$. It is desired to find a function $y$ in the domain of Bessel's operator $B$ such that $B(y) = h$. Let $y = \sum a_n \phi_n$ where the $a_n$ are to be determined. Can you express the $a_n$ in terms of the $c_n$ and thus solve the equation $B(y) = h$? Explain all your steps.

## 6.6 THE VIBRATING DRUMHEAD, THE HEATED DISK, AND THE QUANTUM PARTICLE CONFINED TO A CIRCULAR REGION (THE θ-INDEPENDENT CASE)

We shall apply the previously developed theory to analyze three physical problems: (1) the vibrating drumhead, (2) the heated circular disk, and (3) the quantum particle confined to a circular region $D$ in which the potential $V$ is

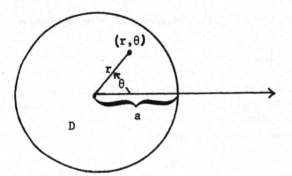

**Figure 6.5** Schematic depiction of the underlying geometry of our three physical problems.

zero. We depict all three situations schematically by the same diagram shown in Figure 6.5.

The wave equation, the heat equation, and Schrödinger's equations with $V = 0$ are

Wave equation $\qquad\qquad \Delta u + w^{-2} \dfrac{\partial^2 u}{\partial t^2} = 0$

Heat equation $\qquad\qquad \Delta u + \alpha^{-2} \dfrac{\partial u}{\partial t} = 0$

Schrödinger's equations $\qquad \begin{cases} \left(\dfrac{\hbar^2}{2m}\right) \Delta \Psi = E \Psi & \text{time independent} \\[3mm] \left(\dfrac{\hbar^2}{2m}\right) \Delta \Psi = i\hbar \dfrac{\partial \Psi}{\partial t} & \text{time dependent} \end{cases}$

In the wave equation, $u = u(r, \theta, t)$ stands for the vertical displacement of the drumhead at position $(r, \theta)$ and time $t$. In the heat equation, $u$ stands for temperature. In Schrödinger's equations, the function $\Psi$ has itself no direct physical interpretation, but $|\Psi|^2 = \Psi\bar{\Psi}$ is the probability density for the location of the quantum particle.

In all four equations, we shall adopt the boundary condition that $u$ (*resp.* $\Psi$) equals zero at the boundary $r = a$ of the disk (see Figure 6.5). For the wave equation, the interpretation of this vanishing boundary condition is that the membrane is clamped at the boundary. For the heat equation, the interpretation is that the edge of the disk is maintained at zero temperature. For Schrödinger's equations, we are assuming that the particle is constrained to the disk of radius $r = a$. Thus $\Psi = 0$ outside this disk. The boundary condition

$\Psi = 0$ *on* the boundary ensures that $\Psi$ will be continuous, a reasonable condition given the meaning of $|\Psi|^2$ as a probability density.

We shall also assume that all quantities are independent of the angle $\theta$. In this case the Laplacian becomes $-\Delta = \partial^2/\partial r^2 + (1/r)\,\partial/\partial r$ (Section 6.1). We apply a method that has served us well in the past, namely separation of variables. For the wave equation and heat equation we write $u(r,t) = R(r)T(t)$, where $R$ depends only on the radial coordinate $r$ and $T$ depends only on the time $t$. For Schrödinger's time-dependent equation, we write also $\Psi(r,t) = R(r)T(t)$. For Schrödinger's time-independent equation, no separation is necessary, as the time does not appear.

Making the indicated substitutions, repeating the usual arguments, and including the boundary condition, we get the following separated equations, the time-dependent cases involving a constant $\lambda$:

Wave equation
$$-\left(R'' + \frac{1}{r}R'\right) = \lambda R \qquad T'' = -\lambda w^2 T \qquad (6.6.1)$$
$$R(a) = 0$$

Heat equation
$$-\left(R'' + \frac{1}{r}R'\right) = \lambda R \qquad T' = -\lambda \alpha^2 T \qquad (6.6.2)$$
$$R(a) = 0$$

Schrödinger's time independent
$$-\left(\Psi'' + \frac{1}{r}\Psi'\right) = \frac{2m}{\hbar^2}E\Psi \qquad (6.6.3)$$
$$\Psi(a) = 0$$

Schrödinger's time dependent
$$-\left(R'' + \frac{1}{r}R'\right) = \frac{2m}{\hbar^2}\lambda R \qquad T' = -\left(\frac{\lambda i}{\hbar}\right)T$$
$$R(a) = 0 \qquad (6.6.4)$$

The Schrödinger equations include the square-integrability condition $\int_S |\Psi|^2\,dA < \infty$, where $D$ represents the disk of radius $a$, $D = \{(r,\theta)\colon 0 \leqslant r \leqslant a, -\pi \leqslant \theta \leqslant \pi\}$, and $dA$ is the differential element of area. In polar coordinates, $dA = r\,dr\,d\theta$. As $\Psi$ is independent of $\theta$, integration over $\theta$ simply multiplies the integral by $2\pi$. Hence the square-integrability condition is $\int_0^a |\Psi(r)|^2 r\,dr < \infty$ (time independent) and $\int_0^a |R(r)|^2 r\,dr < \infty$ (time dependent). This condition reflects the physical interpretation of $|\Psi|^2$ as a probability density. Because the particle must be *somewhere*, we must be able to secure $2\pi\int_0^a |\Psi|^2 r\,dr = 1$. We can achieve this normalization by multiplying a nonzero $\Psi$ by a constant when, and only when, the integral $\int_0^a |\Psi|^2 r\,dr$ is finite. Thus Hilbert space is intrinsic to quantum mechanics; it is a consequence of the physical interpretation of $|\Psi|^2$ as a probability density.

The square-integrability condition is not intrinsic to the wave equation and heat equation problems. However, if we add that condition, namely that $\int_0^a |R(r)|^2 r\, dr < \infty$, we do not restrict the physical applications in any way whatever, and we gain the considerable advantage that we bring to bear the entire machinery of Hilbert space theory.

There is the additional advantage that all three problems, namely wave propagation, heat transfer, and quantum mechanics, are then treated by one uniform method, the Hilbert space method.

Look again at the four spatial equations written directly above. All four represent exactly the same Hilbert space operator, namely Bessel's operator $B$ acting in the Hilbert space

$$L^2(0, a; r) = \left\{ f(r): \int_0^a |f(r)|^2 r\, dr < \infty \right\}$$

that carries the inner product

$$\langle f, g \rangle = \int_0^a f(r)\overline{g(r)} r\, dr$$

Bessel's operator has the differential expression $b(f) = -(f'' + (1/r)f') = -(1/r)(rf')'$ and has domain consisting of functions $f \in L^2(0, a; r)$ that are compatible with $b$ and such that $f(0)$ is finite and $f(a) = 0$. All three problems, in their spatial component, can be interpreted as asking for those scalars $\mu$ for which the equation

$$B(f) = \mu f \tag{6.6.5}$$

has a nonzero solution $f$ in the domain of $B$.

In Section 6.5 we have done a complete spectral analysis of Bessel's operator. It is Hermitian positive definite and has eigenvalues

$$\mu_n = \frac{j_{0,n}^2}{a^2}, \qquad n = 1, 2, \ldots \tag{6.6.6}$$

where $j_{0,n}$ is the $n$th zero of the Bessel function $J_0(x)$. These are *all* the eigenvalues of $B$, and each one is simple (Theorems 6.9 and 6.10). To the $n$th eigenvalue $\mu_n$ belongs the eigenfunction

$$\phi_n(r) = J_0\left( j_{0,n} \frac{r}{a} \right) \tag{6.6.7}$$

The $\phi_n(r)$ form an orthogonal basis of $L^2(0, a; r)$ (Theorem 6.7).

Hence equation (6.6.5) has a nonzero solution $f$ in the domain of $B$ when, and only when, $\mu$ equals one of the values in equation (6.6.6). And when $\mu = \mu_n$, then the only nonzero solutions $f$ to (6.6.5) are the nonzero scalar multiples of the function $\phi_n(r)$ in (6.6.7).

This information about Bessel's operator provides us with the solution to all four of our equations. We shall deal with them one by one.

*Wave equation:* Refer back to equation (6.6.1). If we are to have a nonzero solution $R$, then the constant $\lambda$ must equal one of the eigenvalues of Bessel's operator. These eigenvalues are given in (6.6.6). The corresponding solution $R(r)$ is given by (6.6.7), $R(r) = \phi_n(r)$. Solving the equation $T'' = -\mu_n w^2 T$ for the time factor $T$, we find

$$T_n(t) = A_n \cos \frac{j_{0,n}wt}{a} + B_n \sin \frac{j_{0,n}wt}{a}$$

where $A_n$ and $B_n$ are constants. Hence, for each $n = 1, 2, \ldots$, the function $u_n(r, t) = \phi_n(r) T_n(t)$, or

$$u_n(r, t) = \left( A_n \cos \frac{j_{0,n}wt}{a} + B_n \sin \frac{j_{0,n}wt}{a} \right) J_0\left( j_{0,n} \frac{r}{a} \right)$$

solves the wave equation. Using linearity, we then take as the general $\theta$-independent solution $u(r, t)$ to the wave equation for the circular drumhead clamped at its boundary the infinite sum of all such solutions:

$$u(r, t) = \sum_{n=1}^{\infty} u_n = \sum_{n=1}^{\infty} \left( A_n \cos \frac{j_{0,n}wt}{a} + B_n \sin \frac{j_{0,n}wt}{a} \right) J_0\left( j_{0,n} \frac{r}{a} \right) \qquad (6.6.8)$$

The constants $A_n$ and $B_n$ are determined from the initial position $u(r, 0)$ and the initial velocity $(\partial u/\partial t)(r, 0)$ (Exercise 1).

*Heat equation:* Refer back to equation (6.6.2). If we are to have a nonzero solution $R$, then the constant $\lambda$ must equal one of the eigenvalues of Bessel's operator, given in (6.6.6). Then $R(r) = \phi_n(r)$. Solving the equation $T' = -\mu_n \alpha^2 T$ for the time factor $T$, we find

$$T_n(t) = A_n \exp\left( \frac{-j_{0,n}^2 \alpha^2 t}{a^2} \right)$$

where $A_n$ is a constant. Hence, for each $n = 1, 2, \ldots$ the function $u_n(r, t) = \phi_n(r) T_n(t)$, or

$$u_n(r, t) = A_n \exp\left( \frac{-j_{0,n}^2 \alpha^2 t}{a^2} \right) J_0\left( j_{0,n} \frac{r}{a} \right)$$

solves the heat equation. Using linearity, we then take as the general $\theta$-independent solution $u(r, t)$ to the wave equation for a circular disk with its edge held at zero temperature the infinite sum of all such solutions:

$$u(r, t) = \sum_{n=1}^{\infty} u_n = \sum_{n=1}^{\infty} A_n \exp\left( \frac{-j_{0,n}^2 \alpha^2 t}{a^2} \right) J_0\left( j_{0,n} \frac{r}{a} \right) \qquad (6.6.9)$$

The constants $A_n$ are determined from the initial temperature distribution $u(r, 0)$ (Exercise 4).

In the heat equation, the boundary condition of zero temperature at the edge of the disk, a condition expressed mathematically by $u(a, t) = 0$, can easily be relaxed. For example, suppose the edge of the disk was maintained at 10°C. In this case we recognize that $u = 10$ is a solution of the wave equation $\Delta u + (1/\alpha^2)\,\partial u/\partial t = 0$, the "steady-state" solution. If $u(r, t)$ is the solution of the heat equation that meets the boundary condition $u(a, t) = 10$, then $v(r, t) = u(r, t) - 10$ is a solution that satisfies the homogeneous boundary condition $v(a, t) = 0$. So we solve for $v(r, t)$ by the Hilbert space method just described; then $u = v + 10$ is the final answer.

*Schrödinger's time-independent equation:* Refer back to equation (6.6.3). If we are to have a nonzero solution $\Psi$, then the quantity $(2m/\hbar^2)E$ must equal one of the eigenvalues of Bessel's operator. These eigenvalues are given in (6.6.6). Hence we must have $(2m/\hbar^2)E = j_{0,n}{}^2/a^2$, or $E = E_n$ where

$$E_n = \frac{j_{0,n}{}^2\hbar^2}{2ma^2}, \qquad n = 1, 2, \ldots \tag{6.6.10}$$

Thus the total energy $E$ of a particle (with $V = 0$) restricted to a circular disk of radius $a$ can have only certain discrete values, namely the values $E_n$ given by equation (6.6.1). Thus the total energy is *quantized*, and we see again confirmation of Schrödinger's assertion, quoted many times earlier, that his equation "carries within itself the quantum conditions." If we take for $a$ the Bohr radius $a = 5.2917 \times 10^{-9}$ cm and for $m$ the mass of the electron, $m = 9.1086 \times 10^{-28}$ g, we find for the energy values

$$E_n = j_{0,n}{}^2(13.6)\text{ eV}, \qquad n = 1, 2, \ldots$$

Thus the first energy level for an electron of zero potential energy bound within a circular disk of radius $a = 5.2917 \times 10^{-9}$ cm (no $\theta$ dependence) is $(2.406)^2(13.6) = 79.7$ eV. Compare this with the first energy level of an electron of zero potential energy bound within an *interval* of length $L = 5.2917 \times 10^{-9}$ cm. There the quantized energy values are $n^2h^2/8mL^2$ (Section 5.4, Exercise 1), so $E_n(\text{interval})/E_n(\text{disk}) = n^2\pi^2/j_{0,n}{}^2$. Putting $n = 1$, we get $E_1(\text{interval}) = 134.3$ eV.

*Schrödinger's time-dependent equation:* Refer back to equation (6.6.4). If we are to have a nonzero solution $R$, then the quantity $(2m/\hbar^2)\lambda$ must equal one of the eigenvalues of Bessel's operator, given in (6.6.6). Hence we must have $(2m/\hbar^2)\lambda = j_{0,n}{}^2/a^2$, so for the constant $\lambda$ we must have $\lambda = \lambda_n$ where

$$\lambda_n = \frac{j_{0,n}{}^2\hbar^2}{2ma^2} \tag{6.6.11}$$

Then $R(r) = R_n(r)$ where $R_n(r) = J_0(j_{0,n}r/a)$. Solving the equation $T' = -(\lambda_n i/\hbar)T$ for the time factor $T$, we find $T = T_n$ where

$$T_n(t) = A_n \exp\left(\frac{-ij_{0,n}^2\hbar t}{2ma^2}\right)$$

Hence, for each $n = 1, 2, \ldots$ the function $\Psi_n(r, t) = R_n(r)T_n(t)$, or

$$\Psi_n(r, t) = A_n \exp\left(\frac{-ij_{0,n}^2\hbar t}{2ma^2}\right) J_0\left(j_{0,n}\frac{r}{a}\right)$$

solves Schrödinger's time-dependent equation. Using linearity, we then take the general $\theta$-independent solution $\Psi(r, t)$ to Schrödinger's time-dependent equation for a quantum particle with zero potential energy bound within a circular disk of radius $r = a$ as the infinite sum of all such solutions:

$$\Psi(r, t) = \sum_{n=1}^{\infty} \Psi_n = \sum_{n=1}^{\infty} A_n \exp\left(\frac{-ij_{0,n}^2\hbar t}{2ma^2}\right) J_0\left(j_{0,n}\frac{r}{a}\right) \qquad (6.6.12)$$

The constants $A_n$ are determined by the initial state $\Psi(r, 0)$ (Exercise 7). Using equation (6.6.1) for the energy levels, $E_n = j_{0,n}^2\hbar^2/2ma^2$, we may also write (6.6.12)

$$\Psi(r, t) = \sum_{n=1}^{\infty} A_n \exp\left(\frac{-iE_n t}{\hbar}\right) J_0\left(j_{0,n}\frac{r}{a}\right) \qquad (6.6.13)$$

**Exercises**

1. Show that the constants $A_n$ and $B_n$ in the solution to the wave equation (6.6.8) are

$$A_n = \frac{2}{a^2(J_0'(j_{0,n}))^2} \int_0^a u(r, 0)J_0\left(j_{0,n}\frac{r}{a}\right) r \, dr$$

$$B_n = \frac{2}{awj_{0,n}(J_0'(j_{0,n}))^2} \int_0^a \frac{\partial u}{\partial t}(r, 0)J_0\left(j_{0,n}\frac{r}{a}\right) r \, dr$$

2. (Continuation) (a) If $u(r, 0) = J_0(j_{0,1}r/a)$ and $\partial u/\partial t(r, 0) = 0$, show that equation (6.6.8) reduces to $u(r, t) = \cos(j_{0,1}wt/a)J_0(j_{0,1}r/a)$. Graph this as a function of $x = r/a$, $0 \leqslant x \leqslant 1$, for $t = n\pi a/4wj_{0,1}$, $n = 0, 1, 2, 3, 4$. (You are sketching a cross section of the drumhead along a radius.) Find the frequency $\nu_1$ of vibration in this mode (in terms of $w$ and $a$).
(b) Same as (a) except $u(r, 0) = J_0(j_{0,2}r/a)$. Compute $\nu_2/\nu_1$.
(c) Same as (a) except $u(r, 0) = J_0(j_{0,3}r/a)$ and only for $t = 0$. Compute $\nu_3/\nu_1$.
3. (Continuation) Look down on the drumhead from above. A curve on the drumhead that remains fixed at $u = 0$ for all time $t$ is called a *nodal curve*.

For each of the three solutions of Exercise 2, prove that the nodal curves are circles, and compute their radii. Take $a = 1$ and make a separate sketch of each nodal curve for each solution. Mark on your sketches the radius of each nodal curve.

4. Show that the constants $A_n$ in the solution of the heat equation (6.6.9) are

$$A_n = \frac{2}{a^2(J_0'(j_{0,n}))^2} \int_0^a u(r, 0) J_0\left(j_{0,n}\frac{r}{a}\right) r \, dr$$

5. A circular steel disk of radius $a = 20$ cm and uniform thickness is heated initially to a temperature of $u(r, 0) = 100(1 - (r/20)^2)°C$, $r$ in cm. The disk is insulated, and the temperature at its edge is maintained at $0°C$ for all subsequent time. Show that the temperature $u(r, t)$ in the disk is given by

$$u(r, t) = 800 \sum_{n=1}^\infty \frac{1}{j_{0,n}{}^3(-J_0'(j_{0,n}))} \exp(-1.294 j_{0,n}{}^2 t) J_0\left(j_{0,n}\frac{r}{20}\right)$$

where $r$ is in cm, $t$ in hours, and $u$ in °C. Using four terms, tabulate and plot $u(r, t)$ vs $r$, $0 \leqslant r \leqslant 20$, for selected times $t$. [Hint: Use Exercise 1, Section 6.5, and Table 5.1, Section 5.2.]

6. Show that the normalized time-independent wave function $\Psi_n(r)$ for the $n$th eigenstate of a quantum particle of potential energy $V = 0$ restricted to a circular disk of radius $a$ is

$$\Psi_n(r) = \frac{1}{\sqrt{\pi} \, a J_0'(j_{0,n})} J_0\left(j_{0,n}\frac{r}{a}\right)$$

Plot $|a\Psi_n(r)|^2$ versus $r$, $0 \leqslant r \leqslant a$, for $n = 1, 2$, and 3. Remembering that $|\Psi|^2$ is the probability density for the location of the particle, discuss your three graphs. At what radii does $|\Psi|^2$ have a local maximum? At what radii is the probability density zero?

7. Show that the constants $A_n$ in equation (6.6.13) for the time-dependent wave function $\Psi(r, t)$ are

$$A_n = \frac{2}{a^2(J_0'(j_{0,n}))^2} \int_0^a \Psi(r, 0) J_0\left(j_{0,n}\frac{r}{a}\right) r \, dr$$

Prove that if $\Psi(r, 0)$ is normalized,

$$2\pi \int_0^a |\Psi(r, 0)|^2 r \, dr = 1$$

then $\Psi(r, t)$ remains normalized for every time $t$.

### 6.7  θ DEPENDENCE: BESSEL'S EQUATION AND BESSEL'S OPERATOR OF INTEGRAL ORDER p

In the previous section we studied the problem of the circular vibrating drumhead under the assumption that the displacement $u$ is independent of the angle $\theta$. We look now at the kinds of equations that arise when we separate variables in the general case.

The wave equation in polar coordinates is

$$\frac{\partial^2 u}{\partial r^2} + \frac{1}{r}\frac{\partial u}{\partial r} + \frac{1}{r^2}\frac{\partial^2 u}{\partial \theta^2} = \frac{1}{w^2}\frac{\partial^2 u}{\partial t^2} \quad \text{(equation (6.1.10)} \qquad (6.7.1)$$

We deal with a circular drumhead of radius $r = a$ clamped at the rim, so that

$$u(a, \theta, t) = 0 \qquad \text{for all } \theta, t \qquad (6.7.2)$$

We again seek to solve (6.7.1) by the method of separation of variables, writing $u = R(r)\Theta(\theta)T(t)$. Substituting into (6.7.1) and dividing by $R\Theta T$, we get

$$\frac{R''}{R} + \frac{1}{r}\frac{R'}{R} + \frac{1}{r^2}\frac{\Theta''}{\Theta} = \frac{1}{w^2}\frac{T''}{T}$$

The right side depends only on time $t$; the left side does not. Hence both the left and right sides equal the same constant, say $-\lambda$. Setting the left side equal to $-\lambda$ and multiplying through by $r^2$ we get

$$r^2\frac{R''}{R} + r\frac{R'}{R} + \lambda r^2 = -\frac{\Theta''}{\Theta} \qquad (6.7.3)$$

The right side depends only on $\theta$, the left side only on $r$. Hence both the left side and the right side equal the same constant, say $\gamma$. Hence we have

$$\Theta'' + \gamma\Theta = 0 \qquad (6.7.4)$$

This is an eigenvalue problem with eigenvalue parameter $\gamma$. We seek to set this up as an eigenvalue problem for a Hilbert space operator. The appropriate space is $L^2(C)$, where $C$ is the circle of unit radius obtained by bending the interval $-\pi \leqslant \theta \leqslant \pi$ into a circle and then joining the point $\theta = -\pi$ to the point $\theta = \pi$: $C = \{\theta\colon -\pi \leqslant \theta \leqslant \pi,\ -\pi \equiv \pi\}$. The manifold $C$ has no boundary. Our differential operator $L$ in $L^2(C)$ has differential expression $l(\Theta) = -\Theta''$ and domain consisting of functions in $L^2(C)$ compatible with $l$ and satisfying the necessary smoothness conditions $\Theta(-\pi) = \Theta(\pi)$, $\Theta'(-\pi) = \Theta'(\pi)$ ensuring continuity of $\Theta$ and its derivative at the point $-\pi \equiv \pi$ of $C$ (refer back to Theorem 4.2 of Section 4.1). We have dealt with this Hilbert space operator before. Its eigenvalues are $\gamma = 0, 1, 4, 9, \ldots$. The eigenvalue $\gamma = 0$ is simple, and the other eigenvalues have multiplicity 2. The function 1/2 is a convenient eigenfunction for $\gamma = 0$, and the pair of functions $\cos p\theta$, $\sin p\theta$ span the eigenspace belonging to the eigenvalue $\gamma = p^2$, $p$ a positive integer. What we have here is the classical Fourier series. The main thing we need at the moment is the information that equation (6.7.4), now interpreted as an eigenvalue problem for a Hilbert space operator, has nonzero solutions $\Theta$ when, and only when, $\gamma = p^2$, $p = 0, 1, 2, \ldots$.

Thus equation (6.7.3) becomes $r^2(R''/R) + r(R'/R) + \lambda r^2 = p^2$. Multiply through by $R/r^2$, rearrange terms, include the boundary condition (6.7.2), and you will get

$$-\left(R'' + \frac{1}{r}R'\right) + \frac{p^2}{r^2}R = \lambda R, \qquad R(a) = 0 \qquad (6.7.5)$$

This is an eigenvalue problem, with eigenvalue parameter $\lambda$. We seek to set this up as an eigenvalue problem for a Hilbert space operator. We thus require, first, an appropriate $L^2$ space of functions on $0 \leqslant r \leqslant a$. Our operator will have differential expression $b_p(R) = -(R'' + (1/r)R') + (p^2/r^2)R$. Rewrite this in Sturmian form, $-(1/r)(rR')' + (p^2/r^2)R$, and read off from the Sturmian form the appropriate weight function for our Hilbert space, namely $w(r) = r$. Thus the appropriate Hilbert space is $L^2(0, a; r)$, which is exactly the same space appropriate for Bessel's operator of order zero studied in Section 6.2. Thus this family of operators all act in the same Hilbert space.

**Definition 6.12.** Fix an integer $p$. Bessel's operator $B_p$ has the *differential expression*

$$b_p(f) = -\frac{1}{r}\frac{d}{dr}\left(r\frac{df}{dr}\right) + \frac{p^2}{r^2}f$$

and *domain*$(B_p)$ consisting of functions $f(r)$ in $L^2(0, a; r)$ that satisfy the following three conditions:

1. $f$ is compatible with $b_p$; i.e., both $f$ and $b_p(f) \in L^2(0, a; r)$.
2. $f(a) = 0$.
3. $\lim_{r \to 0+} f(r)$ exists and is finite.

For the functions $f$ in the domain of $B$, and *only* for those functions, Bessel's operator is defined by

$$B_p(f) = b_p(f) = -\frac{1}{r}\frac{d}{dr}\left(r\frac{df}{dr}\right) + \frac{p^2}{r^2}f, \qquad f \in \text{domain}(B_p)$$

We may abbreviate condition 3 by simply saying that $f(0)$ is finite.

Note that when $p = 0$ the operator $B_0$ as defined above is exactly the same as the operator $B$ that we analyzed in prior sections. So the operator in $L^2(0, a; r)$ that we earlier called Bessel's operator now appears as the first operator $B_0$ in a sequence $B_p$, $p = 0, 1, 2, \ldots$, of Hilbert space operators, all acting in $L^2(0, a; r)$.

The three conditions that define the domain of Bessel's operator have been chosen with the purpose of making the operator Hermitian.

The first condition, compatibility, restricts the domain to the functions $f \in L^2(0, a; r)$ such that $b_p(f) = -(1/r)(rf')' + (p^2/r^2)f$ also belongs to $L^2(0, a; r)$. This condition assures us that the operator will always take its values in our Hilbert space $L^2(0, a; r)$; without the compatibility requirement we could not be sure that we would get a Hilbert space operator at all.

As in previous cases, compatibility carries with it certain restrictions on the boundary behavior.

**Theorem 6.13.** *If $f(r)$ is compatible with Bessel's differential expression $b_p(f) = -(1/r)(rf')' + (p^2/r^2)f$, then both $f(a)$ and $f'(a)$ are finite.*

When we also impose condition 3 of Definition 6.12, some strong restrictions on the boundary behavior result:

**Theorem 6.14.** *Let $D$ denote the subspace of $L^2(0, a; r)$ consisting of functions $y(r)$ that satisfy conditions 1 and 3 of Definition 6.12, namely that $y$ is compatible with Bessel's differential expression $b_p(y) = -(1/r)(ry')' + (p^2/r^2)y$ and $y(0)$ is finite. If $p \neq 0$, then, for any functions $f(r)$, $g(r)$ in $D$,*

1. *$f'(r)$ and $f(r)/r$ belong to $L^2(0, a; r)$, and*
2. *$\lim_{r \to 0+} rf'(r)\overline{g(r)} = 0$.*

You are not responsible for the proofs of Theorems 6.13 and 6.14, which are given in Appendix 6.A, but you will need the theorems themselves to prove that Bessel's operator $B_p$ is Hermitian: note that the domain of Bessel's operator consists of all the functions $y(r)$ in the subspace $D$ defined in Theorem 6.14 that satisfy $y(a) = 0$.

**Theorem 6.15.** *Bessel's operator $B_p$ is Hermitian and positive definite, i.e.,*

$$\langle B_p(f), g \rangle = \langle f, B_p(g) \rangle \qquad \text{for all } f, g \text{ in domain}(B_p)$$

*and*

$$\langle B_p(f), f \rangle > 0 \qquad \text{whenever } 0 \neq f \in \text{domain}(B_p)$$

You are asked to supply the proof as Exercise 3.

Recall these definitions: An *eigenfunction* $f$ of $B_p$ is defined by these three characteristics: (1) $f$ belongs to domain($B_p$), (2) $f \neq 0$, and (3) $B_p(f) = \lambda f$ for some scalar $\lambda$. The $\lambda$'s for which this happens are the *eigenvalues of $B_p$*. For a fixed eigenvalue $\lambda$, any nonzero function $f$ in the domain of $B_p$ that satisfies $B_p(f) = \lambda f$ is called an *eigenfunction belonging to the eigenvalue $\lambda$*. Given an eigenvalue $\lambda$, the collection of eigenfunctions that belong to it, together with the zero function, is the *eigenspace* belonging to $\lambda$.

**Corollary 6.16.**   *All eigenvalues of $B_p$ are real and positive, and eigenfunctions belonging to different eigenvalues are orthogonal.*

You are asked to give the proof in Exercise 4.

Corollary 6.16 tells us that, if $B_p$ has any eigenvalues at all, they must be real and positive. We now take up the task of finding the eigenvalues, and corresponding eigenfunctions, for $B_p$. For $p = 0$, we have done this already in Section 6.5. We use the same procedure here. We first find the general solution to the homogeneous differential equation $y'' + (1/r)y' - (p^2/r^2)y + \lambda y = 0$ where $p$ is a fixed integer. We know that the solution space is two-dimensional, so the general solution has the form $y = c_1 f_1 + c_2 f_2$, where $c_1$ and $c_2$ are arbitrary constants and $f_1$ and $f_2$ linearly independent solutions. Then we determine for what values of the constants $c_1$ and $c_2$ and for what values of the parameter $\lambda$ the function $y$ will lie in the domain of $B$.

Our differential equation is

$$\frac{d^2 y}{dr^2} + \frac{1}{r}\frac{dy}{dr} - \frac{p^2}{r^2}y + \lambda y = 0$$

We make the change of variable $x = \sqrt{\lambda}\,r$, which keeps us in the real numbers as we know $\lambda > 0$. Making this substitution, canceling a common factor of $\lambda$, and multiplying through by $x^2$ we get

$$x^2 \frac{d^2 y}{dx^2} + x \frac{dy}{dx} + (x^2 - p^2)y = 0$$

This is called *Bessel's equation of order p*. It is the equation we must solve to find the eigenvalues and eigenfunctions of Bessel's differential operator $B_p$.

**Exercises**

1. (Repeat of Exercise 6, Section 4.1)   Let $p$ be an integer and let $b_p$ denote Bessel's differential expression

$$b_p(y) = -\frac{1}{r}\frac{d}{dr}\left(r\frac{dy}{dr}\right) + \frac{p^2}{r^2}y \qquad \text{in } L^2(0, a; r)$$

Consider the quadratic polynomial $y = a_0 + a_1 r + a_2 r^2$, which obviously belongs to $L^2(0, a; r)$.

(a) Prove that, if $p = 0$, then $y$ is compatible with $b_p$ if, and only if, $a_1 = 0$.

(b) Prove that, if $p = \pm 1$, then $y$ is compatible with $b_p$ if, and only if, $a_0 = 0$.

(c) Prove that, if $|p| \geqslant 2$, then $y$ is compatible with $b_p$ if, and only if, both $a_0 = 0$ and $a_1 = 0$.

2. Let $p$ be a fixed positive integer.

   (a) Prove that $r^p$ is compatible with Bessel's differential expression $b_p$. Prove also that $b_p(r^{-p}) = 0$.

   (b) Prove that $r^\alpha$, $\alpha$ a real number, is compatible with $b_p$ if, and only if, either $\alpha = p$ or $\alpha > 1$.

   (c) Prove that $a^2 r^p - r^{p+2}$ lies in the domain of $B_p$.

3. Suppose $f$ and $g$ belong to the domain of Bessel's operator $B_p$.

   (a) Let $\varepsilon$ be a small positive number, and derive the formula

   $$\int_\varepsilon^a b_p(f) \bar{g} r \, dr = \varepsilon f'(\varepsilon) \bar{g}(\varepsilon) + \int_\varepsilon^a f' \bar{g}' r \, dr + p^2 \int_\varepsilon^a \left(\frac{f}{r}\right)\left(\frac{\bar{g}}{r}\right) r \, dr.$$

   (b) Refer to Theorems 6.13 and 6.14, and pass to the limit as $\varepsilon \to 0+$ to derive the formula

   $$\langle B_p(f), g \rangle = \int_0^a f' \bar{g}' r \, dr + p^2 \int_0^a \left(\frac{f}{r}\right)\left(\frac{\bar{g}}{r}\right) r \, dr$$

   $$= \langle f', g' \rangle + p^2 \left\langle \frac{f}{r}, \frac{g}{r} \right\rangle$$

   for all $f, g$ in the domain of $B_p$.

   (c) In your result in (b), interchange $f$ and $g$, take the complex conjugate of the result, and show therefrom that $B_p$ is Hermitian.

   (d) From your result in (b) deduce that $B_p$ is positive definite.

4. (a) From the fact that $B_p$ is Hermitian, prove that its eigenvalues must be real.

   (b) From the fact that $B_p$ is Hermitian, prove that eigenfunctions belonging to different eigenvalues are orthogonal.

   (c) From the fact that $B_p$ is positive definite, prove that its eigenvalues are $> 0$.

5. Show that for any differential operator $A_p$ with differential expression the same as Bessel's and domain determined by conditions 1 and 3 of Definition 6.12, we have $\langle A_p(f), g \rangle - \langle f, A_p(g) \rangle = a(f(a)\overline{g'(a)} - f'(a)\overline{g(a)})$. Show that if the domain is further restricted by the condition $\alpha y(a) + \beta y'(a) = 0$, where $\alpha$ and $\beta$ are real and not both zero, then the resulting operator $A_p$ is Hermitian.

## 6.8  $J_p(x)$: THE BESSEL FUNCTIONS OF THE FIRST KIND OF INTEGRAL ORDER $p$

In the previous section we expressed the wave equation in polar coordinates *including* the dependence on the angle variable $\theta$. We assumed the boundary condition $u=0$ when $r=a$, separated variables, and were then led to the following equation in the radius variable $r$: $-y''-(1/r)y'+(p^2/r^2)y=\lambda y$ where $p$ is a fixed integer. We interpreted this as an eigenvalue problem for a Hermitian operator $B_p$ which operates in the Hilbert space $L^2(0,a;r)$, has differential expression $b_p(y)=-y''-(1/r)y'+(p^2/r^2)y$, and has domain as described in the previous section. The problem of solving the $r$-variable part of the wave equation is equivalent to finding the eigenvalues of $B_p$ and the corresponding eigenfunctions.

To find the eigenvalues $\lambda$ of the operator $B_p$ we must solve the differential equation $y''+(1/r)y'-(p^2/r^2)y+\lambda y=0$ and then determine when such solutions lie in the domain of our operator. As in the previous section, we make the change of variable $x=\sqrt{\lambda}\,r$ (knowing $\lambda>0$), and *Bessel's differential equation of integral order $p$* results:

$$x^2\frac{d^2y}{dx^2}+x\frac{dy}{dx}+(x^2-p^2)y=0 \qquad (6.8.1)$$

The parameter $\lambda$ has disappeared. Our job is now to solve equation (6.8.1).

Equation (6.8.1) has nonconstant coefficients, so we try a power series solution, a method that has served us well in the past (refer to Section 2.8). We begin by assuming a power series expansion for $y$,

$$y=a_0+a_1x+a_2x^2+a_3x^3+\cdots+a_lx^l+\cdots$$

where the coefficients $a_i$ are to be determined so that $y$ satisfies (6.8.1). We set up the calculation this way:

$$
\begin{array}{llll}
x^2y''= & & 2a_2x^2+3\cdot2a_3x^3+\cdots+l(l-1)a_lx^l+\cdots \\
xy'= & a_1x+ & 2a_2x^2+\quad3a_3x^3+\cdots+ & la_lx^l+\cdots \\
x^2y= & & a_0x^2+\quad a_1x^3+\cdots+ & a_{l-2}x^l+\cdots \\
-p^2y= & -p^2a_0-p^2a_1x- & p^2a_2x^2-\quad p^2a_3x^3-\cdots- & p^2a_lx^l-\cdots
\end{array}
$$

Adding, and using $l(l-1)+l=l^2$, we get

$$0=-p^2a_0+(1^2-p^2)a_1x+((2^2-p^2)a_2+a_0)x^2$$
$$+((3^2-p^2)a_3+a_1)x^3+\cdots+((l^2-p^2)a_l+a_{l-2})x^l+\cdots$$

If $y$ is to satisfy the differential equation (6.8.1), the power series on the right must vanish identically, so all its coefficients must be zero. So our coefficients must be chosen to satisfy the following system of equations:

$$p^2 a_0 = 0$$
$$(1^2 - p^2)a_1 = 0$$
$$(2^2 - p^2)a_2 + a_0 = 0$$
$$(3^2 - p^2)a_3 + a_1 = 0$$
$$(4^2 - p^2)a_4 + a_2 = 0 \qquad\qquad (6.8.2)$$
$$(5^2 - p^2)a_5 + a_3 = 0$$
$$\vdots$$
$$(l^2 - p^2)a_l + a_{l-2} = 0$$
$$\vdots$$

If $p = 0$, then we may choose $a_0 = 1$. We then get $a_1 = a_3 = a_5 = \cdots = 0$, and we arrive back at $J_0(x)$.

Suppose now $p \geqslant 1$. The equations (6.8.2) force $a_0 = a_1 = \cdots = a_{p-1} = 0$. Then the equation $((p+1)^2 - p^2)a_{p+1} + a_{p-1} = 0$ implies $a_{p+1} = 0$. Continuing, we find $a_{p+k} = 0$ for $k$ odd. We may specify $a_p$ at will. The equation $((p+2)^2 a_{p+2} + a_p = 0$ then yields $a_{p+2} = -a_p/((p+2)^2 - p^2)$. The next equation, $((p+4)^2 - p^2)a_{p+4} + a_{p+2} = 0$, gives

$$a_{p+4} = \frac{-a_{p+2}}{(p+4)^2 - p^2} = \frac{a_p}{((p+4)^2 - p^2)((p+2)^2 - p^2)}$$

Continuing, we find

$$a_{p+2l} = \frac{(-1)^l a_p}{((p+2l)^2 - p^2)((p+2(l-1))^2 - p^2)\cdots((p+2)^2 - p^2)}$$

Each term in the denominator is the difference of squares, $a^2 - b^2$. Factoring as $a^2 - b^2 = (a+b)(a-b)$ and collecting first the terms with the $+$ sign, we get

$$(p+2l+p)(p+2l-2+p)\cdots(p+2+p) = (2p+2l)(2p+2l-2)\cdots(2p+2)$$
$$= 2^l(p+l)(p+l-1)\cdots(p+1)$$
$$= \frac{2^l(p+l)!}{p!}$$

The terms with the $-$ sign yield

$$(p+2l-p)(p+2l-2-p)\cdots(p+2-p) = (2l)(2l-2)\cdots(2)$$
$$= 2^l l!$$

Our choice for $a_p$ is $a_p = 1/p!2^p$. Using this normalization, we get

$$a_{p+2l} = \frac{(-1)^l}{2^{p+2l}(p+l)!l!}$$

Thus our power series solution is

$$J_p(x) = \sum_{l=0}^{\infty} \frac{(-1)^l}{l!(p+l)!} \left(\frac{x}{2}\right)^{p+2l} = \left(\frac{x}{2}\right)^p \sum_{l=0}^{\infty} \frac{(-1)^l}{l!(p+l)!} \left(\frac{x}{2}\right)^{2l} \qquad (6.8.3)$$

*the Bessel function of the first kind of integral order p.*

Note that while $J_0(0) = 1$, we have instead $J_p(0) = 0$, $p \geqslant 1$. In fact, $J_p(x)$ has a zero of order $p$ at $x = 0$. The ratio test shows that the series (6.8.3) converges for all values of $x$. Thus $J_p(x)$ is an entire function—one that is analytic at every point of the complex plane. It is mildly surprising that $J_p(x)$ is analytic at $x = 0$ because the leading coefficient in Bessel's equation vanishes at $x = 0$, so we might have expected a singularity there. The singularity does show up in the second linearly independent solution.

We may obtain a second linearly independent solution to Bessel's equation by the methods of Chapter 2: find the Wronskian $W$ of Bessel's equation by Abel's formula; then solve the first order equation $y'J_p - yJ'_p = W$ for the second linearly independent solution $y$. When you do this calculation (Exercise 12) you will find

$$y = J_p(x) \int \frac{dx}{xJ_p^2(x)} \qquad (6.8.4)$$

as a second linearly independent solution to Bessel's equation.

It is convenient for our purpose to recast this solution in a somewhat different form. $J_p(x)$ is analytic everywhere and has, at $x = 0$, a zero of order $p$. Thus $xJ_p^2(x)$ has a zero of order $2p+1$ at $x = 0$, so the Laurent series of $1/xJ_p^2(x)$ begins with a term $(\text{const}) \times x^{-2p-1}$. When we do the integration specified in (6.8.4) and then multiply by $J_p(x)$, we find that the second linearly independent solution $y$ to Bessel's differential equation is a linear combination of two terms, one having the form (a Laurent series that starts with $x^{-p}$), the other $J_p(x) \ln x$. In any case, we conclude that, for any $p \geqslant 0$, $y$ has a singularity at $x = 0$.

If $a$ is a nonzero constant and $c$ any constant, then $ay + cJ_p$ is again a linearly independent solution to Bessel's equation, so we have a rather wide latitude in our normalization of this second solution. The following is a common choice.

$$Y_p(x) = -\frac{1}{\pi}\left(\frac{2}{x}\right)^p \sum_{l=0}^{p-1} \frac{(p-l-1)!}{l!}\left(\frac{x}{2}\right)^{2l} + \frac{2}{\pi}\ln\left(\frac{x}{2}\right)J_p(x)$$

$$-\frac{1}{\pi}\left(\frac{x}{2}\right)^p \sum_{l=0}^{\infty}(-1)^l \frac{\Psi(l+1)+\Psi(p+l+1)}{l!(p+l)!}\left(\frac{x}{2}\right)^{2l} \qquad (6.8.5)$$

where

$$\Psi(1) = -\gamma = -\lim_{n\to\infty}\left[\sum_{k=1}^{n}\frac{1}{k} - \ln n\right] = -0.577215\ldots$$

and

$$\Psi(l) = -\gamma + \sum_{k=1}^{l-1}\frac{1}{k}, \qquad l \geqslant 2$$

The function $Y_p(x)$ is the *Bessel function of the second kind of integral order p*. It is singular at $x=0$, as is clear from equation (6.8.5).

The general solution to Bessel's equation is $c_0 J_p(x) + c_1 Y_p(x)$. What is important for us is the requirement that the solution belong to the domain of Bessel's operator $B_p$. This requirement includes the condition that the solution be bounded at $r=0$, which means also at $x=0$ as $x=\sqrt{\lambda}r$. The general solution $c_0 J_p(x) + c_1 Y_p(x)$ is bounded at $x=0$ if and only if $c_1 = 0$, because $Y_p(x)$ is singular at $x=0$.

The power series for $J_p(x)$,

$$J_p(x) = \left(\frac{x}{2}\right)^p \sum_{l=0}^{\infty}\frac{(-1)^l}{l!(p+l)!}\left(\frac{x}{2}\right)^{2l}$$

$$= \left(\frac{x}{2}\right)^p\left(\frac{1}{p!} - \frac{1}{(p+1)!}\left(\frac{x}{2}\right)^2 + \frac{1}{2!(p+2)!}\left(\frac{x}{2}\right)^4 - \cdots\right) \qquad (6.8.6)$$

is a rapidly convergent alternating series. As it is an alternating series, if we use a finite number of terms to compute $J_p(x)$ we make an error which is an absolute value less than the last term omitted. Hence, if we keep $n$ terms ($l=0$ through $l=n-1$), we will be in error by less than $x^{2n}/n!(p+n)!2^{2n}$. Note that $J_p(x)$ has the same parity as $p$, so we need only compute it for $x \geqslant 0$. The computation of the truncated power series is easily programmed just as for $J_0(x)$.

The asymptotic behavior of $J_p(x)$ is given by

$$J_p(x) \cong \sqrt{\frac{2}{\pi x}}\cos\left(x - \frac{p\pi}{2} - \frac{\pi}{4}\right), \qquad x \to \infty \qquad (6.8.7)$$

The $n$th positive zero of $J_p$ is denoted $j_{p,n}$, $n = 1, 2, \ldots$. Equation (6.8.7) suggests that the large zeros should be approximately given by

$$j_{p,n} - \frac{p\pi}{2} - \frac{\pi}{4} \cong (2n - 1)\frac{\pi}{2}$$

or

$$j_{p,n} \cong \left(n + \frac{p}{2} - \frac{1}{4}\right)\pi \qquad \text{valid for large } n \tag{6.8.8}$$

For $p = 0$, we have seen that equation (6.8.8) gives the zeros of $J_0(x)$ to an accuracy ranging from 2% for $j_{0,1}$ to 0.01% for $j_{0,10}$ (Table 6.2). In Table 6.5 we list $j_{p,n}$ for $1 \leqslant p \leqslant 8$, $1 \leqslant n \leqslant 5$. For large $p$ and small n, equation (6.8.8) is a less good approximation. For example, $j_{8,1} = 12.22509$, while (6.8.8) gives 14.9. And $j_{8,5} = 26.26681$, while (6.8.8) gives 27.5, a 4.7% error.

Table 6.5  Zeros of $J_p(x)$

| $n$ | $j_{1,n}$ | $j_{2,n}$ | $j_{3,n}$ | $j_{4,n}$ |
|---|---|---|---|---|
| 1 | 3.83171 | 5.13562 | 6.38016 | 7.58834 |
| 2 | 7.01559 | 8.41724 | 9.76102 | 11.06471 |
| 3 | 10.17347 | 11.61984 | 13.01520 | 14.37254 |
| 4 | 13.32369 | 14.79595 | 16.22347 | 17.61597 |
| 5 | 16.47063 | 17.95982 | 19.40942 | 20.82693 |

| $n$ | $j_{5,n}$ | $j_{6,n}$ | $j_{7,n}$ | $j_{8,n}$ |
|---|---|---|---|---|
| 1 | 8.77148 | 9.93611 | 11.08637 | 12.22509 |
| 2 | 12.33860 | 13.58929 | 14.82127 | 16.03777 |
| 3 | 15.70017 | 17.00382 | 18.28758 | 19.55454 |
| 4 | 18.98013 | 20.32079 | 21.64154 | 22.94517 |
| 5 | 22.21780 | 23.58608 | 24.93493 | 26.26681 |

Source: M. Abramowitz and I. A. Stegun (Editors), *Handbook of Mathematical Functions*, Applied Math. Series 55, N.B.S., 1965.

## Exercises

1. Is $J_p(x)$ an even function of $x$ or an odd function?
2. Using the ratio test, show that the series for $J_p(x)$ (equation (6.8.3)) converges for all $x$, thus represents an entire function.

The following exercises defines $J_p$ for negative integers $p$.

3. Let $p$ be a nonnegative integer. Extend the definition of the Bessel function of the first kind to negative integral orders by defining $J_{-p}(x)$ as the series resulting from replacing $p$ by $-p$ in the series (6.8.3), subject to the convention

$$\frac{1}{(-1)!} = \frac{1}{(-2)!} = \frac{1}{(-3)!} = \cdots = \frac{1}{(-n)!} = \cdots = 0$$

Show then that

$$J_{-p}(x) = \sum_{l=p}^{\infty} \frac{(-1)^l}{l!(l-p)!} \left(\frac{x}{2}\right)^{2l-p}$$

and that

$$J_{-p}(x) = (-1)^p J_p(x)$$

Show that this latter formula and equation (6.8.3) in the text are now valid for all integers $p$, positive, negative and zero.

4. (Continuation) Let $p$ be an integer, positive, negative, or zero. By differentiating the series

$$x^p J_p(x) = \sum_{l=0}^{\infty} \frac{(-1)^l}{l!(l+p)!2^{2l+p}} x^{2l+2p}$$

term by term, show that $(x^p J_p(x))' = x^p J_{p-1}(x)$. Using this formula, show that

$$xJ_p' + pJ_p = xJ_{p-1} \tag{1}$$

(In particular, $J_0' = J_{-1} = -J_1$.)

5. (Continuation) Show that $(x^{-p}J_p(x))' = -x^{-p}J_{p+1}(x)$. Using this formula, show that

$$xJ_p' - pJ_p = -xJ_{p+1} \tag{2}$$

6. (Continuation) By combining equation (1) of Exercise 4 with equation (2) of Exercise 5, derive the equation $2J_p' = J_{p-1} - J_{p+1}$ and the *recursion formula for Bessel functions*

$$xJ_{p+1} = 2pJ_p - xJ_{p-1}$$

both valid for all integers $p$.

7. Using the convention set down in Exercise 3, show that the evaluations

$$x^{-p}J_p(x)|_{x=0} = \frac{1}{2^p p!}, \qquad x^p J_p(x)|_{x=0} = \frac{(-1)^p}{2^{-p}(-p)!}$$

are valid for all integers $p$.

8. Using Exercises 4 and 5, show that

$$\int x^p J_{p-1}(x)\,dx = x^p J_p(x) + C$$

and

$$\int x^p J_{p+1}(x)\,dx = -x^{-p} J_p(x) + C$$

Show also that, for $\lambda \neq 0$,

$$\int_0^1 t^p J_{p-1}(\lambda t)\,dt = \frac{J_p(\lambda)}{\lambda}, \qquad p \geq 1$$

and

$$\int_0^1 t^p J_{p+1}(\lambda t)\,dt = \frac{1}{\lambda^{p+1}}\left(\frac{1}{2^p p!} - \frac{J_p(\lambda)}{\lambda^p}\right), \qquad p \geq 0$$

## The Generating Function for Bessel Functions

9. Write $h(x,w) = \sum_{p=-\infty}^{\infty} w^p J_p(x)$. Hold $x$ constant, differentiate this equation with respect to $w$, and use the recursion formula (Exercise 6) to show that $h$ satisfies the differential equation $dh/h = (x/2)(1 + 1/w^2)\,dw$. Integrate this equation and then determine the integration constant by substituting an appropriate value of $x$. Thus derive the *generating function for Bessel functions*:

$$\exp\left[\frac{x}{2}\left(w - \frac{1}{w}\right)\right] = \sum_{p=-\infty}^{\infty} w^p J_p(x)$$

10. Show that

$$\sum_{p=-\infty}^{\infty} w^p J_p(x) = J_0(x) + \sum_{p=1}^{\infty}\left[w^p + \left(\frac{-1}{w}\right)^p\right] J_p(x)$$

Derive the formula

$$1 = \sum_{p=-\infty}^{\infty} J_p(x) = J_0(x) + 2\sum_{p=1}^{\infty} J_{2p}(x)$$

11. Derive the formulas below and explain how they can be valid simultaneously

$$e^{\alpha x} = \sum_{p=-\infty}^{\infty} (\alpha + \sqrt{\alpha^2 + 1})^p J_p(x)$$

$$= \sum_{p=-\infty}^{\infty} (\alpha - \sqrt{\alpha^2 + 1})^p J_p(x)$$

12. Complete the calculation outlined in this section to derive the formula

$$y = J_p(x) \int \frac{dx}{x J_p^2(x)}$$

for a second linearly independent solution to Bessel's equation.

13. Show that $\cos x = \sum\limits_{-\infty}^{\infty} (-1)^n \, J_{2n}(x)$ and $\sin x = \sum\limits_{-\infty}^{\infty} (-1)^n \, J_{2n+1}(x)$

## 6.9 THE EIGENVALUES AND EIGENFUNCTIONS OF BESSEL'S OPERATOR OF INTEGRAL ORDER $p$

We can now determine the eigenvalues and eigenfunctions of Bessel's operator $B_p$ which has the differential expression $b_p(y) = -(1/r)(ry')' + (p^2/r^2)y$ and domain consisting of compatible functions $y$ in $L^2(0, a; r)$ such that $y(0)$ is finite and $y(a) = 0$.

Any eigenfunction $y$ of $B_p$ must satisfy the differential equation $r^2 y'' + ry' - p^2 y + \lambda r^2 y = 0$, which becomes Bessel's equation $x^2 y'' + xy' + (x^2 - p^2)y = 0$ after the substitution $x = \sqrt{\lambda} r$. The general solution of Bessel's equation is $c_0 J_p(x) + c_1 Y_p(x)$, where $J_p$ and $Y_p$ are the Bessel functions of integral order $p$ of the first and second kind, respectively. Replacing $x$ by $\sqrt{\lambda} r$, we find that the general solution to $r^2 y'' + ry' - p^2 y + \lambda r^2 y = 0$ (which is equivalent to the equation $b_p(y) = \lambda y$) is $c_0 J_p(\sqrt{\lambda} r) + c_1 Y_p(\sqrt{\lambda} r)$. And the question before us now is: *for what values of the constants $c_0$, $c_1$ and for what values of the parameter $\lambda$ does the function $f(r) = c_0 J_p(\sqrt{\lambda} r) + c_1 Y_p(\sqrt{\lambda} r)$ lie in the domain of $B_p$?* To get into the domain of $B_p$, the function $f(r)$ must pass through various gates. We list them one by one.

$f(0)$ *is finite*: This forces $c_1 = 0$ as $Y_p(\sqrt{\lambda} r)$ has a singularity at $r = 0$. So $f(r) = c_0 J_p(\sqrt{\lambda} r)$.

*Both $f(r)$ and $b_p(f(r))$ must belong to $L^2(0, a; r)$*: As $f(r)$ satisfies $b_p(f) = \lambda f$, it is enough to show that $f$ belongs to our $L^2$ space. But as $f(r) = c_0 J_p(\sqrt{\lambda} r)$ is entire, it is bounded on every finite interval and therefore certainly square integrable there against the weight function $r$. Thus $f(r) = c_0 J_p(\sqrt{\lambda} r)$ belongs to $L^2(0, a; r)$ and is compatible with $b_p$.

$f(a) = 0$: We must have $c_0 \neq 0$ lest $f$ equal the trivial zero solution, so $f(a) = 0$ means $J_p(\sqrt{\lambda} a) = 0$, which requires

$$\sqrt{\lambda} a = j_{p,n}$$

for some zero $j_{p,n}$, $n = 1, 2, \ldots$, of $J_p$. Thus

$$\lambda_{p,n} = \frac{1}{a^2} j_{p,n}^2, \qquad n = 1, 2, \ldots \tag{6.9.1}$$

are eigenvalues of Bessel's operator $B_p$. To each of these eigenvalues belongs the eigenfunction

$$\phi_{p,n}(r) = J_p\left(j_{p,n}\frac{r}{a}\right). \tag{6.9.2}$$

where we have used the convenient normalization $c_0 = 1$. Thus we have

$$B_p(\phi_{p,n}) = \lambda_{p,n}\phi_{p,n} \quad \text{or} \tag{6.9.3}$$

$$B_p\left(J_p\left(j_{p,n}\frac{r}{a}\right)\right) = \frac{1}{a^2}j_{p,n}^2 J_p\left(j_{p,n}\frac{r}{a}\right)$$

for $n = 1, 2, \ldots$, where $B_p$ is Bessel's operator.

**Theorem 6.17.** *Bessel's operator $B_p$ (refer to Definition 6.12) has eigenvalues*

$$\lambda_{p,n} = \frac{1}{a^2}j_{p,n}^2, \qquad n = 1, 2, \ldots$$

*where $j_{p,n}$ is the nth zero of $J_p(x)$, the Bessel function of the first kind of integral order p. To each of these eigenvalues belongs the eigenfunction*

$$\phi_{p,n}(r) = J_p\left(j_{p,n}\frac{r}{a}\right), \qquad n = 1, 2, \ldots$$

*These eigenfunctions form an orthogonal family in $L^2(0, a; r)$; i.e.,*

$$\langle\phi_{p,m}, \phi_{p,n}\rangle = \int_0^a \phi_{p,m}\phi_{p,n}r\,dr = \int_0^a J_p\left(j_{p,m}\frac{r}{a}\right)J_p\left(j_{p,n}\frac{r}{a}\right)r\,dr = 0$$

*when $m \neq n$.*

The orthogonality comes from the fact that the $\phi_{p,n}$ are eigenfunctions of a Hermitian operator. If we make the substitution $x = r/a$, the orthogonality relation becomes

$$\int_0^1 J_p(j_{p,m}x)J_p(j_{p,n}x)x\,dx = 0, \qquad m \neq n$$

a fact that is difficult to verify directly. But in our approach it is an immediate consequence of the principle that, for a Hermitian operator in Hilbert space, eigenfunctions belonging to different eigenvalues are orthogonal.

Next, we shall determine the norm (squared) of the eigenfunctions $\phi_{p,n}$. This means evaluating the integrals

$$\|\phi_{p,n}(r)\|^2 = \int_0^a J_p\left(j_{p,n}\frac{r}{a}\right)^2 r\,dr$$

Make the substitution $x = j_{p,n}r/a$ to get

$$\|\phi_{p,n}(r)\|^2 = \left(\frac{a}{j_{p,n}}\right)^2 \int_0^{j_{p,n}} J_p(x)^2 x \, dx$$

To evaluate this integral, we shall make use of the following result:

**Lemma 6.18.** *Any solution $y$ of Bessel's differential equation $x^2 y'' + xy' + (x^2 - p^2)y = 0$ satisfies the antiderivative equation*

$$2 \int xy^2 \, dx = x^2(y')^2 + (x^2 - p^2)y^2 + C$$

*Proof.* Multiply Bessel's differential equation by $y'$ to get $x^2 y'' y' + x(y')^2 + (x^2 - p^2)yy' = 0$. Observe that $[x^2(y')^2]' = 2x^2 y'' y' + 2x(y')^2$, so the preceding equation may be written $[x^2(y')^2]' + 2(x^2 - p^2)yy' = 0$. Observe next that $[(x^2 - p^2)y^2]' = 2(x^2 - p^2)yy' + 2xy^2$, so the preceding equation can be written $2xy^2 = [x^2(y')^2]' + [(x^2 - p^2)y^2]'$. Integrate this last equation to derive the conclusion of the lemma.

As $y = J_p(x)$ satisfies Bessel's equation, we can use Lemma 6.18 to evaluate our integral:

$$2 \int_0^{j_{p,n}} x J_p(x)^2 \, dx = x^2 [J_p'(x)]^2 + (x^2 - p^2)J_p(x)^2 \big|_0^{j_{p,n}}$$

$$= j_{p,n}^2 J_p'(j_{p,n})^2$$

Hence

$$\|\phi_{p,n}(r)\|^2 = \frac{a^2}{2} J_p'(j_{p,n})^2 \tag{6.9.4}$$

Hence, to know these norms, one needs to know the value of the derivative of $J_p$ at its zeros.

**Theorem 6.19.** *For each fixed nonnegative integer $p$ the functions*

$$\phi_{p,n}(r) = J_p\left(j_{p,n}\frac{r}{a}\right), \qquad n = 1, 2, \ldots$$

*form an orthogonal basis of $L^2(0, a; r)$.*

For the proof of this result we refer again to J. R. Higgins' book *Basis Properties of Sets of Special Functions* (Cambridge University Press, 1977, §2.3).

Theorem 6.19 provides infinitely many orthogonal bases of the Hilbert space $L^2(0, a; r)$, one for each integer $p = 0, 1, 2, \ldots$! That fact is generally a source of wonder for students, and the question frequently arises: How do we know which basis to use? There are two answers.

First, theoretically, it makes no difference which basis you use. Any one orthogonal basis of $L^2(0, a; r)$ is as good, theoretically, as any other orthogonal basis. When I say "theoretically," I am referring to the four key properties that we have stressed time and time again, most recently after Theorem 6.7. Let's run over these four properties, once again:

1.  For each fixed nonnegative integer $p$, we have

    $$f(r) = \sum_{n=1}^{\infty} c_n \phi_{p,n}(r) \tag{6.9.5}$$

    valid for *any* $f(r)$ in $L^2(0, a; r)$, where the uniquely determined Fourier-Bessel coefficients $c_n$ are given by $c_n = \langle f, \phi_{p,n} \rangle / \langle \phi_{p,n}, \phi_{p,n} \rangle$. The series (6.9.5) converges in the mean-square metric.
2.  The Fourier-Bessel coefficients $c_n$ make the $N$th partial sum of the series on the right of (6.9.5) the best mean-square approximation to $f$.
3.  The expansion (6.9.5) is valid in the mean-square sense for *every* $f(r)$ in $L^2(0, a; r)$. The *only* requirement on a function $f(r)$ that it have a Fourier-Bessel series (6.9.4) convergent to it in the mean-square metric is the square-integrability requirement:

    $$\int_0^a |f(r)|^2 r \, dr < \infty$$

4.  We may take the inner product of both sides of the Fourier-Bessel series (6.9.5) with any function $g$ in $L^2(0, a; r)$ and we will get a valid numerical series.

All four of those properties hold for any one of the bases $\{\phi_{p,n}: n = 1, 2, \ldots\}$. That is what I mean when I say that, theoretically, it doesn't matter what basis you use. Any orthogonal basis has those four properties.

The second answer is that, practically, there are various advantages that one basis will hold over another. For example, the basis $\{\phi_{6,n}: n = 1, 2, \ldots\}$ diagonalizes the operator $B_6$. Therefore, if you are working with the operator $B_6$, then use as orthogonal basis its eigenfunctions, namely the $\{\phi_{6,n}: n = 1, 2, \ldots\}$. Also, if you are analyzing a physical problem that leads to $B_6$, you should use the orthogonal basis $\{\phi_{6,n}: n = 1, 2, \ldots\}$.

If it is a matter of simply expanding a given function $f$ in a Fourier-Bessel series, then you want to choose the basis that gives the most rapid convergence, so you can take just a few terms of your series to get a good approximation. A reasonably reliable rule of thumb for making such a choice is this: compare the shape of the graph of the function you wish to expand to that of the first few basis functions. The closer the qualitative agreement, the more accurate will be a few-term partial sum. For example, to expand $f(r) = 1 - r^2$ in $L^2(0, 1; r)$ use the $p = 0$ basis because $J_0(0) = 1$ while $J_p(x)$ has, at $x = 0$, a zero of order $p$. So $J_0(j_{0,1}r)$ most closely resembles the shape of your function $f(r) = 1 - r^2$. We have seen in Exercise 1b of Section 6.5 that indeed the first four terms of the $p = 0$ series for $f(c) = 1 - r^2$ does give an excellent approximation to the function. Likewise, for $f(r) = r^p(1 - r^2)$, use the $J_p$ basis.

**Theorem 6.20.** *Select and fix a nonnegative integer $p$. If the function $f(r)$ belongs to $L^2(0, a; r)$ and is orthogonal to every $\phi_{p,n}(r)$, $n = 1, 2, \ldots$, then $f = 0$.*

The proof uses the fact that the $\phi_{p,n}$ are an orthogonal basis of $L^2(0, a; r)$. You are asked to give the proof in Exercise 1.

**Theorem 6.21.** *Bessel's operator $B_p$ has no eigenvalues other than $\lambda_{p,n} = (1/a^2)j_{p,n}^2$, $n = 1, 2, \ldots$.*

*Proof.* Exercise 2.

**Theorem 6.22.** *Each of the eigenvalues of Bessel's operator is simple.*

*Proof.* Exercise 3.

**Theorem 6.23.** *Suppose $f(r)$ lies in the domain of Bessel's operator $B_p$. Then the equation obtained by applying $B_p$ term by term to the Fourier-Bessel expansion for $f(r)$ is valid. Put another way, if $f = \sum_{n=1}^{\infty} c_n \phi_{p,n}$ where $c_n = \langle f, \phi_{p,n} \rangle / \langle \phi_{p,n}, \phi_{p,n} \rangle$ and if $f \in \text{domain}(B_p)$, then*

$$B_p(f) = \sum_{n=1}^{\infty} c_n B_p(\phi_{p,n}) = \frac{1}{a^2} \sum_{n=1}^{\infty} j_{p,n}^2 c_n \phi_{p,n}$$

*Proof.* Exercise 4.

### Exercises

1. Prove Theorem 6.20.
2. Prove Theorem 6.21.
3. Prove Theorem 6.22.
4. Prove Theorem 6.23.
5. What is the kernel of Bessel's operator $B_p$? Explain.
6. Fix a nonnegative integer $p$. Suppose $h$ is a given function in $L^2(0, a; r)$ and thus has a known Fourier-Bessel expansion $h = \sum_{n=1}^{\infty} c_n \phi_{p,n}$, $c_n = \langle h, \phi_{p,n} \rangle / \langle \phi_{p,n}, \phi_{p,n} \rangle$. It is desired to find a function $y$ in the domain of Bessel's operator $B_p$ such that $B_p(y) = h$. Let $y = \sum a_n \phi_{p,n}$ where the $a_n$ are to be determined. Can you express the $a_n$ in terms of the $c_n$ and thus solve the equation $B_p(y) = h$? Explain all your steps.

## 6.10  PROJECT ON BESSEL FUNCTIONS OF
##       NONINTEGRAL ORDER

The gamma function, $\Gamma(x)$, is defined by the integral

$$\Gamma(x) = \int_0^\infty t^{x-1} e^{-t}\, dt$$

convergent for $x > 0$.

1. Show that $\Gamma(1)=1$. Using integration by parts, show that $\Gamma(x+1)=x\Gamma(x)$, $x>0$. Using this latter formula and mathematical induction, show that $\Gamma(n+1) = n!$ for every integer $n = 0,1,2,\ldots$.

2. Show that $\Gamma(\tfrac{1}{2})=2\int_0^\infty e^{-u^2}\, du=\sqrt{\pi}$. Show by induction that

$$\Gamma\left(n+\frac{1}{2}\right) = \frac{(2n)!\sqrt{\pi}}{4^n n!}, \qquad n = 0,1,2,\ldots$$

3. On some calculators, the "$x!$" button computes $\Gamma(x+1)$ for real $x>0$. Thus to compute $\Gamma(n+\tfrac{1}{2})$, enter $n-\tfrac{1}{2}$ and hit $x!$. If your calculator does this, complete the table below. (A simple program will compute $(2n)!\sqrt{\pi}/4^n n!$.)

| $n$ | $(n-\tfrac{1}{2})!$ | $(2n)!\sqrt{\pi}/4^n n!$ |
|---|---|---|
| 0 | | |
| 1 | | |
| 2 | | |
| 3 | | |
| 4 | | |

Thus the gamma function interpolates the factorial function to general—in fact complex—arguments. In the book *Complex Variables* by Polya and Latta, the following facts are established. The gamma function, $\Gamma(z)$, when analytically continued, is analytic everywhere except at the points $z=0, -1, -2,\ldots$, where it has simple poles. $\Gamma(z)$ is never zero; thus $1/\Gamma(z)$ is entire. The function $1/\Gamma(z)$ has simple zeros at the points $z=0, -1, -2,\ldots$.

4. In the formula for the Bessel function

$$J_\nu(x) = \left(\frac{x}{2}\right)^\nu \sum_{l=0}^\infty \frac{(-1)^l}{l!(\nu+l)!}\left(\frac{x}{2}\right)^{2l}$$

replace $(v+l)!$ by $\Gamma(v+l+1)$ and thereby define $J_v$ for *all* values of $v$, not just $v=p$, $p=0,1,2,\ldots$. Prove: if $v$ is not an integer, then $J_v$ and $J_{-v}$ are linearly independent. Is this true if $v$ is an integer?

5. Assume $v$ is not an integer. Show by direct substitution that $y=J_v(x)$ satisfies Bessel's equation $x^2y''+xy'+(x^2-v^2)y=0$. Conclude (without substitution) that $J_{-v}$ also satisfies it. What is the general solution of Bessel's equation when $v$ is not an integer?

6. Using the results of Exercises 2 and 4, show that

$$J_{-1/2}(x)=\left(\frac{2}{\pi x}\right)^{1/2}\cos x \quad\text{and}\quad J_{1/2}(x)=\left(\frac{2}{\pi x}\right)^{1/2}\sin x$$

7. Using the recursion formula for Bessel functions (Section 6.8, Exercise 6) which is valid also for general $v$, prove that the Bessel functions of half-integral order are elementary.

## APPENDIX 6.A  MATHEMATICAL THEORY OF BESSEL'S OPERATOR

This optional appendix presents some details of the mathematical theory underlying Bessel's operator. The intent and spirit of our discussion here, as well as its level, are the same as those of Appendix 4.A, which provided the theory of Legendre's and Hermite's operators. I suggest that the reader refer back to the first few pages of that appendix for the concept of absolute continuity and for the basic properties of the (Lebesgue) integral, as we will again be using that material here.

We are dealing in this case with the Hilbert space $L^2(0,a;r)$ consisting of functions $f(r)$, $0\leqslant r\leqslant a$, such that

$$\int_0^a |f(r)|^2 r\,dr < \infty$$

The inner product in this space is

$$\langle f,g\rangle = \int_0^a f(r)\overline{g(r)}r\,dr$$

In this chapter we have seen how Bessel's differential expression

$$b_p(y) = -\frac{1}{r}\frac{d}{dr}\left(r\frac{dy}{dr}\right)+\frac{p^2}{r^2}y \tag{6.A.1}$$

arises naturally in the mathematical representation of two-dimensional physical problems when the governing equation involves the Laplacian expressed in polar coordinates. Our goal is to construct a Hermitian differential operator in $L^2(0, a; r)$ from Bessel's differential expression. We shall refer to this as a "Bessel's operator" and denote it $B_p$.

The construction of this operator follows the principles laid down in Section 4.1 and which we followed in the construction of Legendre's operator and Hermite's operator. It boils down to specifying the domain for $B_p$. This domain can be no larger than the subspace of $L^2(0, a; r)$ consisting of functions that are compatible with the differential expression $b_p$, that is, functions $y \in L^2(0, a; r)$ such that $b_p(y)$ also belongs to $L^2(0, a; r)$. This compatibility condition on the domain ensures that our operator will always take its values in the Hilbert space $L^2(0, a; r)$; without this condition there is no assurance that we would get a Hilbert space operator at all. So we begin with a study of compatibility and its consequences. For simplicity, we shall assume all functions are real.

**Definition 6.24.** Select and fix an integer $p$. A function $y(r)$ is *compatible* with Bessel's differential expression $b_p(y) = -(1/r)(ry')' + (p^2/r^2)y$ acting in the Hilbert space $L^2(0, a; r)$ if the following four conditions are fulfilled:

1. $y(r) \in L^2(0, a; r)$, i.e., $\int_0^a |y(r)|^2 r \, dr < \infty$.
2. $y(r)$ is differentiable with finite derivative $y'(r)$ for every $r$ in the open interval $0 < r < a$.
3. The function $ry'(r)$ is absolutely continuous in the open interval $0 < r < a$. This means that for any $r_0$, $0 < r_0 < a$, we have

$$ry'(r) - r_0 y'(r_0) = \int_{r_0}^r (ty'(t))' \, dt, \qquad 0 < r < a$$

4. $b_p(y) \in L^2(0, a; r)$, i.e.,

$$\int_0^a \left| -\frac{1}{r}(ry')' + \frac{p^2}{r^2} y \right|^2 r \, dr < \infty$$

As we have seen in our earlier studies in Appendix 4.A, functions that satisfy the compatibility requirement often turn out to be restricted in other, unexpected, ways. Think back, for example, to Hermite's differential expression, where compatibility carried with it the boundary conditions $e^{-x^2/2} f'(x)\overline{g(x)} \to 0$ as $x \to \pm\infty$. This was enough to make the "maximal" operator associated with Hermite's differential expression, namely the operator with the largest possible domain, Hermitian. This will not turn out to be the case with Bessel's operator—other conditions beside compatibility are needed. So our program will be first to investigate the consequences of compatibility, then to see what additional conditions are required (other than the physical boundary condition $f(a) = 0$) to make our operator Hermitian.

The mathematics involved when $p=0$ turns out to be somewhat different from that involved when $|p| \geqslant 1$, so we investigate these two cases separately.

### 6.A.1 The Case $p = 0$

**Lemma 6.25.** *If $y(r)$ is compatible with Bessel's differential expression $b_0(y) = -(1/r)(ry')'$, then*

1. *both $\lim_{r \to a-} y(r)$ and $\lim_{r \to a-} y'(r)$ exist and are finite,* and
2. *$\lim_{r \to 0+} ry'(r)$ exists and is finite.*

*Proof.* In the case $p=0$, condition 4 of Definition 6.24 asserts that $b_0(y) = -(1/r)(ry')'$ belongs to $L^2(0, a; r)$. Now use the fact that a function square integrable over a finite interval is integrable there (use Schwarz's inequality). Thus $b_0(y)$ is integrable over $[0, a]$ so that for some fixed $r_0$, $0 < r_0 < a$, both of the following limits exist:

$$\lim_{r \to 0+} \int_r^{r_0} (xy'(t))' \, dt \quad \text{and} \quad \lim_{r \to a-} \int_{r_0}^r (ty'(t))' \, dt$$

By 3 of Definition 6.24 these integrals are, respectively,

$$r_0 y'(r_0) - ry'(r) \quad \text{and} \quad ry'(r) - r_0 y'(r_0)$$

Hence $\lim ry'(r)$ exists at both ends, $r \to 0+$ and $r \to a-$. In particular, $\lim_{r \to 0+} ry'(r)$ exists and is finite, which proves 2.

As $\lim(1/r) = 1/a > 0$ as $r \to a-$, we get that the $\lim y'(r) = \lim(1/r)(ry'(r)) = (1/a) \lim ry'(r)$ exists as $r \to a-$. That proves the last half of 1. The first half now follows from the fact that, for any fixed $r_0$, $0 < r_0 < a$, the function $y'(r)$ is continuous on the closed interval $[r_0, a]$, so by the fundamental theorem of calculus,

$$y(a) = \int_0^a y'(t) \, dt + y(r_0)$$

is finite. That completes the proof of Lemma 6.25.

Lemma 6.25 tells us to what extent functions are put on their good behavior by the proscription of compatibility. Any function $y(r)$ compatible with Bessel's differential expression $b_0$ is quite well behaved at the right-hand end point $r = a$; both its value $y(a)$ and its derivative $y'(a)$ are finite there. Also, at the left-hand end point $r = 0$, a weak kind of boundary condition is forced, namely the finiteness of $\lim ry'(r)$ as $r \to 0+$.

Compatibility does not force much beyond what is stated in Lemma 6.25. For example, the function $y(r) = A + Cr^2$, where $A$ and $C$ are arbitrary real constants, is compatible with $b_0$, yet $y(a) = A + Ca^2$ and $y'(a) = 2Ca$ may be given any values we wish. Hence compatibility, by itself, imposes no restriction at all on the values $y(a)$ and $y'(a)$ aside from the fact that they are finite. Further, $y(r) = \ln r$, $0 < r \leqslant a$ (and $y(0) = 0$) is compatible with $b_0$, yet $y(r) \rightarrow -\infty$ as $r \rightarrow 0+$ and $\lim ry'(r) = 1$ as $r \rightarrow 0+$. So compatibility does not force $y(r)$ to be bounded in the neighborhood of $r = 0$, nor does it restrict the value of limit of $ry'(r)$ at $r = 0$, aside from the fact that it is finite.

Keep in mind the goal of this appendix. We want to select a subspace of $L^2(0, a; r)$ such that the operator that has that subspace as domain, and has as its "law of action" the Bessel differential expression $b_0$, is Hermitian. For any pair of functions $f(r)$, $g(r)$ that are compatible with $b_0$, both inner products $\langle b_0(f), g \rangle$ and $\langle f, b_0(g) \rangle$ can be formed (are finite). Our goal is to narrow the domain to the extent necessary to secure the equality of these numbers. And we don't want to narrow it excessively so as to limit the range of applications of our operator.

The natural question arises: Is the "maximal" operator associated with Bessel's differential expression, namely the operator whose domain consists of *all* compatible functions, Hermitian? The answer is "no," even if the domain is further restricted by the boundary condition $f(a) = 0$. See Exercises 2 and 3 of Section 6.2. Some further restriction of the domain is necessary if we are to secure a Hermitian operator.

To see what kind of additional restriction is required, we apply the standard method, the same one we have used right along: integration by parts. Select $\varepsilon > 0$, $\varepsilon < a$. Then for any $f, g$ compatible with $b_0$,

$$\int_\varepsilon^a b_0(f)gr \, dr = -\int_\varepsilon^a \frac{1}{r}(rf')'gr \, dr = -\int_\varepsilon^a g \, d(rf')$$

$$= -rf'g|_\varepsilon^a + \int_\varepsilon^a rf'g' \, dr$$

$$\int_\varepsilon^a b_0(f)gr \, dr = \varepsilon f'(\varepsilon)g(\varepsilon) - af'(a)g(a) + \int_\varepsilon^r rf'g' \, dr \tag{6.A.2}$$

As $\varepsilon \rightarrow 0+$, the left side of (6.A.2) tends to $\langle b_0(f), g \rangle$. The right side therefore will necessarily tend to the same finite limit as $\varepsilon \rightarrow 0+$. But we don't know how the first and third terms behave *individually* as $\varepsilon \rightarrow 0+$. The following result shows how to get rid of the term $\varepsilon f'(\varepsilon)g(\varepsilon)$.

**Lemma 6.26.** *If $y(r)$ is compatible with Bessel's differential expression $b_0(y) = -(1/r)(ry')'$, and $\lim_{r \to 0+} y(r)$ exists and is finite, then $\lim_{r \to 0+} ry'(r) = 0$.*

Note what is being asserted here. Lemma 6.25 established that if $y$ is compatible with $b_0$ then $\lim r y'(r)$ exists and is finite as $r \to 0+$. Lemma 6.26 asserts that if we add the requirement that $y(r)$ has a finite limit at $r = 0$, then $\lim r y'(r) = 0$.

*Proof of Lemma 6.26.* In equation (6.A.2) put $g = f$. As $\varepsilon \to 0+$, the left side approaches $\langle b_0(f), f \rangle$. By Lemma 6.25 and our assumption that $f(\varepsilon)$ has a finite limit as $\varepsilon \to 0+$, it follows that $\varepsilon \, f'(\varepsilon) f(\varepsilon)$ has a finite limit as $\varepsilon \to 0+$. Hence

$$\lim_{\varepsilon \to 0+} \int_\varepsilon^a |f'(r)|^2 r \, dr$$

exists and is finite. Thus $f'(r) \in L^2(0, a; r)$. Now we know from Lemma 6.25 that $\lim_{\varepsilon \to 0+} r f'(r) = c$, a finite number. Suppose $c \neq 0$. Then, when $r$ is small, we have $|r f'(r)| \cong |c|$, which implies that for sufficiently small $\delta > 0$ we have

$$r |f'(r)| > \tfrac{1}{2} |c| > 0, \qquad 0 < r < \delta$$

Thus $|f'(r)|^2 > |c|^2 / 4 r^2$, $0 < r < \delta$, whence

$$\int_0^a |f'(r)|^2 r \, dr \geqslant \int_0^\delta |f'(r)|^2 r \, dr > \frac{|c|^2}{4} \int_0^\delta \frac{dr}{r} = \infty$$

which contradicts the fact that $f' \in L^2(0, a; r)$. The proof of Lemma 6.26 is complete. Lemma 6.26 still is valid if we replace the assumption that $\lim f(r)$ exists by the assumption that $f(r)$ is bounded. Almost the same proof works for this case.

Using Lemmas 6.25 and 6.26, we may now prove that Bessel's operator $B_0$ (Definition 6.1 of Section 6.2) is Hermitian and positive definite. See Exercise 4 of Section 6.2.

### 6.A.2 The Case $|p| \geqslant 1$

Bessel's differential expression $b_p(y) = -(1/r)(r y')' + (p^2/r^2) y$ is the sum of two expressions:

$$b_p = b_0 + M_{p^2/r^2}$$

where $b_0(y) = -(1/r)(r y')'$ and $M_{p^2/r^2}(y) = (p^2/r^2) y$ is multiplication by $p^2/r^2$. Noting this, it is tempting to define an operator for $b_p$ by specifying its domain as the intersection of the domains of $B_0$ and $M_{p^2/r^2}$. A difficulty with this approach is that the function $y = r$ is not compatible with $b_0$ or with $M_{1^2/r^2}$, but is compatible with $b_1$:

$$b_0(r) = -\left(\frac{1}{r}\right)(r r')' = -\left(\frac{1}{r}\right) r' = -\frac{1}{r}$$

so

$$\int_0^a (b_0(r))^2 r \, dr = \int_0^a \frac{1}{r^2} r \, dr = \int_0^a \frac{dr}{r} = \infty$$

and

$$M_{1/r^2}(r) = \frac{1}{r^2} r = \frac{1}{r}$$

so by the same argument, $M_{1/r^2}(r) \notin L^2(0, a; r)$. But

$$b_1(r) = -\frac{1}{r} + \frac{1}{r} = 0$$

so $y = r$ is compatible with $b_1$.

So in attempting to define a domain for Bessel's expression $b_p$ by assigning domains individually to $b_0$ and the multiplication operator $M_{p^2/r^2}$ you miss, in the case $p=1$, the function $f(r)=r$, which is compatible with $b_1$ but compatible with neither $b_0$ nor $M_{1^2/r^2}$. We should therefore treat Bessel's differential expression $b_p$ as a single entity rather than the sum of two expressions.

**Lemma 6.27.** *If $y(r)$ is compatible with Bessel's differential expression $b_p(y) = -(1/r)(ry')' + (p^2/r^2)y$, then both $y(r)$ and $y'(r)$ have finite limits at $r = a$.*

Hence, at the right-hand end point $r = a$, compatibility with $b_p$ enforces the same restrictions as does compatibility with $b_0$—refer back to Lemma 6.25.

*Proof of Lemma 6.27.* We shall be brief here, as the proof of Lemma 6.27 follows closely that of Lemma 6.25. Any function that belongs to $L^2(0, a; r)$ also belongs to $L^2(a/2, a; r)$. Hence if $y$ is compatible with $b_p$, then $y \in L^2(a/2, a; r)$ so also $(p^2/r^2)y \in L^2(a/2, a; r)$ as $r^2$ is bounded away from zero on $[a/2, a]$. Thus $b_0(y) = b_p(y) - (p^2/r^2)y$ belongs to $L^2(a/2, a; r)$. Now follow the same pattern of proof as in Lemma 6.25.

Thus any function $y(r)$ compatible with $b_p$ will have a finite value $y(a)$ and a finite derivative $y'(a)$ at the right-hand end point of the interval $[0, a]$. Easy examples show that $y(a)$ and $y'(a)$ may be given any values we wish. Hence compatibility, by itself, imposes no restrictions at all on the values $y(a)$ and $y'(a)$ aside from the fact that they are finite.

Now look at the situation at $r=0$. If $y$ is compatible with $b_p$, then both $y$ and $b_p(y)$ belong to $L^2(0, a; r)$. Hence the integral $\int_0^a b_p(y)yr \, dr$ is finite (use

Schwarz's inequality). Select $t > 0$, $t < a$. Then

$$\int_t^a b_p(y) yr\, dr = \int_t^a \left( -\frac{1}{r}(ry')' + \frac{p^2}{r^2} y \right) yr\, dr$$

$$= -\int_t^a (ry')' y\, dr + p^2 \int_t^a \left( \frac{y}{r} \right)^2 r\, dr$$

Do the usual integration by parts on the first term.

$$= -ry'(r)y(r)|_t^a + \int_t^a (y')^2 r\, dr + p^2 \int_t^a \left( \frac{y}{r} \right)^2 r\, dr$$

So

$$\int_t^a b_p(y) yr\, dr = ty'(t)y(t) - ay'(a)y(a) + \int_t^a (y')^2 r\, dr$$

$$+ p^2 \int_t^a \left( \frac{y}{r} \right)^2 r\, dr \tag{6.A.3}$$

As already pointed out, the limit on the left of equation (6.A.3), the limit being taken as $t \to 0+$, exists (in fact, has the value $\langle b_p(y), y \rangle$). Hence the corresponding limit on the right must also exist. Both integrals on the right have nonnegative integrands, so if either limit is not finite, then it must be $+\infty$. Hence, to balance, we must have $ty'(t)y(t) \to -\infty$ as $t \to 0+$. Now $ty'(t)y(t) = \frac{1}{2}t(y^2(t))'$, so, given any positive $M$ however large, there is a $\delta > 0$ such that $\frac{1}{2}t(y^2(t))' < -M$ for $0 < t < \delta$, whence $(y^2(t))' < -2M/t$ for $0 < t < \delta$. Integrating,

$$\int_u^\delta (y^2(t))'\, dt < -2M \int_u^\delta \frac{dt}{t}$$

so

$$y^2(\delta) - y^2(u) < -2M(\ln \delta - \ln u) \qquad \text{for any } u$$

such that $0 < u < \delta$. Multiplying by $-1$ we get

$$y^2(u) - y^2(\delta) > 2M(\ln \delta - \ln u) \tag{6.A.4}$$

As $-\ln(u) \to +\infty$ as $u \to 0$, we arrive at this conclusion: *if either of the integrals on the right side of 6.A.3 diverges as $t \to 0+$, then $y^2(t) \to \infty$ as $t \to 0+$.* We can restate the conclusion this way: *if $y(t)$ is bounded as $t \to 0+$, then both integrals on the right of (6.A.3) converge.* Noting that the convergence of both is tantamount to the pair of requirements that both $y'$ and $y/r$ belong to $L^2(0, a; r)$, we are in a position to frame a precise result. Remember we are assuming $|p| \geq 1$.

**Lemma 6.28.** *If $y(r)$ is compatible with Bessel's differential expression $b_p(y) = -(1/r)(ry')' + (p^2/r^2)y$, and if $y(r)$ is bounded at $y = 0$, then both $y'(r)$ and $y(r)/r$ belong to $L^2(0, a; r)$.*

**Lemma 6.29.** *Let $D$ denote the subspace of $L^2(0, a; r)$ consisting of functions $y(r)$ that satisfy these two conditions:*

1. *$y$ is compatible with Bessel's differential expression $b_p(y) = -(1/r)(ry')' + (p^2/r^2)y$, $p \neq 0$.*
2. *$y(r)$ is bounded at $r = 0$.*

*Then, for any pair of functions $f(r)$, $g(r)$ in $D$,*

$$\lim_{r \to 0+} rf'(r)g(r) = 0$$

*Proof.* We shall use the following fact repeatedly in this proof: if $f(r)$ and $g(r)$ both belong to $L^2(0, a; r)$, then the integral $\int_0^a f(r)g(r)r\,dr$ is finite. This follows directly from Schwarz's inequality.

Select functions $f$, $g$ in $D$, and choose a small positive number $\varepsilon$. Then

$$\int_\varepsilon^a b_p(f)gr\,dr = \int_\varepsilon^a \left( -\frac{1}{r}(rf')' + \frac{p^2}{r^2}f \right)gr\,dr$$

$$= -\int_\varepsilon^a (rf')'g\,dr + p^2 \int_\varepsilon^a \left(\frac{f}{r}\right)\left(\frac{g}{r}\right)r\,dr$$

Use parts on the first integral:

$$\int_\varepsilon^a (rf')'g\,dr = \int_\varepsilon^a g\,d(rf') = rf'g|_\varepsilon^a - \int_\varepsilon^a f'g'r\,dr$$

$$= af'(a)g(a) - \varepsilon f'(\varepsilon)g(\varepsilon) - \int_\varepsilon^a f'g'r\,dr$$

Combine:

$$\int_\varepsilon^a b_p(f)gr\,dr = \varepsilon f'(\varepsilon)g(\varepsilon) - af'(a)g(a) + \int_\varepsilon^a f'g'r\,dr + p^2 \int_\varepsilon^a \left(\frac{f}{r}\right)\left(\frac{g}{r}\right)r\,dr$$

As both $b_p(f)$ and $g$ belong to $L^2(0, a; r)$, the integral on the left converges as $\varepsilon \to 0+$. So do both integrals on the right—use Lemma 6.28. Hence $\lim \varepsilon f'(\varepsilon)g(\varepsilon)$ exists and is finite, say $\lim \varepsilon f'(\varepsilon)g(\varepsilon) = c$. We must show $c = 0$.

Assume $c > 0$; then, for sufficiently small positive $\delta$, $rf'(r)g(r) > c/2$, $0 < r \leqslant \delta$. So $f'(r)g(r) > c/2r$, which implies $(f')(g/r) > c/2r^2$, $0 < r \leqslant \delta$. Integrate this last inequality:

$$\int_\varepsilon^\delta (f')\left(\frac{g}{r}\right)r\,dr > \frac{c}{2}\int_\varepsilon^\delta \frac{1}{r^2}r\,dr = \frac{c}{2}(\ln \delta - \ln \varepsilon)$$

As $\varepsilon \to 0+$, the right side goes to infinity, which forces the left side to go to infinity. But this is a contradiction—the integral on the left must be finite, because both $f'$ and $g/r$ belong to $L^2(0, a; r)$ by Lemma 6.28. Hence $c > 0$ is excluded. A similar argument excludes $c < 0$, so we must have $c = 0$, and Lemma 6.29 is proved.

Lemmas 6.28 and 6.29 are the key to proving that Bessel's operator $B_p$ is Hermitian and positive definite. See Exercise 3 of Section 6.7.

# 7

# Eigenvalues of the Laplacian, with Applications

## 7.1 THE LAPLACIAN AS A HILBERT SPACE OPERATOR

In the previous chapter we studied the wave equation $\Delta u + (1/w^2)\, \partial^2 u/\partial t^2 = 0$ for a disk (drumhead) $D$ of radius $a$. The wave amplitude $u$ was assumed to satisfy the condition $u = 0$ on the boundary of $D$. Our procedure in that study relied on separation of variables. We wrote $u(r, \theta, t) = R(r)\Theta(\theta)T(t)$, separating the time variable $t$ and the two space variables $r$ and $\theta$ (see Section 6.7).

Let us redo that procedure in a more general way, by separating out only the time variable $t$. Let $p$ symbolize the space variables, for example, $(x, y)$ in Cartesian coordinates or $(r, \theta)$ in polar. Write $u = P(p)T(t)$ expressing the wave amplitude $u$ as the product of two functions, $P$ and $T$, $P$ depending only on the space variables $p$ and $T$ only on the time $t$. As the Laplacian $\Delta$ operates only on the space variables, we have $\Delta u = \Delta(PT) = (\Delta P)T$. Also $\partial^2 u/\partial t^2 = PT''$. Substituting $u = PT$ into the wave equation and dividing by $PT$, we get

$$\frac{\Delta P}{P} = -\frac{1}{w^2}\frac{T''}{T} \qquad (7.1.1)$$

As the left side of (7.1.1) depends only on position and the right side only on time, the left side is independent of position (because the right side is), so must be a constant, say $\lambda$. Thus we get this equation in the space variables

$$\Delta(P) = \lambda P \qquad (7.1.2)$$

402

which carries the additional condition that $P = 0$ on the boundary of $D$, as is dictated by the boundary condition on $u$.

Equation (7.1.2) is an eigenvalue problem for the Laplacian operator $\Delta$; $\lambda$ is the eigenvalue parameter. Equation (7.1.2) is the focus of this chapter. Note that the eigenvalue problem (7.1.2) is, formally, exactly the same as the eigenvalue problems we studied in Chapters 4 and 6, even though (7.1.2) is based on a partial differential expression in two variables. Those earlier problems all had the form $L(y) = \lambda y$, where $L$ represented a Hilbert space operator, namely the one-dimensional Laplacian, or Legendre's or Hermite's operator (Chapter 4), or Bessel's operator (Chapter 6). To bring (7.1.2) under our general theory we must study the two-dimensional Laplacian $\Delta$ as a Hilbert space operator in its own right. That is our project in this chapter.

What is the procedure for constructing a Hilbert space operator from a given differential expression? I hope you remember. It goes like this. First, specify the $L^2$ space in which your operator will act; second, define the domain of your operator as a subspace of that $L^2$ space, and third, prove that the resulting operator is Hermitian.

For the Laplacian $\Delta$, the underlying differential expression is a *partial* differential expression in two variables, namely $-\partial^2/\partial x^2 - \partial^2/\partial y^2$, rather than an ordinary differential expression in one variable. While this fact introduces a few technical difficulties, it does not change this basic procedure for the construction of the Hilbert space operator $\Delta$. Exactly the same Hilbert space concepts apply to this case as to the earlier ones.

To study the two-dimensional Laplacian $\Delta$ as a Hilbert space operator in its own right, we begin with the Hilbert space in which the operator acts, which is the space of square-integrable functions on the disk $D$:

$$L^2(D) = \left\{ f : \iint_D |f|^2 \, dA < \infty \right\} \tag{7.1.3}$$

The symbol $dA$ is the differential element of area; in cartesian coordinates $dA = dx \, dy$, in polar $dA = r \, dr \, d\theta$. Just as in Section 3.9, Theorem 3.16, one can prove that $L^2(D)$ is a vector space; in particular, if $f_1$ and $f_2$ belong to $L^2(D)$, so does $f_1 + f_2$. And, just as in the one-variable case, we speak of a function $f$ as "essentially zero" when $\iint_D |f|^2 \, dA = 0$ (see Section 3.6). Note that a function can be nonzero on a curve in $D$, yet still be essentially zero, because the value of a function on a one-dimensional subset of $D$ won't affect the area integral. (We're talking about *actual* functions here, not mythical delta functions.)

Using exactly the same arguments that established Theorem 3.17 of Section 3.9, we can prove that the formula

$$\langle f, g \rangle = \iint_D f \bar{g} \, dA \tag{7.1.4}$$

defines an inner product on $L^2(D)$. That is to say, the expression (7.1.4) has the three characteristic properties

Conjugate bilinear:
$$\begin{cases} \langle af+bg,h\rangle = a\langle f,h\rangle + b\langle g,h\rangle \\ \langle f,ag+bh\rangle = \bar{a}\langle f,g\rangle + \bar{b}\langle f,h\rangle \end{cases}$$

Hermitian symmetric:    $\langle f,g\rangle = \overline{\langle g,f\rangle}$

Positive definite:
$\langle f,f\rangle \geqslant 0$, and $\langle f,f\rangle = 0$ implies
$f$ is essentially zero

Therefore the Schwarz inequality and all other results that depend only on these three properties remain valid for $L^2(D)$.

Having now secured the appropriate Hilbert space, we will complete the specification of the Laplacian $\Delta$ as a Hilbert space operator by defining its domain as an appropriate subspace of $L^2(D)$. We shall follow the same prescription that we have used right along:

Hilbert space differential operator
= differential expression + domain

We shall use the symbol $\delta$ for the Laplacian differential expression; $\delta(f) = -\partial^2 f/\partial x^2 - \partial^2 f/\partial y^2$ in cartesian coordinates. (Don't confuse this with Dirac's delta function.)

**Definition 7.1 (Provisional)**   Let $D = \{(x,y): x^2+y^2 \leqslant a^2\}$ denote the disk of radius $a > 0$ and $\partial D = \{(x,y): x^2+y^2 = a^2\}$ its boundary. The *Laplacian operator* $\Delta$ has the differential expression

$$-\delta(f) = \begin{cases} \dfrac{\partial^2 f}{\partial x^2} + \dfrac{\partial^2 f}{\partial y^2} & \text{in cartesian coordinates} \\[4mm] \dfrac{\partial^2 f}{\partial r^2} + \dfrac{1}{r}\dfrac{\partial f}{\partial r} + \dfrac{1}{r^2}\dfrac{\partial^2 f}{\partial \theta^2} & \text{in polar coordinates (Section 6.1)} \end{cases}$$

and domain consisting of those functions $f$ in $L^2(D)$ that satisfy the following two conditions:

1.   $f$ is compatible with $\delta$; i.e., $\delta(f) \in L^2(D)$.
2.   $f = 0$ on $\partial D$.

(This definition is a special case of the forthcoming Definition 7.4, which applies to a more general region than a disk and includes other possible boundary conditions; thus the label "provisional.")

According to Definition 7.1, the Laplacian $\Delta$ is defined for the functions, and *only* those functions, that satisfy conditions 1 and 2. For such functions $f$,

$\Delta(f)=\delta(f)$. The differential expression $\delta$ by itself will in general apply to a much larger class of functions. But the *operator* $\Delta$ needs both parts, the differential expression *and* the domain. Thus $\Delta(f)$ is defined only when

1. Both $\iint_D |f|^2\, dA < \infty$ and $\iint_D |\delta(f)|^2\, dA < \infty$ (compatibility).
2. $f = 0$ on $\partial D$ (boundary condition).

The compatibility condition (1) guarantees that our operator, when applied to a function in its domain, produces a function in our Hilbert space $L^2(D)$. Without compatibility we could not be sure that we would be defining a Hilbert space operator at all.

The boundary condition (2), which is known as the Dirichlet boundary condition, reflects the physical origin of our problem—the vibrating drumhead clamped at the rim. Other boundary conditions can be used as well (see Section 7.3).

The compatibility condition contains implicitly a requirement on the continuity of the first partial derivatives. Using very much the same reasoning as in Theorem 4.2 in Section 4.1, we reach the same conclusion:

**Theorem 7.2.** *A necessary condition that a function f be compatible with the Laplacian differential expression is that f and its first partial derivatives be continuous at each interior point of the disk D.*

(The interior points of $D$ are $\{(x, y): x^2 + y^2 < a^2\}$, which are the points of $D$ not on its boundary circle $\partial D = \{(x, y): x^2 + y^2 = a^2\}$.)

For example, suppose $f(x, y) = |x|$. Then $\partial f/\partial x \equiv \pm 1$ according as $x \gtrless 0$. And $\partial f/\partial y = 0$. Thus $\partial^2 f/\partial x^2 = \delta(x)$, Dirac's delta function (see Section 2.10), and $\partial^2 f/\partial y^2 = 0$. So $\delta(f) = -\partial^2 f/\partial x^2 - \partial^2 f/\partial y^2 = -\delta(x)$. (The symbol $\delta(f)$ on the left stands for the Laplacian differential expression $\delta$ applied to the function $f$; the symbol $\delta(x)$ on the right is Dirac's delta function.) Then, from condition 1,

$$\iint_D |\delta(f)|^2\, dA = \int_{-a}^{a} dy \int_{-\sqrt{a^2-y^2}}^{\sqrt{a^2-y^2}} (\delta(x))^2\, dx = \infty$$

(see Section 4.1, equation (4.1.4)). Thus $f(x, y) = |x|$ is not compatible with the Laplacian differential expression, owing to the discontinuity in $\partial f/\partial x$ which produces a delta function in $\partial^2 f/\partial x^2$.

This necessary condition that $f$ and its first partial derivatives be continuous refers to cartesian coordinates. However, polar coordinates (shown in Figure 7.1) are by far the most useful coordinates for the disk. When one defines a function $f(r, \theta)$ in terms of polar coordinates, one must pay careful attention to the behavior of $f$ near $r = 0$ and near $\theta = \pm \pi$ to make

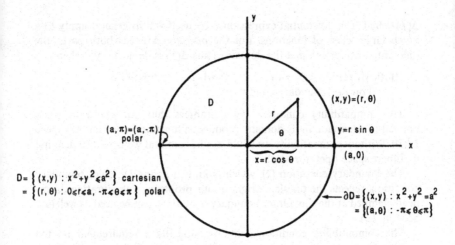

**Figure 7.1**   Polar coordinates and the disk $D$.

$f$ and its first partials continuous on the disk. We'll go into some detail on that matter now.

Polar coordinates are depicted in Figure 7.1. The transformation from polar coordinates $(r, \theta)$ to cartesian coordinates $(x, y)$ is given by

$$x = r \cos \theta, \qquad y = r \sin \theta \tag{7.1.5}$$

with the restrictions

$$r \geqslant 0, \qquad -\pi \leqslant \theta \leqslant \pi \tag{7.1.6}$$

In polar coordinates, all pairs $(0, \theta)$, $-\pi \leqslant \theta \leqslant \pi$, represent the same point in 2-space, the origin. And for each fixed $r > 0$, the two pairs $(r, \pi)$ and $(r, -\pi)$ represent the same point. We can describe the assignment of polar coordinates on the disk in a manner similar to that we used in describing coordinates on the circle. Begin with a rectangle $R = \{(r, \theta): 0 \leqslant r \leqslant a, -\pi \leqslant \theta \leqslant \pi\}$ (see Figure 7.2). Bend the rectangle into a cylinder by joining the top edge to the bottom edge. In this process the left edge of the rectangle becomes a circle, the left circular edge of the cylinder, and the right edge becomes the circular edge of the cylinder at its right end. Now take the left circular edge of the cylinder you have constructed, and shrink it to a point. This will give a cone (kind of). Finally, flatten the cone to make the disk $D$. This process is also described in Figure 7.2.

In practice, we often define a function $f(r, \theta)$ on the disk by some simple formula valid in the rectangle $R$ of Figure 7.2. For example, we might set

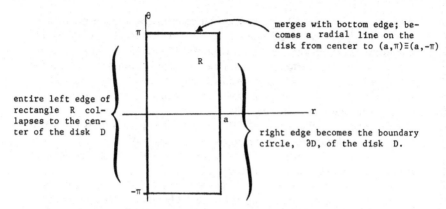

**Figure 7.2** When the rectangle $R = \{(r, \theta): 0 \leqslant r \leqslant a, -\pi \leqslant \theta \leqslant \pi\}$ is mapped via polar coordinates to the disk $D$, the entire left edge collapses to a single point, the center. The bottom and top edges are identified and become a radial line, and the right edge becomes the boundary circle.

$g(r, \theta) = \cos \theta$ or $h(r, \theta) = r \cos(\theta/2)$. Functions like this are quite well behaved in the interior of the rectangle $R$. The problem occurs at the edges because of the identifications we have just described. Thus, if $f(r, \theta)$ is to represent a continuous function *on the disk* $D$, then $f(0, \theta)$ must be independent of $\theta$ because $f(0, \theta)$ is just the value of $f$ at the center of the disk. Moreover, for $\varepsilon > 0$ and small, $f(\varepsilon, \theta)$ must remain close to that constant value. Furthermore, we must have $f(r, \pi) = f(r, -\pi)$, $0 \leqslant r \leqslant a$, because this common value is just the value of $f$ on the radial line on the disk from its center to the point $(a, \pi) \equiv (a, -\pi)$. And, for $\varepsilon > 0$ and small, we must have $f(r, \pi - \varepsilon) \approx f(r, -\pi + \varepsilon)$, $0 \leqslant r \leqslant a$, to ensure continuity of $f$ at this radial line.

In the first example cited, $g(r, \theta) = \cos \theta$, we have $g(0, \theta) = \cos \theta$ which takes all values between $-1$ and $+1$, so there is no hope of assigning some value to $g(r, \theta) = \cos \theta$ at the center of the disk to make it continuous there. Thus $g(r, \theta) = \cos \theta$ does *not* belong to the domain of the Laplacian $\Delta$.

The second example, $h(r, \theta) = r \cos(\theta/2)$, is continuous at the center of the disk and is continuous along the radial line $\{(r, \theta) \equiv (r, -\pi): 0 \leqslant r \leqslant a\}$ because $h(r, \pi) = h(r, -\pi) = 0$, $0 \leqslant r \leqslant a$, and $h(r, \pi - \varepsilon) \approx h(r, -\pi + \varepsilon)$ for $\varepsilon > 0$ and small (owing to the continuity of cosine). But $\partial h / \partial \theta = -(r/2) \sin(\theta/2)$ is not continuous along this radial line (Exercise 9), so $h(r, \theta) = r \cos(\theta/2)$ does not belong to the domain of the Laplacian either.

The continuity of the partial derivatives $\partial f / \partial x$ and $\partial f / \partial y$ is equivalent to the continuity of $\partial f / \partial r$ and $\partial f / \partial \theta$ at every point except $r = 0$. This equivalence follows from the equations $\partial f / \partial x = \cos \theta \, \partial f / \partial r - (1/r) \sin \theta \, \partial f / \partial \theta$ and $\partial f / \partial y = \sin \theta \, \partial f / \partial r + (1/r) \cos \theta \, \partial f / \partial \theta$ (equation (6.1.9) of Section 6.1).

However, $r=0$ is a singular point for polar coordinates, and it is possible for $\partial f/\partial r$ and $\partial f/\partial \theta$ to be continuous at $r=0$ while $\partial f/\partial x$ and $\partial f/\partial y$ are discontinuous there, and vice versa (Exercises 10 and 11). So, at $r=0$, either one needs to use cartesian coordinates to check continuity of partial derivatives, or one may simply visualize the surface $z=f(r,\theta)$ and check whether or not planes containing the $z$ axis cut this surface in curves whose slope is continuous at $r=0$ (Exercise 10).

Remember that the continuity of $f$ and its first partial derivatives is a *necessary* condition for $f$ to belong to the domain of $\Delta$. If $f$ fails these conditions, it cannot belong. But if it passes, you still must test whether $\delta(f)\in L^2(D)$ and whether $f=0$ on $\partial D$.

These fussy considerations apply when you want to test whether a function $f(r,\theta)$ defined on the rectangle $R$ of Figure 7.2 belongs to the *domain* of the Laplacian $\Delta$. If you are only testing whether $f$ belongs to the Hilbert space $L^2(D)$, you can simply define $f(r,\theta)$ in the interior of the rectangle $R$ of Figure 7.2, disregarding its values at the edges. The edges collapse to points and lines in the disk $D$, and the behavior of $f$ on these one-dimensional subsets of the disk can't affect the area integral.

In terms of cartesian and polar coordinates, the integral (7.1.4) is given by

$$\langle f,g \rangle = \begin{cases} \int_{-a}^{a} dy \int_{-\sqrt{a^2-y^2}}^{\sqrt{a^2-y^2}} dx\, f(x,y)\overline{g(x,y)} = \int_{-a}^{a} dx \int_{-\sqrt{a^2-x^2}}^{\sqrt{a^2-x^2}} dy\, f(x,y)\overline{g(x,y)} \\[2mm] \int_{0}^{a} r\,dr \int_{-\pi}^{\pi} d\theta\, f(r,\theta)\overline{g(r,\theta)} = \int_{-\pi}^{\pi} d\theta \int_{0}^{a} r\,dr\, f(r,\theta)\overline{g(r,\theta)} \end{cases}$$

That concludes our introduction to the two-dimensional Laplacian $\Delta$. We have set up the Hilbert space for our operator and have specified its domain. Now comes the important question: Is $\Delta$ Hermitian? Same question: Do we have $\langle \Delta f,g \rangle = \langle f,\Delta g \rangle$ for all $f,g$ in the domain of $\Delta$? In the next two sections we shall generalize the Laplacian to more general planar regions than disks, shall admit boundary conditions more general than the Dirichlet, and shall prove that, even with the broader definition, the Laplacian $\Delta$ is Hermitian.

## Exercises

In all exercises, $D$ stands for the disk of Figure 7.1.

1. In all three cases, determine $\|f\|^2 = \langle f,f \rangle$:

(a) $f(x,y) = \begin{cases} 1 & \text{if } y=0 \\ 0 & \text{if } y\neq 0 \end{cases}$
(b) $f(x,y) = \begin{cases} 1 & \text{if } y=x^2 \\ 0 & \text{if } y\neq x^2 \end{cases}$

(c) $f(r,\theta) = \begin{cases} 1 & \text{if } \theta=0 \\ 0 & \text{if } \theta\neq 0 \end{cases}$

2. Show that, if $|f(x, y)| \leqslant M$ for all $(x, y)$ in $D$, then $\|f\|^2 \leqslant \pi a^2 M^2$. Conclude that any bounded function on the disk belongs to $L^2(D)$.

3. Prove the inequalities
   (a) $(\iint_D |f| |g| \, dA)^2 \leqslant (\iint_D |f|^2 \, dA)(\iint_D |g|^2 \, dA)$ for any $f, g$ in $L^2(D)$.
   (b) $(\iint_D |f| \, dA)^2 \leqslant \pi a^2 \iint_D |f|^2 \, dA$ for any $f$ in $L^2(D)$.

4. Of the following six functions, three belong to $L^2(D)$ and three do not. Determine which is which (prove your answer) and, for the three functions that belong to $L^2(D)$, calculate $\|f\|^2 = \langle f, f \rangle$. (Put $a = 1$.)

   (a) $f(x, y) = \dfrac{1}{x^2 + y^2}$        (b) $f(x, y) = \dfrac{xy}{x^2 + y^2}$

   (c) $f(x, y) = |x|^{-1/2}$        (d) $f(r, \theta) = r^{-3/4}$

   (e) $f(r, \theta) = \dfrac{\cos \theta}{r}$        (f) $f(r, \theta) = r^{-1/2} \ln r$.

5. Let $\gamma_n(r, \theta) = r^n \cos n\theta$, $n = 0, 1, 2, \ldots$, and let $\sigma_n(r, \theta) = r^n \sin n\theta$, $n = 1, 2, \ldots$. Show that, in $L^2(D)$,
   (a) the $\gamma_n$ form an orthogonal family,
   (b) the $\sigma_n$ form an orthogonal family, and
   (c) every $\gamma_n$ is orthogonal to every $\sigma_n$.
   Also compute $\|\gamma_n\|^2$ and $\|\sigma_n\|^2$.

6. If $f_1(r, \theta) = g_1(r)h_1(\theta)$ and $f_2(r, \theta) = g_2(r)h_2(\theta)$ both belong to $L^2(D)$, show that $\langle f_1, f_2 \rangle = \langle g_1, g_2 \rangle_B \langle h_1, h_2 \rangle_C$ where $\langle \cdot, \cdot \rangle_B$ is the inner product in $L^2(0, a; r)$ (Section 6.2) and $\langle \cdot, \cdot \rangle_C$ that in $L^2(C)$, where $C$ is the unit circle (Section 4.5).

7. (a) Using the result of Exercise 2, show that $f(x, y) = |y|$ belongs to $L^2(D)$.
   (b) Is $f(x, y) = |y|$ compatible with the Laplacian differential expression? Prove your answer.

8. Determine the points in $D$ where $f(r, \theta) = r|\cos \theta|$ fails to have continuous first partial derivatives. Is this function compatible with the Laplacian differential expression? Prove your answer.

9. Define a function $h$ on the disk $D$ by $h(r, \theta) = r \cos(\theta/2)$. Show that $h$ is continuous along the radial line $\{(r, \pi) \equiv (r, -\pi): 0 \leqslant r \leqslant a\}$ but $\partial h/\partial \theta$ is not. Is $h$ compatible with the Laplacian differential expression?

10. Let $f(x, y) = \sqrt{x^2 + y^2} = r$.
    (a) Show that $\partial f/\partial r$ and $\partial f/\partial \theta$ are continuous in $D$.
    (b) Show that $\partial f/\partial x$ and $\partial f/\partial y$ are discontinuous at $(0, 0)$.
    (c) Describe the shape of the surface $z = \sqrt{x^2 + y^2}$ in the neighborhood of $(0, 0)$ and use your description to explain the apparent discrepancy in (a) and (b).

11. Let $f(r,\theta)=r\cos\theta$. Show that $\partial f/\partial r$ is discontinuous at $r=0$ but that $\partial f/\partial x$ and $\partial f/\partial y$ are continuous everywhere. Is $f$ compatible with the Laplacian differential expression?

12. For what values of $\alpha$ do we have $\delta(r^\alpha)\in L^2(D)$? ($\delta$ stands for the Laplacian differential expression.)

13. Let $f(r,\theta)=(a^2-r^2)^{-1/4}$.
    (a) Show that $f$ and its first partial derivatives are continuous at each interior point of $D$.
    (b) Show that $f\in L^2(D)$.
    (c) Prove that $f$ is *not* compatible with the Laplacian differential expression.

14. Prove that $\gamma_n(r,\theta)=r^n\cos n\theta$, $n=0,1,2,\ldots$, and $\sigma_n(r,\theta)=r^n\sin n\theta$, $n=1,2,\ldots$, are all compatible with the Laplacian differential expression $\delta$. Do any of these functions belong to the domain of the Laplacian $\Delta$ (Definition 7.1)?

15. For what values of $\mu$ is $f(r,\theta)=r^\mu\cos\mu\theta$ compatible with $\delta$?

16. Set $a=1$ (unit disk). Let $f_{p,m}(r,\theta)=\cos p\theta J_p(j_{p,m}r)$, $p=0,1,2,\ldots$, $m=1,2,\ldots$; and $g_{q,n}(r,\theta)=\sin q\theta J_q(j_{q,n}r)$, $q=1,2,\ldots$, $n=1,2,\ldots$.
    (a) Using the result of Exercise 2, show, without evaluating any integrals, that all $f_{p,m}$ and $g_{q,n}$ belong to $L^2(D)$.
    (b) Using the method of Exercise 6, show that $\langle f_{p,m},f_{q,n}\rangle=\langle g_{p,m},g_{q,n}\rangle$ $=0$ unless both $p=q$ and $m=n$. Show also that $\langle f_{p,m},g_{q,n}\rangle=0$ for all $p$, $q$, $m$, $n$.
    (c) Show that $\delta(f_{p,m})=j_{p,m}^2 f_{p,m}$ and $\delta(g_{q,n})=j_{q,n}^2 g_{q,n}$ where $\delta$ is the Laplacian differential expression.
    (d) Prove that all $f_{p,m}$ and $g_{q,n}$ lie in the domain of the Laplacian operator $\Delta$.
    (e) Explain how parts (c) and (d) supply a sequence of eigenvalues of the Laplacian $\Delta$. List the first seven of the eigenvalues in this sequence ordered according to magnitude, and mark each with its multiplicity.

## 7.2 DIFFERENTIAL FORMS, THE STOKES THEOREM, AND INTEGRATION BY PARTS IN TWO VARIABLES

In the previous section we dealt with a geometric configuration that consisted of the disk $D$ and its boundary $\partial D$, which is a circle. Generalize this configuration: let $C$ stand for a simple closed curve—i.e., a curve in the plane that forms a loop without self-intersections—and let $D$ be the region enclosed by $C$, together with $C$ itself. The general situation is depicted in Figure 7.3. The curve $C$ is the boundary of $D$, and we shall usually write $\partial D$ instead of $C$.

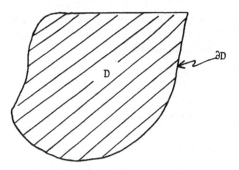

**Figure 7.3**   General region in the plane.

As Figure 7.3 suggests, $\partial D$ need not be a smoothly turning curve, but may have sharp corners.

We can construct the Hilbert space $L^2(D)$ for this general region just as we did when $D$ was the unit disk:

$$L^2(D) = \left\{ f : \iint_D |f|^2 \, dA < \infty \right\}$$

We can define the Laplacian operator $\Delta$ in $L^2(D)$ just as before: it has the same differential expression, $-\delta(f) = \partial^2 f/\partial x^2 + \partial^2 f/\partial y^2$, and its domain consists of functions in $L^2(D)$ that are compatible with $\delta$ and vanish on $\partial D$. And we can ask the same question: Is $\Delta$ Hermitian? That is, do we have $\langle \Delta f, g \rangle = \langle f, \Delta g \rangle$ for all $f, g$ in the domain of $\Delta$? Same question, put in the form of an equation:

$$\iint_D (\Delta f)\bar{g} \, dA \stackrel{?}{=} \iint_D f(\overline{\Delta g}) \, dA \qquad \text{for all } f, g \in \text{domain}(\Delta)$$

or:

$$\iint_D \left( \frac{\partial^2 f}{\partial x^2} + \frac{\partial^2 f}{\partial y^2} \right) \bar{g} \, dA \stackrel{?}{=} \iint_D f \left( \frac{\partial^2 \bar{g}}{\partial x^2} + \frac{\partial^2 \bar{g}}{\partial y^2} \right) dA \qquad \text{for all } f, g \text{ in domain}(\Delta)$$

All our earlier one-variable proofs of Hermitianness have used integration by parts. Here we shall use it again. But this time it will be the two-variable form of integration by parts. And that is the goal of this section: to develop the formula for integration by parts in two variables.

I feel that the clearest way to do this is by use of differential forms. Differential forms occur throughout elementary calculus of several variables, but, while their theory is simple and straightforward, only a few texts develop that theory systematically. So I shall just outline the basic algorithms needed here. They are very easy to understand and to apply. For a deeper exposition of differential forms, one may consult the following texts:

H. M. Edwards, *Advanced Calculus* (Krieger, 1980)

R. C. Buck, *Advanced Calculus* (McGraw-Hill, 1965), Chapter 7

R. T. Seeley, *Calculus of Several Variables* (Scott Foresman, 1970), §5.8

J. Marsden and A. Tromba, *Vector Calculus*, 3rd ed. (W. H. Freeman, 1988), §8.6.

An ordinary function of two variables, $f(x, y)$, we call a *0-form*. A *1-form* is an expression $f(x, y) dx + g(x, y) dy$ that occurs as an integrand in a line integral. A *2-form* is an expression $f(x, y) dx\, dy$ that occurs as the integrand in an area integral. The symbol $\omega$ is frequently used for 1- and 2-forms.

For example,

$e^{-x} \sin y$ is a 0-form.

$\omega = e^{-x} \sin y\, dx + e^{-x} \cos y\, dy$ is a 1-form.

$\omega = e^{-(x^2 + y^2)} dx\, dy$ is a 2-form.

You should be familiar from your calculus course with the integration of 1-forms over curves and the integration of 2-forms over planar regions.

We can add and subtract forms of the same type. The sum of two 0-forms means the same thing as the sum of two functions. For 1-forms, we simply add them term by term. For example, if $\omega_1 = y\, dx - x\, dy$ and $\omega_2 = x\, dx + y\, dy$, then $\omega_1 + \omega_2 = (y + x) dx + (-x + y) dy$. We handle 2-forms the same way: if, for example, $\omega_1 = (x^2 + y^2) dx\, dy$ and $\omega_2 = -2xy\, dx\, dy$, then $\omega_1 + \omega_2 = (x^2 - 2xy + y^2) dx\, dy = (x - y)^2 dx\, dy$. Addition and subtraction of forms of different types are not defined.

The *product* of differential forms is determined by the simple rules

$$dx\, dy = -dy\, dx, \qquad dx\, dx = dy\, dy = 0$$

together with usual distributive and associative laws and the commutativity of functions with differentials: $f(x, y) dx = dx\, f(x, y)$ and $f(x, y) dy = dy\, f(x, y)$. For example, if $\omega_1 = y\, dx - x\, dy$ and $\omega_2 = x\, dx + y\, dy$, then

$$\omega_1 \omega_2 = (y\, dx - x\, dy)(x\, dx + y\, dy)$$

$$= (y\, dx)(x\, dx) + (y\, dx)(y\, dy) - (x\, dy)(x\, dx) - (x\, dy)(y\, dy)$$

$$= xy\, dx\, dx + y^2\, dx\, dy - x^2\, dy\, dx - xy\, dy\, dy$$

$$= 0 + y^2\, dx\, dy + x^2\, dx\, dy - 0$$

$$= (x^2 + y^2)\, dx\, dy$$

Thus, the product of two 1-forms is a 2-form, the product of a 0-form (function) with a form of any type is a form of the same type, and the product of a 1-form with a 2-form is zero (Exercise 2).

Note that the product of forms is noncommutative; the order of factors matters. This is already evident in the rule $dx\, dy = -dy\, dx$.

The last operation we shall introduce is the *exterior derivative* of differential forms. The operation of taking the exterior derivative is denoted "$d$" just as in ordinary differentiation. The exterior derivative of a 0-form (function) is just its total differential. Thus, for a function $f = f(x, y)$, its exterior derivative is

$$df = \frac{\partial f}{\partial x} dx + \frac{\partial f}{\partial y} dy$$

For example,

$$d(e^{-x}\sin y) = \frac{\partial}{\partial x}(e^{-x}\sin y)\, dx + \frac{\partial}{\partial y}(e^{-x}\sin y)\, dy$$

$$= -e^{-x}\sin y\, dx + e^{-x}\cos dy$$

$$= e^{-x}(\cos y\, dy - \sin y\, dx)$$

Thus the exterior derivative of a 0-form is a 1-form.

The exterior derivative of a 1-form, say $\omega = f\, dx + g\, dy$, is defined by the rule

$$d\omega = df\, dx + dg\, dy$$

where the products $df\, dx$ and $dg\, dy$ are calculated using the multiplication rules introduced earlier in this section. Thus

$$df\, dx = \left(\frac{\partial f}{\partial x} dx + \frac{\partial f}{\partial y} dy\right) dx$$

$$= \frac{\partial f}{\partial x} dx\, dx + \frac{\partial f}{\partial y} dy\, dx = -\frac{\partial f}{\partial y} dx\, dy$$

and

$$dg\, dy = \left(\frac{\partial g}{\partial x} dx + \frac{\partial g}{\partial y} dy\right) dy$$

$$= \frac{\partial g}{\partial x} dx\, dy + \frac{\partial g}{\partial y} dy\, dy = \frac{\partial g}{\partial x} dx\, dy$$

So $d\omega = (\partial g/\partial x - \partial f/\partial y)\, dx\, dy$ when $\omega = f\, dx + g\, dy$. We see that the exterior derivative of a 1-form is a 2-form. The exterior derivative of a 2-form, say $\omega = f\, dx\, dy$, is defined to be $d\omega = df\, dx\, dy$. In the 2-dimensional case, which is the one we are exclusively concerned with here, the exterior derivative of a 2-form is always zero (Exercise 11a).

Recall from your calculus course Green's theorem in the plane:

$$\int_{\partial D} f \, dx + g \, dy = \iint_D \left(\frac{\partial g}{\partial x} - \frac{\partial f}{\partial y}\right) dx \, dy \tag{7.2.1}$$

where $D$ is a planar region and $\partial D$ its boundary, just as depicted in Figure 7.3. The line integral on the left is done counterclockwise, in the direction that keeps the region $D$ on your left. The usual hypotheses that guarantee the validity of Green's theorem are that the functions $f$ and $g$ and their first partial derivatives are continuous at each point of $D$, including $\partial D$.

Denote the 1-form that occurs under the integral sign on the left side of (7.2.1) by $\omega$; $\omega = f \, dx + g \, dy$. The exterior derivative $d$ of this 1-form was computed in the previous paragraph: $d\omega = (\partial g/\partial x - \partial f/\partial y) \, dx \, dy$. This is the 2-form that occurs on the right side of Green's formula (7.2.1). Hence, using the notation of differential forms, we may restate Green's theorem in the following neat way:

$$\int_{\partial D} \omega = \iint_D d\omega \tag{7.2.2}$$

Under suitable hypotheses, equation (7.2.2) holds not only for planar regions $D$ but also for surfaces such as a hemisphere (in this case $\partial D$ is the equator curve). Equation (7.2.2) also holds in dimensions greater than 2. In its more general formulation, $D$ is an $n$-dimensional region, $\partial D$ its $(n-1)$-dimensional boundary, and $\omega$ an $(n-1)$-form. For example, $D$ might be a ball in 3-dimensional space; then $\partial D$ would be its spherical surface, and $\omega$ would be a 2-form. In this case, equation (7.2.2) says that (under suitable hypotheses) the 2-dimensional integral of $d\omega$ over the spherical surface equals the 3-dimensional integral of the exterior derivative, $d\omega$, over the solid ball enclosed by that surface. In this extended form, equation (7.2.2), when suitably specialized, includes the classical theorems of Gauss, Green, and Stokes. For this reason, it is often called the *general Stokes theorem*. This one compact and easy to remember formula includes, as special cases, those three major theorems. Part of the value of the theory of differential forms traces to the fact that it is so well adapted to the statement of the general Stokes theorem (7.2.2).

But we are not concerned here with the higher-dimensional application of the Stokes theorem (7.2.2). We want to apply it to the configuration shown in Figure 7.3, that of a region $D$ in the plane bounded by a simple closed curve,

$\partial D$. Our objective? To derive the 2-dimensional formula for integration by parts.

If $f$ is a function and $\omega$ a 1-form, then the product $f\omega$ is again a 1-form. The exterior derivative of $f\omega$ is given by

$$d(f\omega) = (df)\omega + f\,d\omega \tag{7.2.3}$$

The proof, a matter of direct calculation (Exercise 10(iii)). Formula (7.2.3) corresponds in almost every respect with the product rule for differentiation, except that, in (7.2.3), one has to maintain the order of the terms because multiplication of forms is noncommutative.

The forms occurring in (7.2.3) are all 2-forms; integrate them over the region $D$, and rearrange terms:

$$\iint_D f\,d\omega = \iint_D d(f\omega) - \iint_D (df)\omega \tag{7.2.4}$$

Apply the Stokes theorem to the first term on the right, and you have the *2-dimensional formula for integration by parts*:

$$\boxed{\iint_D f\,d\omega = \int_{\partial D} f\omega - \iint_D (df)\omega} \tag{7.2.5}$$

Note the very close parallel with the formula for integration by parts in one variable:

$$\int_a^b u\,dv = uv|_a^b - \int_a^b v\,du$$

The one-variable formula involves functions and their derivatives; formula (7.2.5) involves differential forms and exterior derivatives. In the next section we shall apply formula (7.2.5) to prove that the Laplacian is Hermitian.

### Exercises

1. Each of the following describes a region $D$. In each case sketch and shade the region $D$ and clearly mark $\partial D$.
   (a) $D = \{(x, y): x^2/a^2 + y^2/b^2 \leqslant 1\}$, $a, b > 0$
   (b) $D = \{(r, \theta): 0 \leqslant r \leqslant 2(1 + \cos\theta)\}$.
   (c) $D = \{(x, y): |x| + |y| \leqslant 1\}$.
   (d) $D = \{(r, \theta): 0 \leqslant r \leqslant 2\cos\theta\}$.
   (e) $D = \{(x, y): -1 \leqslant x \leqslant 1, -1 \leqslant y \leqslant 1\}$.
2. (a) If $\omega_1 = x\,dx + y\,dy$ and $\omega_2 = y\,dx - x\,dy$, compute $\omega_1 + \omega_2$, $\omega_1 - \omega_2$, $\omega_1\omega_2$, and $\omega_2\omega_1$.

(b) If $\omega_1 = f_1\,dx + g_1\,dy$, $\omega_2 = f_2\,dx + g_2\,dy$, compute $\omega_1\omega_2$ and $\omega_2\omega_1$. Thus show $\omega_1\omega_2 = -\omega_2\omega_1$.

(c) Prove that the product of a 1-form $\omega = f\,dx + g\,dy$ and a 2-form $\sigma = h\,dx\,dy$ is always zero: $\omega\sigma = \sigma\omega = 0$.

3. Compute $df$ for the following functions:

    (a) $f = e^x \cos y$        (c) $f = \arctan(y/x)$

    (b) $f = x^2 + y^2$      (d) $f = \sin^2 y$

4. Compute the exterior derivative of the functions $\partial f/\partial x$ and $\partial f/\partial y$.

5. Let $f = f(x,y)$ and $g = g(x,y)$ be functions and $c$ a constant. Prove the following rules for the exterior derivative:

    (i) $d(cf) = c\,df$ homogeneous $\left.\vphantom{\begin{matrix}a\\b\end{matrix}}\right\}$ linear

    (ii) $d(f + g) = df + dg$ additive

    (iii) $d(fg) = (df)g + f\,dg$ product rule

    (iv) $d(f/g) = (g\,df - (dg)f)/g^2$ quotient rule

    (v) $d(f^\alpha) = \alpha f^{\alpha-1}\,df$ power rule

6. Show that the exterior derivative of a function $f(x)$ of one variable is $df = f'(x)\,dx$, the ordinary differential. Use this fact to recompute Exercise 3d. Use this, together with the rule $d(f+g) = df + dg$, to recompute Exercise 3b. Use this, together with the product rule 5(iii), to recompute Exercise 3a.

7. Show: If $\phi(\cdot)$ is a function of one variable, and if $f(x,y) = \phi(g(x,y))$, then $df = \phi'(g)\,dg$. Use this, and the quotient rule 5(iv), to recompute Exercise 3(c).

8. Compute the exterior derivative of each of the following 1-forms:

    (a) $\omega = \frac{1}{2}(-y\,dx + x\,dy)$     (c) $\omega = (x^2 + y^2)\,dy$

    (b) $\omega = (\partial f/\partial y)\,dx - (\partial f/\partial x)\,dy$     (d) $\omega = (x\,dy - y\,dx)/(x^2 + y^2)$

9. The calculus of differential forms yields another benefit; it automatically computes the element of area in a change of coordinates.

    (a) Prove the formula $dx\,dy = r\,dr\,d\theta$ for the element of area in polar coordinates by computing $dx$ and $dy$ for $x = r\cos\theta$, $y = r\sin\theta$ and then multiplying.

    (b) For a general change of coordinates $x = x(u,v)$, $y = y(u,v)$, prove the formula

$$dx\,dy = \begin{vmatrix} \dfrac{\partial x}{\partial u} & \dfrac{\partial x}{\partial v} \\[2mm] \dfrac{\partial y}{\partial u} & \dfrac{\partial y}{\partial v} \end{vmatrix} du\,dv$$

    which involves the *Jacobian determinant*.

10. Let $f = f(x,y)$ be a function, $\omega = g\,dx + h\,dy$, $\sigma = p\,dx + q\,dy$ 1-forms, and $c$ a constant. Prove the following rules for the exterior derivative:

    (i) $d(c\omega) = c\,d\omega$   homogeneous $\Big\}$ linear
    (ii) $d(\omega + \sigma) = d\omega + d\sigma$   additive
    (iii) $d(f\omega) = (df)\omega + f\,d\omega$ $\Big\}$ product rule
    (iv) $d(\omega f) = (d\omega)f - \omega\,df$
    (v) $d(\omega/f) = (f\,d\omega - (df)\omega)/f^2$   quotient rule

11. In the following steps, prove that $d^2 = 0$.
    (a) Show that for any 2-form $\omega = f\,dx\,dy$, $d\omega = 0$. Deduce that $d^2\omega = d(d\omega) = 0$ for any 1-form $\omega$.
    (b) Show that $d^2 f = 0$ for any function $f = f(x, y)$. On what does this fact rely?
    (c) Deduce from (a) and (b) that $d^2 = 0$ always.

12. A 1-form $\omega$ is *closed* if $d\omega = 0$; *exact* if $\omega = df$ for some function $f$. A 2-form $\sigma$ is *exact* if $\sigma = d\omega$ for some 1-form $\omega$. Prove that any exact form is closed (use Exercise 11c).

13. (Continuation) Which of the following forms is closed?
    (a) $\omega = y\,dx + x\,dy$         (c) $\omega = x\,dx + dy$
    (b) $\omega = xy\,dx + \frac{1}{2}(x^2 + y^2)\,dy$     (d) $\omega = df$, $f$ a function

14. (Continuation) When is $\omega = f\,dx + g\,dy$ closed?

15. (Continuation) By finding in each case a 1-form $\omega$ such that $\sigma = d\omega$, show that each of the following 2-forms is exact:
    (a) $\sigma = 2\,dx\,dy$       (b) $\sigma = 2(x + y)\,dx\,dy$

16. (Continuation) By finding a 1-form $\omega$ such that $\sigma = d\omega$, show that the 2-form

$$\sigma = \left(\frac{\partial^2 f}{\partial x^2} + \frac{\partial^2 f}{\partial y^2}\right) dx.dy$$

is exact.

17. By converting to polar coordinates, show that
    (a) $-y\,dx + x\,dy = r^2\,d\theta$
    (b) $\dfrac{\partial f}{\partial y}\,dx - \dfrac{\partial f}{\partial x}\,dy = \dfrac{1}{r}\dfrac{\partial f}{\partial \theta}\,dr - r\dfrac{\partial f}{\partial r}\,d\theta$

18. Let $\partial D$ be the boundary of the square with vertices $(0, 0)$, $(1, 0)$, $(1, 1)$, and $(0, 1)$ and let $\omega = -y\,dx + x\,dy$.
    (a) Compute $\int_{\partial D} \omega$
    (b) Compute $\int_D d\omega$
    (c) Compare your answers to (a) and (b) and thus check the general Stokes theorem (equation (7.2.2)) in this particular case.

19. Same as Exercise 18 with $\partial D = $ unit circle. (Use polar coordinates.)

20. Show that the area of the general region $D$ depicted in Figure 7.3 is given by the integral

$$\frac{1}{2}\int_{\partial D} -y\,dx + x\,dy$$

Use this formula to show that the area of the ellipse whose boundary is parametrized by $x = a\cos\theta$, $y = b\sin\theta$, $-\pi \leqslant \theta \leqslant \pi$, equals $\pi ab$.

21. Derive the formula

$$\iint_D \left(\frac{\partial^2 f}{\partial x^2} + \frac{\partial^2 f}{\partial y^2}\right) dx\,dy = \int_{\partial D} -\frac{\partial f}{\partial y}\,dx + \frac{\partial f}{\partial x}\,dy$$

22. The general Stokes theorem

$$\int_{\partial D} \omega = \iint_D d\omega \qquad\qquad\qquad (7.2.2)$$

holds in any dimension. Interpret it in the following dimension 1 case. Let $D$ be the interval $a \leqslant x \leqslant b$. Then $\partial D$, the boundary to $D$, consists of just two points: $x = a$ and $x = b$. Let $\omega$ be a 0-form, $\omega = f(x)$. What does formula (7.2.2) become in this case?

23. (Continuation)  Let $D$ be a *curve* in the $x$–$y$ plane from the point $(x_0, y_0)$ to the point $(x_1, y_1)$. By interpreting (7.2.2) in this case, derive the formula

$$\int_D \frac{\partial f}{\partial x}\,dx + \frac{\partial f}{\partial y}\,dy = f(x_1, y_1) - f(x_0, y_0)$$

for a (well-behaved) function $f(x, y)$. In particular, what is the integral on the left when $(x_0, y_0) = (x_1, y_1)$, i.e., when curve $D$ is closed?

## 7.3  THE LAPLACIAN IS HERMITIAN

In the previous section we traveled a detour to develop some of the basic calculus of two variables. This side trip was necessary because certain topics we needed, such as differential forms, the general Stokes theorem, and especially the formula for integration by parts in several variables (equation (7.2.5)), are not normally part of the calculus sequence. We developed this theory for the purpose of proving that the Laplacian is Hermitian. Let us apply it to that task now.

Consider the general configuration described at the beginning of the previous section: $D$ a region in the plane bounded by a simple closed curve $\partial D$. Figure 7.4 is a repeat of Figure 7.3 from the preceding section depicting that configuration. For emphasis, we also repeat here the construction of the Hilbert space $L^2(D)$.

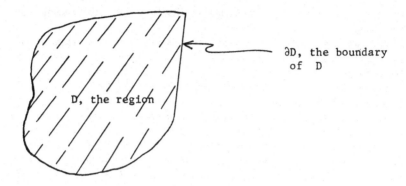

∂D, the boundary
of D

D, the region

**Figure 7.4**   General region in the plane.

The Hilbert space $L^2(D)$ consists of those functions that are square integrable over $D$:

$$L^2(D) = \left\{ f : \iint_D |f|^2 \, dx \, dy < \infty \right\}$$

We have used cartesian coordinates, $dA = dx \, dy$, to be concrete. In practice, you should use whatever coordinate system is most convenient. The choice will depend mainly on the shape of $D$. If $D$ is a rectangle, then use cartesian coordinates; if $D$ is a disk, use polar. In this section we will stick with cartesian coordinates.

The Hilbert space $L^2(D)$ carries the inner product

$$\langle f, g \rangle = \iint_D f\bar{g} \, dx \, dy$$

which one checks easily (as in the case of the disk) has the three characteristic properties: conjugate bilinear, Hermitian symmetric, and positive definite. The elements of this Hilbert space are functions of two variables. The Hilbert spaces we considered in earlier chapters consisted of functions of one variable. Does this confuse us? Not us! We *know* that once we have established the three key properties, conjugate bilinear, Hermitian symmetric, and positive definite, then *everything* we proved in Chapter 3 holds automatically here. Those proofs, which depended *only* on those three properties, hold in all particular cases. Nothing has to be reproved. So, for example, the inequality in Exercise 3 of Section 7.1 is just Schwarz's inequality specialized to this particular case. Exercise 3 didn't require a proof (did you notice that?). Our Hilbert space theory provides a single framework to tie together formulas and constructions that might not otherwise appear connected.

Now we come to the definition of the Laplacian. Our earlier definition of the Laplacian, Definition 7.1, was special in two respects: first, it applied to the particular region $D =$ unit disk; second, it used the Dirichlet boundary condition (that $f = 0$ on the unit circle $\partial D$). We want to broaden our definition in both respects.

We can easily go from the unit disk to the general domain $D$, such as pictured in Figure 7.4. Just form the Hilbert space $L^2(D)$, use the same differential expression $\delta$, and begin with the "maximal" operator whose differential expression is $\delta$ and whose domain consists of functions $f$ in $L^2(D)$ that are compatible with $\delta$, i.e., for which $\delta(f) \in L^2(D)$. It is then a matter of restricting the domain further through the imposition of boundary conditions. Each different boundary condition will produce a different Laplacian operator because, as we well remember, operators with different domains are different Hilbert space operators notwithstanding that they share the same differential expression.

As to the boundary conditions, the particular boundary condition used in any individual case depends on the physical problem being modeled. The Dirichlet boundary condition, namely that $f = 0$ on $\partial D$, is the mathematical representation of either the clamped rim of the vibrating drumhead, the zero-temperature edge of the heated disk, or the continuity of the wave function $\Psi$ for the quantum particle confined to a circular region. Other physical situations will call for other boundary conditions.

For example, consider the heated disk with insulated rim. The insulated rim means that no heat can enter or leave the disk through its edge. (In this 2-dimensional model, heat flows only in the plane of the disk. No heat flows vertically.) In our discussion of heat flow (Chapter 5) we stated one of the two basic principles of heat flow as follows: "The quantity of heat flowing across a unit area per second is proportional to the directional derivative of the temperature normal to that area, and the flow is in the direction of decreasing temperature." For a heat flow problem in a region $D$ whose boundary $\partial D$ is insulated, the boundary condition is therefore that the directional derivative of the temperature $u(x, y, t)$, in the direction normal to the boundary curve $\partial D$, vanishes on that boundary.

Suppose the boundary curve is described parametrically by the smooth functions $x = x(t)$, $y = y(t)$ or, equivalently, by the vector equation $\bar{r}(t) = x(t)\bar{i} + y(t)\bar{j}$, where $\bar{i} = (0, 1), \bar{j} = (0, 1)$ are unit vectors in the positive $x$ and $y$ directions, respectively. Then the vector $\bar{r}'(t) = x'(t)\bar{i} + y'(t)\bar{j}$ is tangent to the curve $\partial D$. The vector $\bar{n}(t) = -y'(t)\bar{i} + x'(t)\bar{j}$ is orthogonal to the tangent vector because their dot product is zero: $\bar{r} \cdot \bar{n} = -x'y' + x'y' = 0$. Hence $\bar{n}$ is normal to the boundary curve. The directional derivative of the temperature function

$u(x, y, t)$ in the direction $\bar{n}$ is $\nabla u \cdot \bar{n}/\|\bar{n}\|$, where $\nabla u$ is the gradient of $u$: $\nabla u = (\partial u/\partial x)\bar{i} + (\partial u/\partial y)\bar{j}$. We have

$$\nabla u \cdot \bar{n} = -\frac{\partial u}{\partial x}\frac{dy}{dt} + \frac{\partial u}{\partial y}\frac{dx}{dt}$$

Thus, for a heat flow problem in a region $D$ with insulated boundary $\partial D$, the mathematical boundary condition that expresses the fact that no heat flows across the boundary is $\nabla u \cdot \bar{n} = 0$ on $\partial D$. This, in turn, implies that $(\nabla u \cdot \bar{n}) dt = 0$ on $\partial D$. But $(\nabla u \cdot \bar{n}) dt = -(\partial u/\partial x) dy + (\partial u/\partial y) dx$, a differential 1-form. Hence we arrive at the *Neumann boundary condition*, namely that the differential 1-form $(\partial u/\partial y) dx - (\partial u/\partial x) dy$ vanishes on $\partial D$. This is equivalent to the condition that the derivative of the temperature $u$, in the direction $\bar{n}$ normal to $\partial D$, equals zero on $\partial D$.

The Neumann boundary condition also applies to the vibrating drumhead with free boundary. "Free boundary" means that the drumhead is held in tension by a mechanism at its edge that allows it to move up and down freely, without friction. Hence there can be no vertical component of the tension at the edge of the drumhead. This means that the drumhead must remain flat at its edge; i.e., its slope there, normal to the boundary, must be zero. That is the Neumann boundary condition.

We summarize our discussion in a formal definition.

**Definition 7.3.** Let $\partial D$ be a simple closed curve in the plane and let $D$ stand for the region interior to this curve together with the curve itself. A function defined in $D$ is said to satisfy the

*Dirichlet boundary condition if* $f = 0$ on $\partial D$
*Neumann boundary condition if* $(\partial f/\partial y) dx - (\partial f/\partial x) dy = 0$ on $\partial D$

Mixed boundary conditions are also possible where $f = 0$ on one portion of $\partial D$ and $(\partial f/\partial y) dx - (\partial f/\partial x) dy = 0$ on the remaining portion. Such a mixed boundary condition would apply, for example, to the heated unit disk where one semicircular section of the boundary was held at zero temperature and the remaining semicircular section was insulated.

We define the Laplacian Hilbert space operator in $L^2(D)$ by first forming the maximal operator whose differential expression is $\delta$ and whose domain consists of those functions $f$ in $L^2(D)$ for which $\delta(f)$ also belongs to $L^2(D)$ (functions compatible with $\delta$). Then the Laplacian is obtained by restricting the domain of the maximal operator by application of the Dirichlet, Neumann, or mixed boundary conditions. Because different domains mean different operators, each different boundary condition determines a different operator. We shall nonetheless follow custom and use the single name "Laplacian," and the symbol $\Delta$, for any Hilbert space operator based on the differential expression $\delta(f) = -\partial^2 f/\partial x^2 - \partial^2 f/\partial y^2$. This is a useful ambiguity in terminology. When necessary, we shall make clear which Laplacian is meant.

**Definition 7.4.** Let $\partial D$ be a simple closed curve in the plane, and let $D$ stand for the region interior to this curve together with the curve itself. The *Laplacian* operator $\Delta$, which acts in the Hilbert space $L^2(D)$, has the differential expression $\delta$ given by $-\delta(f) = \partial^2 f/\partial x^2 + \partial^2 f/\partial y^2$ and domain consisting of functions $f$ in $L^2(D)$ that are compatible with $\delta$ and satisfy either the Dirichlet, Neumann, or mixed boundary conditions on $\partial D$.

Our objective here is to prove that $\Delta$ is Hermitian, i.e., that $\langle \Delta(f), g \rangle = \langle f, \Delta(g) \rangle$ for all functions $f, g$ in the domain of $\Delta$. The identity that we establish in the next theorem will be the key. To prove it, we apply the formula for integration by parts in several variables (equation (7.2.5) of Section 7.2).

**Theorem 7.5.** *For all functions $f, g$ in the domain of the Laplacian $\Delta$, we have*

$$\iint_D \Delta(f)\bar{g}\,dx\,dy = \iint_D \left( \frac{\partial f}{\partial x}\frac{\partial \bar{g}}{\partial x} + \frac{\partial f}{\partial y}\frac{\partial \bar{g}}{\partial y} \right) dx\,dy$$

*Equivalently,*

$$\langle \Delta(f), g \rangle = \left\langle \frac{\partial f}{\partial x}, \frac{\partial g}{\partial x} \right\rangle + \left\langle \frac{\partial f}{\partial y}, \frac{\partial g}{\partial y} \right\rangle$$

*for all $f, g$ in domain($\Delta$). These formulas are valid for any boundary condition, Dirichlet, Neumann, or mixed.*

**Proof.** We begin our proof by computing the exterior derivative of the 1-form $\omega = (\partial f/\partial y)\,dx - (\partial f/\partial x)\,dy$:

$$\begin{aligned}
d\omega &= d\left(\frac{\partial f}{\partial y}\right)dx - d\left(\frac{\partial f}{\partial x}\right)dy \\
&= \left( \frac{\partial}{\partial x}\left(\frac{\partial f}{\partial y}\right)dx + \frac{\partial}{\partial y}\left(\frac{\partial f}{\partial y}\right)dy \right)dx - \left( \frac{\partial}{\partial x}\left(\frac{\partial f}{\partial x}\right)dx + \frac{\partial}{\partial y}\left(\frac{\partial f}{\partial x}\right)dy \right)dy \\
&= \frac{\partial^2 f}{\partial x\,\partial y}\,dx\,dx + \frac{\partial^2 f}{\partial y^2}\,dy\,dx - \frac{\partial^2 f}{\partial x^2}\,dx\,dy + \frac{\partial^2 f}{\partial y\,\partial x}\,dy\,dy \\
&= -\left( \frac{\partial^2 f}{\partial x^2} + \frac{\partial^2 f}{\partial y^2} \right)dx\,dy \\
&= \Delta(f)\,dx\,dy \qquad \text{when } f \in \text{domain}(\Delta)
\end{aligned}$$

Thus

$$\iint_D \Delta(f)\bar{g}\,dx\,dy = \iint_D \bar{g}\,d\omega \qquad \text{for } f, g \in \text{domain}(\Delta) \tag{7.3.1}$$

To the integral on the right we apply integration by parts:

$$\iint_D \bar{g}\,d\omega = \int_{\partial D} \bar{g}\omega - \iint_D (d\bar{g})\,\omega \tag{7.3.2}$$

Now compute the integrand on the far right of (7.3.2):

$$(d\bar{g})\omega = \left(\frac{\partial \bar{g}}{\partial x}\,dx + \frac{\partial \bar{g}}{\partial y}\,dy\right)\left(\frac{\partial f}{\partial y}\,dx - \frac{\partial f}{\partial x}\,dy\right)$$

$$= \frac{\partial \bar{g}}{\partial x}\frac{\partial f}{\partial y}\,dx\,dx - \frac{\partial \bar{g}}{\partial x}\frac{\partial f}{\partial x}\,dx\,dy + \frac{\partial \bar{g}}{\partial y}\frac{\partial f}{\partial y}\,dy\,dx - \frac{\partial \bar{g}}{\partial y}\frac{\partial f}{\partial x}\,dy\,dy$$

$$= -\left(\frac{\partial f}{\partial x}\frac{\partial \bar{g}}{\partial x} + \frac{\partial f}{\partial y}\frac{\partial \bar{g}}{\partial y}\right)dx\,dy$$

Insert this and the formula for $\omega$ into (7.3.2), and combine with (7.3.1):

$$\iint_D \Delta(f)\bar{g}\,dx\,dy = \int_{\partial D} \bar{g}\left(\frac{\partial f}{\partial y}\,dx - \frac{\partial f}{\partial x}\,dy\right) + \iint_D \left(\frac{\partial f}{\partial x}\frac{\partial \bar{g}}{\partial x} + \frac{\partial f}{\partial y}\frac{\partial \bar{g}}{\partial y}\right)dx\,dy$$

$$\tag{7.3.3}$$

Now consider the first integral on the right of equation (7.3.3). It is an integral of the 1-form $\bar{g}((\partial f/\partial y)\,dx - (\partial f/\partial x)\,dy)$ around the boundary of $D$. The functions $f$ and $g$ in equation (7.3.3) lie in the domain of $\Delta$. If we are using the Dirichlet boundary condition, then both $f$ and $g$ will be zero on $\partial D$ so, in particular, $\bar{g}((\partial f/\partial y)\,dx - (\partial f/\partial x)\,dy) = 0$ on $\partial D$; thus the integral of this form over the boundary is zero. If we are using the Neumann boundary condition, then both $(\partial f/\partial y)\,dx - (\partial f/\partial x)\,dy = 0$ and $(\partial g/\partial y)\,dx - (\partial g/\partial x)\,dy = 0$ on $\partial D$. So, in this case too, the integrand is zero on $\partial D$; thus the first integral on the right of (7.3.3) is again zero. For the mixed boundary condition, write the integral over $\partial D$ as the sum of two integrals, one over the portion where the functions vanish, the other over the remaining portion where the 1-form vanishes. Each integral is zero because the integrand is zero. Once again, the first integral on the right of equation (7.3.3) is zero. Thus it is zero for either the Dirichlet, Neumann, or mixed boundary conditions, and we have proved Theorem 7.5.

Notice that there is much less attention to mathematical niceties in the proof of Theorem 7.5 than there was in the corresponding proofs in Chapters 4 and 6 of the Hermitian character of the one-variable Legendre, Hermite, and Bessel operators. I have relaxed the level of mathematical rigor here in the 2-variable case because the 2-variable rigorous analysis becomes so complex it would overwhelm the discussion. Thus, in equation (7.3.3), in the first term on the right of the equal sign, we are taking for granted that the functions involved are sufficiently regular along the boundary of $D$ so that the product $\bar{g}((\partial f/\partial y)\,dx - (\partial f/\partial x)\,dy)$ goes to zero on $\partial D$ if either $\bar{g}=0$ or the 1-form is zero. That is, we are assuming that there is no singularity in one factor that cancels the zero of the other. In the one-variable case we proved rigorously (in appendices) that compatibility forced strong restrictions on the boundary behavior of the functions involved. In this, the 2-variable case, we are omitting the corresponding rigorous arguments. Thus, we are "thinking like engineers or physicists." This can be interpreted as meaning that we are thinking clearly, because we are leaving aside, for the present, excessive details that might muddy the waters and threaten our clear perception of the main features of the Laplacian operator.

From Theorem 7.5 we deduce our main result.

**Theorem 7.6.** *The Laplacian is Hermitian and positive semidefinite (under either the Dirichlet, Neumann, or mixed boundary conditions). With the Dirichlet boundary condition, it is also positive definite.*

You are asked to supply the proof in Exercise 6.

### Exercises

1. Show that the function $f(x,y)=9xy-3xy^3-3x^3y+x^3y^3$ satisfies the Neumann boundary condition on the square $D=\{(x,y):\ -1\leqslant x\leqslant 1,\ -1\leqslant y\leqslant 1\}$. [Hint: On the top edge of the square $y=1$ and $dy=0$, so $f_y\,dx-f_x\,dy=f_y(x,1)\,dx$. Similarly for the other three edges.]

2. Show that in polar coordinates

$$\frac{\partial f}{\partial y}\,dx - \frac{\partial f}{\partial x}\,dy = \frac{1}{r}\frac{\partial f}{\partial \theta}\,dr - r\frac{\partial f}{\partial r}\,d\theta \quad \text{(see Exercise 17b of Section 7.2)}$$

   Thus express the Neumann boundary condition in polar coordinates.

3. (a) Using the result in Exercise 2, show that $f(r,\theta)=(3r-r^3)g(\theta)$ satisfies the Neumann boundary condition on the unit disk for any $g(\theta)$.

   (b) Show that for any disk of radius $a$, $f(r,\theta)$ satisfies the Neumann boundary condition if, and only if,

$$\frac{\partial f}{\partial r}(a,\theta) = 0$$

   for all $\theta$.

4. Show that, on the unit disk, all the functions

$$\cos m\theta J_0(j_{1,n}r), \qquad m = 0, 1, 2, \ldots, n = 1, 2, \ldots$$
$$\sin m\theta J_0(j_{1,n}r), \qquad m = 1, 2, \ldots, n = 1, 2, \ldots$$

satisfy the Neumann boundary condition. (See Exercise 4 of Section 6.8.)

5. Consider the region $D$ bounded by the ellipse

$$\partial D = \left\{(x, y): \frac{x^2}{a^2} + \frac{y^2}{b^2} = 1\right\}$$

Show that $f(x, y) = 3g - g^3$ satisfies the Neumann boundary condition, where $g = g(x, y) = (x^2/a^2 + y^2/b^2)^{1/2}$.

6. Start with the formula from Theorem 7.5: for all $f, g$ in the domain of the Laplacian $\Delta$ we have

$$\langle \Delta f, g \rangle = \left\langle \frac{\partial f}{\partial x}, \frac{\partial g}{\partial x} \right\rangle + \left\langle \frac{\partial f}{\partial y}, \frac{\partial g}{\partial y} \right\rangle$$

   (a) Using this formula, prove that $\Delta$ is Hermitian.
   (b) Using this formula, prove that $\Delta$ is positive semidefinite, i.e., $\langle \Delta f, f \rangle \geqslant 0$ for all $f$ in the domain of $\Delta$.
   (c) Assuming the Dirichlet boundary condition, prove that $\Delta$ is positive definite; i.e., prove that if $f$ belongs to the domain of $\Delta$ and $\langle \Delta f, f \rangle = 0$ then $f = 0$.

7. What two conclusions about its eigenvalues and eigenfunctions can you deduce from the fact that Laplacian is Hermitian? What additional information about its eigenvalues can you deduce from the fact that $\Delta$ is positive semidefinite? What other information about the eigenvalues is provided by the Dirichlet boundary condition? Prove your answers.

8. When is 0 an eigenvalue of $\Delta$ and when not? Prove your answer, which will depend on the boundary conditions.

9. (Continuation) Suppose 0 is an eigenvalue of $\Delta$. Find a corresponding eigenfunction, and prove that 0 must be simple (multiplicity 1).

10. In the proof of Theorem 7.5 we showed that if $\omega = (\partial f/\partial y) dx - (\partial f/\partial x) dy$ then $d\omega = \delta(f) dx\, dy$ where $\delta(f)$ is the Laplacian differential expression in Cartesian coordinates. Express $\omega$ in polar coordinates; then compute the exterior derivative of this expression to show that $d\omega = \delta(f)r\, dr\, d\theta$ where $\delta(f)$ is the Laplacian differential expression in polar coordinates (Section 6.1).

## 7.4 GENERAL FACTS ABOUT THE EIGENVALUES OF THE LAPLACIAN

Where do we stand? We begin with a simple closed curve in the plane, and we consider the region enclosed by the curve together with the curve itself. We use the letter $D$ to symbolize this (closed) region and denote its boundary, which is our original curve, by $\partial D$. The way we have set things up, $\partial D$ is part of $D$. We shall refer to such a region $D$, constructed in this way, as a *two-dimensional manifold with boundary*.

Next we form the Hilbert space $L^2(D)$ consisting of all functions that are square integrable over the region $D$. Acting in this Hilbert space is the Laplacian operator $\Delta$ with differential expression $\delta(f) = -\partial^2 f/\partial x^2 - \partial^2 f/\partial y^2$ and domain consisting of functions in $L^2(D)$ that are compatible with $\delta$ and satisfy either the Dirichlet, the Neumann, or the mixed boundary conditions. In the last section (Theorem 7.6) we proved that the Laplacian is Hermitian and positive semidefinite (positive definite under the Dirichlet boundary conditions). This implies that, if the Laplacian has any eigenvalues, these eigenvalues must be real and nonnegative. Whether or not the Laplacian *has* any eigenvalues we haven't settled yet. While not every Hermitian operator has eigenvalues, most interesting ones do—as we have seen—and it turns out that the Laplacian is no exception. We will come to that in a moment.

First let's review, one more time, the definition of eigenvalue and eigenfunction. The scalar $\lambda$ is said to be an *eigenvalue* of the Laplacian operator $\Delta$ if there is a nonzero function $f$ in the domain of $\Delta$ such that $\Delta(f) = \lambda f$. Any nonzero $f$ that satisfies this equation is called *an eigenfunction belonging to the eigenvalue* $\lambda$. I want to stress that this is exactly the same definition we have used many, many times before. Exactly the same Hilbert space concepts apply to this case as to the earlier ones, even though the Laplacian differential expression involves partial derivatives and the earlier cases involved only ordinary derivatives. The same overlying theoretical framework covers all these cases.

The Laplacian differential expression, $\delta(f) = -\partial^2 f/\partial x^2 - \partial^2 f/\partial y^2$, has constant coefficients. As was pointed out in Section 6.1, it corresponds to the one-variable expression $l(f) = -d^2 f/dx^2$. The two-variable Laplacian, like its one-variable counterpart, is a very simple second order expression. But when one uses the two-variable Laplacian expression to define the Laplacian operator $\Delta$ in $L^2(D)$, a new and interesting feature enters, namely the *shape* of the planar region $D$. This geometric feature has no counterpart in the one-variable case. For each different region $D$, there will be a different Hilbert space $L^2(D)$ and a different operator $\Delta$ in that space, notwithstanding the same differential expression shared by all such operators. As the operators are different, their eigenvalues will in general be different. To each region $D$ we can thus associate the eigenvalues of the Laplacian operator $\Delta$ in $L^2(D)$. It is natural to take the point of view that the eigenvalues belong to the region $D$. In each case, though, we must also specify the boundary conditions. For example, we might speak of the "eigenvalues of the disk, Dirichlet boundary conditions," "eigenvalues of the rectangle, Neumann boundary conditions."

Only for a few very simple planar regions are the eigenvalues explicitly known. We shall do the computation for two regions, the rectangle and the disk, in the next sections. But for most regions, even simple ones like triangles, we cannot compute the eigenvalues explicitly. The method that we used in the one-variable case becomes impractical for two variables. In the one-variable case, we followed—by and large—the following method: we first found the general solution $y = c_1 f_1(x) + c_2 f_2(x)$ to the differential equation $l(y) - \lambda y = 0$. The functions $f_1$ and $f_2$ contained the parameter $\lambda$. We then determined for what values of $\lambda$ and for what values of the constants $c_1$ and $c_2$ the function $y$ belonged to the domain of our operator. In the two-variable case, the partial differential equation $\Delta f - \lambda f = 0$ has too complicated a general solution to make this method practical. (In the case of the rectangle and the disk, we use separation of variables to reduce to the one-variable case.)

So, in the absence of explicit solutions, one looks to obtain qualitative information about the eigenvalues. Much general information is known, and I shall summarize the basic facts in Theorems 7.7, 7.8, and 7.9 that follow. Roughly, the essential facts are these: the eigenvalues of the Laplacian form an infinite sequence that increases to $+\infty$, each eigenvalue has finite multiplicity, and the corresponding family of eigenfunctions form an orthogonal basis of $L^2(D)$.

That should all sound familiar to you because it simply says that the situation as regards the two-dimensional Laplacian is exactly the same as the situation for the Legendre, Hermite, and Bessel operators and the one-dimensional Laplacian, $l(y) = -y''$. Those operators also had an infinite sequence of eigenvalues increasing to $+\infty$, and the associated eigenfunctions formed an orthogonal basis of the underlying Hilbert space. Once again, we see the value of our Hilbert space theory. It gives us one framework into which all our examples fit and are thus unified. It gives us one language to describe these examples. The Laplacian is just one more example of a Hermitian operator in Hilbert space whose eigenfunctions form an orthogonal basis of that space. We may apply to the Laplacian exactly the same general methods we have used for the other operators: Legendre, Hermite, Bessel, and the one-dimensional Laplacian. So the basic facts that I am about to state in formal theorems fit easily into our program—they are understandable and useful. As to their detailed proofs, I shall make some remarks later bearing on that. But, generally, we treat these theorems as we have treated the assertion that the Legendre polynomials form an orthogonal basis of $L^2(-1, 1)$ and similar assertions—we shall leave their proofs to later stages in our career, content now to understand and apply them.

**Theorem 7.7.** *Let $D$ stand for the region comprising that part of the plane enclosed by a simple closed curve $\partial D$ together with the curve itself, and let $\Delta$ symbolize the Laplacian operator in the Hilbert space $L^2(D)$ under the Dirichlet, Neumann, or mixed boundary conditions. Then, under suitable smoothness conditions on the boundary $\partial D$, the Laplacian $\Delta$ has an infinite sequence of eigenvalues*

$$(0 = \lambda_0) < \lambda_1 < \lambda_2 < \cdots < \lambda_n < \lambda_{n+1} < \cdots$$

*($\lambda_0 = 0$ is a simple eigenvalue for the Neumann boundary condition but not an eigenvalue for either the Dirichlet or mixed boundary conditions.) We have $\lambda_n \to \infty$ as $n \to \infty$. Each eigenvalue has finite multiplicity. If $\lambda_0 = 0$ is an eigenvalue, then it is simple. If $\lambda_0 = 0$ is not an eigenvalue, then $\lambda_1$ is simple.*

Recall what is meant by the *eigenspace* belonging to the eigenvalue $\lambda$: it consists of all functions $f$ in the domain of $\Delta$ that satisfy $\Delta f = \lambda f$. This is a subspace of $L^2(D)$ (in fact, a subspace of the domain of $\Delta$) that we shall denote $E(\lambda)$. An alternative description: $E(\lambda) = \ker(\Delta - \lambda I)$. Every *nonzero* function in $E(\lambda)$ is an eigenfunction belonging to the single eigenvalue $\lambda$. The dimension of $E(\lambda)$ is called the *multiplicity* of $\lambda$. In the one-variable case, we remember well that the multiplicity of a second order ordinary differential operator is either 1 or 2. This is no longer true in two variables: for the Laplacian, we only know that each $E(\lambda_i)$ has finite dimension.

Denote the dimension of the eigenspace $E(\lambda_i)$ by $d(i)$: $d(i) = \dim E(\lambda_i)$. When $i \neq j$, then $\lambda_i \neq \lambda_j$, so every nonzero function in $E(\lambda_i)$ will be orthogonal to every nonzero function in $E(\lambda_j)$. Why? Because, for a Hermitian operator, eigenfunctions belonging to different eigenvalues are orthogonal (see Exercise 7 of Section 7.3). However, in the eigenspace $E(\lambda_i)$, every nonzero function is an eigenfunction belonging to the single eigenvalue $\lambda_i$, so certainly not every pair of functions from $E(\lambda_i)$ will be orthogonal in general. For example, choose $f$ and $2f$, $0 \neq f \in E(\lambda_i)$. But we can use the Gram-Schmidt method (from our linear algebra course) to produce an orthogonal basis of the finite-dimensional space $E(\lambda_i)$.

Start with the first eigenspace, which according to Theorem 7.7 always has dimension 1. In the Neumann case the first eigenspace will be $E(\lambda_0)$ corresponding to the eigenvalue $\lambda_0 = 0$. We write $d(0) = \dim(E(\lambda_0)) = 1$. Any nonzero constant function will span $E(\lambda_0)$; we may take $\phi_0 = 1$ as a convenient eigenfunction (Exercise 9 of Section 7.3). In either the Dirichlet or mixed case, the first eigenspace is $E(\lambda_1)$ (which also has dimension 1 according to Theorem 7.7). In this case we write $d(1) = \dim(E(\lambda_1)) = 1$ and denote a spanning eigenfunction by $\phi_1$.

Now pass to the second eigenspace, which will be either $E(\lambda_1)$ in the Neumann case or $E(\lambda_2)$ in the Dirichlet and mixed cases. To be concrete, I will stick now with the Dirichlet case and let you fill in the parallel details for the case when $\lambda_0 = 0$ is an eigenvalue. So $E(\lambda_2)$ is our second eigenspace, and it has dimension $d(2)$. Gram-Schmidt produces $d(2)$ orthogonal functions that span $E(\lambda_2)$. Label these eigenfunctions in succession $\phi_2$, $\phi_3, \ldots, \phi_{1+d(2)}$. Gram-Schmidt produces $d(3)$ orthogonal functions that span $E(\lambda_3)$. Label these $\phi_{1+d(2)+1}, \phi_{1+d(2)+2}, \ldots, \phi_{1+d(2)+d(3)}$. Continuing in this way, we get a sequence $\phi_1, \phi_2, \ldots$ of orthogonal eigenfunctions of the Laplacian such that the first $d(1)$ eigenfunctions belong to the eigenvalue $\lambda_1$ and span the eigenspace $E(\lambda_1)$, the next block of $d(2)$ eigenfunctions belong to $\lambda_2$ and span $E(\lambda_2)$, etc.

**Theorem 7.8.** *As constructed in the previous paragraph, the eigenfunctions of the Laplacian $\phi_n$, $n = 1, 2, \ldots$, form an orthogonal basis of $L^2(D)$. Hence, every $f$ in $L^2(D)$ has a general Fourier expansion*

$$f = \sum_{i=1}^{\infty} c_i \phi_i \tag{7.4.1}$$

*convergent to $f$ in the mean-square metric,*

$$\left\| f - \sum_{i=1}^{N} c_i \phi_i \right\|^2 = \iint_D \left| f - \sum_{i=1}^{N} c_i \phi_i \right|^2 dA \to 0 \qquad \text{as} \qquad N \to \infty \tag{7.4.2}$$

*The coefficients $c_i$ are given by the usual formula, $c_i = \langle f, \phi_i \rangle / \langle \phi_i, \phi_i \rangle$.*

I have stated Theorem 7.8 for the case where $\lambda_0 = 0$ is not an eigenvalue. If $\lambda_0 = 0$ is an eigenvalue, start with $\phi_0 = 1$ (or $\phi_0 =$ some convenient nonzero constant).

Let us review briefly some of the detailed information provided by Theorem 7.8.

First, to say that the $\phi_n$, $n = 1, 2, \ldots$, form an orthogonal basis of $L^2(D)$ means that the series (7.4.1) is valid in the sense of mean-square convergence (7.4.2) assuming *only* that

$$\iint_D |f|^2 dA < \infty$$

That is the *only* requirement necessary to ensure the validity of (7.4.1). We don't need to assume any continuity of $f$, any differentiability of $f$, or anything else. The series (7.4.1) is valid for *every* $f$ in $L^2(D)$. This is all the more surprising when we remember that the $\phi_n$ all belong to the domain of the

Laplacian $\Delta$ which is a proper subspace of $L^2(D)$. Every *finite* sum, $\Sigma_{i=1}^n c_i \phi_i$, will belong to the domain of $\Delta$, so if we use only finite sums then we can represent only functions in the domain of $\Delta$. But, when we pass to infinite series (7.4.1), then we get *every* function in $L^2(D)$!

But bear in mind that mean-square convergence does not imply pointwise convergence at every point. Hence the series (7.4.1) may converge pointwise at several points in $D$ to values different from the values of $f$ at those points.

Finally, recall that we may take the inner product on both sides of (7.4.1) with any $g$ in $L^2(D)$ and get a valid numerical series

$$\langle f, g \rangle = \sum_{i=1}^{\infty} c_i \langle \phi_i, g \rangle$$

In particular, with $g = f$, we have $\langle \phi_i, f \rangle = \overline{\langle f, \phi_i \rangle} = \bar{c}_i \langle \phi_i, \phi_i \rangle = \bar{c}_i \| \phi_i \|^2$, so

$$\langle f, f \rangle = \sum_{i=1}^{\infty} |c_i|^2 \| \phi_i \|^2 \tag{7.4.3}$$

which is the *Parseval equality* for (7.4.1).

For several purposes, it is convenient to list the eigenvalues of the Laplacian by repeating each eigenvalue a number of times equal to its multiplicity. Denote the eigenvalues in this list by $\mu_j$. In the case of the Neumann boundary condition, when $\lambda_0 = 0$ is an eigenvalue, start the list with $\mu_0 = 0$, so then the list runs $\mu_j, j = 0, 1, 2, \ldots$. In the other two cases, when $\lambda_0 = 0$ is not an eigenvalue, start with $\mu_1 = \lambda_1$, so then the list runs $\mu_j$, $j = 1, 2, \ldots$. To simplify the explanation, I shall assume now that we are in the Dirichlet or mixed case when $\lambda_0 = 0$ is not an eigenvalue and leave it to you to supply the details for the Neumann case. (All you will need to change is to begin with $\mu_0 = 0$ instead of $\mu_1 = \lambda_1$.)

For example, suppose $\lambda_1$ has multiplicity 1 (as it must), $\lambda_2$ has multiplicity 3, and $\lambda_3$ has multiplicity 2; then the initial segments of the listings would compare as follows:

$$0 < \lambda_1 < \qquad \lambda_2 \qquad < \qquad \lambda_3 \qquad < \cdots$$

$$0 < \mu_1 < \overbrace{\mu_2 = \mu_3 = \mu_4} < \overbrace{\mu_5 = \mu_6} \quad < \cdots$$

In general

$$0 < \lambda_1 = \mu_1 < \lambda_2 = \mu_2 = \mu_3 = \cdots = \mu_{1+d(2)} < \lambda_3 = \mu_{1+d(2)+1}$$

$$= \cdots = \mu_{1+d(2)+d(3)} < \lambda_4 \cdots$$

If $\phi_j, j = 1, 2, \ldots$, is the orthogonal basis of $L^2(D)$ described in the paragraph preceding Theorem 7.8, then $\Delta(\phi_j) = \mu_j \phi_j, j = 1, 2, \ldots$.

**Theorem 7.9.** *If* $\mu_n$, $n = 1, 2, \ldots$, *is a listing of the eigenvalues of the Laplacian according to their multiplicity, then*

$$\lim_{n \to \infty} \frac{\mu_n}{n} = \frac{4\pi}{\text{area}(D)}$$

This asymptotic formula, established by the mathematician Hermann Weyl in 1911, shows that the asymptotic behavior of the eigenvalues of the Laplacian depends only on the size (area) of the region $D$ and not on the shape of $D$.

Theorem 7.9 is somewhat advanced, and I shall not even discuss its proof here. I have included it because it is a famous old interesting result, and it will provide us with a kind of yardstick to set against the sequence of eigenvalues of the rectangle and disk that we will compute in the next sections.

As to Theorems 7.7 and 7.8, I will discuss the essential features of the method of proof but not give a detailed proof. Theorem 7.7 asserts the existence of an infinite sequence of eigenvalues of the Laplacian. As I mentioned earlier, we are able to compute the eigenvalues explicitly for only a very few special regions $D$. So, in the proof of Theorem 7.7, it is a matter of proving the existence of eigenvalues without actually finding the numbers. The common techniques in problems of this sort are called *variational methods*. I shall discuss these here not only because they can be used to prove Theorems 7.7 and 7.8 but also because they provide a general means for approximating eigenvalues when you cannot actually compute them.

We'll begin with a review of some of the material from Section 4.4. Let $H$ be a Hilbert space with inner product $\langle f, g \rangle$, norm $\|f\| = \langle f, f \rangle^{1/2}$, and intrinsic metric $d(f, g) = \|f - g\| = \langle f - g, f - g \rangle^{1/2}$.

**Definition 7.10.** A subspace $M$ of the Hilbert space $H$ is said to be *dense* in $H$ if, given any $f \in H$ and given any positive real number $\varepsilon$, however small, one can then find an $m \in M$ so that $\|f - m\| < \varepsilon$.

We can put this definition in capsule form thus: $M$ is dense in $H$ if there are elements in $M$ arbitrarily close to every $f$ in $H$. For example, on the real number line, not every real number is rational, that is, of the form $a/b$ for integers $a$ and $b$. For example, neither $\sqrt{2}$ nor $\pi$ is rational. Yet the rationals are dense in the real number line, meaning that for every real number, whether rational or not, one can find a rational number as close to it as you please. (Proof: Choose a very large positive integer $N$, then mark down the rationals, $0$, $\pm 1/N$, $\pm 2/N$, etc. These form a net of rationals on the real number line, and every real number is within distance $1/N$ from one of them.

As $N$ can be made as large as you please, you can get as close as you please to every real number.)

With reference to Definition 7.10, remember that words like "close" refer to the intrinsic metric of the Hilbert space $H$. That is the exclusive measure of distance that we have been using right along. In the Hilbert space $H = L^2(D)$, for example, to say that $\|f - m\| < \varepsilon$ means that

$$\left[ \iint_D |f(x, y) - m(x, y)|^2 \, dx \, dy \right]^{1/2} < \varepsilon.$$

If our subspace $M$ is all of $H$, $M = H$, then clearly $M$ will be dense in $H$. We can take in every case $m = f$ and get $\|f - m\| = 0 < \varepsilon$.

The Legendre, Hermite, and Bessel operators that we studied in Chapters 4 and 6 all had their domains dense in their respective Hilbert spaces. For example, the domain of the Legendre operator, which is the subspace of $L^2(-1, 1)$ consisting of those functions $f$ in $L^2(-1, 1)$ compatible with Legendre's differential expression $l(y) \equiv -((1-x^2)y')'$ and such that $f(\pm 1)$ are finite, is dense in $L^2(-1, 1)$. We did not dwell on this fact in Chapter 4 because we didn't need it then. But all the operators we have considered have dense domains. In particular, the *domain of the Laplacian is dense in $L^2(D)$*. We shall not prove this here, in accordance with our program of just discussing the main ideas in the proof of Theorem 7.7. To go into a detailed proof that the domain of the Laplacian is dense in $L^2(D)$—or that the domains of any of the operators we have considered previously are dense in their respective Hilbert spaces—would involve us in a lot of detailed mathematical analysis, which would divert us from our program of concentrating on the principal concepts of Hilbert space theory and their application to science and engineering, a program we have adhered to right from the beginning.

**Lemma 7.11.** *Suppose $M$ is a dense subspace of the Hilbert space $H$. If $f \in H$, and $\langle f, m \rangle = 0$ for all $m \in M$, then $f = 0$.*

I shall outline the proof here and leave the details for you in Exercise 5. First note that if $M = H$, then the proof is easy. In that case, since we have $\langle f, m \rangle = 0$ for all $m \in M$ this is in particular true with $m = f$. So $\langle f, f \rangle = 0$. But for an inner product, $\langle f, f \rangle = 0$ happens only for $f = 0$.

So the case we need to consider is when $M$ is dense in $H$ but not equal to it, and $f \notin M$. By hypothesis, for any $m \in M$, $\langle f, m \rangle = 0$; therefore

$$\langle f, f \rangle = \langle f, f \rangle - 0 = \langle f, f \rangle - \langle f, m \rangle = \langle f, f - m \rangle$$

for any $m \in M$. Now take absolute values and apply a famous inequality—see Exercise 5.

We come now to the variational method, so called, for proving the existence of eigenvalues.

**Theorem 7.12.** *Suppose T is a Hermitian operator with domain dense in a Hilbert space H. Consider the ratio*

$$\frac{\langle Tf, f \rangle}{\langle f, f \rangle} \tag{7.4.4}$$

*where f is a nonzero function in the domain of T. Suppose there is a nonzero function $f_1$ in the domain of T such that the ratio (7.4.4) takes on its smallest value when $f = f_1$. Then: (1) this smallest value, call it $\mu_1$ ($\mu_1 = \langle Tf_1, f_1 \rangle / \langle f_1, f_1 \rangle$), is an eigenvalue of T and (2) $f_1$ is an eigenfunction belonging to $\mu_1$.*

Before the proof, a discussion: the ratio (7.4.4) is always a *real* number, never complex. The denominator is real. So is the numerator, because, if $\alpha = \langle Tf, f \rangle$, then $\bar{\alpha} = \overline{\langle Tf, f \rangle} = \langle f, Tf \rangle = \langle Tf, f \rangle = \alpha$, using the fact that $T$ is Hermitian. Thus the ratio (7.4.4) is always a real number. If $T$ is also positive semidefinite, then the ratio will be $\geqslant 0$; if $T$ is positive definite, then the ratio will be $> 0$. But, while all our applications will be to positive semidefinite operators, we don't need that hypothesis to prove the theorem.

We are assuming that

$$\frac{\langle Tf, f \rangle}{\langle f, f \rangle} \geqslant \mu_1 \tag{7.4.5}$$

for all nonzero $f$ in the domain of $T$ and that we have equality when $f = f_1$. We need to prove that $Tf_1 = \mu_1 f_1$; that is exactly the statement that $\mu_1$ is an eigenvalue and $f_1$ a corresponding eigenfunction.

The inequality (7.4.5) is equivalent to

$$\langle Tf, f \rangle - \mu_1 \langle f, f \rangle \geqslant 0 \qquad \text{for all nonzero } f \in \text{domain}(T) \tag{7.4.6}$$

Since (7.4.6) is true for all $f$ in the domain of $T$, it will be true for $f = f_1 + \alpha g$ where $f_1$ is the particular function whose existence we have explicitly assumed in the statement of the theorem, $g$ is any function in the domain of $T$, and $\alpha$ is any scalar (real or complex number). Substitute $f = f_1 + \alpha g$ in (7.4.6) to get

$$\langle Tf_1, f_1 \rangle + \bar{\alpha} \langle Tf_1, g \rangle + \alpha \langle T\alpha, f_1 \rangle + |\alpha|^2 \langle Tg, g \rangle$$
$$- \mu_1 [\langle f_1, f_1 \rangle + \bar{\alpha} \langle f_1, g \rangle + \alpha \langle g, f_1 \rangle + |\alpha|^2 \langle g, g \rangle] \geqslant 0 \tag{7.4.7}$$

By hypothesis, $\langle Tf_1, f_1 \rangle - \mu_1 \langle f_1, f_1 \rangle = 0$. Also, using the fact that $T$ is Hermitian, we have $\alpha \langle Tg, f_1 \rangle = \alpha \langle g, Tf_1 \rangle = \alpha \overline{\langle Tf_1, g \rangle}$. Hence

$$\bar{\alpha} \langle Tf_1, g \rangle + \alpha \langle Tg, f_1 \rangle = \bar{\alpha} \langle Tf_1, g \rangle + \overline{\bar{\alpha} \langle Tf_1, g \rangle}$$
$$= 2 \operatorname{Re}(\bar{\alpha} \langle Tf_1, g \rangle)$$

Similarly, $\bar{\alpha}\langle f_1, g\rangle + \alpha\langle g, f_1\rangle = 2\,\mathrm{Re}(\bar{\alpha}\langle f_1, g\rangle)$. Using these facts, reduce (7.4.7) to

$$2\,\mathrm{Re}[\bar{\alpha}(\langle Tf_1, g\rangle - \mu_1\langle f_1, g\rangle)] + |\alpha|^2(\langle Tg, g\rangle - \mu_1\langle g, g\rangle) \geqslant 0 \qquad (7.4.8)$$

Inequality (7.4.8) is valid for *all* scalars $\alpha$ and all $g$ in the domain of $T$. Note that, by hypothesis, $\langle Tg, g\rangle - \mu_1\langle g, g\rangle \geqslant 0$; this is just (7.4.6) with $g = f$. Let $s = \langle Tg, g\rangle - \mu_1\langle g, g\rangle$, $s \geqslant 0$, and let $r = \langle Tf_1, g\rangle - \mu_1\langle f_1, g\rangle$. Then, fixing $g$, we can rewrite (7.4.8) as follows:

$$2\,\mathrm{Re}(\bar{\alpha}r) + |\alpha|^2 s \geqslant 0 \qquad \text{for all scalars } \alpha \qquad (7.4.9)$$

This forces $r = 0$. Because if $r \neq 0$, then write the complex number $r$ in its polar form, $r = |r|e^{i\theta}$, and choose $\alpha = \varepsilon e^{i\theta}$ for real $\varepsilon$. Then $\bar{\alpha}r = \varepsilon e^{-i\theta}|r|e^{i\theta} = \varepsilon|r|$, which is real, so $2\,\mathrm{Re}(\bar{\alpha}r) + |\alpha|^2 s = 2\varepsilon|r| + \varepsilon^2 s = \varepsilon(2|r| + \varepsilon s) \geqslant 0$ for all real $\varepsilon$. If $s = 0$, then $\varepsilon = -1$ produces a contradiction. If $s > 0$, take $\varepsilon = -|r|/s$. Then $\varepsilon(2|r| + \varepsilon s) = -|r|^2/s < 0$, another contradiction. The source of this contradiction is our assumption that $r = \langle Tf_1, g\rangle - \mu_1\langle f_1, g\rangle \neq 0$. Hence, we conclude that the validity of (7.4.8) for all scalars $\alpha$ implies that $\langle Tf_1, g\rangle - \mu_1\langle f_1, g\rangle = 0$ for all $g$ in the domain of $T$. We can rewrite this as

$$\langle Tf_1 - \mu_1 f_1, g\rangle = 0 \qquad \text{for all } g \text{ in the domain of } T \qquad (7.4.10)$$

Then, because the domain of $T$ is dense, Lemma 7.11 tells us that $Tf_1 - \mu_1 f_1 = 0$, or $Tf_1 = \mu_1 f_1$, which is what we wanted to prove.

Theorem 7.12 tells us this: to determine whether a Hermitian, domain-dense operator $T$ has an eigenvalue, form the ratio $\langle Tf, f\rangle/\langle f, f\rangle$. If you then find a nonzero function $f_1$ in the domain of $T$ such that, when $f = f_1$, that ratio takes on its smallest possible value, then that value is an eigenvalue and $f_1$ is an eigenfunction belonging to that eigenvalue. If you succeed in this, you will have found the *smallest* eigenvalue (Exercise 8).

There is a subtle point here that we ought to look at carefully. Suppose our operator $T$ is positive semidefinite. Then the ratio $\langle Tf, f\rangle/\langle f, f\rangle$ will always be nonnegative. So it would seem quite reasonable to assume that one could always find a function $f_1$ which would give this ratio its smallest (necessarily nonnegative) value. But this is not the case. Consider the following analogous situation: the function $f(x) = x$ defined on the *open* interval $0 < x < 1$. Clearly $f(x) > 0$ for all such $x$ and $f(x)$ can be made as close to zero as we please. But there is no $x$ in the domain of $f$ for which $f(x) = 0$. So the assumption in Theorem 7.12 that there is a function $f_1$ for which the ratio $\langle Tf, f\rangle/\langle f, f\rangle$ actually *equals* its smallest value is a crucial assumption.

How does Theorem 7.12 apply to the Laplacian? The Laplacian is Hermitian (Section 7.3). It also has dense domain—a fact we are taking on faith here. Thus Theorem 7.12 would guarantee us one eigenvalue if we could find a function $f_1$ that gave the ratio $\langle \Delta f, f\rangle/\langle f, f\rangle$ its smallest value.

Apply Theorem 7.5:

$$\langle \Delta f, f \rangle = \left\langle \frac{\partial f}{\partial x}, \frac{\partial f}{\partial x} \right\rangle + \left\langle \frac{\partial f}{\partial y}, \frac{\partial f}{\partial y} \right\rangle = \iint_D \left( \left| \frac{\partial f}{\partial x} \right|^2 + \left| \frac{\partial f}{\partial y} \right|^2 \right) dx \, dy$$

Hence, for the Laplacian,

$$\frac{\langle \Delta f, f \rangle}{\langle f, f \rangle} = \frac{\iint_D (|\partial f/\partial x|^2 + |\partial f/\partial y|^2) \, dx \, dy}{\iint_D |f|^2 \, dx \, dy} \tag{7.4.11}$$

so it is a matter of finding a function $f_1(x, y)$, in the domain of $\Delta$, that gives the expression on the right its minimum value.

This is an old problem, made famous in the work of the great German mathematician Bernhard Riemann, whose name you may remember from your calculus course—the Riemann integral. (Incidentally, his last name is pronounced "*Ree*-maan".) Riemann was born in 1826 and died in 1866 at the age of 40. In his doctoral dissertation (1851), titled "Foundations of a General Theory of Functions of a Complex Variable", Riemann considers an integral of the same general type as in (7.4.11) and writes that the integral "always for one of these functions has a minimum value" (§16 of his paper). Later, in another paper, he refers again to this result, calls it a "principle," and attributes it to his teacher, Dirichlet. Riemann is generally regarded as one of the greatest and most original mathematicians who ever lived. His published research, though small in quantity, shows its profound influence throughout much of modern mathematics and physics: calculus, number theory, complex variables, differential geometry, and general relativity. Yet, in the matter of the "Dirichlet principle," he had overstated the case. A valid proof of the Dirichlet principle came only 30 years after Riemann's death. We can perhaps draw a few conclusions: (1) everyone—even the best—makes mistakes (that should make you feel better); (2) science is a human enterprise carried on by fallible human beings, and the advance of science is messier and more erratic than our textbooks would have us believe; and (3) the existence of the eigenvalues of the Laplacian is a difficult and sophisticated matter.

As to the latter conclusion, there is a full and detailed proof of Theorems 7.7 and 7.8 in the book *Variational Methods for Eigenvalue Problems* by S. H. Gould (2nd ed., Univ. of Toronto Press, 1966) which is the most elementary treatment I know. The classic treatise *Methods of Mathematical Physics*, Vol. I, by R. Courant and D. Hilbert (Interscience, 1953) also has a proof, but they assume the validity of the Dirichlet principle, and refer to Vol. II for the proof of that principle. In any case, we shall drop the discussion of the proof of Theorems 7.7 and 7.8 at this point. Our purpose is to understand and use these results.

Return for the moment to Theorem 7.12. That theorem tells us how to find the *smallest* eigenvalue and a corresponding eigenfunction $f_1$. What about the next eigenvalue? Here is what you do. You consider the same ratio

$\langle Tf, f \rangle / \langle f, f \rangle$, but only for those nonzero functions $f$ in the domain of $T$ that also satisfy $\langle f, f_1 \rangle = 0$. If there is an $f_2$ in that class that gives the ratio its minimum value, say $\mu_2$, then $\mu_2$ is an eigenvalue and $f_2$ a corresponding eigenfunction. At the next step, you consider the ratio $\langle Tf, f \rangle / \langle f, f \rangle$ once again, this time for those nonzero $f$ in the domain of $T$ that satisfy *both* $\langle f, f_1 \rangle = 0$ and $\langle f, f_2 \rangle = 0$, etc.; see exercises 7 through 10.

As I have already mentioned, only for the simplest regions $D$ can we compute the eigenvalues of the Laplacian exactly. And the rate of convergence of Weyl's limit relation stated in Theorem 7.9, namely $\lim(\mu_n/n) = 4\pi/\text{area}(D)$, is quite slow and can be badly off even for large values of $n$. Hence one needs a method for approximating the eigenvalues of $\Delta$ in situations where they are not exactly calculable. It is already useful to have a method for estimating $\lambda_1$. (Let's stick with the Dirichlet boundary condition.) Theorem 7.12 is useful here too. To use this result to estimate $\lambda_1$, the trick is to choose a function $f$ in the domain of $\Delta$ that puts the ratio $\langle Tf, f \rangle / \langle f, f \rangle$ close to its minimum value. If you knew an eigenfunction belonging to $\lambda_1$, then this eigenfunction would give $\lambda_1$ exactly. Our limited experience with nodal curves (Exercise 3 of Section 6.6) suggests that the first eigenfunction has only the boundary as a nodal curve, so a natural choice for $f$ is a function in the domain of $\Delta$ that vanishes only on $\partial D$. Some rather simple guesses for $f$ can produce quite accurate estimates (Exercises 11 and 13).

The Laplacian is linear: for any $f, g$ in its domain and for any scalars $a$ and $b$, $\Delta(af + bg) = a\,\Delta(f) + b\,\Delta(g)$. This automatically entails linearity for any finite number of functions: $\Delta(c_1 f_1 + \cdots + c_n f_n) = c_1\,\Delta(f_1) + \cdots + c_n\,\Delta(f_n)$. But linearity does not in general carry over to infinite series. We need to know something more about $\Delta$, for example, that it is bounded or that it is Hermitian. Well, the Laplacian is not bounded (Exercise 2), but it is Hermitian. So we have

**Theorem 7.13.** *Suppose $f$ lies in the domain of the Laplacian $\Delta$ (under either the Dirichlet, Neumann, or mixed boundary conditions). Then the equation obtained by applying $\Delta$ term by term to the general Fourier expansion for $f$ is valid. Put another way, if $f = \Sigma_{n=1}^{\infty} c_n \phi_n$ where $c_n = \langle f, \phi_n \rangle / \langle \phi_n, \phi_n \rangle$, and if $f \in \text{domain}(\Delta)$, then*

$$\Delta(f) = \sum_{n=1}^{\infty} c_n \Delta(\phi_n) = \sum_{n=1}^{\infty} c_n \mu_n \phi_n$$

*(Under the Neumann boundary condition, start with $\mu_0 = 0$.)*

The proof, which is assigned as Exercise 7, uses the fact that $\Delta$ is Hermitian.

Theorems 7.7 and 7.8 give us a way to solve, at least in principle, the wave equation and the heat equation for any (decent) planar region $D$, also to solve Schrödinger's equations for a quantum particle confined to such a region $D$ in which the potential is zero. See Exercises 15 through 17.

This has been a long section, but one that is easily summarized: the Laplacian operator is a Hermitian operator in Hilbert space, and its eigenvalues and eigenfunctions behave pretty much like those of the Legendre, Hermite, and Bessel operators that we learned to know and love in earlier chapters. When we speak Hilbert space, we speak the universal language of these operators.

### Exercises

1. Suppose $\phi_1, \phi_2, \ldots, \phi_n, \ldots$ constitute an orthogonal basis for a Hilbert space $H$ whose inner product we denote $\langle \cdot, \cdot \rangle$, as usual.
   (a) Prove that it is impossible to expand $\phi_1$ in terms of the remaining $\phi_i$'s; i.e., prove that there are no constants $c_2, c_3, \ldots$, so that
   
   $$\phi_1 = c_2\phi_2 + c_3\phi_3 + \cdots + c_n\phi_n + \cdots$$
   
   (remember that by definition an orthogonal basis cannot contain the zero function).
   (b) Generalize your result in (a): show that no $\phi_m$ can be expanded in terms of the remaining $\phi_i$'s.
   (c) Using your result from (b), show that if we delete one function from an orthogonal basis of a Hilbert space then the remaining functions no longer constitute an orthogonal basis of that space.
2. An operator $L$ in a Hilbert space is said to be *bounded* if there is a positive real number $M$ so that $\|Lf\| \leqslant M\|f\|$ for all $f$ in the domain of $M$. Using Theorem 7.7, show that the Laplacian $\Delta$ is not bounded. [Hint: Belonging to each eigenvalue $\lambda_n$ there is at least one eigenfunction $\Psi_n$, and, by definition of eigenfunction, $\|\Psi_n\| \neq 0$. Refer to Section 4.4.]
3. When we write a numerical series $a = \Sigma a_i$, where the $a_i$ are real or complex numbers, we mean that

$$\left| a - \sum_{i=1}^{n} a_i \right| \to 0 \qquad \text{as } n \to \infty$$

If $\phi_1, \phi_2, \ldots$ is an orthogonal basis for a Hilbert space $H$ and $f \in H$ has the general Fourier expansion $f = \Sigma c_i\phi_i$, prove that for any $g \in H$ we have

$$\langle f, g \rangle = \sum_{i=1}^{\infty} c_i \langle \phi_i, g \rangle$$

as a numerical series. [Hint: You must prove that

$$\left| \langle f, g \rangle - \sum_{i=1}^{n} c_i \langle \phi_i, g \rangle \right| \to 0 \qquad \text{as } n \to \infty$$

Show first that

$$\left| \langle f, g \rangle - \sum_{i=1}^{n} c_i \langle \phi_i, g \rangle \right| = \left| \left\langle f - \sum_{i=1}^{n} c_i \phi_i, g \right\rangle \right|$$

then use a famous inequality.]

4. Let $N(\mu)$ be the number of eigenvalues of the Laplacian, counted with their multiplicity, that are $\leqslant \mu$. Show that the asymptotic relation of Theorem 7.9 has this equivalent formulation.

$$\lim_{\mu \to \infty} \frac{N(\mu)}{\mu} = \frac{\text{area}(D)}{4\pi}$$

5. (a) Suppose $\alpha$ is a nonnegative real number such that $\alpha \leqslant \varepsilon$ for *every* positive real number $\varepsilon$. What is the only possible value that $\alpha$ can have? Prove your answer.

   (b) Suppose that $M$ is a dense subspace of a Hilbert space $H$. Using your result in (a), prove that $M^{\perp} = 0$.

6. Refer to the discussion preceding Theorem 7.8. Prove: $f \perp E(\lambda_2)$ if, and only if, $\langle f, \phi_i \rangle = 0, 2 \leqslant i \leqslant 1 + d(2)$. ($f \perp M$, or $f \in M^{\perp}$, means $\langle f, m \rangle = 0$ for all $m \in M$.)

7. Prove Theorem 7.13: Suppose $\mu_n$, $n = 1, 2, \ldots$, are the eigenvalues of the Laplacian listed according to their multiplicity and $\phi_n$ are the corresponding orthogonal eigenfunctions as constructed in the discussion preceding Theorem 7.8. Every $f \in L^2(D)$ has an expansion $f = \sum_{n=1}^{\infty} c_n \phi_n$ (Theorem 7.8). Prove: If also $f \in \text{domain}(\Delta)$, then

$$\Delta(f) = \sum_{n=1}^{\infty} \mu_n c_n \phi_n$$

[Hint: If $f \in \text{domain}(\Delta)$, then $\Delta(f) \in L^2(D)$ so $\Delta f = \sum a_n \phi_n$.]

In Exercises 8, 9, and 10 assume that $\mu_0 = 0$ is not an eigenvalue of $\Delta$.

8. (Continuation) (a) Show that, for any $f$ in the domain of $\Delta$,

$$\langle \Delta(f), f \rangle = \sum_{i=1}^{\infty} \mu_i |c_i|^2 \langle \phi_i, \phi_i \rangle$$

(b) Using the result in part (a), show that $\langle\Delta(f),f\rangle\geqslant\lambda_1\langle f,f\rangle$ for any $f$ in the domain of $\Delta$, where $\lambda_1$ is the first eigenvalue of $\Delta$. Thus conclude for any nonzero $f$ in the domain of $\Delta$

$$\frac{\langle\Delta(f),f\rangle}{\langle f,f\rangle}\geqslant\lambda_1$$

(c) Prove: If $0\neq f\in E(\lambda_1)$, then $\langle\Delta(f),f\rangle/\langle f,f\rangle=\lambda_1$.

9. (Continuation) Prove the converse of Exercise 8c: If $0\neq f\in\text{domain}(\Delta)$ and $\langle\Delta(f),f\rangle/\langle f,f\rangle=\lambda_1$, then $f\in E(\lambda_1)$.

10 (Continuation) Show that if $0\neq f\in\text{domain}(\Delta)$ and $\langle f,\phi_1\rangle=0$, then

$$\frac{\langle\Delta f,f\rangle}{\langle f,f\rangle}\geqslant\mu_2$$

with equality when $f=\phi_2$. Generalize.

11. Show that the function $f(r)=1-r^2$ is in the domain of the Laplacian on the unit disk, Dirichlet boundary conditions. Use this function in Theorem 7.12 to compute an upper bound for the first eigenvalue $\lambda_1$ of the disk. How close are you to the correct value? (See Exercise 16 of Section 7.1.)

Exercises 12 through 14 estimate the first Dirichlet eigenvalue $\lambda_1$ of the triangle $T$ determined by the points $(0,0)$, $(1,0)$, and $(0,1)$.

12. (a) In a paper published in the *Proceedings of the American Mathematical Society*, vol. 100 (1987), pp. 175–182, Robert Brooks and Peter Waksman proved the inequality

$$\lambda_1\geqslant\frac{(L+\sqrt{4\pi A})^2}{16A^2}$$

for the first Dirichlet eigenvalue $\lambda_1$ of any triangle with area $A$ and perimeter $L$. Compute this lower bound for the triangle $T$.

(b) Polya has conjectured that, for the Dirichlet boundary condition, the Weyl asymptotic formula (Theorem 7.9) is always low: $\mu_n\geqslant4\pi n/\text{area}(D)$, $n=1,2,\ldots$. Assuming the truth of this conjecture, compute another lower bound for $\lambda_1$ of our triangle $T$.

13. Show that the function $f(x,y)=xy(x+y-1)$ lies in the domain of the Laplacian $\Delta$ on the triangle $T$, Dirichlet boundary conditions. Using the inequality $\lambda_1\leqslant\langle\Delta f,f\rangle/\langle f,f\rangle$, $0\neq f\in\text{domain}(\Delta)$ (see Theorem 7.12 or Exercise 8), compute an upper bound for $\lambda_1$ of $T$. [Ans.: $\langle\Delta f,f\rangle=1/90$, $\langle f,f\rangle=1/5040$.]

14. (a) Show that $f(x,y)=\sin\pi x\sin 2\pi y+\sin 2\pi x\sin\pi y$ satisfies the Dirichlet boundary condition for the triangle $T$.

   (b) For the function $f(x, y)$ in (a), show that $\Delta f = \lambda f$. Find $\lambda$. Prove that $f \in \text{domain}(\Delta)$.

   (c) Without any further calculation, use (a) and (b) to find $\langle \Delta f, f \rangle / \langle f, f \rangle$. You have found the first Dirichlet eigenvalue of the triangle $T$.

15. A membrane is stretched over a planar rim-curve $\partial D$ to form a drumhead $D$. Let $u = u(x, y, t)$ stand for the vertical displacement of the drumhead at position $(x, y)$ and time $t$. Show how Theorems 7.7 and 7.8 enable us to find $u$, and derive an expression for $u$.

   [Ans.: $u = \sum_{n=1}^{\infty} (a_n \cos w\sqrt{\mu_n}\, t + b_n \sin w\sqrt{\mu_n}\, t)\phi_n(x, y).$]

   Determine the constants $a_n$ and $b_n$ in terms of the initial position $u(x, y, 0)$ and initial velocity $(\partial u/\partial t)(x, y, 0)$.

16. Let $D$ represent a metal plate with insulated boundary $\partial D$. Let $u = u(x, y, t)$ stand for the temperature of the plate at position $(x, y)$ and time $t$. Show how Theorems 7.7 and 7.8 enable us to find $u$, and derive an expression for $u$.

   [Ans.: $u = \sum_{n=0}^{\infty} c_n \phi_n(x, y) \exp(-\alpha^2 \mu_n t).$]

   Determine the constants $c_n$ in terms of the initial temperature $u(x, y, 0)$. Find $\lim_{t \to \infty} u(x, y, t)$.

17. Let $D$ represent a region in the plane inside which is confined a quantum particle of mass $m$ under a potential $V = 0$. Let $\Psi = \Psi(x, y, t)$ stand for the probability density for finding the particle at position $(x, y)$ and time $t$. Show how Theorems 7.7 and 7.8 enable us to find $\Psi$, and derive an expression for $\Psi$.

   [Ans.: $\Psi = \sum_{n=1}^{\infty} c_n \phi_n(x, y) \exp(-i\mu_n \hbar t/2m).$]

   Determine the constants $c_n$ in terms of the initial probability density $\Psi(x, y, 0)$.

## 7.5 EIGENVALUES OF THE RECTANGLE

In this section I shall use the Dirichlet boundary condition exclusively, leaving computations for other boundary conditions to the exercises.

Take $a$ as the length of one side of the rectangle $D$, use $b$ to stand for the length of the other side, and locate $D$ in the $x$–$y$ plane as in Figure 7.5.

Our objective is to find the eigenvalues of the rectangle, Dirichlet boundary conditions. Which is to say, find all scalars $\lambda$ such that $\Delta f = \lambda f$ has a nonzero solution $f$ in the domain of $\Delta$.

Our differential equation is

$$\delta(f) = -\frac{\partial^2 f}{\partial x^2} - \frac{\partial^2 f}{\partial y^2} = \lambda f \tag{7.5.1}$$

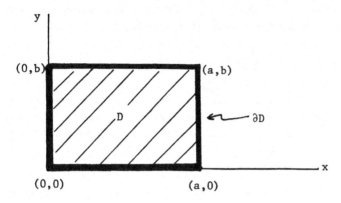

**Figure 7.5**  The rectangle.

We know $\lambda > 0$ as the Laplacian is positive definite. We seek all nonzero functions $f(x, y)$ in $L^2(D)$ that solve (7.5.1) and satisfy $f = 0$ on $\partial D$. Note that if $f \in L^2(D)$ solves (7.5.1), then $\delta(f) = \lambda f \in L^2(D)$ also, so $f$ is automatically compatible with $\delta$, thus in the domain of the Laplacian operator $\Delta$.

We do what has worked for us in the past: separate variables. Set $f(x, y) = X(x)Y(y)$. As

$$\iint_D |f(x, y)|^2 \, dx \, dy = \left( \int_0^a |X(x)|^2 \, dx \right) \left( \int_0^b |Y(y)|^2 \, dy \right)$$

we have $f \in L^2(D)$ if, and only if, both $X \in L^2(0, a)$ and $Y \in L^2(0, b)$ ($f \neq 0$). Substitute $f = XY$ into (7.5.1), divide by $XY$, and rearrange to get $-X''/X = Y''/Y + \lambda$. This equation contains only the variable $x$ on the left, only the variable $y$ on the right. Thus the left side does not in fact depend on $x$ (because the right side does not), hence is a constant, say $\gamma$. We get $-X'' = \gamma X$ and $-Y'' = (\lambda - \gamma)Y$. The Dirichlet boundary condition, $f(x, y) = 0$ on $\partial D$, is equivalent to both $X(0) = X(a) = 0$ and $Y(0) = Y(b) = 0$ for the separated functions $X$ and $Y$. Hence we are now faced with two Hilbert space eigenvalue problems, each for the one-variable Laplacian (Section 4.5):

$$-X'' = \gamma X, \qquad X \in L^2(0, a), \ X(0) = X(a) = 0 \qquad (7.5.2)$$

$$-Y'' = (\lambda - \gamma)Y, \qquad Y \in L^2(0, b), \ Y(0) = Y(b) = 0 \qquad (7.5.3)$$

We are back in familiar territory. The eigenvalues of the problem (7.5.2) are $\gamma = m^2\pi^2/a^2$, $m = 1, 2, \ldots$. These are all simple, and corresponding eigenfunctions are $X_m(x) = \sin(m\pi x/a)$. The eigenvalues for problem (7.5.3) are given by $\lambda - m^2\pi^2/a^2 = n^2\pi^2/b^2$, $n = 1, 2, \ldots$. These are also simple, and corresponding eigenfunctions are $Y_n = \sin(n\pi y/b)$. The functions

$$f_{m,n}(x, y) = \sin\frac{m\pi x}{a}\sin\frac{n\pi y}{b}$$

obviously lie in $L^2(D)$, satisfy $\delta(f_{m,n}) = (m^2/a^2 + n^2/b^2)\pi^2 f_{m,n}$ so are compatible with $\delta$, and $f_{m,n} = 0$ on $\partial D$. Hence we have found a sequence of scalars $\lambda = (m^2/a^2 + n^2/b^2)\pi^2$ such that $\Delta f = \lambda f$ has a nonzero solution $f = f_{m,n}$ in the domain of $\Delta$. We shall assume that separation of variables, which has never failed us before, has come through again and has given us *all* the eigenvalues of the rectangle:

$$\lambda = \left(\frac{m^2}{a^2} + \frac{n^2}{b^2}\right)\pi^2, \qquad m, n = 1, 2, \ldots \tag{7.5.4}$$

with their corresponding eigenfunctions:

$$f_{m,n}(x, y) = \sin\frac{m\pi x}{a}\sin\frac{n\pi y}{b}, \qquad m, n = 1, 2, \ldots \tag{7.5.5}$$

Let us consider in detail the eigenvalue sequence (7.5.4) for the unit square, $a = b = 1$. We wish to arrange these eigenvalues in increasing order, following the general pattern described in Section 7.4, and to determine the multiplicity of each eigenvalue. When $a = b = 1$, then $\lambda = (m^2 + n^2)\pi^2$, $m, n = 1, 2, \ldots$. The eigenfunction corresponding to the pair $(m, n)$ is $f_{m,n}(x, y) = \sin m\pi x \sin n\pi y$. If $m^2 + n^2 \neq p^2 + q^2$, then we know that the eigenfunctions $f_{m,n}$ and $f_{p,q}$ will be orthogonal in $L^2(D)$ because they are eigenfunctions of a Hermitian operator belonging to different eigenvalues. But more than this is true: $\langle f_{m,n}, f_{p,q} \rangle = 0$ *unless both $m = p$ and $n = q$.* (Remember that $m, n, p, q$ are all $\geq 1$.) We can prove this italicized statement by a direct calculation: $4\langle f_{m,n}, f_{p,q} \rangle = \langle \sin mt, \sin pt \rangle_C \langle \sin nt, \sin qt \rangle_C$, which expresses the $L^2(D)$-inner product on the left in terms of the inner product $\langle f, g \rangle_C = (1/\pi)\int_{-\pi}^{\pi} f\bar{g}\,dt$ in $L^2(C)$, where $C$ is the unit circle. We studied this in Section 3.6 and showed there that the inner product on the right is zero if either $m \neq p$ or $n \neq q$. Let us summarize what we have just proved in a formal statement:

**Lemma 7.14.** *Let $m, n$ be integers $\geq 1$, and consider the functions $f_{m,n}(x, y) = \sin m\pi x \sin n\pi y$, which are the eigenfunctions of the Laplacian on the unit square $D = \{(x, y): 0 \leq x \leq 1, 0 \leq y \leq 1\}$ (Dirichlet boundary conditions). If either $m \neq p$ or $n \neq q$, then $f_{m,n}$ is orthogonal to $f_{p,q}$.*

Hence, if $m^2 + n^2 = p^2 + q^2$, so that the same eigenvalue $\lambda = (m^2 + n^2)\pi^2 = (p^2 + q^2)\pi^2$ corresponds to the pair $(m, n)$ and the pair $(p, q)$, but if $m \neq p$, then $f_{m,n}$ and $f_{p,q}$ will be orthogonal eigenfunctions both belonging to $\lambda$. As orthogonal eigenfunctions are certainly linearly independent, the multiplicity of this $\lambda$ must be at least 2.

For example, $1^2 + 2^2 = 2^2 + 1^2 = 5$, and 5 cannot be written as the sum of two positive integral squares in any other way. So $\lambda = 5\pi^2$ has multiplicity 2. On the other hand, $1^2 + 1^2 = 2$, so $\lambda = 2\pi^2$ has multiplicity 1. In general, $\lambda = r\pi^2$ *has multiplicity equal to the number of different ways the integer r can be written* $r = m^2 + n^2$ *for integers* $m, n \geq 1$. So the determination of the multiplicity of the eigenvalues of the unit square reduces to a problem in number theory! The multiplicity does exceed 2. For example, $65 = 4^2 + 7^2 = 7^2 + 4^2 = 1^2 + 8^2 = 8^2 + 1^2$, so the multiplicity of $\lambda = 65\pi^2$ is 4. Also $1105 = 4^2 + (33)^2 = 9^2 + (32)^2 = (12)^2 + (31)^2 = (23)^2 + (24)^2$ plus the reversal of these, so the multiplicity of $\lambda = 1105\pi^2$ is 8. The multiplicity, while always finite of course, does become arbitrarily large. This is proved in the classic treatise *An Introduction to the Theory of Numbers* by G. H. Hardy and E. M. Wright, (Oxford Univ. Press, 5th ed., 1979) §16.9. See also Exercises 1 and 2.

Go back now from the square to the general rectangle of sides $a$ and $b$, for which the eigenvalue sequence is given by (7.5.4). Rewrite (7.5.4) this way: $\lambda = (1/ab)(m^2(b/a) + n^2(a/b))\pi^2$. Now construct a rectangle having the same area as the unit square, so $ab = 1$, but with sides determined by $b/a = 1 + \varepsilon$ where $\varepsilon = \sqrt{2}/10^4 = 1.41421356\ldots \times 10^{-4} = 0.000141421356\ldots$. Then $b = a(1 + \varepsilon)$, so $1 = ab = a^2(1 + \varepsilon)$, hence $a = (1 + \varepsilon)^{-1/2} = 0.99992930\ldots$, and $b = 1/a = (1 + \varepsilon)^{1/2} = 1.00007071\ldots$. This rectangle differs imperceptibly from the unit square. Its eigenvalue sequence is given by $\lambda = (m^2(1 + \varepsilon) + n^2(1 + \varepsilon)^{-1})\pi^2$, $m, n = 1, 2, \ldots$. What are the multiplicities in this case? If $m^2(1+\varepsilon) + n^2(1+\varepsilon)^{-1} = p^2(1+\varepsilon) + q^2(1+\varepsilon)^{-1}$, then $(m^2 - p^2)(1+\varepsilon)^2 = (q^2 - n^2)$. If $m^2 \neq p^2$, then we would be able to get $1 + 2\varepsilon + \varepsilon^2 = (q^2 - n^2)/(m^2 - p^2)$. As $\varepsilon^2 = 2/10^8$ this would lead to $2\sqrt{2} = 10^4((q^2 - n^2)/(m^2 - p^2) - 1 - 2/10^8)$, which says that $\sqrt{2}$ is a rational number, which is not so. Hence we must have $m^2 = p^2$, hence $m = p$ and $q = n$. Thus there are no repetitions in the sequence of eigenvalues, and each one is simple. We have used $\varepsilon = \sqrt{2}/10^4$, but could just as well have used $\varepsilon = \sqrt{2}/10^{10}$ or an even much smaller irrational number and obtained the same result. By altering imperceptibly the sides of our unit square, leaving its area constant at 1, we have rendered all its eigenvalues simple, even though the square itself had eigenvalues of arbitrarily large multiplicity. While the multiplicities of the eigenvalues changed abruptly, the eigenvalues themselves have changed very little as shown in Table 7.1, where I have listed, according to multiplicity, the first 33 eigenvalues $\mu_n$ of the unit square and some of the "perturbed" unit square. According to Weyl's limit relation, $\mu_n/n \to 4\pi/\text{area}(D)$. In our case area $(D) = 1$, so $\mu_n/n\pi^2 \to 4/\pi = 1.273\ldots$. As you can see in Table 7.1, the numbers $\mu_n/n\pi^2$ are drifting toward $4/\pi$ but the convergence is so slow that this limit relation has no practical utility in this range.

**Table 7.1**  Eigenvalues $\mu_n$ of the Unit Square and the Perturbed Unit Square Listed According to Their Multiplicity (Dirichlet Boundary Condition)

| $n$ | $p$ | $q$ | $\dfrac{\mu_n}{\pi^2} = p^2 + q^2$ | $\dfrac{\mu_n}{\pi^2 n}$ | Perturbed square |
|----|----|----|----|----|----|
|    |    |    |    |    | $\dfrac{\mu_n}{\pi^2} = p^2(1+\varepsilon) + q^2(1+\varepsilon)^{-1}$ |
| 1  | 1  | 1  | 2  | 2     | 2.00000002 |
| 2  | 1  | 2  | 5  | 2.5   | 4.999575817 |
| 3  | 2  | 1  | 5  | 1.7   | 5.000424283 |
| 4  | 2  | 2  | 8  | 2.0   | 8.00000008 |
| 5  | 1  | 3  | 10 | 2.0   | 9.998868812 |
| 6  | 3  | 1  | 10 | 1.7   | 10.00113139 |
| 7  | 2  | 3  | 13 | 1.9   | 12.99929308 |
| 8  | 3  | 2  | 13 | 1.6   | 13.00070719 |
| 9  | 1  | 4  | 17 | 1.9   | 16.99787900 |
| 10 | 4  | 1  | 17 | 1.7   | 17.00212134 |
| 11 | 3  | 3  | 18 | 1.6   | etc. |
| 12 | 2  | 4  | 20 | 1.7   | All these eigenvalues |
| 13 | 4  | 2  | 20 | 1.5   | are simple. |
| 14 | 3  | 4  | 25 | 1.8   | |
| 15 | 4  | 3  | 25 | 1.7   | |
| 16 | 1  | 5  | 26 | 1.6   | |
| 17 | 5  | 1  | 26 | 1.53  | |
| 18 | 2  | 5  | 29 | 1.61  | |
| 19 | 5  | 2  | 29 | 1.53  | |
| 20 | 4  | 4  | 32 | 1.60  | |
| 21 | 3  | 5  | 34 | 1.62  | |
| 22 | 5  | 3  | 34 | 1.55  | |
| 23 | 1  | 6  | 37 | 1.61  | |
| 24 | 6  | 1  | 37 | 1.54  | |
| 25 | 2  | 6  | 40 | 1.60  | |
| 26 | 4  | 2  | 40 | 1.54  | |
| 27 | 4  | 5  | 41 | 1.52  | |
| 28 | 5  | 4  | 41 | 1.46  | |
| 29 | 3  | 6  | 45 | 1.55  | |
| 30 | 6  | 3  | 45 | 1.50  | |
| 31 | 1  | 7  | 50 | 1.61  | |
| 32 | 5  | 5  | 50 | 1.56  | |
| 33 | 7  | 1  | 50 | 1.52  | |
|    |    |    |    | $4/\pi = 1.273\ldots$ | |

In Exercises 3 and 4, you will find the eigenvalues of the rectangle under other boundary conditions.

Once we have the eigenvalues of the rectangle, we are in a position to simply "write down" the solution of the wave equation—see Exercise 15 of the previous section. Using the Dirichlet boundary condition, the solution of $\Delta u + (1/w^2)\, \partial^2 u/\partial t^2 = 0$ on the unit square is

$$u(x, y, t) = \sum_{n=1}^{\infty} (a_n \cos w\sqrt{\mu_n}\, t + b_n \sin w\sqrt{\mu_n}\, t) \sin p\pi x \sin q\pi y \qquad (7.5.6)$$

If the initial velocity is zero, then $b_n = 0$, $n = 1, 2, \ldots$. The $\mu_n$'s are the eigenvalues listed according to their multiplicity. These are given in Table 7.1. If we use the distinct eigenvalues $\lambda_1 = 2\pi^2$, $\lambda_2 = 5\pi^2$, $\lambda_3 = 8\pi^2, \ldots$ and if we take all $b_n = 0$, then we may rewrite (7.5.6) as

$$u(x, y, t) = \sum_{n=1}^{\infty} \left( \sum_{j=1}^{d(n)} a_{nj} \sin p\pi x \sin q\pi y \right) \cos w\sqrt{\lambda_n}\, t \qquad (7.5.7)$$

where $d(n) = \dim(E(\lambda_n))$ is the multiplicity of $\lambda_n$ and the interior sum is taken over all integer pairs $(p, q)$, $p \geqslant 1$, $q \geqslant 1$, such that $\lambda_n = (p^2 + q^2)\pi^2$. The first few terms of the series (7.5.7) are

$$u = (a_{11} \sin \pi x \sin \pi y) \cos w\sqrt{2\pi}\, t + (a_{21} \sin \pi x \sin 2\pi y$$

$$+ a_{22} \sin 2\pi x \sin \pi y) \cos w\sqrt{5\pi}\, t$$

$$+ (a_{31} \sin 2\pi x \sin 2\pi y) \cos w\sqrt{8\pi}\, t + \cdots \qquad (7.5.8)$$

The term $a_{21} \sin \pi x \sin 2\pi y + a_{22} \sin 2\pi x \sin \pi y$ is the "general eigenfunction" belonging to the eigenvalue $\lambda_2 = 5\pi^2$, meaning that every eigenfunction in $E(\lambda_2)$ has that form for suitable constants $a_{21}$, $a_{22}$ not both zero. These eigenfunctions have physical significance. For example, if we take the initial position of our square drumhead as $u(x, y, 0) = 2 \sin\pi x \sin 2\pi y + \sin 2\pi x \sin \pi y$ and take its initial velocity as zero, then (7.5.6) reduces simply to $u = (2 \sin \pi x \sin 2\pi y + \sin 2\pi x \sin \pi y) \cos w\sqrt{5\pi}\, t$. If we were to sprinkle iron filings on our square drumhead and then vibrate it in this mode, the iron filings would fly off except on the curve determined by $2 \sin \pi x \sin 2\pi y + \sin 2\pi x \sin \pi y = 0$ because this curve remains fixed at zero amplitude for all time. Such a curve is called a *nodal curve*.

**Definition 7.15.** Let $f(x, y)$ be an eigenfunction of the Laplacian on a region $D$. The curves in $D$, excluding $\partial D$, determined by the equation $f(x, y) = 0$ are called the *nodal curves* of $f$. The nodal curves divide $D$ into subregions called the *nodal regions* of $f$.

For example, let's sketch the nodal curve $f(x, y) = 2 \sin \pi x \sin 2\pi y +$

$\sin 2\pi x \sin \pi y = 0$. The curve is the locus of *all* pairs $(x, y)$ in $D$ for which this equation holds. On the boundary $\partial D$ of the square $D$ *every* eigenfunction vanishes because we are using the Dirichlet boundary condition. We have agreed to exclude this. Hence the interesting part of the curve occurs when $0 < x < 1$ and $0 < y < 1$. In this range $\sin \pi x \neq 0$ and $\sin \pi y \neq 0$. Using the relation $\sin 2\theta = 2 \sin \theta \cos \theta$, we get $f(x,y) = 2 \sin \pi x \sin \pi y (2 \cos \pi y + \cos \pi x)$. As $\sin \pi x \sin \pi y \neq 0$, $f = 0$ is equivalent to $2 \cos \pi y + \cos \pi x = 0$, or $\cos \pi y = -\frac{1}{2} \cos \pi x$. For $0 < x < 1$ we have $-\frac{1}{2} < -\frac{1}{2} \cos \pi x < \frac{1}{2}$. For each such $x$, there is one and only one $y$ with $0 < y < 1$ and $\cos \pi y = -\frac{1}{2} \cos \pi x$. A program for computing $y$ as a function of $x$ is easily written.

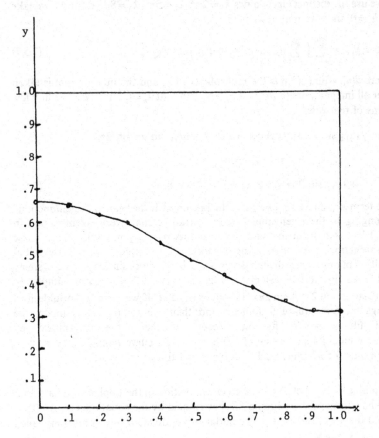

**Figure 7.6** Nodal curve for the eigenfunction $f(x, y) = 2 \sin \pi x \sin 2\pi y + \sin 2\pi x \sin \pi y = 0$ corresponding to the eigenvalue $\lambda_2 = 5\pi^2$ of the unit square.

When we "perturb" the square, all its eigenfunctions become simple, so then the eigenfunction belonging to a given eigenvalue is unique up to a nonzero multiplicative constant. Thus, for the perturbed square, we can speak of the nodal curve belonging to a given eigenvalue. As you see from Table 7.1, the eigenvalue $\lambda_2 = 5\pi^2$, which has multiplicity 2 for the square, splits into two different eigenvalues when we perturb the square, $\lambda_2 = 4.999575817\pi^2$ and $\lambda_3 = 5.000424283\pi^2$, both simple. In the perturbed case, the nodal curve for $\lambda_2$ is $\sin(\pi x/a)\sin(2\pi y/b) = 0$ and for $\lambda_3$ is $\sin(2\pi x/a)\sin(\pi y/b) = 0$ where $a = 0.99992930\ldots$ and $b = 1.00004771\ldots$. These are simply the lines $y = \frac{1}{2}$ for $\lambda_2$ and $x = \frac{1}{2}$ for $\lambda_3$. These various nodal curves for the square and perturbed square are graphed in Figures 7.6 and 7.7.

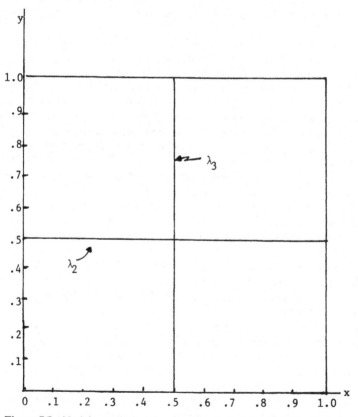

**Figure 7.7** Nodal curves for the eigenfunction $\sin(\pi x/a)\sin(2\pi y/b)$ belonging to $\lambda_2 = 4.999575817\pi^2$ and for the eigenfunction $\sin(2\pi x/a)\sin(\pi y/b)$ belonging to $\lambda_3 = 5.00042483\pi^2$ of the "perturbed" unit square.

Once we know the eigenvalues of the rectangle, we are in a position to write down the solution not only of the wave equation but also of the heat equation and special cases of the Schrödinger equations—see Exercises 15, 16, and 17 of Section 7.4. Exercises 7 and 8 at the end of this section deal with the heat equation for a square steel plate.

### Exercises

1. (a) Prove the identity $(a^2 + b^2)(c^2 + d^2) = (ac + bd)^2 + (ad - bc)^2$.
   (b) Using the identity in (a), prove: If the integers $r, s$ can each be written as the sum of two integral squares, so can their product $rs$.
   (c) Given that $5 = 1^2 + 2^2$ and $13 = 2^2 + 3^2$, use your result in (b) to write $(5)(13) = 65$ as the sum of two integral squares. Do this in four different ways.

2. A *prime* is a positive integer $\geqslant 2$ divisible only by 1 and itself. It is a fact that any prime of the form $4k + 1$ is the sum of two integral squares. See the treatise by Hardy and Wright referred to in the text, or the book *The Higher Arithmetic* by H. Davenport (5th ed., Camb. Univ. Press, 1982). For example $5 = 1^2 + 2^2, 13 = 2^2 + 3^2, 17 = 1^2 + 4^2, 29 = 2^2 + 5^2, 37 = 1^2 + 6^2$, etc. Using the method in Exercise 1, write $1105 = (65)(17)$ as a sum of two squares, in eight different ways. With a bit of grim determination, you can write $(1105)(29) = 32,045$ as the sum of two squares in 16 different ways. What is the multiplicity of the eigenvalue $\lambda = 32,045\pi^2$ of the unit square?

3. Find the eigenvalues and corresponding eigenfunctions for the rectangle shown in Figure 7.5 under Neumann boundary conditions. In the special case of the unit square, $a = b = 1$, show how the determination of the multiplicity of each eigenvalue reduces to a number-theoretic problem. For the unit square, list the numbers $\mu_n/\pi^2$, where $\mu_n$ are the eigenvalues of the square, listed according to their multiplicity (Neumann boundary condition), $0 \leqslant n \leqslant 29$. List also $\mu_n/\pi^2 n$. Separately, list the discrete eigenvalues $\lambda_n/\pi^2$ in increasing order of magnitude, $0 \leqslant n \leqslant 15$, and give the multiplicity of each. Do these multiplicities become arbitrarily large? Justify your answer.

4. Find the eigenvalues and corresponding eigenfunctions for the rectangle shown in Figure 7.5 under these mixed boundary conditions: Dirichlet on the bottom, Neumann on the remaining three sides. In the special case of the unit square, $a = b = 1$, show that the multiplicity of the eigenvalue $r\pi^2$ equals the number of different ways we can write $4r = (2p)^2 + (2q - 1)^2$, $p = 0, 1, 2, \ldots, q = 1, 2, \ldots$. Do these multiplicities increase without limit? Justify your answer. For the unit square, list the numbers $\mu_n/\pi^2$, where $\mu_n$ are the eigenvalues of the square, listed according to their multiplicity, $0 \leqslant n \leqslant 16$. Also list $\mu_n/\pi^2 n$.

5. The functions $f(x, y) = \sin \pi x \sin 2\pi y \pm \sin 2\pi x \sin \pi y$ are eigenfunctions

belonging to the eigenvalue $\lambda_2 = 5\pi^2$ of the unit square, Dirichlet boundary conditions. Graph these two nodal curves.

6. (Continuation) Refer back to Exercise 14 of Section 7.4, which dealt with the first Dirichlet eigenvalue of the triangle $T$ determined by $(0,0)$, $(1,0)$, and $(0,1)$. Is the first Dirichlet eigenvalue of $T$ the second Dirichlet eigenvalue of the unit square? Why?

7. A square metal plate, 1 meter on a side, is heated to an initial temperature $u(x, y, 0)$, $0 \leqslant x \leqslant 1$, $0 \leqslant y \leqslant 1$. At time $t = 0$ the source of heat is removed, and all four edges are put to 0°C and held at that temperature. The plate is perfectly insulated on its flat surfaces so that heat flows only parallel to the $x$–$y$ plane. Show that the subsequent temperature $u(x, y, t)$ of the plate is

$$u(x, y, t) = \sum_{n=1}^{\infty} c_n \sin p\pi x \sin q\pi y e^{-\alpha^2 \mu_n t}$$

where $\mu_n = (p^2 + q^2)\pi^2$, $p \geqslant 1$, $q \geqslant 1$, are the eigenvalues of the square listed in Table 7.1, and $\alpha^2 = \kappa/\sigma\rho$, $\kappa =$ thermal conductivity, $\sigma =$ specific heat, and $\rho =$ density. What are the correct units for $\alpha^2$ here? Find a formula for the coefficients $c_n$.

8. (Continuation) Suppose the plate in Exercise 7 is made of steel and is heated to an initial temperature

$$u(x, y, 0) = 100(1 - y)°C, \qquad 0 \leqslant x \leqslant 1, 0 \leqslant y \leqslant 1$$

Keeping terms $n = 1$ through $n = 17$ (there will be 10 nonzero terms), graph $u(0.5, y, t)$ versus $y$ and $u(x, 0.2, t)$ versus $x$ for $t = 0$, 0.1, 0.5, and 1 hour. At $t = 0$ plot the function $u(x, y, 0) = 100(1 - y)$; don't attempt to use your series at $t = 0$. [Hints for an HP-15C program: Put time $t$ in register 0 (hours). Punch in xENTERy ($x$, $y$ in meters). The following program takes it from there.

```
fLBLA
gπ × STO2                          puts πy in Reg 2
x≷y                                interchanges x and y
gπ × STO1                          puts πx in Reg 1
.51086 RCL × 0 CHS STO3            puts −α²π²t in Reg 3
0 ENTER, ENTER, ENTER             clears the stack
RCL1 SIN    ⎫
RCL2 SIN ×  ⎬                      computes first term of series
2 RCL × 3 eˣ × ⎭
RCL1 SIN         ⎫
2 RCL × 2 SIN ×  ⎬                 computes the second term
5 RCL × 3 eˣ 2 ÷ ⎭
+                                  adds the first two terms
etc.
```

With 10 nonzero terms, the HP-15C takes about 30 seconds per value.]

## 7.6   EIGENVALUES OF THE DISK

In this section we shall use the Dirichlet boundary conditions exclusively.

Take $a$ as the radius of the disk $D$, and locate the disk as in Figure 7.8. Our objective is to find the eigenvalues of the disk, Dirichlet boundary conditions. Which is to say, find all scalars $\lambda$ such that $\Delta f = \lambda f$ has a nonzero solution $f$ in the domain of $\Delta$.

Our differential equation is

$$\delta(f) = -\frac{\partial^2 f}{\partial r^2} - \frac{1}{r}\frac{\partial f}{\partial r} - \frac{1}{r^2}\frac{\partial^2 f}{\partial \theta^2} = \lambda f \tag{7.6.1}$$

We know $\lambda > 0$ as the Laplacian is positive definite. We seek all nonzero functions $f(r, \theta)$ in $L^2(D)$ that solve (7.6.1) and satisfy $f = 0$ on $\partial D$. Note that if $f \in L^2(D)$ solves (7.6.1), then $\delta(f) = \lambda f \in L^2(D)$ also, so $f$ is automatically compatible with $\delta$, thus in the domain of the Laplacian operator $\Delta$.

We have already solved this problem in Section 6.9. The eigenvalues are

$$\lambda = \frac{j_{p,q}^2}{a^2}, \qquad p = 0, 1, 2\ldots, \qquad q = 1, 2, \ldots$$

where $j_{p,q}$ is the $q$th zero of the Bessel function $J_p(x)$. The eigenvalues with $p = 0$ are simple; those with $p \geqslant 1$ have multiplicity 2. The eigenfunction $J_0(j_{0,q}r/a)$ spans the eigenspace for $\lambda = j_{0,q}^2/a^2$, $q = 1, 2, \ldots$. The pair of eigenfunctions $\cos p\theta\, J_p(j_{p,q}r/a)$ and $\sin p\theta\, J_p(j_{p,q}r/a)$ are orthogonal in $L^2(D)$ and span the eigenspace for $\lambda = j_{p,q}^2/a^2$, $p = 1, 2, \ldots, q = 1, 2, \ldots$. In Table 7.2 I have listed, according to multiplicity, the first 30 eigenvalues $\mu_n$ of the unit disk ($a = 1$). According to Weyl's limit relation $\mu_n/n \to 4\pi/\mathrm{area}(D)$. In our case

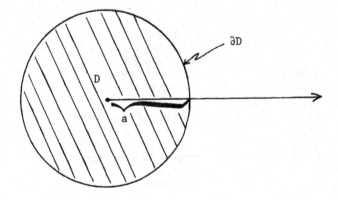

**Figure 7.8**   The disk.

area$(D) = \pi(1)^2 = \pi$, so $\mu_n/n \to 4$. As you can see in Table 7.2, the numbers $\mu_n/n$ are drifting toward 4, but the convergence is slow and erratic. In Table 7.2 I have also listed the first 17 discrete eigenvalues $\lambda_n$ of the unit disk, arranged in increasing magnitude, together with the multiplicity of each.

**Table 7.2** Eigenvalues $\mu_n = j_{p,q}^2$ of the Unit Disk, Listed According to Their Multiplicity (Dirichlet Boundary Condition). Also Listed: Discrete Eigenvalues $\lambda_n$ with Their Multiplicity

| $n$ | $p$ | $q$ | $\mu_n = j_{p,q}^2$ | $\dfrac{\mu_n}{n}$ | $n$ | $\lambda_n$ | Mult. |
|---|---|---|---|---|---|---|---|
| 1 | 0 | 1 | 5.78319 | 5.78 | 1 | 5.78319 | 1 |
| 2 | 1 | 1 | 14.6820 | 7.34 | 2 | 14.6820 | 2 |
| 3 | 1 | 1 | 14.6820 | 4.89 | 3 | 26.3746 | 2 |
| 4 | 2 | 1 | 26.3746 | 6.59 | 4 | 30.4713 | 1 |
| 5 | 2 | 1 | 26.3746 | 5.27 | 5 | 40.7064 | 2 |
| 6 | 0 | 2 | 30.4713 | 5.08 | 6 | 49.2185 | 2 |
| 7 | 3 | 1 | 40.7064 | 5.82 | 7 | 57.5829 | 2 |
| 8 | 3 | 1 | 40.7064 | 5.09 | 8 | 70.8499 | 2 |
| 9 | 1 | 2 | 49.2185 | 5.47 | 9 | 74.8870 | 1 |
| 10 | 1 | 2 | 49.2185 | 4.92 | 10 | 76.9389 | 2 |
| 11 | 4 | 1 | 57.5829 | 5.23 | 11 | 95.2775 | 2 |
| 12 | 4 | 1 | 57.5829 | 4.80 | 12 | 98.7263 | 2 |
| 13 | 2 | 2 | 70.8499 | 5.45 | 13 | 103.499 | 2 |
| 14 | 2 | 2 | 70.8499 | 5.06 | 14 | 122.428 | 2 |
| 15 | 0 | 3 | 74.8870 | 4.99 | 15 | 122.908 | 2 |
| 16 | 5 | 1 | 76.9389 | 4.81 | 16 | 135.021 | 2 |
| 17 | 5 | 1 | 76.9389 | 4.53 | 17 | 139.040 | 1 |
| 18 | 3 | 2 | 95.2775 | 5.29 | | | |
| 19 | 3 | 2 | 95.2775 | 5.01 | | | |
| 20 | 6 | 1 | 98.7263 | 4.94 | | | |
| 21 | 6 | 1 | 98.7263 | 4.70 | | | |
| 22 | 1 | 3 | 103.499 | 4.70 | | | |
| 23 | 1 | 3 | 103.499 | 4.50 | | | |
| 24 | 4 | 2 | 122.428 | 5.10 | | | |
| 25 | 4 | 2 | 122.428 | 4.90 | | | |
| 26 | 7 | 1 | 122.908 | 4.73 | | | |
| 27 | 7 | 1 | 122.908 | 4.55 | | | |
| 28 | 2 | 3 | 135.021 | 4.82 | | | |
| 29 | 2 | 3 | 135.021 | 4.66 | | | |
| 30 | 0 | 4 | 139.040 | 4.63 | | | |
| | | | | $\downarrow$ | | | |
| | | | | 4.0 | | | |

Once we know the eigenvalues of the disk, we are in a position to write down the solutions to the wave and heat equations in the disk (Exercise 2). We can also solve the Schrödinger equations for a quantum particle confined to a disk inside which the potential is zero (Exercise 3).

## Exercises

1. Let $D$ be the disk of radius $a$.
   (a) Show that, in $L^2(D)$, $\|J_0(j_{0,q}r/a)\|^2 = \pi a^2 (J_0'(j_{0,q}))^2$.
   (b) Show that, in $L^2(D)$, with $p \geq 1$,

   $$\|\cos p\theta \, J_p(j_{p,q}r/a)\|^2 = \|\sin p\theta \, J_p(j_{p,q}r/a)\|^2 = \frac{\pi a^2}{2}(J_p'(j_{p,q}))^2$$

   (c) Show that, for $p \geq 1$, $\cos p\theta \, J_p(j_{p,q}r/a)$ is orthogonal to $\sin p\theta \, J_p(j_{p,q}r/a)$ in $L^2(D)$.

2. Show that vertical displacement $u(r, \theta, t)$ of the circular drumhead of radius $r = a$, tension $T$, and density $\rho$ can be written

   $$u(r, \theta, t) = \sum_{n=1}^{\infty} \left[ (a_n \cos p\theta + b_n \sin p\theta) \cos \sqrt{\mu_n} \frac{w}{a} t \right.$$
   $$\left. + (c_n \cos p\theta + d_n \sin p\theta) \sin \sqrt{\mu_n} \frac{w}{a} t \right] J_p(\sqrt{\mu_n} \, r/a)$$

   where the $\mu_n = j_{p,q}^2$ are the eigenvalues of the unit disk, Dirichlet boundary conditions, listed according to multiplicity in increasing order of magnitude; where for each $n$, the appropriate corresponding value of $p$ is taken as in Table 7.2; and where $w = (T/\rho)^{1/2}$. Give formulas for computing the coefficients $a_n$, $b_n$, $c_n$, $d_n$ in terms of the initial position $u(r, \theta, 0)$ and initial velocity $(\partial u/\partial t)(r, \theta, 0)$, distinguishing the cases $p = 0$ and $p \neq 0$.

3. An electron is confined to a circular region of the Bohr radius $a = \hbar^2/me^2$ inside which the potential energy $V = 0$. Show that the possible total energies of the electron are $E_n = (me^4/2\hbar^2)\mu_n = 13.6 \, \mu_n \, \text{eV}$, where $\mu_n$, $n = 1, 2, \ldots$, are the eigenvalues of the unit disk. Compute $E_1$ in eV.

## 7.7  THE LAPLACIAN ON THE SPHERE

In the previous two sections we computed the eigenvalues of the Laplacian on the rectangle and on the disk. In this section and the next we do the same for the Laplacian on the sphere.

Denote the sphere by $S$. In cartesian coordinates $S = \{(x, y, z): x^2 + y^2 + z^2 = a^2\}$, where the positive constant $a$ is the radius of the sphere; see Figure 7.9. Distinguish the sphere $S$ which is a surface (two-dimensional) from

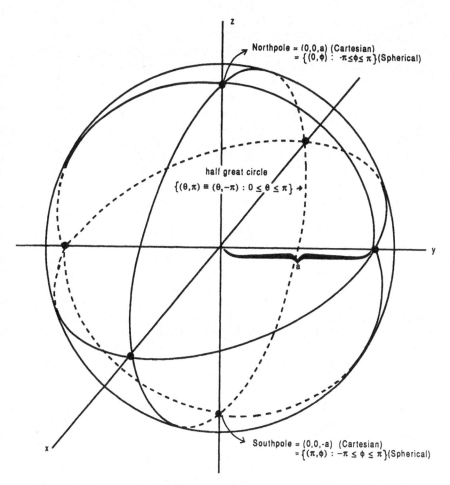

**Figure 7.9** The sphere, $S$. $S = \{(x, y, z): x^2 + y^2 + z^2 = a^2\}$ (Cartesian); $S = (\theta, \phi)$: $0 \leqslant \theta \leqslant \pi, \ -\pi \leqslant \phi \leqslant \pi\}$ (spherical; see Figure 7.10).

the solid ball $B = \{(x, y, z): x^2 + y^2 + z^2 \leqslant a^2\}$ (three-dimensional). The sphere is the boundary of the ball: $S = \partial B$.

While the sphere $S$ is two-dimensional, like the rectangle and disk, it differs from the rectangle and disk in two essential respects. First, $S$ is a curved surface. It requires a three-dimensional ambient space, unlike the rectangle and disk, which lie in a plane. Second, $S$ has no boundary: $\partial S = \varnothing$. The sphere shares this property with the circle. The circle is an example of a one-dimensional manifold without boundary; the sphere, an example of a two-dimensional manifold without boundary. Extending the common terminology wherein we describe the circle as a closed curve, we can describe the sphere as a closed surface (closes on itself, therefore no boundary).

Even though our computations will be exclusively for the sphere, it will be worth our while to broaden our perspective at the beginning and state the basic definitions and facts for the Laplacian on general closed surfaces. This broader view accords with that we took in Section 7.3, where we discussed the Laplacian in a general planar region (see Figure 7.3 or 7.4), even though our computations were for the rectangle and disk.

The kind of general closed surface we have in mind is exemplified by the following: the sphere, the surface of an ellipsoid, or the surface of a torus (doughnut). We won't get any more precise than that, and we just ask that you keep before you some sort of mental picture of the general closed surface. Let us denote the general closed surface by $M$.

As a two-dimensional object, the surface $M$ is generally described by expressing the $x$, $y$, and $z$ coordinates in terms of two independent parameters, say $u$ and $v$: $x = x(u, v)$, $y = y(u, v)$, $z = z(u, v)$. The Laplacian differential expression $\delta$ on $M$ is then obtained by changing the variables in $\delta(f) = -\partial^2 f/\partial x^2 - \partial^2 f/\partial y^2 - \partial^2 f/\partial z^2$ to $u$, $v$ by applying the chain rule for partial differentiation. We shall do all this for the sphere in a moment.

And what holds for the sphere holds for the general closed surface: $M$ must be curved—it cannot lie in a plane. In this sense, $M$ is more complicated than the planar surfaces we considered in Section 7.3. On the other hand, the general closed surface $M$ is simpler than the general planar region in that $M$ has no boundary. And, since it has no boundary, the definition of the Laplacian on $M$ will not involve any boundary conditions.

We shall also ask that $M$ be smooth—that is, have no sharp corners or edges.

To define the Laplacian $\Delta$ as a Hilbert space operator on the closed surface $M$, proceed just as in the planar case. First, construct the Hilbert space $L^2(M)$ consisting of all square-integrable functions on the surface:

$$L^2(M) = \left\{ f : \int_M |f|^2 \, dA < \infty \right\} \tag{7.7.1}$$

where $dA$ is the differential element of area on the surface. Just as in Section 3.9, Theorem 3.16, one can prove that $L^2(M)$ is a vector space; in particular, if $f_1$ and $f_2$ belong to $L^2(M)$, so does $f_1 + f_2$. Using exactly the same arguments that established Theorem 3.16, we can prove that the formula

$$\langle f, g \rangle = \int_M f\bar{g} \, dA \tag{7.7.2}$$

defines an inner product on $L^2(M)$. That is to say, the expression (7.7.2) has the three characteristic properties: conjugate bilinear, Hermitian symmetric, and positive definite.

With this preparation, we are in a position to define the Laplacian operator and to state its principal properties.

**Definition 7.16.** Let $M$ denote a smooth closed surface in $\mathbf{R}^3$ such as the sphere, surface of an ellipsoid, or surface of a torus, and let $L^2(M)$ denote the Hilbert space of all square-integrable functions on $M$. The *Laplacian operator* $\Delta$ has the differential expression $\delta$ obtained from the Cartesian Laplacian $-\partial^2 f/\partial x^2 - \partial^2 f/\partial y^2 - \partial^2 f/\partial z^2$ by expressing $x$, $y$, and $z$ in terms of the two parameters describing the surface; $\Delta$ has domain consisting of functions $f$ in $L^2(M)$ such that $\delta(f)$ also belongs to $L^2(M)$ (compatibility). As $M$ has no boundary, there are no boundary conditions.

The Laplacian on such closed surfaces has the same properties as the Laplacian on planar surfaces (Section 7.4).

**Theorem 7.17.** *The Laplacian $\Delta$ on a closed surface $M$, as defined in Definition 7.16, is a positive semidefinite Hermitian operator on the Hilbert space $L^2(M)$. $\Delta$ has an infinite sequence of eigenvalues*

$$0 = \lambda_0 < \lambda_1 < \lambda_2 < \cdots < \lambda_n < \lambda_{n+1} < \cdots$$

*such that $\lambda_n \to \infty$ as $n \to \infty$. Each $\lambda_n$ has finite multiplicity. The eigenvalue $\lambda_0 = 0$ is simple, and the corresponding eigenspace is spanned by any nonzero constant.*

As $\Delta$ is Hermitian, eigenfunctions belonging to different eigenvalues will be orthogonal. As the eigenspace belonging to each eigenvalue $\lambda$ is finite-dimensional, we can use the Gram-Schmidt process to construct a finite family of orthogonal eigenfunctions, equal in number to the multiplicity of the eigenvalue $\lambda$, that span $\lambda$'s eigenspace. Put all these eigenfunctions together in one sequence, as described in the discussion following Theorem 7.7.

**Theorem 7.18.** *The complete set of eigenfunctions of the Laplacian, listed as described in the previous paragraph, form an orthogonal basis of the Hilbert space $L^2(M)$.*

If we count the eigenvalues according to their multiplicity (see Section 7.4), we have Weyl's limit relation.

**Theorem 7.19.** *If $\mu_n$, $n = 0, 1, 2, \ldots$, is a listing of the eigenvalues of the Laplacian according to their multiplicity, then $\lim_{n \to \infty} \mu_n/n = 4\pi/\mathrm{area}(M)$.*

As to the proofs of Theorems 7.17, 7.18, and 7.19, we shall prove Theorem 7.17 for the sphere only. The proof of the general assertion of Theorem 7.17, together with the proofs of Theorems 7.18 and 7.19, shall be left to a future stage of your career, as was the case with the corresponding assertions in Section 7.4.

Now, having set the stage with an exposition of the facts concerning the Laplacian on a general closed surface, we shall, for the remainder of this section, specialize to the sphere.

Spherical coordinates are the appropriate coordinates for the sphere. We shall use the physicist's notation for spherical coordinates, as depicted in Figure 7.10. These probably differ from the notation you used in calculus in that the symbols $\theta$ and $\phi$ have been interchanged, and $r$ replaces $\rho$. The transformation from spherical coordinates $(r, \theta, \phi)$ to cartesian coordinates $(x, y, z)$ is given by

$$x = r \sin \theta \cos \phi, \qquad y = r \sin \theta \sin \phi, \qquad z = r \cos \theta \tag{7.7.3}$$

with the restrictions

$$r \geqslant 0, \qquad 0 \leqslant \theta \leqslant \pi, \qquad -\pi \leqslant \phi \leqslant \pi \tag{7.7.4}$$

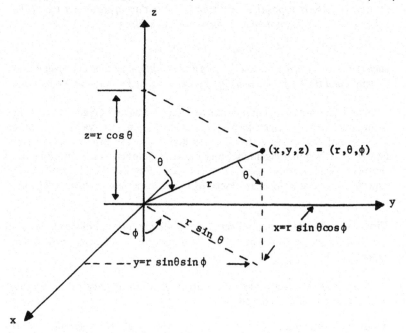

**Figure 7.10**  Spherical coordinates.

In spherical coordinates, all triples $(0, \theta, \phi)$ represent the same point in 3-space, the origin. And, for each fixed $r = a > 0$, all triples $(a, 0, \phi)$, $-\pi \leqslant \phi \leqslant \pi$, represent the same point, namely the north pole of the sphere of radius $a$; and all triples $(a, \pi, \phi)$, $-\pi \leqslant \phi \leqslant \pi$, represent the south pole of the sphere of radius $a$. Further, for each fixed $a > 0$ and each fixed $\theta$, $0 < \theta < \pi$, the two triples $(a, \theta, -\pi)$ and $(a, \theta, \pi)$ represent the same geometric point.

We are interested in coordinates on the sphere $S$, namely $\{(a, \theta, \phi),$ $0 \leqslant \theta \leqslant \pi, -\pi \leqslant \phi \leqslant \pi\}$, where $a$ is a fixed positive constant, the radius of the sphere. We'll abbreviate these coordinates simply as $(\theta, \phi)$. We can describe this assignment of coordinates on the sphere in a manner similar to that we used in describing coordinates on the circle. In the case of the circle, we began with an interval $\{x: a \leqslant x \leqslant b\}$, bent it round, then joined the point $x = a$ to the point $x = b$ (refer to Sections 3.2 and 4.5). In the case of the sphere, begin with a rectangle $R = \{(\theta, \phi): 0 \leqslant \theta \leqslant \pi, -\pi \leqslant \phi \leqslant \pi\}$. (See Figure 7.11.) Bend the rectangle into a cylinder by joining the top edge to the bottom edge. In this process the left edge of the rectangle becomes a circle, the left circular edge of the cylinder, and the right edge becomes the circular edge of the cylinder at its

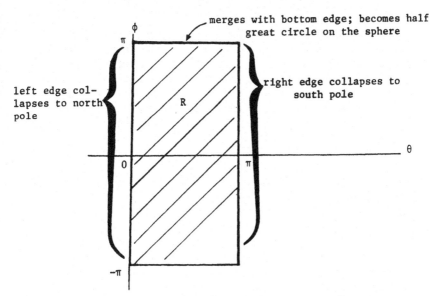

**Figure 7.11** When the rectangle $R = \{(\theta, \phi): 0 \leqslant \theta \leqslant \pi, -\pi \leqslant \phi \leqslant \pi\}$ is mapped via spherical coordinates to the sphere, the entire left edge collapses to a single point, the north pole; the right edge, to the south pole. The bottom and top edges are identified, $(\theta, -\pi) \equiv (\theta, \pi)$, $0 \leqslant \theta \leqslant \pi$, and become one curve on the sphere, a half great circle from the north pole to the south pole—see Figure 7.9.

right end. Then shrink the left circular edge of the cylinder to a point, the north pole of the sphere, and shrink the right circular edge to a point, the south pole.

In practice, we usually define a function $f(\theta, \phi)$ on the sphere by some simple formula valid in the rectangle $R$ of Figure 7.11. For example, we might set $g(\theta, \phi) = \sin\theta \cos(\phi/2)$ or $h(\theta, \phi) = \cos\theta \sin\phi$. Functions like this are quite well behaved in the interior of the rectangle $R$. The problem occurs at the edges, because of the identifications we have just described (which are also depicted in Figure 7.11).

You need to pay special attention to these identifications when testing functions for membership in the domain of the Laplacian operator $\Delta$. According to Definition 7.16, the domain of $\Delta$ consists of functions $f$ that are compatible with the Laplacian differential expression $\delta$; i.e., both $f$ and $\delta(f)$ must belong to $L^2(S)$. Using much the same reasoning as in Theorem 4.2 in Section 4.1 and Theorem 7.2 in Section 7.1, we reach the same conclusion:

**Theorem 7.20.** *A necessary condition that a function $f$ be compatible with the Laplacian differential expression on the sphere $S$ is that $f$ and its first partial derivatives be continuous at every point of $S$.*

In connection with Theorem 7.20, note these two facts. First, as $S$ has no boundary, every point is an interior point. Thus the continuity conditions apply at every point. Second, the north and south poles of the sphere ($\theta = 0$ and $\theta = \pi$) are singular points for spherical coordinates, just as $r = 0$ is a singular point for polar coordinates (Section 7.1). So, in checking continuity of partial derivatives, check $\partial f/\partial\theta$ and $\partial f/\partial\phi$ at every point except $\theta = 0$ and $\theta = \pi$. At the north pole and south pole, use cartesian coordinates, $x = u$, $y = v$, $z = \pm(a^2 - u^2 - v^2)^{1/2}$, and check continuity of $\partial f/\partial u$ and $\partial f/\partial v$ at $(0, 0, \pm a)$.

Thus if a function $f(\theta, \phi)$, defined in the rectangle $R$ of Figure 7.11, is to represent a continuous function *on the sphere* $S$, then $f(0, \phi)$ must be constant, independent of $\phi$ (this constant value is the value of $f$ at the north pole of the sphere); and, when $\theta$ is small, all values $f(\theta, \phi)$, $-\pi \leqslant \phi \leqslant \pi$, must remain close to that constant value. Similarly, $f(\pi, \phi)$ must be constant, and $f(\pi - \varepsilon, \phi) \approx f(\pi, \phi)$ for small positive $\varepsilon$, and $-\pi \leqslant \phi \leqslant \pi$. Furthermore, if such an $f(\theta, \phi)$, defined in $R$, is to represent a continuous function on $S$, then it must satisfy $f(\theta, \pi) = f(\theta, -\pi)$, $0 < \theta < \pi$, because these common values are the values of $f$ at the half great circle on $S$ that results from identifying the top edge of $R$ with its bottom edge. Also, we must have $f(\theta, -\pi + \varepsilon) \approx f(\theta, \pi)$ and $f(\theta, \pi - \varepsilon) \approx f(\theta, \pi)$, $0 < \theta < \pi$, for $\varepsilon$ small and positive.

Consider the two examples introduced earlier: $g(\theta, \phi) = \sin\theta \cos(\phi/2)$ and $h(\theta, \phi) = \cos\theta \sin\phi$. In the first example, $g(0, \phi) = g(\pi, \phi) = 0$ independent of $\phi$, and, when $\theta$ is near 0 or $\pi$, then $g(\theta, \phi)$ is close to 0 because $|g(\theta, \phi)| = |\sin\theta|$

$\times |\cos(\phi/2)| \leqslant |\sin \theta|$. Thus $g$ is continuous at the north and south poles. Furthermore, $g(\theta, \pi) = g(\theta, -\pi) = 0$, and $g(\theta, \pi - \varepsilon) \approx g(\theta, -\pi + \varepsilon)$, $0 < \theta < \pi$, for $\varepsilon$ small and positive. Hence $g(\theta, \phi) = \sin \theta \cos(\phi/2)$ defines a continuous function on the sphere $S$. On the other hand, for $h(\theta, \phi) = \cos \theta \sin \phi$, $h(0, \phi) = \sin \phi$ is not independent of $\phi$ but takes all values between $-1$ and $1$, so there is no hope of assigning some value to $h$ at the north pole to make it continuous there.

While $g$ is continuous on the sphere, $\partial g/\partial \phi = -\frac{1}{2} \sin \theta \sin(\phi/2)$ has the value $-\frac{1}{2} \sin \theta$ at the top edge of the rectangle but the value $+\frac{1}{2} \sin \theta$ at the bottom edge, so there is no hope of making $\partial g/\partial \phi$ into a continuous function on the sphere. Thus $g$ does not belong to the domain of the Laplacian.

Another example: $f(\theta, \phi) = \sin \theta \cos \phi$. In this case, $f$ and $\partial f/\partial \phi$ are continuous at every point of the sphere, and $\partial f/\partial \theta$ is continuous at each point of the half great circle $\{(\theta, \pi) \equiv (\theta, -\pi): 0 < \theta < \pi\}$. But $\partial f/\partial \theta = \cos \theta \cos \phi$ appears to be discontinuous at $\theta = 0$ because $f_\theta(0, \phi) = \cos \phi$ is not independent of $\phi$. But this apparent discontinuity is due to the fact that the north and south poles are singular points for spherical coordinates. As I pointed out earlier, to check continuity of partial derivatives at $\theta = 0$ and $\theta = \pi$, use cartesian coordinates $x = u$, $y = v$, $z = \pm(a^2 - u^2 - v^2)^{1/2}$. In these coordinates, $f(\theta, \phi) = \sin \theta \cos \phi = u/a$, so $\partial f/\partial u = 1/a$, $\partial f/\partial v = 0$. As these are both continuous at $(0, 0, \pm a)$, $f(\theta, \phi) = \sin \theta \cos \phi$ has continuous first partial derivatives at every point of the sphere $S$, thus satisfies the necessary condition of Theorem 7.20.

Remember that the continuity of $f$ and its first partial derivatives is a *necessary* condition for $f$ to belong to the domain of $\Delta$. If $f$ fails these conditions, it cannot belong. But, if it passes, you must still test whether $\delta(f) \in L^2(S)$.

These fussy considerations apply when you want to test whether a function $f(\theta, \phi)$ defined on the rectangle $R$ of Figure 7.11 belongs to the domain of the Laplacian $\Delta$. If you are only testing whether $f$ belongs to the Hilbert space $L^2(S)$, then you can simply define $f(\theta, \phi)$ in the interior of $R$, disregarding its values at the edges. The edges collapse to points and a half great circle on the sphere $S$, and the behavior of $f$ on these zero- and one-dimensional subsets of $S$ can't affect the area integral. (If you change the value of $f$ at a few points, or along a curve, and $g$ is the changed function, then $g$ is "essentially equal" to $f$: $\|g - f\| = 0$.)

We shall need the differential element of volume, $dV$, in spherical coordinates and the differential element of area on the sphere, $dA$, also in spherical coordinates. I shall ask you to do these calculations as Exercises 11 and 15 at the end of this section. The method you will use is the calculus of differential forms in three variables, whose rules and definitions are an easy extension of those for two variables (which were explained in Section 7.2). The

basic definitions and rules for differential forms in three variables run as
follows; you will need this material to do several of the exercises.

A 0-*form* is an ordinary function $f(x, y, z)$ of three variables. A 1-*form* is an
expression $f\,dx + g\,dy + h\,dz$ that occurs as an integrand in a line integral. A
2-*form* is an expression $f\,dy\,dz + g\,dz\,dx + h\,dx\,dy$ that occurs as an integrand
in a surface integral. A 3-*form* is an expression $f\,dx\,dy\,dz$ that occurs as an
integrand in a volume integral.

We can add and subtract forms of the same type. The product of
forms is determined by the simple rules $dy\,dz = -dz\,dy$, $dz\,dx = -dx\,dz$,
$dx\,dy = -dy\,dx$, $dx\,dx = dy\,dy = dz\,dz = 0$, together with the usual distributive
and associative laws and the commutativity of functions with differentials.

The exterior derivative of a 0-form (function) $f$ is just its total differential:
$df = (\partial f/\partial x)\,dx + (\partial f/\partial y)\,dy + (\partial f/\partial z)\,dz$. The exterior derivative of a 1-form
$\omega = f\,dx + g\,dy + h\,dz$ is defined by the rule $d\omega = df\,dx + dg\,dy + dh\,dz$, where
$df, dg, dh$ are the total differentials and the products $df\,dx$, $dg\,dy$, and $dh\,dz$ are
calculated using the exterior multiplication explained in the previous paragraph. Thus we see that the exterior derivative of a 1-form is a 2-form. The
exterior derivative of a 2-form, defined similarly, is a 3-form. The exterior
derivative of a 3-form is zero.

The differential element of area on the sphere $S$ of radius $a$ expressed in
spherical coordinates is $dA = a^2 \sin\theta\,d\theta\,d\phi$ (Exercise 15). Hence, for functions
$f(\theta, \phi)$ expressed in spherical coordinates, the square-integrability condition
that characterizes membership in the Hilbert space $L^2(S)$ is

$$\|f\|^2 = a^2 \int_{-\pi}^{\pi} d\phi \int_0^{\pi} d\theta \sin\theta\,|f(\theta, \phi)|^2 = a^2 \int_0^{\pi} d\theta \sin\theta \int_{-\pi}^{\pi} d\phi\,|f(\theta, \phi)|^2 < \infty$$

$$(7.7.5)$$

The inner product is

$$\langle f, g \rangle = a^2 \int_{-\pi}^{\pi} d\phi \int_0^{\pi} d\theta \sin\theta\,f(\theta, \phi)\overline{g(\theta, \phi)}$$

$$= a^2 \int_0^{\pi} d\theta \sin\theta \int_{-\pi}^{\pi} d\phi\,f(\theta, \phi)\overline{g(\theta, \phi)}. \qquad (7.7.6)$$

Exercise 4 provides practice with integrals of this type.

To convert the Laplacian differential expression $\delta(u) = -\partial^2 u/\partial x^2 -$
$\partial^2 u/\partial y^2 - \partial^2 u/\partial z^2$ to spherical coordinates, we follow the same pattern as we
did with polar coordinates (Section 6.1). Equations (7.7.3), or Figure 7.10,
yield $r^2 = x^2 + y^2 + z^2$. Take the partial derivative with respect to ·$x$ to get

$2r\,\partial r/\partial x=2x$. Do the same with $y$ and $z$; then use (7.7.3). You will find

$$\frac{\partial r}{\partial x}=\frac{x}{r}=\sin\theta\cos\phi,\qquad \frac{\partial r}{\partial y}=\frac{y}{r}=\sin\theta\sin\phi,\qquad \frac{\partial r}{\partial z}=\frac{z}{r}=\cos\theta \qquad (7.7.7)$$

Equations (7.7.3) also yield $r^2\sin^2\theta=x^2+y^2$. Take the partial derivative of this equation with respect to $x$, then $y$. Differentiate $z=r\cos\theta$ with respect to $z$. Manipulate your results using (7.7.7), and derive thereby the following equations:

$$\frac{\partial\theta}{\partial x}=\frac{\cos\theta\cos\phi}{r},\qquad \frac{\partial\theta}{\partial y}=\frac{\cos\theta\sin\phi}{r},\qquad \frac{\partial\theta}{\partial z}=-\frac{\sin\theta}{r} \qquad (7.7.8)$$

Finally, differentiate $\tan\phi=y/x$ with respect to $x$, $y$ and $z$; then use (7.7.7) and (7.7.8) to obtain

$$\frac{\partial\phi}{\partial x}=-\frac{\sin\phi}{r\sin\theta},\qquad \frac{\partial\phi}{\partial y}=\frac{\cos\phi}{r\sin\theta},\qquad \frac{\partial\phi}{\partial z}=0 \qquad (7.7.9)$$

The chain rule tells us that

$$\frac{\partial u}{\partial x}=\frac{\partial u}{\partial r}\frac{\partial r}{\partial x}+\frac{\partial u}{\partial\theta}\frac{\partial\theta}{\partial x}+\frac{\partial u}{\partial\phi}\frac{\partial\phi}{\partial x}$$

and similarly for $\partial u/\partial y$ and $\partial u/\partial z$. Using the chain rule together with (7.7.7), (7.7.8), and (7.7.9), we get

$$\frac{\partial u}{\partial x}=\sin\theta\cos\phi\,\frac{\partial u}{\partial r}+\frac{\cos\theta\cos\phi}{r}\frac{\partial u}{\partial\theta}-\frac{\sin\phi}{r\sin\theta}\frac{\partial u}{\partial\phi}$$

$$\frac{\partial u}{\partial y}=\sin\theta\sin\phi\,\frac{\partial u}{\partial r}+\frac{\cos\theta\sin\phi}{r}\frac{\partial u}{\partial\theta}+\frac{\cos\phi}{r\sin\theta}\frac{\partial u}{\partial\phi} \qquad (7.7.10)$$

$$\frac{\partial u}{\partial z}=\cos\theta\,\frac{\partial u}{\partial r}-\frac{\sin\theta}{r}\frac{\partial u}{\partial\theta}$$

which express the *first* cartesian partials of $u$ in terms of partial derivatives with respect to spherical coordinates. We must now repeat this procedure to express the *second* cartesian partials of $u$ in spherical coordinates. Use the chain rule again:

$$\frac{\partial^2 u}{\partial x^2}=\frac{\partial}{\partial x}\left(\frac{\partial u}{\partial x}\right)=\frac{\partial}{\partial r}\left(\frac{\partial u}{\partial x}\right)\frac{\partial r}{\partial x}+\frac{\partial}{\partial\theta}\left(\frac{\partial u}{\partial x}\right)\frac{\partial\theta}{\partial x}+\frac{\partial}{\partial\phi}\left(\frac{\partial u}{\partial x}\right)\frac{\partial\phi}{\partial x}$$

then use (7.7.7), (7.7.8), and (7.7.9) to get

$$\frac{\partial^2 u}{\partial x^2}=\frac{\partial}{\partial r}\left(\sin\theta\cos\phi\,\frac{\partial u}{\partial x}\right)+\cos\theta\,\frac{\partial}{\partial\theta}\left(\frac{\cos\phi}{r}\frac{\partial u}{\partial x}\right)-\sin\phi\,\frac{\partial}{\partial\phi}\left(\frac{1}{r\sin\theta}\frac{\partial u}{\partial x}\right)$$

$$(7.7.11)$$

Do the same for $\partial^2 u/\partial y^2$ and $\partial^2 u/\partial z^2$, add the resulting three equations, then use (7.7.10). The work is assigned as Exercise 17. You will get the *Laplacian differential expression in spherical coordinates*

$$-\delta(u) = \frac{1}{r^2}\left[\frac{\partial}{\partial r}\left(r^2\frac{\partial u}{\partial r}\right) + \frac{1}{\sin\theta}\left(\frac{\partial}{\partial\theta}\left(\sin\theta\frac{\partial u}{\partial\theta}\right) + \frac{1}{\sin\theta}\frac{\partial^2 u}{\partial\phi^2}\right)\right] \qquad (7.7.12)$$

That is the Laplacian in three dimensions. On the surface of the sphere $r = a = $ constant. Thus the *Laplacian differential expression on the sphere of radius a*:

$$\delta(u) = \frac{-1}{a^2\sin\theta}\left(\frac{\partial}{\partial\theta}\left(\sin\theta\frac{\partial u}{\partial\theta}\right) + \frac{1}{\sin\theta}\frac{\partial^2 u}{\partial\phi^2}\right) \qquad (7.7.13)$$

We come now to the proof that the Laplacian of Definition 7.16 is a positive semidefinite Hermitian operator on $L^2(S)$. This is part of the statement of Theorem 7.17 for the particular case of the sphere. What we have to prove is this: $\langle\Delta(f), g\rangle = \langle f, \Delta(g)\rangle$ for all $f, g$ in the domain of $\Delta$, where $\langle\cdot,\cdot\rangle$ is the inner product defined by (7.7.6). Just as in the planar case, this proof relies on integration by parts in two variables (equation (7.2.5) of Section 7.2), which is a consequence of the general Stokes theorem (equation (7.2.2) of Section 7.2). As the sphere $S$ has no boundary, $\partial S = \varnothing$, the integral over the boundary vanishes, and Stokes' theorem takes the form $\iint_S d\omega = 0$, so that integration by parts becomes

$$\iint_S f\,d\omega = -\iint_S (df)\,\omega \qquad (7.7.14)$$

You will use (7.7.14) to prove that the Laplacian on the sphere is Hermitian positive semidefinite (Exercise 18).

## Exercises

In Exercises 1 through 3, $M$ stands for a general closed surface like the sphere, surface of an ellipsoid, or surface of a torus.

1. In both cases, explain why $\|f\|^2 = \iint_M |f|^2\,dA = 0$:
    (a) $f = 0$ except at one point of $M$.
    (b) $f = 0$ except on a curve in $M$.
2. Show that, if $|f| \leqslant B$ at all points of $M$, then $\|f\|^2 \leqslant \text{area}(M)B^2$. Conclude that any bounded function on $M$ belongs to $L^2(M)$.
3. Prove the inequalities
    (a) $(\iint_M |f||g|\,dA)^2 \leqslant (\iint_M |f|^2\,dA)(\iint_M |g|^2\,dA)$ for any $f, g$ in $L^2(M)$.
    (b) $(\iint_M |f|\,dA)^2 \leqslant \text{area}(M)\iint_M |f|^2\,dA$ for any $f$ in $L^2(M)$.

In all the following exercises, $S$ stands for the sphere of radius $a$.

4. Of the following six functions, three belong to $L^2(S)$ and three do not. Determine which is which (prove your answer) and, for the three functions that belong to $L^2(S)$, calculate $\|f\|^2 = \langle f, f \rangle$.
    (a) $f(\theta, \phi) = \cos \phi / \sqrt{\sin \theta}$
    (b) $f(\theta, \phi) = \cos \phi / \sin \theta$
    (c) $f(\theta, \phi) = 1 / \sqrt{|\cos \phi|}$
    (d) $f(\theta, \phi) = 1$
    (e) $f(\theta, \phi) = \sqrt{|\cos \theta|} \ln(\sin \theta) / \sqrt{\sin \theta}$
    (f) $f(\theta, \phi) = 1/(\theta - \phi) \sqrt{\sin \theta}$
5. Show that $f(\theta, \phi) = \sin \theta \cos \theta \cos \phi$ and $g(\theta, \phi) = \sin \theta \cos \phi$ are orthogonal in $L^2(S)$.
6. If $f_1(\theta, \phi) = g_1(\theta)h_1(\phi)$ and $f_2(\theta, \phi) = g_2(\theta)h_2(\phi)$ both belong to $L^2(S)$, show that $\langle f_1, f_2 \rangle = a^2 \langle g_1, g_2 \rangle_p \langle h_1, h_2 \rangle_c$ where

$$\langle g_1, g_2 \rangle_p = \int_{-1}^{+1} g_1 \overline{g_2} \, dz, \qquad z = \cos \theta$$

is the inner product in $L^2(-1, 1)$, and $\langle h_1, h_2 \rangle_c$ is the inner product in $L^2(C)$, $C$ the unit circle.
7. Let $S'$ be the sphere with its north and south poles removed. Show that the equation $f(\theta, \phi) = \cos \phi$, $0 < \theta < \pi$, defines a continuous function on $S'$. What values, if any, can you assign to $f$ at the north and south poles so that $f$ becomes a continuous function on $S$?
8. Let $f(\theta, \phi) = (\pi^2 - \phi^2) \sin \theta$. Show that $f$ is continuous on $S$ but $\partial f / \partial \phi$ is not. Does $f$ belong to the domain of the Laplacian operator $\Delta$? Give reasons for your answer.
9. Let $f(\theta, \phi) = \sin \theta \sin \phi$.
    (a) Show that $f$ is continuous on $S$.
    (b) Show that $f$ has continuous first partial derivatives at each point of $S$. How do you reconcile this with the fact that $\partial f / \partial \theta = \cos \theta \sin \phi$ is a discontinuous function on $S$ at the north and south poles?
    (c) Using the conclusion of Exercise 2, show that $f$ and $\delta(f)$ belong to $L^2(S)$. Thus show that $f$ belongs to the domain of $\Delta$ and, in fact, is an eigenfunction of the Laplacian on $S$. To what eigenvalue does $f$ belong?
10. Prove: If $f(\theta, \phi) = g(\theta)h(\phi)$ lies in the domain of the Laplacian on the sphere $S$, then
    (a) $g(0) = g(\pi) = 0$ unless $h(\phi)$ is constant.
    (b) If $h(\phi)$ is a nonzero constant, then both $\lim_{\theta \to 0+} g(\theta)$ and $\lim_{\theta \to \pi-} g(\theta)$ exist and are finite.
    (c) $h(\pi) = h(-\pi)$ and $h'(\pi) = h'(-\pi)$ unless $g(\theta)$ is identically zero. (Assume that $g(\theta)$ is nicely behaved in $0 < \theta < \pi$, and same for $h(\phi)$, $-\pi < \phi \langle \pi$.)

Exercises 11 through 16 should be done in sequence.

11. (a) Using the first of equations (7.7.3), show that

$$dx = \sin\theta \cos\phi \, dr + r \cos\theta \cos\phi \, d\theta - r \sin\theta \sin\phi \, d\phi$$

Using the other two equations from (7.7.3), derive similar expressions for $dy$ and $dz$.

   (b) Using the expressions derived in part (a), show that

$$dy \, dz = -r \sin\phi \, dr \, d\theta - r \sin\theta \cos\theta \cos\phi \, dr \, d\phi$$
$$+ r^2 \sin^2\theta \cos\phi \, d\theta \, d\phi$$

   (exterior product).

   (c) Exterior multiply the expression for $dx$ from part (a) by that for $dy \, dz$ from part (b) to derive the formula for the *volume element in spherical coordinates*:

$$dV = dx \, dy \, dz = r^2 \sin\theta \, dr \, d\theta \, d\phi$$

12. (a) On the sphere $r = a$, $dr = 0$. By specializing your results from Exercise 11a, derive the following expressions for $dx$, $dy$, and $dz$ on the sphere:

$$dx = a \cos\theta \cos\phi \, d\theta - a \sin\theta \sin\phi \, d\phi$$
$$dy = a \cos\theta \sin\phi \, d\theta + a \sin\theta \cos\phi \, d\phi$$
$$dz = -a \sin\theta \, d\theta$$

   (b) Using ordinary algebra (not exterior algebra) compute $(dx)^2$, $(dy)^2$, and $(dz)^2$ from part (a); thus derive the *differential element of arc length on the sphere of radius a*

$$(ds)^2 = (dx)^2 + (dy)^2 + (dz)^2 = a^2((d\theta)^2 + \sin^2\theta(d\phi)^2)$$

   (This is also called the *first fundamental form* or the *Riemannian metric* for the sphere.)

   (c) Rederive the formula $(ds)^2 = (a \, d\theta)^2 + (a \sin\theta \, d\phi)^2$ by applying the Pythagorean theorem to an infinitesimal right triangle on the sphere $S$.

13. Using the formula $ds = a\sqrt{(d\theta)^2 + \sin^2\theta(d\phi)^2}$ (Exercise 12b and c) and performing the integration $\int ds$, compute the length of the curve cut from the sphere $S$ by a plane parallel to the $xy$ plane at height $b$ above it, $0 \leqslant b \leqslant a$. Check your result by elementary geometry.

14. In Exercise 12c, make the infinitesimal triangle into an infinitesimal rectangle to suggest that $dA = a^2 \sin\theta \, d\theta \, d\phi$ is the differential element of area on the sphere $S$.

15. (a) Compute the exterior products $dy \, dz$, $dz \, dx$, and $dx \, dy$ on the sphere $S$ of radius $a$.

(b) Using ordinary algebra (not exterior algebra) compute $(dy\,dz)^2$, $(dz\,dx)^2$ and $(dx\,dy)^2$ from part (a); thus derive the *differential element of area on the sphere of radius a*.

$$(dA)^2 = (dy\,dz)^2 + (dz\,dx)^2 + (dx\,dy)^2 = (a^2 \sin\theta\,d\theta\,d\phi)^2$$

16. Let a surface $M$ be parametrized by variables $u$, $v$, and suppose the Riemannian metric for $M$ (see Exercise 12b) is given by $(ds)^2 = g_{11}(du)^2 + g_{22}(dv)^2$. Then it is a fact that the element of surface area on $M$ is $dA = \sqrt{g}\,du\,dv$, and the Laplacian differential expression for $M$ is

$$\delta(f) = \frac{-1}{\sqrt{g}}\left[\frac{\partial}{\partial u}\left(\frac{\sqrt{g}}{g_{11}}\frac{\partial f}{\partial u}\right) + \frac{\partial}{\partial v}\left(\frac{\sqrt{g}}{g_{22}}\frac{\partial f}{\partial v}\right)\right]$$

where $g = g_{11}g_{22}$. Compute both for the sphere $S$ in spherical coordinates $u = \theta$, $v = \phi$.

17. In the following steps, derive equations (7.7.12) and (7.7.13).
   (a) Compute equations for $\partial^2 u/\partial y^2$ and $\partial^2 u/\partial z^2$ corresponding to equation (7.7.11) of the text.
   (b) Adding equation (7.7.11) to the two equations from (a), show that

$$-\delta(u) = \frac{\partial}{\partial r}\left(\sin\theta\cos\phi\,\frac{\partial u}{\partial x} + \sin\theta\sin\phi\,\frac{\partial u}{\partial y} + \cos\theta\,\frac{\partial u}{\partial z}\right)$$

$$+ \cos\theta\,\frac{\partial}{\partial\theta}\left(\frac{\cos\phi}{r}\frac{\partial u}{\partial x} + \frac{\sin\phi}{r}\frac{\partial u}{\partial y}\right) - \sin\theta\,\frac{\partial}{\partial\theta}\left(\frac{1}{r}\frac{\partial u}{\partial z}\right)$$

$$- \sin\phi\,\frac{\partial}{\partial\phi}\left(\frac{1}{r\sin\theta}\frac{\partial u}{\partial x}\right) + \cos\phi\,\frac{\partial}{\partial\phi}\left(\frac{1}{r\sin\theta}\frac{\partial u}{\partial y}\right)$$

   (c) Use equations (7.7.10) to finish up.
18. (a) Define a 1-form $\omega$ by

$$\omega = \frac{1}{\sin\theta}\frac{\partial f}{\partial\phi}\,d\theta - \sin\theta\frac{\partial f}{\partial\theta}\,d\phi$$

   Show that $d\omega = \delta(f)\,dA$ where $dA = a^2 \sin\theta\,d\theta\,d\phi$.
   (b) Using (a) and equation (7.7.14), show that

$$\langle \Delta f, g\rangle = \frac{1}{a^2}\left\langle \frac{\partial f}{\partial\theta}, \frac{\partial g}{\partial\theta}\right\rangle + \frac{1}{a^2}\left\langle \frac{1}{\sin\theta}\frac{\partial f}{\partial\phi}, \frac{1}{\sin\theta}\frac{\partial g}{\partial\phi}\right\rangle$$

   valid for any $f$, $g$ in the domain of $\Delta$.
   (c) Using (b), prove that the Laplacian on the sphere is Hermitian positive semidefinite.

## 7.8 EIGENVALUES OF THE SPHERE; SPHERICAL HARMONICS

In this section we complete our task of finding the eigenvalues of the sphere $S$. We seek all numbers $\lambda$ such that the equation

$$\Delta(f) = \lambda f \tag{7.8.1}$$

has a nonzero solution $f$ that lies in the domain of the spherical Laplacian operator $\Delta$.

We have advance information about these eigenvalues and their corresponding eigenfunctions. As the Laplacian on the sphere is Hermitian positive semidefinite (Exercise 18 of Section 7.7), all its eigenvalues are nonnegative. Further, Theorem 7.17 tells us that the eigenvalues form a sequence tending to $+\infty$ and that each eigenvalue has finite multiplicity. Moreover, the complete set of eigenfunctions of $\Delta$ form an orthogonal basis of the Hilbert space $L^2(S)$ (Theorem 7.18).

Our job now is to find the eigenvalues precisely, to compute their multiplicities, and to ascertain their eigenfunctions.

Every solution of the operator equation $\Delta(f) = \lambda f$ is a solution of the partial differential equation

$$\frac{-1}{a^2 \sin\theta}\left(\frac{\partial}{\partial\theta}\left(\sin\theta\frac{\partial f}{\partial\theta}\right) + \frac{1}{\sin\theta}\frac{\partial^2 f}{\partial\phi^2}\right) = \lambda f \tag{7.8.2}$$

It will be enough to solve (7.8.2) with $a = 1$, because if $\lambda$ is an eigenvalue of the unit sphere ($a = 1$), then $\lambda/a^2$ is an eigenvalue of the sphere of radius $a$, as we see easily from (7.8.2). So we shall assume $a = 1$.

Separate variables, $f(\theta, \phi) = g(\theta)h(\phi)$, and perform a few manipulations to obtain

$$\sin^2\theta\frac{g''}{g} + \sin\theta\cos\theta\frac{g'}{g} + \lambda\sin^2\theta = -\frac{h''}{h} \tag{7.8.3}$$

The right side of (7.8.3) depends only on $\phi$, the left side only on $\theta$; they are therefore equal to a common constant, say $\gamma$. We get two equations involving this constant, one of which is

$$-h'' = \gamma h \tag{7.8.4}$$

As $f(\theta, \phi) = g(\theta)h(\phi)$ belongs to the domain of the Laplacian on the sphere $S$, $\partial f/\partial\phi = g(\theta)h'(\phi)$ and $\partial^2 f/\partial\phi^2 = g(\theta)h''(\phi)$ are both continuous on $S$. Unless $g(\theta)$ is identically zero, this implies that $h(\phi)$ and $h'(\phi)$ are continuous on the unit circle $C$ that we get by bending the interval $\{\phi: -\pi \leqslant \phi \leqslant \pi\}$ into a circle and identifying the point $\phi = -\pi$ with $\phi = \pi$. In particular, $h(\pi) = h(-\pi)$ and

$h'(\pi) = h'(-\pi)$ (see Exercise 10 of Section 7.7). Furthermore, $h(\phi)$ belongs to $L^2(C)$ (Exercise 1). Equation (7.8.4) then implies that $h$ is compatible with the one-dimensional Laplacian on the unit circle $C$. The upshot: *equation (7.8.4) is equivalent to the eigenvalue problem for the one-dimensional Laplacian $\Delta_c$ on the unit circle $C$* (see Definition 4.13):

$$\Delta_C(h) = \gamma h \tag{7.8.5}$$

We know very well that these eigenvalues are $\gamma = m^2$, $m = 0, 1, 2, \ldots$, and that the corresponding eigenfunctions comprise the classical Fourier series. The eigenvalue $m = 0$ is simple, and we usually take $h = \frac{1}{2}$ as a convenient eigenfunction. The eigenvalues $m^2$, $m = 1, 2, \ldots$, are each double, and the pair of eigenfunctions $\cos m\phi$, $\sin m\phi$ span the eigenspace for $\gamma = m^2$.

The second equation that we get from (7.8.3) is

$$\sin^2 \theta \frac{g''}{g} + \sin \theta \cos \theta \frac{g'}{g} + \lambda \sin^2 \theta = m^2 \tag{7.8.6}$$

Make the substitution $z = \cos \theta$ and do a few manipulations to get

$$-\frac{d}{dz}\left( (1 - z^2) \frac{dg}{dz} \right) + \frac{m^2}{1 - z^2} g = \lambda g \qquad \text{(Exercise 2)} \tag{7.8.7}$$

The variable $z = \cos \theta$ has the range $-1 \leqslant z \leqslant 1$ because $0 \leqslant \theta \leqslant \pi$. The differential expression on the left of (7.8.7) is already in Sturmian form and shows a weight function $r(z) = 1$. Thus we are dealing with the Hilbert space $L^2(-1, 1)$ and a differential operator on $L^2(-1, 1)$ which has the differential expression $l_m(g) \equiv -((1 - z^2)g')' + (1 - z^2)^{-1}m^2 g$ with an appropriately defined domain. Because $f(\theta, \phi) = g(\theta)h(\phi)$ belongs to $L^2(S)$, $g(z)$ ($z = \cos \theta$) belongs to $L^2(-1, 1)$ (Exercise 1). Moreover, because $f(\theta, \phi)$ belongs to the domain of the Laplacian, $g(\theta)$ is finite at $\theta = 0$ and $\theta = \pi$, thus at $z = +1$, $z = -1$ for $z = \cos \theta$. Indeed, if $h(\phi)$ is not constant, $g = 0$ at $z = \pm 1$ (see Exercise 10 of Section 7.7).

These remarks lead to this conclusion: *equation (7.8.7) is equivalent to the eigenvalue problem*

$$L_m(g) = \lambda g \tag{7.8.8}$$

*for a differential operator $L_m$ acting in the Hilbert space $L^2(-1, 1)$. The operator $L_m$ has the differential expression $l_m(g) \equiv -((1 - z^2)g')' + (1 - z^2)^{-1}m^2 g$, where $m$ is a fixed integer, and has domain consisting of the functions $g$ in $L^2(-1, 1)$ that are compatible with $l_m$ and are such that both values $g(\pm 1)$ are finite.*

When $m = 0$ (which corresponds to $h = \frac{1}{2}$), the operator $L_0$ is just Legendre's operator that we studied very thoroughly in Sections 4.6, 4.7, and 4.8. Its eigenvalues are $n(n + 1)$, $n = 0, 1, 2, \ldots$, and each eigenvalue is simple.

*Thus, the numbers $n(n+1)$, $n=0,1,2,\ldots$, are eigenvalues of the unit sphere.* Do we pick up other eigenvalues when $m \neq 0$? The answer is "no," contrary to what our experience with Bessel's operator $B_p$ might lead us to expect, because the operator $B_p$, based on the differential expression $b_p(f) \equiv -r^{-1}(rf')' + r^{-2}p^2 f$, has different eigenvalues for different values of the integral parameter $p$ (Sections 6.7 through 6.9).

**Definition 7.21.** Fix a nonnegative integer $m$. *Legendre's operator $L_m$* has the differential expression $l_m(g) \equiv -((1-z^2)g')' + (1-z^2)^{-1}m^2 g$ and domain consisting of functions $g(z)$ in $L^2(-1,1)$ that are compatible with $l_m$ and such that $g(-1)$ and $g(1)$ are both finite.

**Theorem 7.22.** *Fix a nonnegative integer $m$. Legendre's operator $L_m$ is Hermitian positive semidefinite. It has eigenvalues $\lambda = n(n+1)$, $n = m$, $m+1$, $m+2,\ldots$. Each eigenvalue is simple and has the corresponding eigenfunction*

$$P_n^m(z) = (1 - z^2)^{m/2} \frac{d^m}{dz^m} P_n(z)$$

*where $P_n(z)$ is the nth Legendre polynomial. For each fixed nonnegative integer $m$, the "associated Legendre functions" $P_n^m(z)$, $n = m, m+1, m+2, \ldots$, constitute an orthogonal basis of $L^2(-1,1)$.*

The proof of Theorem 7.22, which will rely mainly on the exercises and on Appendix 7.A, shall be taken up shortly. But first we should get to the heart of the matter.

**Theorem 7.23.** *The eigenvalues of the sphere $S$ of radius $a$ are $\lambda_n = 4\pi n(n+1)/A$, $n = 0, 1, 2, \ldots$, where $A = 4\pi a^2$ is the area of the sphere. The nth eigenvalue, $\lambda_n = 4\pi n(n+1)/A$, has multiplicity $2n+1$; its eigenspace is spanned by the $2n+1$ orthogonal functions*

$$\cos m\phi \; P_n^m(\cos \theta), \qquad m = 0, 1, 2, \ldots, n$$
$$\sin m\phi \; P_n^m(\cos \theta), \qquad m = 1, 2, \ldots, n$$

*The totality of these functions constitute an orthogonal basis of $L^2(S)$.*

    *Proof.* We shall use the handy notation

$$Y_n^m(\theta, \phi) = \begin{Bmatrix} \cos m\phi \\ \sin m\phi \end{Bmatrix} P_n^m(\cos \theta)$$

Let's begin by checking that all the $Y_n^m$ satisfy the necessary conditions for compatibility with the Laplacian differential expression, namely that they, together with their first partial derivatives, are continuous on the sphere $S$.

The functions $Y_n^m(\theta, \phi)$ are quite well behaved in the interior of the rectangle $R = \{(\theta, \phi): 0 < \theta < \pi, \ -\pi < \phi < \pi\}$, so it is just a matter of checking the behavior at the seam $(\theta, -\pi) \equiv (\theta, \pi)$, $0 < \theta < \pi$, and at the north and south poles $\theta = 0$ and $\theta = \pi$. (For all this, refer to the previous section.) Clearly $Y_n^m(\theta, -\pi) = Y_n^m(\theta, \pi)$, and the same for $\partial Y_n^m / \partial \theta$, $\partial Y_n^m / \partial \phi$, owing to the fact that $\cos(-m\pi) = \cos m\pi$ and $\sin(-m\pi) = \sin m\pi$. It is also clear that $Y_n^m(\theta, -\pi + \varepsilon) \approx Y_n^m(\theta, \pi - \varepsilon)$ for small positive $\varepsilon$, and same for $\partial Y_n^m / \partial \theta$, $\partial Y_n^m / \partial \phi$, because of the continuity of $\cos m\phi$ and $\sin m\phi$. So the $Y_n^m(\theta, \phi)$, together with their first partial derivatives, are all continuous along the half great circle $\{(\theta, -\pi) \equiv (\theta, \pi): 0 < \theta < \pi\}$.

At the north and south poles, $\theta = 0$ and $\theta = \pi$, we have $z = \cos \theta = \pm 1$. If $m > 0$, $P_n^m(z) = (1 - z^2)^{m/2}(P_n(z))^{(m)}$ is zero at $\pm 1$; and, if $m = 0$, then $Y_n^0(z) = P_n^0(z) = P_n(z)$ is simply a polynomial, thus continuous at $z = \pm 1$. So in all cases $Y_n^m(\theta, \phi)$ is continuous at $\theta = 0, \pi$. To check the continuity of the partial derivatives at $\theta = 0$ and $\theta = \pi$ we need to use $\partial Y_n^m / \partial x$ and $\partial Y_n^m / \partial y$ owing to the fact that the points $\theta = 0, \pi$ are singular for spherical coordinates. I'll do $\partial Y_n^m / \partial x$; the calculations for $Y_n^m / \partial y$ are virtually the same. Use the chain rule: $\partial f / \partial x = (\partial f / \partial \theta)(\partial \theta / \partial x) + (\partial f / \partial \phi)(\partial \phi / \partial x)$. We have $\partial \theta / \partial x = \cos \phi / \cos \theta$ and $\partial \phi / \partial x = -\sin \phi / \sin \theta$. Consider the one case $Y_n^m(\theta, \phi) = \cos m\phi \ P_n^m(z)$, $z = \cos \theta$; the case involving $\sin m\phi$ is handled exactly the same way. We have $\partial Y_n^m / \partial \phi = -m \sin m\phi \ P_n^m(z)$. And $\partial Y_n^m / \partial \theta = \cos m\phi \ dP_n^m / d\theta$. The variable $z = \cos \theta$ is more convenient: $dP_n^m / d\theta = (dP_n^m / dz)(dz / d\theta) = -\sin \theta (dP_n^m / dz) = -\sqrt{1 - z^2} \ dP_n^m / dz$. A routine calculation shows that

$$\frac{dP_n^m}{dz} = (1 - z^2)^{(m/2) - 1} \left( -mz \frac{d^m P_n(z)}{dz^m} + (1 - z^2) \frac{d^{m+1} P_n(z)}{dz^{m+1}} \right)$$

Assemble all the pieces, do a few patient manipulations, and you will get

$$\frac{\partial Y_n^m}{\partial x} = m(1 - z^2)^{(m-1)/2} \cos(m - 1)\phi \frac{d^m P_n(z)}{dz^m}$$

$$- \cos m\phi (1 - z^2)^{(m+1)/2} \frac{d^{m+1} P_n(z)}{dz^{m+1}}$$

The north pole corresponds to $z = 1$, the south pole to $z = -1$. When $m = 0$, $\partial Y_n^m / \partial x = -\sqrt{1 - z^2} \ P_n'(z)$, which is independent of $\phi$ and has the value 0 at $z = \pm 1$. When $m = 1$, $\partial Y_n^m / \partial x = P_n'(z) - \cos \phi(1 - z^2)P_n''(z)$, which equals $P_n'(\pm 1)$ at $z = \pm 1$ independent of $\phi$. When $m > 1$, $\partial Y_n^m / \partial x = 0$ at $z = \pm 1$ independent of $\phi$. Thus $\partial Y_n^m / \partial x$ is continuous at the north and south poles of the sphere $S$. And similarly for $\partial Y_n^m / \partial y$. Thus all the $Y_n^m$, together with their first partial derivatives, are continuous at *every* point of the sphere $S$.

The $Y_n^m(\theta, \phi)$ are bounded, thus all belong to $L^2(S)$ (Exercise 2 of Section

7.7). We also need to show that the $Y_n^m$ are compatible with the Laplacian $\Delta$, i.e., that $\Delta(Y_n^m) \in L^2(S)$. But this will follow if we show that $\Delta(Y_n^m) = \lambda Y_n^m$. And we need to establish this equation anyway to prove that the $Y_n^m$ are eigenfunctions of the Laplacian. As the expression says, we shall kill two birds with one stone:

$$\Delta(Y_n^m) = \frac{-1}{a^2 \sin \theta} \left( \frac{\partial}{\partial \theta} \left( \sin \theta \frac{\partial Y_n^m}{\partial \theta} \right) + \frac{1}{\sin \theta} \frac{\partial^2 Y_n^m}{\partial \phi^2} \right)$$

In terms of the variable $z = \cos \theta$, $\partial/\partial \theta = -\sin \theta \, \partial/\partial z$, so

$$\Delta(Y_n^m) = \frac{-1}{a^2 \sin \theta} \left( \sin \theta \frac{\partial}{\partial z} \left( \sin^2 \theta \frac{\partial Y_n^m}{\partial z} \right) + \frac{1}{\sin \theta} \frac{\partial^2 Y_n^m}{\partial \phi^2} \right)$$

$$= \frac{-1}{a^2} \left( \frac{\partial}{\partial z} \left( (1 - z^2) \frac{\partial Y_n^m}{\partial z} \right) + \frac{1}{1 - z^2} \frac{\partial^2 Y_n^m}{\partial \phi^2} \right)$$

It will be sufficient to do the case $Y_n^m = \cos m\phi \, P_n^m(z)$ because the calculation for $Y_n^m = \sin m\phi \, P_n^m(z)$ is virtually the same.

$$\Delta(Y_n^m) = \frac{\cos m\phi}{a^2} \left( -\frac{d}{dz} \left( (1 - z^2) \frac{dP_n^m(z)}{dz} \right) + \frac{m^2}{1 - z^2} P_n^m(z) \right)$$

$$= \frac{\cos m\phi}{a^2} L_m(P_n^m(z)) = \frac{\cos m\phi}{a^2} n(n + 1) P_n^m(z)$$

$$= \frac{n(n + 1)}{a^2} Y_n^m = \frac{4\pi n(n + 1)}{A} Y_n^m$$

In the second to the last step, we have used the fact that the $P_n^m$ are eigenfunctions of Legendre's operator $L_m$ belonging to the eigenvalue $n(n + 1)$, $n \geqslant m$ (Theorem 7.22). We have solved equation (7.8.1):

$$\Delta(Y_n^m) = \frac{4\pi n(n + 1)}{A} Y_n^m, \qquad 0 \leqslant m \leqslant n, \; n = 0, 1, 2, \ldots \qquad (7.8.9)$$

where $A = 4\pi a^2$ is the area of the sphere $S$. For each fixed $n$, there are $2n + 1$ functions $Y_n^m$ that satisfy (7.8.9), namely $\cos m\phi \, P_n^m(\cos \theta)$, $0 \leqslant m \leqslant n$, and $\sin m\phi \, P_n^m(\cos \theta)$, $1 \leqslant m \leqslant n$. These $2n + 1$ functions are orthogonal (Exercise 11), so they are certainly linearly independent. Thus each $\lambda_n = 4\pi n(n + 1)/A$ is an eigenvalue of $\Delta$ on $S$ of multiplicity at least $2n + 1$.

Next we come to the assertion that the totality of functions $\cos m\phi \, P_n^m(\cos \theta)$, $0 \leqslant m \leqslant n$; $\sin m\phi \, P_n^m(\cos \theta)$, $1 \leqslant m \leqslant n$; $n = 1, 2, \ldots$, constitute an orthogonal basis of $L^2(S)$. At this point we shall again rely on J. R. Higgins's book, *Completeness and Basis Properties of Sets of Special Functions*, that we have referred to many times before. See Section 2.2 (including Problem 2.10) in Higgins' book for the proof that these functions constitute an orthogonal basis of $L^2(S)$.

Knowing that the $Y_n^m$ constitute an orthogonal basis of $L^2(S)$, we can then infer that the $\lambda_n = 4\pi n(n+1)/A$ are *all* the eigenvalues of the sphere and that, for each fixed $n$, $\lambda_n$ has multiplicity *exactly* $2n+1$ (Exercise 12). Thus, if $f(\theta, \phi)$ is a function on $S$ that satisfies $\Delta(f) = \lambda f$, then $\lambda = 4\pi n(n+1)/A$ for some $n = 0, 1, 2, \ldots$, and $f$ must be a linear combination of the $2n+1$ functions $\cos m\phi \, P_n^m(\cos \theta)$, $0 \leqslant m \leqslant n$, and $\sin m\phi \, P_n^m(\cos \theta)$, $1 \leqslant m \leqslant n$.

Except for the pending proof of Theorem 7.22, that completes the proof of Theorem 7.23, thus completes our project of this section and the last section: to determine the eigenvalues of the sphere.

For the same reasons of convenience of normalization that applied to Fourier series, we shall insert a factor $\frac{1}{2}$ in the $m = 0$ harmonics and take as orthogonal basis of $L^2(S)$ the *spherical harmonics*

$$\frac{1}{2} P_n(\cos \theta) \qquad\qquad n = 0, 1, 2, \ldots$$

$$\cos m\phi \, P_n^m(\cos \theta), \qquad 1 \leqslant m \leqslant n, \, n = 1, 2, \ldots \qquad (7.8.10)$$

$$\sin m\phi \, P_n^m(\cos \theta), \qquad 1 \leqslant m \leqslant n, \, n = 1, 2, \ldots$$

Thus *every* $f(\theta, \phi)$ in $L^2(S)$ has an expansion

$$f(\theta, \phi) = \frac{1}{2} a_{00} + \sum_{n=1}^{\infty} \left( \frac{1}{2} a_{0n} P_n(\cos \theta) + \sum_{m=1}^{n} (a_{mn} \cos m\phi + b_{mn} \sin m\phi) P_n^m(\cos \theta) \right)$$

$$(7.8.11)$$

The expansion coefficients are given by the usual formulas:

$$\frac{1}{2} a_{0,n} = \frac{\langle f, P_n(\cos \theta) \rangle}{\langle P_n(\cos \theta), P_n(\cos \theta) \rangle}, \qquad n = 0, 1, 2, \ldots \qquad (7.8.12)$$

and

$$a_{mn} = \frac{\langle f, \cos m\phi \, P_n^m(\cos \theta) \rangle}{\langle \cos m\phi \, P_n^m(\cos \theta), \cos m\phi \, P_n^m(\cos \theta) \rangle}, \qquad 1 \leqslant m \leqslant n, \, n = 1, 2, \ldots \quad (7.8.13)$$

$$b_{mn} = \frac{\langle f, \sin m\phi \, P_n^m(\cos \theta) \rangle}{\langle \sin m\phi \, P_n^m(\cos \theta), \sin m\phi \, P_n^m(\cos \theta) \rangle}, \qquad 1 \leqslant m \leqslant n, \, n = 1, 2, \ldots \quad (7.8.14)$$

The inner product is that of $L^2(S)$; refer to equation (7.7.6) of Section 7.7. To put this expansion to practical use, we must evaluate the norms of the spherical harmonics. These norms (squared) are the denominators in equations (7.8.12), (7.8.13), and (7.8.14).

$$\langle P_n(\cos \theta), P_n(\cos \theta) \rangle = \int_{-\pi}^{\pi} d\phi \int_{0}^{\pi} d\theta \sin \theta \, P_n(\cos \theta)^2 = 2\pi \int_{-1}^{1} P_n(z)^2 \, dz$$

$$= \frac{4\pi}{2n+1}, \qquad n = 0, 1, 2, \ldots$$

So

$$a_{0n} = \frac{2n+1}{2\pi} \int_{-\pi}^{\pi} d\phi \int_{0}^{\pi} d\theta \sin \theta f(\theta, \phi) P_n(\cos \theta), \qquad n = 0, 1, 2, \ldots$$

Also, for $m \geqslant 1$

$$\langle {\textstyle{\cos \atop \sin}} m\phi \, P_n^m(\cos \theta), {\textstyle{\cos \atop \sin}} m\phi \, P_n^m(\cos \theta) \rangle = \int_{-\pi}^{\pi} ({\textstyle{\cos \atop \sin}} m\phi)^2 \, d\phi \int_{-1}^{1} (P_n^m(z))^2 \, dz$$

$$= \pi \int_{-1}^{1} (P_n^m(z))^2 \, dz = \frac{2\pi}{2n+1} \frac{(n+m)!}{(n-m)!}$$

(We shall come to the derivation of the very last formula in a minute.) Hence

$$\left. \begin{matrix} a_{mn} \\ b_{mn} \end{matrix} \right\} = \frac{2n+1}{2\pi} \frac{(n-m)!}{(n+m)!} \int_{-\pi}^{\pi} d\phi \int_{0}^{\pi} d\theta \sin \theta f(\theta, \phi) {\textstyle{\cos \atop \sin}} m\phi \, P_n^m(\cos \theta) \qquad (7.8.15)$$

Formula (7.8.15) is valid for *all $m, n$, $0 \leqslant m \leqslant n$, $n = 0, 1, 2, \ldots$* (that was the point of including the factor $\frac{1}{2}$).

Let's review some of the facts about the expansion (7.8.11) in spherical harmonics.

First, the coefficients given by equation (7.8.15) make the mean-square error

$$\left\| f - \left( \frac{a_{00}}{2} + \sum_{n=1}^{N} \left( \frac{a_{0n}}{2} P_n(\cos \theta) + \sum_{m=1}^{n} (a_{mn} \cos m\phi + b_{mn} \sin m\phi) P_n^m(\cos \theta) \right) \right) \right\|$$

$$(7.8.16)$$

an absolute minimum; any alteration of these coefficients will increase this error. Of course, the norm is that of $L^2(S)$; see equation (7.7.5) of Section 7.7.

Second, with the coefficients given by (7.8.15), the expression (7.8.16), which is the norm of the difference between $f(\theta, \phi)$ and the $N$th partial sum of its expansion in spherical harmonics, tends to zero as $N \to \infty$. This is true for any $f(\theta, \phi) \in L^2(S)$. It is in this sense that the series (7.8.11) is valid. If you alter the function $f(\theta, \phi)$ at a few points on the sphere, or even on a curve, you won't affect the coefficients $a_{mn}, b_{mn}$ because changing $f$ on a set of zero area won't affect the integrals in (7.8.15). Therefore, the series in (7.8.11) may converge at a few points to values different from the values of $f(\theta, \phi)$. Remember the meaning of the equal sign in (7.8.11). It means exactly that the norm of the difference as given in equation (7.8.16) goes to zero as $N \to \infty$. This "$L^2$ convergence" has the substantial advantage that it holds for any square-integrable $f(\theta, \phi)$ without the necessity of assuming differentiability, continuity, etc. To say that the expansion (7.8.11) is valid for every $f$ in $L^2(S)$

means the same thing as saying that the spherical harmonics form an orthogonal basis of $L^2(S)$.

To complete this story we must prove Theorem 7.22, which involves Legendre's operator $L_m$. This operator has the differential expression $l_m(g) = -((1-z^2)g')' + m^2(1-z^2)^{-1}g$ and has domain consisting of the functions in $L^2(-1, 1)$ that are compatible with $l_m$ and are finite at $\pm 1$. Exercise 3, which uses material developed in Appendix 7.A, provides the proof that $L_m$ is Hermitian positive semidefinite (positive definite if $m \geq 1$). The next step is to determine the eigenvalues of $L_m$, i.e., the nonnegative numbers $\lambda$ such that the differential equation $-((1-z^2)g')' + m^2(1-z^2)^{-1}g = \lambda g$ has a nonzero solution $g$ in the domain of $L_m$.

**Theorem 7.24.**   *If $0 \leq m \leq n$, the function*

$$g(z) = (1 - z^2)^{m/2} \frac{d^m P_n(z)}{dz^m}$$

*is a nonzero solution of the equation*

$$-\frac{d}{dz}\left((1 - z^2)\frac{dg}{dz}\right) + \frac{m^2}{1 - z^2} g = n(n + 1)g$$

The proof of Theorem 7.24 is assigned as Exercises 4 and 5. The *associated Legendre functions*

$$P_n^m(z) = (1 - z^2)^{m/2}\frac{d^m P_n(z)}{dz^m}, \qquad n = m, m + 1, m + 2, \ldots \qquad (7.8.17)$$

lie in the domain of Legendre's operator $L_m$; thus the operator $L_m$ has eigenvalues $m(m+1)$, $(m+1)(m+2)$, $(m+2)(m+3), \ldots$, with corresponding eigenfunctions $P_m^m, P_{m+1}^m, P_{m+2}^m, \ldots$ (Exercise 7). At this point we do not know if these are all the eigenvalues, nor do we know that each one is simple. But, once we have ascertained the fact that, for each fixed nonnegative integer $m$, the associated Legendre functions $P_m^m(z)$, $P_{m+1}^m(z)$, $P_{m+2}^m(z), \ldots$ form an orthogonal basis of $L^2(-1, 1)$, it will follow that the numbers $\lambda = n(n+1)$, $n = m$, $m+1$, $m+2, \ldots$ are *all* the eigenvalues of Legendre's operator $L_m$ and that each of the corresponding eigenfunctions $P_n^m(z)$, $n = m$, $m+1$, $m+2, \ldots$, is simple (Exercise 9). Thus the proof of Theorem 7.22, and so that of Theorem 7.23, which is our main result in this section, hinges on the proof that, for each fixed nonnegative integer $m$, the associated Legendre functions $P_n^m(z)$, $n = m$, $m+1$, $m+2, \ldots$, constitute an orthogonal basis of $L^2(-1, 1)$.

For this proof, which requires mathematical theory more advanced than we wish to develop here, we refer once again to the splendid little book by J. R. Higgins, *Completeness and Basis Properties of Sets of Special Functions*; see Section 2.1, particularly Problem 2.8.

The associated Legendre functions satisfy recursion relations similar to those satisfied by Legendre's polynomials. We will also need the normalization constant that we have already used in equation (7.8.15). Both formulas use the following result.

**Lemma 7.25.** *The nth Legendre polynomial $P_n(z)$ satisfies the equations*

$$-(1-z^2)\frac{d^{m+2}P_n(z)}{dz^{m+2}} + 2(m+1)z\frac{d^{m+1}P_n(z)}{dz^{m+1}}$$

$$+ (m(m+1)-n(n+1))\frac{d^m P_n(z)}{dz^m} = 0, \quad 0 \leqslant m \leqslant n$$

The proof is assigned as Exercise 5. This relation is used to prove Theorem 7.24 and the following two theorems.

**Theorem 7.26.** *The associated Legendre functions $P_n^m(z) = (1 - z^2)^{m/2}\, d^m P_n(z)/dz^m$ satisfy the following recurrence in m:*

$$P_n^{m+2}(z) - \frac{2(m+1)z}{\sqrt{1-z^2}}\, P_n^{m+1}(z) + (n(n+1)$$

$$- m(m+1))P_n^m(z) = 0$$

*and the following recurrence in n:*

$$(n-m+1)P_{n+1}^m(z) - (2n+1)zP_n^m(z)$$

$$+ (n+m)P_{n-1}^m(z) = 0$$

The proof is assigned as Exercises 13 and 14.

**Theorem 7.27**

$$\int_{-1}^{1} [P_n^m(z)]^2\, dz = \frac{2}{2n+1}\frac{(n+m)!}{(n-m)!}, \quad 0 \leqslant m \leqslant n,\ n = 0, 1, 2, \ldots$$

The proof is assigned as Exercise 15.

In previous chapters of this book, whenever we have introduced an orthogonal basis of a particular Hilbert space, we have provided explicit examples of expansions of various interesting functions in that Hilbert space. We should do the same for the spherical harmonics and the Hilbert space $L^2(S)$. There *is* one particularly interesting example of such an expansion. Let $(\theta_0, \phi_0)$ denote a fixed point on the sphere $S$, and let $\gamma$ represent the angle between the vector from the origin to $(\theta_0, \phi_0)$ and one from the origin to the general point $(\theta, \phi)$. (See Section 7.7, particularly Figures 7.9 and 7.10.) Then we have

$$\cos \gamma = \sin \theta \sin \theta_0 \cos(\phi - \phi_0) + \cos \theta \cos \theta_0 \qquad (7.8.18)$$

(Exercise 16).   Considered as a function of $\theta$ and $\phi$, $\cos \gamma$ is a function on the sphere $S$. It and its first partial derivatives are continuous there. This is really evident from the geometric way that the angle $\gamma$ is defined, because if $\theta_0 = 0$, then $\gamma = \theta$ so $\cos \gamma = \cos \theta = z/a$, which is clearly a smooth function at every point of the sphere. In general, the angle $\gamma$ has really the same geometric meaning as the angle $\theta$, except that $\theta$ refers to a vector from the origin to the north pole, while $\gamma$ refers to a vector from the origin to a point $(\theta_0, \phi_0)$. But the sphere looks the same from any radius. Hence $\cos \gamma$ will have the same smoothness properties as $\cos \theta$. In fact, $\cos \gamma$, as well as all $P_n(\cos \gamma)$, where the $P_n$ are Legendre's polynomials, belongs to the domain of the Laplacian (Exercise 18). Certainly then, all $P_n(\cos \gamma)$, $n = 0, 1, 2, \ldots$, lie in $L^2(S)$. Our example of a specific spherical harmonics expansion is that for $P_n(\cos \gamma)$.

**Theorem 7.28, Addition Theorem for Legendre Polynomials** (Legendre, 1782).   *Let* $\cos \gamma = \sin \theta \sin \theta_0 \cos(\phi - \phi_0) + \cos \theta \cos \theta_0$. *Then*

$$P_n(\cos \gamma) = P_n(\cos \theta) P_n(\cos \theta_0)$$

$$+ 2 \sum_{m=1}^{n} \frac{(n-m)!}{(n+m)!} P_n^m(\cos \theta) P_n^m(\cos \theta_0) \cos m(\phi - \phi_0), \qquad n = 1, 2, \ldots.$$

The proof of the addition theorem is somewhat tricky. Our methods do yield the general form of the expansion for $P_n(\cos \gamma)$, but I don't know any particularly easy way to determine the numerical coefficients. For a course at our level, it would seem appropriate to leave it at that officially and allow those interested to take up the details as a project (see Exercises 19 through 21).

### Exercises

In this set of exercises, we continue the notation of Section 7.7. $S$ denotes the sphere of radius $a$, and $\theta, \phi$ denote spherical coordinates.

1. Show that $f(\theta, \phi) = g(\theta)h(\phi)$ belongs to $L^2(S)$ if and only if both $h(\phi)$ belongs to $L^2(C)$, $C$ the unit circle, and $g(z)$, $z = \cos \theta$, belongs to $L^2(-1, 1)$. (Assume $\|f\| \neq 0$.)

2. Show that the substitution $z = \cos \theta$ converts the differential equation

$$-\sin^2 \theta \frac{d^2 g}{d\theta^2} - \sin \theta \cos \theta \frac{dg}{d\theta} + m^2 g = \lambda \sin^2 \theta g$$

to

$$-\frac{d}{dz}\left((1-z^2)\frac{dg}{dz}\right)+\frac{m^2}{1-z^2}g=\lambda g$$

3. (a) Derive the equation

$$\int_r^s l_m(g)\overline{h}\,dz=(1-r^2)g'(r)\overline{h(r)}-(1-s^2)g'(s)\overline{h(s)}$$

$$+\int_r^s(1-z^2)g'(z)\overline{h'(z)}\,dz+m^2\int_r^s\frac{g(z)\overline{h(z)}}{1-z^2}\,dz$$

valid for any $g,h$ in the domain of Legendre's operator $L_m$, $-1<r<s<1$.

(b) Using (a) and Theorem 7.35, derive the equation

$$\langle L_m(g),h\rangle=\langle\sqrt{1-z^2}\,g'(z),\sqrt{1-z^2}\,h'(z)\rangle$$

$$+m^2\left\langle\frac{g(z)}{\sqrt{1-z^2}},\frac{h(z)}{\sqrt{1-z^2}}\right\rangle$$

valid for any pair of functions $g$, $h$ in the domain of $L_m$.

(c) Using (b), prove that $L_m$ is Hermitian positive definite if $m\geqslant 1$.

Exercises 4 and 5 constitute the proof of Theorem 7.24.

4. (a) Prove: If $g(z)=P_n^m(z)=(1-z^2)^{m/2}\,d^mP_n(z)/dz^m$ satisfies the equation

$$-\frac{d}{dz}\left((1-z^2)\frac{dg}{dz}\right)+\frac{m^2}{1-z^2}g=n(n+1)g$$

then the $n$th Legendre polynomial $P_n(z)$ satisfies the equation

$$-(1-z^2)\frac{d^{m+2}}{dz^{m+2}}P_n(z)+2(m+1)z\frac{d^{m+1}}{dz^{m+1}}P_n(z)$$

$$+(m(m+1)-n(n+1))\frac{d^m}{dz^m}P_n(z)=0$$

(b) Reverse your steps in (a): If the $n$th Legendre polynomial $P_n(z)$ satisfies the second equation in (a), then $P_n^m(z)=(1-z^2)^{m/2}\,d^mP_n(z)/dz^m$ satisfies the first.

5. (Continuation) Prove that the $n$th Legendre polynomial $P_n(z)$ satisfies the second equation of Exercise 4(a) for $0\leqslant m\leqslant n$.

6. Compute the associated Legendre functions $P_n^m(z) = (1-z^2)^{m/2} d^m P_n(z)/dz^m$, $0 \leqslant m \leqslant n \leqslant 3$. Also express each function as a linear combination of $\cos k\theta$ and $\sin k\theta$, where $z = \cos \theta$. (Answers below.)

| $n$ | $m$ | $P_n^m(z)$ | $P_n^m(\theta)$ |
|---|---|---|---|
| 0 | 0 | 1 | 1 |
| 1 | 0 | $z$ | $\cos \theta$ |
| 1 | 1 | $(1-z^2)^{1/2}$ | $\sin \theta$ |
| 2 | 0 | $(1/2)(3z^2 - 1)$ | $(1/4)(3 \cos 2\theta + 1)$ |
| 2 | 1 | $3z(1-z^2)^{1/2}$ | $(3/2) \sin 2\theta$ |
| 2 | 2 | $3(1-z^2)$ | $(3/2)(1 - \cos 2\theta)$ |
| 3 | 0 | $(1/2)(5z^3 - 3z)$ | $(1/8)(5 \cos 3\theta + 3 \cos \theta)$ |
| 3 | 1 | $(3/2)(5z^2 - 1)(1-z^2)^{1/2}$ | $(3/8)(5 \sin 3\theta + \sin \theta)$ |
| 3 | 2 | $15z(1-z^2)$ | $(15/4)(\cos \theta - \cos 3\theta)$ |
| 3 | 3 | $15(1-z)^{3/2}$ | $(15/4)(3 \sin \theta - \sin 3\theta)$ |

7. Fix a nonnegative integer $m$. Verify that the associated Legendre functions $P_n^m(z)$ are continuous with continuous first derivative in $-1 < z < 1$, are bounded at $z = \pm 1$ (in fact, are zero there when $m \geqslant 1$), and all belong to $L^2(-1, 1)$. Combining Exercises 4 and 5, show that $l_m(P_n^m) = n(n+1)P_n^m$, $n \geqslant m$, where $l_m$ is Legendre's differential expression. Prove that the $\lambda = n(n+1)$ are eigenvalues of Legendre's operator $L_m$ with corresponding eigenfunctions $P_n^m$, $n = m, m+1, m+2....$

8. (Continuation, $m$ still fixed)   May we assert that

$$\int_{-1}^{1} P_k^m(z)P_l^m(z) \, dz = 0 \qquad \text{when } k \neq l, \, k, \, l \geqslant m$$

without doing any calculations whatever? If so, explain why.

9. Fix a nonnegative integer $m$. Take as known the fact that the associated Legendre functions $P_n^m(z)$, $n = m, m+1, m+2,...$, are an orthogonal basis of $L^2(-1, 1)$. Using this fact, prove that the $\lambda = n(n+1)$, $n = m, m+1, m+2,...$, are all the eigenvalues of Legendre's operator $L_m$ and that each one is simple.

10. May we assert, without doing any calculations, that $\langle Y_n^l, Y_p^m \rangle = 0$ when $n \neq p$ for any $l, m$ satisfying $0 \leqslant l \leqslant n, 0 \leqslant m \leqslant p$? (Here the $Y$'s are spherical harmonics and $\langle \cdot, \cdot \rangle$ is the inner product in $L^2(S)$, $S$ the sphere.) If so, explain why.

11. Fix a nonnegative integer $n$. Prove that the $2n+1$ functions

$$\cos m\phi\, P_n^m(\cos\theta), \qquad 0 \leqslant m \leqslant n$$
$$\sin m\phi\, P_n^m(\cos\theta), \qquad 1 \leqslant m \leqslant n$$

are orthogonal in $L^2(S)$, $S$ the sphere. (See Exercise 6 of Section 7.7.)

12. Take as known the fact that the spherical harmonics are an orthogonal basis of $L^2(S)$. Using this fact, prove that the $\lambda = n(n+1)$, $n = 0, 1, 2, \ldots$, are *all* the eigenvalues of the Laplacian on the unit sphere and that the eigenvalue $n(n+1)$ has multiplicity exactly $2n+1$.

13. Prove that the associated Legendre functions satisfy the following recursion in $m$:

$$P_n^{m+2}(z) - \frac{2(m+1)z}{\sqrt{1-z^2}} P_n^{m+1}(z) - (m(m+1) - n(n+1))P_n^m(z) = 0$$

[Hint: use the second equation of Exercise 4a or Lemma 7.25.]

14. (a) Derive the equation $(2n+1)P_n = P_{n+1}' - P_{n-1}'$ for Legendre's polynomial $P_n$. [Hint: combine the recursion formula (Theorem 4.36 in Section 4.8) with the formula $P_{n+1}' - zP_n' = (n+1)P_n$ from Exercise 10 in Section 4.8.]

(b) Prove that the associated Legendre functions satisfy the following recursion in $n$:

$$(n-m+1)P_{n+1}^m(z) = (2n+1)zP_n^m(z) - (n+m)P_{n-1}^m(z)$$

15. (a) Let

$$I_{m,n} = \int_{-1}^{1} [P_n^m(z)]^2\, dz = \int_{-1}^{1} \left[ (1-z^2)^m \frac{d^m P_n}{dz^m} \right] d\left( \frac{d^{m-1} P_n}{dz^{m-1}} \right)$$

Using integration by parts and Lemma 7.25, show that $I_{m,n} = (m+n)(n-m+1)I_{m-1,n}$, $1 \leqslant m \leqslant n$.

(b) Using your result from (a) together with the known formula $I_{0,n} = 2/(2n+1)$ (Theorem 4.35, Section 4.8), show that

$$I_{m,n} = \int_{-1}^{1} [P_n^m(z)]^2\, dz$$

$$= \frac{2}{2n+1} \frac{(n+m)!}{(n-m)!}, \qquad 0 \leqslant m \leqslant n,\ n = 0, 1, 2, \ldots$$

16. Show that, if $\gamma$ is the angle between the vectors $x\bar{i} + y\bar{j} + z\bar{k}$ and $x_0\bar{i} + y_0\bar{j} + z_0\bar{k}$ where $x = \sin\theta\cos\phi$, $y = \sin\theta\sin\phi$, $z = \cos\theta$, and $\bar{i}, \bar{j}, \bar{k}$ are the unit vectors in the positive $x$, $y$, and $z$ directions, respectively, then

$$\cos\gamma = \sin\theta\sin\theta_0\cos(\phi - \phi_0) + \cos\theta\cos\theta_0$$

17. Suppose $f(\theta, \phi)$ belongs to the domain of the unit spherical Laplacian $\Delta$ and satisfies $\Delta(f) = n(n + 1)f$. Explain why the spherical harmonics expansion (7.8.11) for $f$ then takes the form

$$f(\theta, \phi) = \tfrac{1}{2}a_{00} \qquad \text{when } n = 0$$

$$f(\theta, \phi) = \tfrac{1}{2}a_{0n}P_n(\cos \theta)$$

$$+ \sum_{m=1}^{n} (a_{mn} \cos m\phi + b_{mn} \sin m\phi)P_n^m(\cos \theta) \qquad \text{when } n \geqslant 1$$

18. Let $w = \cos \gamma = \sin \theta \sin \theta_0 \cos(\phi - \phi_0) + \cos \theta \cos \theta_0$.

   (a) Show that

   $$\Delta(P_n(w)) = -\left(\left(\frac{\partial w}{\partial \theta}\right)^2 + \frac{1}{\sin^2 \theta}\left(\frac{\partial w}{\partial \phi}\right)^2\right)\frac{\partial^2 P_n}{\partial w^2}$$

   $$- \left(\frac{\partial^2 w}{\partial \theta^2} + \frac{\cos \theta}{\sin \theta}\frac{\partial w}{\partial \theta} + \frac{1}{\sin^2 \theta}\frac{\partial^2 w}{\partial \phi^2}\right)\frac{\partial P_n}{\partial w}$$

   where $\Delta$ is the unit spherical Laplacian.

   (b) Prove the identities

   $$\left(\frac{\partial w}{\partial \theta}\right)^2 + \frac{1}{\sin^2 \theta}\left(\frac{\partial w}{\partial \phi}\right)^2 = 1 - w^2,$$

   $$\frac{\partial^2 w}{\partial \theta^2} + \frac{\cos \theta}{\sin \theta}\frac{\partial w}{\partial \theta} + \frac{1}{\sin^2 \theta}\frac{\partial^2 w}{\partial \phi^2} = -2w$$

   (c) Combine (a) and (b) to prove that $P_n(\cos \gamma)$ is an eigenfunction of the Laplacian belonging to the eigenvalue $n(n+1)$. (Thus, in particular, $P_n(\cos \gamma)$ belongs to the domain of $\Delta$.)

Exercises 19 through 21 bear on Legendre's addition theorem. Consider these three exercises as optional.

19. Combining Exercises 17 and 18, show that the spherical harmonics expansion of $P_n(\cos \gamma)$ has the form

$$P_n(\cos \gamma) = \tfrac{1}{2}a_{0n}(\theta_0, \phi_0)P_n(\cos \theta) + \sum_{m=1}^{n} a_{mn}(\theta_0, \phi_0) \cos m\phi \, P_n^m(\cos \theta)$$

$$+ \sum_{m=1}^{n} b_{mn}(\theta_0, \phi_0) \sin m\phi \, P_n^m(\cos \theta)$$

   in which, as indicated, the coefficients depend on the point $(\theta_0, \phi_0)$.

20. (Continuation) Argue that the expansion in Exercise 19 holds equally

well if we interchange the pairs $(\theta, \phi)$ and $(\theta_0, \phi_0)$. Comparing both expansions, show that

$$P_n(\cos \gamma) = \tfrac{1}{2}\alpha_{0n} P_n(\cos \theta_0) P_n(\cos \theta)$$

$$+ \sum_{m=1}^{n} \alpha_{mn} \cos m\phi_0 \cos m\phi \, P_n^m(\cos \theta_0) P_n^m(\cos \theta)$$

$$+ \sum_{m=1}^{n} \beta_{mn} \sin m\phi_0 \sin m\phi \, P_n^m(\cos \theta_0) P_n^m(\cos \theta)$$

where the $\alpha$'s and $\beta$'s are numerical constants that are independent of all four variables $\theta$, $\phi$, $\theta_0$, $\phi_0$.

21. (Continuation)   In Exercise 20, show that

$$\alpha_{mn} = \beta_{mn} = 2\frac{(n-m)!}{(n+m)!}$$

and thus prove Theorem 7.28, the addition theorem for Legendre polynomials.

22. From the addition theorem (7.28), deduce the following two formulas:

$$1 = P_n(z)^2 + 2 \sum_{m=1}^{n} \frac{(n-m)!}{(n+m)!} P_n^m(z)^2, \qquad -1 \leqslant z \leqslant 1, \; n = 1, 2, \ldots$$

$$\int_0^{2\pi} P_n(\cos \gamma) \, d\phi = 2\pi P_n(\cos \theta) P_n(\cos \theta_0)$$

23. From the addition theorem (7.28), deduce the following two formulas:

$$P_n^m(\cos \theta_0) \cos m\phi_0 = \frac{2n+1}{4\pi} \int_{-\pi}^{\pi} d\phi \int_0^{\pi} d\theta \sin \theta \, P_n(\cos \gamma) \cos m\phi \, P_n^m(\cos \theta)$$

$$P_n^m(\cos \theta_0) \sin m\phi_0 = \frac{2n+1}{4\pi} \int_{-\pi}^{\pi} d\phi \int_0^{\pi} d\theta \sin \theta \, P_n(\cos \gamma) \sin m\phi \, P_n^m(\cos \theta)$$

24. Let $S$ be the sphere of radius $a$ and area $A = 4\pi a^2$. Let $\mu_n$, $n = 0, 1, 2, \ldots$, be the eigenvalues of $S$ listed according to their multiplicity.
    (a) Write down the $\mu_n$ for $0 \leqslant n \leqslant 24$. [Hint: The list begins $\mu_0 = 0 = 4\pi(0)(1)/A$, $\mu_1 = \mu_2 = \mu_3 = 4\pi(1)(2)/A$, $\mu_4 = \mu_5 = \mu_6 = \mu_7 = \mu_8 = 4\pi(2)(3)/A$, etc.]
    (b) Show that

$$\lim_{n \to \infty} \frac{\mu_n}{n} = \frac{4\pi}{A}$$

[Hint: $\lim_{n \to \infty} \mu_n/n = \lim_{n \to \infty} \mu_{g(n)}/g(n)$ where $g(n)$ is any subsequence.]

## 7.9  THE HYDROGEN ATOM

In Schrödinger's series of papers that created wave mechanics (the series of papers that form the basis for our Chapter 5), the very first paper consisted primarily of the solution of his wave equation for the hydrogen atom. As the simplest atom, the hydrogen atom was the touchstone to test any theory purporting to explain the discreteness of spectral lines. We'll take up the analysis of the hydrogen atom now.

Schrödinger's time-independent equation has the form

$$\frac{\hbar^2}{2m}\Delta\Psi + V\Psi = E\Psi \tag{7.9.1}$$

(equation (6.1.4) of Section 6.1). Look at equation (7.9.1) this way: there is a Hilbert space operator $H$ whose differential expression is $(\hbar^2/2m)\Delta + V$. Our job is to find the eigenvalues and eigenfunctions of $H$, which is to say, to solve the equation

$$H(\Psi) = E\Psi \tag{7.9.2}$$

This point of view accords with that expounded in Section 5.9. The total energy of the system, which is an observable quantity, is matched to the operator $H$, and if there are discrete (quantized) values of the total energy, they will be the eigenvalues of $H$, namely the values $E$ for which equation (7.9.2) has a nonzero solution $\Psi$ in the domain of $H$.

In setting up the details of this analysis, we shall follow the prescription of Section 5.8: (1) define the physical system, (2) form the configuration space, (3) identify the states and the state space, and (4) match observables to Hermitian operators.

(1) Our *physical system* consists of an electron of mass $m$ and a proton, which attract each other with a force of magnitude $e^2/r^2$, where $e$ is the electronic charge in appropriate units and $r$ is the distance between proton and electron. As the proton is roughly 2000 times heavier than the electron, we shall simplify the problem by taking the proton's mass as infinite and taking the proton's location as the origin of our coordinate system. Thus our idealized physical system consists of a single electron of mass $m$ moving in three-dimensional space under the influence of a force of magnitude $e^2/r^2$ directed toward the origin. There are no other constraints on the electron's motion. The potential energy $V(r)$ of the electron, as a function of distance $r$ from the origin, is $V(r) = -e^2/r$, with reference $V(\infty) = 0$ (see Section 5.3).

(2) The *configuration space* of our idealized physical system, the Cartesian space sufficient to specify the position of our unconstrained single electron, is $\mathbf{R}^3$, ordinary three-dimensional space. Owing to the spherical symmetry

inherent in our problem, spherical coordinates are the coordinates of choice here.

(3) The *state space* for our idealized hydrogen atom is the Hilbert space $L^2(\mathbf{R}^3)$ consisting of all real- or complex-valued functions $\Psi(x, y, z)$ that satisfy

$$\iiint_{-\infty}^{+\infty} |\Psi(x, y, z)|^2 \, dx \, dy \, dz < \infty$$

$L^2(\mathbf{R}^3)$ carries the usual inner product. In terms of the more convenient spherical coordinates, the square-integrability condition is

$$\|\Psi\|^2 = \int_0^\infty r^2 \, dr \int_0^\pi \sin\theta \, d\theta \int_{-\pi}^\pi d\phi |\Psi(r, \theta, \phi)|^2 < \infty \qquad (7.9.3)$$

and the inner product is

$$\langle \Psi_1, \Psi_2 \rangle = \int_0^\infty r^2 \, dr \int_0^\pi \sin\theta \, d\theta \int_{-\pi}^\pi d\phi \, \Psi_1(r, \theta, \phi)\overline{\Psi_2(r, \theta, \phi)} \qquad (7.9.4)$$

(4) As for *matching observables to Hermitian operators* on the state space $L^2(\mathbf{R}^3)$, the total energy $E$ of our hydrogen atom is the observable of interest in this problem. Its operator $H$ has the differential expression $(\hbar^2/2m)\Delta + V$ (see Section 5.9) with an appropriately defined domain. (Here $\Delta$ is the Laplacian differential expression on $L^2(\mathbf{R}^3)$, $m$ is the mass of the electron, and $V$ represents multiplication by $V(r) = -e^2/r$.) We seek the eigenvalues of $H$, i.e., all numbers $E$ such that the equation $H(\Psi) = E\Psi$ has a nonzero solution in the domain of $H$. This leads us back to Schrödinger's equation (7.9.1) and its equivalent form (7.9.2).

Expressed in spherical coordinates, our operator $H$ has the differential expression

$$l(\Psi) \equiv \frac{-\hbar^2}{2mr^2} \left[ \frac{\partial}{\partial r}\left(r^2 \frac{\partial \Psi}{\partial r}\right) + \frac{2me^2}{\hbar^2} r\Psi + \frac{1}{\sin\theta}\left(\frac{\partial}{\partial\theta}\left(\sin\theta \frac{\partial \Psi}{\partial\theta}\right) + \frac{1}{\sin\theta}\frac{\partial^2 \Psi}{\partial\phi^2}\right)\right]$$

$$(7.9.5)$$

Remember our basic equation: operator = differential expression + domain. So we must specify the domain of $H$ to complete the definition of this operator. The domain of $H$ will be a subspace of $L^2(\mathbf{R}^3)$ that will consist of functions $\Psi$ in $L^2(\mathbf{R}^3)$ that are compatible with the differential expression (7.9.5), that is, for which $l(\Psi)$ also belongs to $L^2(\mathbf{R}^3)$. In addition to the compatibility requirement, membership in the domain of $H$ will be further restricted by a boundary condition at $r = 0$. The point $r = 0$ is a boundary point for our problem because of the singular potential $V = -e^2/r$. (This singular potential corresponds to the term $-(\hbar^2/2mr^2)(2me^2r\Psi/\hbar^2) = -e^2\Psi/r$

in (7.9.5). The other terms in (7.9.5) that contain $1/r^2$ and $1/\sin\theta$ do not correspond to real singularities of the physical problem but rather reflect the fact that spherical coordinates are singular on the $z$ axis.) So our job is to specify this domain, check that the resulting operator $H$ is Hermitian, then ascertain the numbers $E$ for which the equation $H(\Psi) = E\Psi$ has a nonzero solution $\Psi$ belonging to the domain of $H$.

To avoid the rather intricate and involved mathematics of working out the implications of compatibility for the partial differential expression (7.9.5), I shall simply *assume* that we have successfully completed the project just outlined, namely that of specifying a large-as-possible domain to combine with the differential expression (7.9.5) to create the Hermitian operator $H$. This assumption is partially justified by the fact that Schrödinger's differential expression (7.9.5) equals the sum of the Laplacian plus the operation of multiplication by a real function, both of which determine Hermitian operators. Also, one can directly define a domain for (7.9.5), somewhat artificially restricted to be sure, but on which (7.9.5) defines a Hermitian operator (Exercises 1, 2, and 3).

On the basis of this assumption that a suitable domain for the operator $H$ has been specified, we shall now seek the eigenvalues and eigenfunctions of this operator. It will turn out that we can do this even though we don't know the domain of $H$ completely. It will also turn out that the search for the eigenvalues and eigenfunctions of $H$ will yield, after the fact, information on its domain.

So, with these preparations, our problem has boiled down to this: we seek all numbers $E$ such that the partial differential equation (7.9.5), $l(\Psi) = E\Psi$, has a nonzero solution $\Psi$ in the domain of the Hermitian operator $H$ determined by $l$, a domain not yet completely known. Yet, as that domain is a subspace of the state space $L^2(\mathbf{R}^3)$, certainly every function in that domain will belong to $L^2(\mathbf{R}^3)$. If we find an $L^2$ solution $\Psi$ to $l(\Psi) = E\Psi$, such a $\Psi$ will automatically be compatible with the differential expression $l$ (which is given in (7.9.5)), because $l(\Psi) = E\Psi$ says that $l(\Psi)$ is a scalar multiple of $\Psi$, thus certainly in $L^2(\mathbf{R}^3)$ if $\Psi$ is. Thus any $L^2$ function that satisfies $l(\Psi) = E\Psi$ will automatically fulfill the compatibility requirement, which is the major requirement for membership in the domain of $H$. In this way we bring in the domain restriction even though we don't have a complete specification of that domain.

To solve the differential equation $l(\Psi) = E\Psi$ we separate variables; $\Psi(r, \theta, \phi) = R(r)F(\theta, \phi)$. Let us, using (7.9.5), rewrite $l(\Psi) = E\Psi$ as

$$\frac{-1}{r^2}\left[\frac{d}{dr}\left(r^2\frac{dR}{dr}\right)F + \frac{2me^2}{\hbar^2}rRF + \frac{R}{\sin\theta}\left(\frac{\partial}{\partial\theta}\left(\sin\theta\frac{\partial F}{\partial\theta}\right) + \frac{1}{\sin\theta}\frac{\partial^2 F}{\partial\phi^2}\right)\right]$$

$$= \frac{2mE}{\hbar^2}RF \tag{7.9.6}$$

Divide by $RF$ and do a little rearranging to get

$$\frac{1}{R}\frac{d}{dr}r^2\frac{dR}{dr} + \frac{2mr}{\hbar^2}(rE + e^2) = \frac{1}{F}\delta(F) \tag{7.9.7}$$

where $\delta$ is the Laplacian differential expression on the surface of the unit sphere (equation (7.7.13) of Section 7.7 with $a=1$). As the left side of (7.9.7) depends on $r$ alone and the right side does not, the left side does not in fact vary with $r$, hence is a constant, say $\mu$. So from (7.9.7) we get two equations, one of which is

$$\delta(F) = \mu F \tag{7.9.8}$$

As $\Psi = RF$ belongs to $L^2(\mathbf{R}^3)$, $F$ belongs to $L^2(S)$ (Exercise 4). Hence (7.9.8) really asks for the eigenvalues of the Laplacian operator $\Delta$ on the surface of the unit sphere, i.e., the numbers $\mu$ such that the equation $\Delta(F) = \mu F$ has a nonzero solution $F(\theta, \phi)$ that lies in the domain of the spherical Laplacian $\Delta$. We know these eigenvalues from Section 7.8. They are $\mu = l(l+1)$, $l = 0, 1, 2, \ldots$. The eigenvalue $l(l+1)$ has multiplicity $2l+1$. The eigenfunctions are the spherical harmonics. Here, we shall use the exponential notation for the spherical harmonics, as that notation is more convenient for this application to the hydrogen atom. Hence we shall take the $2l+1$ orthogonal eigenfunctions belonging to the eigenvalue $l(l+1)$ as

$$Z_l^m(\theta, \phi) = P_l^{|m|}(\cos\theta)e^{im\phi}, \qquad -l \leqslant m \leqslant l$$

These functions serve as an alternative to the functions defined in (7.8.10) of Section 7.8 because they also span the eigenspace belonging to the eigenvalue $l(l+1)$. We shall also refer to the $Z_l^m$ as spherical harmonics. The functions $Z_l^m$, $-l \leqslant m \leqslant l$, $l = 0, 1, 2, \ldots$, also of course form an orthogonal basis of $L^2(S)$, $S$ the unit sphere: for any $f(\theta, \phi)$ in $L^2(S)$ we have

$$f(\theta, \phi) = \sum_{l=0}^{\infty} \sum_{m=-l}^{l} c_{lm} Z_l^m(\theta, \phi)$$

where

$$c_{lm} = \frac{\langle f, Z_l^m \rangle}{\langle Z_l^m, Z_l^m \rangle}$$

$$= \frac{2l+1}{4\pi}\frac{(l-|m|)!}{(l+|m|)!}\int_{-\pi}^{\pi}d\phi\int_0^{\pi}d\theta\sin\theta f(\theta, \phi)P_l^{|m|}(\cos\theta)e^{-im\phi}$$

Thus one half of (7.9.7) is solved. The other half can be written

$$-\frac{1}{r^2}\frac{d}{dr}\left(r^2\frac{dR}{dr}\right) + \left(\frac{l(l+1)}{r^2} - \frac{2me^2}{\hbar^2 r}\right)R = \frac{2mE}{\hbar^2}R \tag{7.9.9}$$

As $\Psi = RF$ belongs to $L^2(\mathbf{R}^3)$, we have

$$\int_0^\infty |R(r)|^2 r^2 dr < \infty \tag{7.9.10}$$

(Exercise 4).

As a first step in the solution of (7.9.9), we shall convert to a dimensionless independent variable. This is always a good idea. The variable $s = 2me^2r/\hbar^2$ is dimensionless. In terms of $s$, equation (7.9.9) becomes

$$-\frac{1}{s^2}\frac{d}{ds}\left(s^2\frac{dR}{ds}\right) + \left(\frac{l(l+1)}{s^2} - \frac{1}{s}\right)R = \lambda R \tag{7.9.11}$$

where the eigenvalue parameter $\lambda = E\hbar^2/2me^4$ is also dimensionless. The square-integrability condition remains the same:

$$\int_0^\infty |R(s)|^2 s^2 \, ds < \infty \tag{7.9.12}$$

Equation (7.9.11) asks for the eigenvalues $\lambda$ of the operator in the Hilbert space $L^2(0, \infty; s^2)$ determined by the differential expression on the left of (7.9.11). This differential expression does determine a Hermitian operator on $L^2(0, \infty; s^2)$—I have analyzed this operator and specified its domain in Appendix 7.A. But in our analysis here we shall maintain the practical philosophy adopted in this section and shall bypass the mathematical details involved in specifying the domain. We shall instead concentrate on getting the actual eigenvalues and eigenfunctions of (7.9.11).

Determining these eigenvalues and eigenfunctions of (7.9.11) (with (7.9.12)) is not an easy matter. We shall use as a guide the quantum linear oscillator (Section 5.5). There we transferred Hilbert spaces, from $L^2(-\infty, \infty)$ to $L^2(-\infty, \infty; e^{-x^2/2})$, and ended up with the Hermite polynomials in the latter space as our eigenfunctions. It is reasonable to try the same sort of device here. As the Hilbert space $L^2(0, \infty; s^2)$ certainly contains no polynomials, a transfer of Hilbert spaces is called for. We accordingly substitute

$$R(s) = s^k e^{-\beta s} y(s) \tag{7.9.13}$$

into (7.9.11) and (7.9.12). The real constants $k$ and $\beta$ in (7.9.13) will be adjusted to simplify the subsequent equations. We will want $\beta > 0$ in any event, so that our new Hilbert space, $L^2(0, \infty; s^{2k+2}e^{-2\beta s})$, whose weight function we get by substituting (7.9.13) into (7.9.12),

$$\int_0^\infty |y(s)|^2 s^{2k+2}e^{-2\beta s} \, ds < \infty \tag{7.9.14}$$

will contain all polynomials.

Now substitute (7.9.13) into (7.9.11). Keep the new equation in Sturmian form. This substitution requires only straightforward calculus but some rather intense bookkeeping. A guided procedure is provided in Exercises 5 and 6. After the dust has settled, you will get the following equation in $y(s)$:

$$\frac{-1}{s^{2k+2}e^{-2\beta s}}\frac{d}{ds}\left(s^{2k+2}e^{-2\beta s}\frac{dy}{ds}\right) + \frac{l(l+1)-k(k+1)}{s^2}y$$

$$+ \frac{2\beta(k+1)-1}{s}y = (\lambda + \beta^2)y \tag{7.9.15}$$

Next comes the selection of the constants $k$ and $\beta$ to attempt to turn (7.9.15) into something reasonable. The choice of $k$ is easy: putting $k=l$ eliminates the $y/s^2$ term. Put $k=l$ in (7.9.15):

$$\frac{-1}{s^{2l+2}e^{-2\beta s}}\frac{d}{ds}\left(s^{2l+2}e^{-2\beta s}\frac{dy}{ds}\right) + \frac{2\beta(l+1)-1}{s}y = (\lambda + \beta^2)y \tag{7.9.16}$$

Selection of the positive constant $\beta$ is next. It is advantageous first to introduce a new independent variable,

$$x = 2\beta s \tag{7.9.17}$$

With this change of variable, (7.9.16) becomes

$$\frac{-1}{x^{2l+2}e^{-x}}\frac{d}{dx}\left(x^{2l+2}e^{-x}\frac{dy}{dx}\right) + \left(l+1-\frac{1}{2\beta}\right)\frac{y}{x} = \frac{1}{4}\left(\frac{\lambda}{\beta^2}+1\right)y \tag{7.9.18}$$

The accompanying Hilbert space is determined by

$$\int_0^\infty |y(x)|^2 x^{2l+2}e^{-x}\,dx < \infty \tag{7.9.19}$$

Finally, we turn to the choice of $\beta$. Our eigenvalue parameter is $\lambda = E\hbar^2/2me^4$. From our experience with the Bohr atom (Section 5.3), we expect the total energy $E$, thus also $\lambda$, to be negative. Hence if we choose $\beta = \sqrt{-\lambda}$ (positive square root), we will eliminate the term on the right side of (7.9.18). So, substitute $\beta = \sqrt{-\lambda}$ into (7.9.18), multiply through by $x$, and introduce the parameters $\alpha = 2l+1$ and $\nu = 1/2\sqrt{-\lambda}-(l+1)$. You will get

$$\frac{-1}{x^\alpha e^{-x}}\frac{d}{dx}\left(x^{\alpha+1}e^{-x}\frac{dy}{dx}\right) = \nu y \tag{7.9.20}$$

or, written out,

$$xy'' + (\alpha + 1 - x)y' + \nu y = 0 \tag{7.9.20'}$$

Note carefully: by putting the eigenvalue parameter $(\lambda/\beta^2 + 1)/4$ to zero in (7.9.18), then multiplying through by $x$, we have changed the operator and its associated Hilbert space but have not changed the unknown function $y(x)$. The new Hilbert space, and one matched to the operator whose differential expression sits on the left side of (7.9.20), is given by

$$\int_0^\infty |y(x)|^2 x^\alpha e^{-x} \, dx < \infty \qquad (7.9.21)$$

The weight function in (7.9.21) differs from that in (7.9.19) by a factor of $x$ (remember that $\alpha = 2l + 1$).

Now we are at the end of the road. Equation (7.9.20) (or 7.9.20′)) combined with (7.9.21) constitutes the eigenvalue problem for the classical *operator of Laguerre*, whose eigenvalues are $\nu = 0, 1, 2, \ldots$. Each of these eigenvalues is simple; the corresponding eigenfunctions are the *Laguerre polynomials*, denoted $L_\nu^{(\alpha)}(x)$, $\nu = 0, 1, 2, \ldots$. The Laguerre polynomials form an orthogonal basis of the Hilbert space $L^2(0, \infty; x^\alpha e^{-x})$.

The theory of Laguerre's operator and the Laguerre polynomials parallels in every detail the theory of Legendre's operator and the Legendre polynomials (Sections 4.6 through 4.8) and that of Hermite's operator and the Hermite polynomials (Sections 4.9 through 4.11). The Laguerre theory is worked out in the next section, Section 7.10, which is set up as a student project, the results being stated in a series of exercises. While I strongly recommend that you do this project eventually, both to reinforce your understanding of the Legendre and Hermite cases and to prove the results we shall be using here, you need not do project 7.10 now. You can simply refer as needed to the facts stated there. I shall use these facts from 7.10 without explicit reference to the particular exercise or theorem.

Our conclusion from Laguerre theory is now this: equation (7.9.20) (or (7.9.20′)) has a nonzero solution $y(x)$ that lies in the Hilbert space $L^2(0, \infty; x^{2l+1}e^{-x})$ and for which $x^{l+1/2}y(x)$ is finite at $x = 0$ only when $\nu$ is a nonnegative integer, $\nu = 0, 1, 2, \ldots$. For a fixed integer $\nu$ there is, up to scalar multiples, only one such solution, namely the Laguerre polynomial $L_\nu^{(2l+1)}(x)$, which is a polynomial of degree $\nu$. (Remember that, in (7.9.20) and (7.9.20′), $\alpha = 2l + 1$.)

Let's now work backward to get information on the energy eigenvalues $E$ of our original equation (7.9.2) $H(\Psi) = E\Psi$. From $\nu = 1/2\sqrt{-\lambda} - (l+1)$ we get $\sqrt{-\lambda} = 1/2n$, or $\lambda = -1/4n^2$, where $n = \nu + (l+1)$. As $\nu \geq 0$ and $l \geq 0$, the integer $n$ has the range $n = 1, 2, \ldots$. Remembering that $\lambda = E\hbar^2/2me^4$ (refer to the sentence following equation (7.9.11)), we get

$$E = -\frac{me^4}{2\hbar^2 n^2}, \qquad n = 1, 2, \ldots \qquad (7.9.22)$$

as *energy eigenvalues of Schrödinger's operator H for the hydrogen atom.* These discrete total energies for the hydrogen atom agree with those of the Bohr theory (Section 5.3).

What are the eigenfunctions $\Psi_n(r, \theta, \phi)$ that belong to a given energy eigenvalue $E_n = -me^4/2\hbar^2 n^2$, $n$ fixed? It turns out that, for each fixed integer $n$, there are $n^2$ orthogonal eigenfunctions in $L^2(\mathbf{R}^3)$ that belong to the eigenvalue $E_n$. They arise as follows. We solved the eigenvalue equation $H(\Psi) = E\Psi$ by separating variables: $\Psi(r, \theta, \phi) = R(r)F(\theta, \phi)$. We proved that, if $\Psi = R(r)F(\theta, \phi)$ were an eigenfunction of $H$, then $F$ would have to be an eigenfunction of $\Delta$, the Laplacian on the unit sphere. The integer $l$ entered at this point: as an eigenfunction of $\Delta$, $F(\theta, \phi)$ would belong to one of the eigenvalues $l(l+1)$ of the sphere. And, for each fixed $l$, there would be $2l+1$ orthogonal eigenfunctions, namely the spherical harmonics

$$Z_l^m(\theta, \phi) = P_l^{|m|}(\cos \theta)e^{im\phi}, \qquad -l \leqslant m \leqslant l \tag{7.9.23}$$

(We are using the alternative complex form of the spherical harmonics which was introduced earlier in this section.)

Once $l$ is fixed, so is $v$: $v = n - (l+1)$. To the integer $v$ belongs an eigenfunction $R(r)$ in the radius variable. To see what this is, we'll have to retrace our steps through the various changes of dependent and independent variables. First, we changed from $r$ to the dimensionless variable $s = (2me^2/\hbar^2)r$. We shall keep everything in terms of the dimensionless $s$. Then we substituted $R(s) = s^k \exp(-\beta s)y(s)$. Remembering that $k = l$ and that $\beta = \sqrt{-\lambda} = 1/2n$, we get $R(s) = s^l \exp(-s/2n)y(s)$. To determine $y$ we changed to the variable $x = 2\beta s = s/n$ and found $y(x) = L_v^{(2l+1)}(x) = L_v^{(2l+1)}(s/n)$. Combining these, we get for our radial eigenfunction $R$

$$R(s) = s^l e^{-s/2n} L_v^{(2l+1)}\left(\frac{s}{n}\right) \tag{7.9.24}$$

Multiply (7.9.23) by (7.9.24) to get the $n^2$ *orthogonal eigenfunctions that belong to the energy eigenvalue* $E_n = -me^4/2\hbar^2 n^2$:

$$\Psi_n^{l,m}(s, \theta, \phi) = s^l e^{-s/2n} L_v^{(2l+1)}\left(\frac{s}{n}\right) Z_l^m(\theta, \phi),$$

$$l = 0, 1, 2, \ldots, n-1, \ -l \leqslant m \leqslant l, \ v = n - (l+1) \tag{7.9.25}$$

Why are there $n^2$ of these? To each value of $l$ there correspond $2l+1$ spherical harmonics. Add these up to get the total number:

$$\sum_{l=0}^{n-1} (2l+1) = 2\sum_{l=0}^{n-1} l + n = 2\frac{(n-1)(n)}{2} + n = n^2$$

Thus the eigenvalue $E_n$ has multiplicity $n^2$.

The integer $n$ is called the *total quantum number*. It determines the total energy $E_n$ of the bound electron in the hydrogen atom, $E_n = -me^4/2\hbar^2 n^2$. We also refer to the $E_n$ as the *energy levels* of the hydrogen atom. Thus the total energy of the bound electron in the hydrogen atom is quantized; it has discrete, not continuous, values. Quantum theory *predicts* this decisive fact. Quoting Schrödinger again, "the theory furnishes a method of quantization which is absolutely free from arbitrary postulates that this or that quantity must be an integer."

The integer $l$, whose existence is also predicted by Schrödinger's theory, is often referred to as the *orbital quantum number*. It and an integer $m$, $|m| \leqslant l$, determine the angular dependence of the wave function $\Psi(r, \theta, \phi) = R(r)Z_l^m(\theta, \phi)$. When $l = 0$, then $m = 0$ and $Z_0^0(\theta, \phi) = P_0(\cos \theta) = 1$. Thus, in the case $l = 0$, and only in this case, the wave function $\Psi$ is spherically symmetric—depends only on the radius $r$. The eigenfunctions (7.9.25) correspond to *eigenstates* of the hydrogen atom (refer to Section 5.8). The $l = 0$ states, which are the spherically symmetric ones, are called $s$ *states*. Those with $l = 1$, $p$ *states*; $l = 2$, $d$ *states*, etc. Thus, for a fixed total quantum number $n$ (and thus for a fixed total energy $E$), the electron would have one possible $s$ state, three possible $p$ states, five possible $d$ states, etc. up to $2(n-1) + 1 = 2n - 1$ possible states corresponding to the largest value of $l$, $l = n - 1$.

The integer $v = n - (l + 1)$ is sometimes called the *radial quantum number*. When $v = 0$, the radial part (7.9.24) of the eigenfunctions (7.9.25) is never zero for $s > 0$. When $v > 0$, this radial part does have zeros, or "nodes." The probability density, $|\Psi|^2$, of finding the electron at those particular radii from the proton is zero.

In Table 7.3, I have listed the (nonnormalized) eigenfunctions for the hydrogen atom for $n = 1, 2$, and 3.

To complete the story, we must normalize the eigenfunctions $\Psi_n^{l,m}$ by multiplying each by an appropriate constant $c$ (dependent on $n$, $l$, and $m$) so that $\|c\Psi_n^{l,m}\| = 1$. When that is done, the quantity $|c\Psi_n^{l,m}|^2$ will be the probability that an electron, in the state determined by $\Psi_n^{l,m}$, will lie in the differential volume element $dV = r^2 \sin \theta \, dr \, d\theta \, d\phi$. The constant $c$ is $\|\Psi_n^{l,m}\|^{-1}$. From (7.9.25) we have

$$\|\Psi_n^{l,m}\|^2 = \int_0^\infty r^2 \, dr \int_0^\pi d\theta \sin \theta \int_{-\pi}^\pi d\phi \Psi_n^{l,m} \overline{\Psi_n^{l,m}}$$

$$= \int_0^\infty r^2 \, dr \int_0^\pi d\theta \sin \theta \int_{-\pi}^\pi d\phi \, s^{2l} e^{-s/n} \left( L_v^{(\alpha)} \left( \frac{s}{n} \right) \right)^2 \left( P_l^{|m|}(\cos \theta) \right)^2$$

$$= 2\pi \left[ \int_0^\infty r^2 \, dr \, (s^{2l}) e^{-s/n} \left( L_v^{(\alpha)} \left( \frac{s}{n} \right) \right)^2 \right] \left[ \int_0^\pi d\theta \sin \theta (P_l^{|m|}(\cos \theta))^2 \right]$$

$$\|\Psi_n^{l,m}\|^2 = \frac{4\pi(l + |m|)!}{(2l+1)(l - |m|)!} \int_0^\infty r^2 \, dr \, (s^{2l}) e^{-s/n} \left( L_v^{(\alpha)} \left( \frac{s}{n} \right) \right)^2 \tag{7.9.26}$$

**Table 7.3** The Nonnormalized Eigenfunctions $\Psi_n^{l,m}$ of the Hydrogen Atom for $n=1,2,3^a$

| $n$ | $l$ | $m$ | | |
|---|---|---|---|---|
| 1 | 0 | 0 | $(v=0)$ | $\Psi_1^{0,0} = e^{-s/2}$ |
| 2 | 0 | 0 | $(v=1)$ | $\Psi_2^{0,0} = e^{-s/4}(2-s/2)$ |
| | | $-1$ | | $\Psi_2^{1,-1} = se^{-s/4}\sin\theta\, e^{-i\phi}$ |
| | 1 | 0 | $(v=0)$ | $\Psi_2^{1,0} = se^{-s/4}\cos\theta$ |
| | | 1 | | $\Psi_2^{1,1} = se^{-s/4}\sin\theta\, e^{i\phi}$ |
| 3 | 0 | 0 | $(v=2)$ | $\Psi_3^{0,0} = e^{-s/6}(3-s+s^2/18)$ |
| | | $-1$ | | $\Psi_3^{1,-1} = se^{-s/6}(4-s/3)\sin\theta\, e^{-i\phi}$ |
| | 1 | 0 | $(v=1)$ | $\Psi_3^{1,0} = se^{-s/6}(4-s/3)\cos\theta$ |
| | | 1 | | $\Psi_3^{1,1} = se^{-s/6}(4-s/3)\sin\theta\, e^{i\phi}$ |
| | | $-2$ | | $\Psi_3^{2,-2} = s^2e^{-s/6}(3/2)(1-\cos 2\theta)e^{-2i\phi}$ |
| | | $-1$ | | $\Psi_3^{2,-1} = s^2e^{-s/6}(3/2)\sin 2\theta\, e^{-i\phi}$ |
| | 2 | 0 | $(v=0)$ | $\Psi_3^{2,0} = s^2e^{-s/6}(1/4)(3\cos 2\theta+1)$ |
| | | 1 | | $\Psi_3^{2,1} = s^2e^{-s/6}(3/2)\sin 2\theta\, e^{i\phi}$ |
| | | 2 | | $\Psi_3^{2,2} = s^2e^{-s/6}(3/2)(1-\cos 2\theta)e^{2i\phi}$ |

$^a$The integer $n$ is the total quantum number; the corresponding total energy of the electron is $E_n = -me^4/2\hbar^2n^2$. The integer $l$ is the orbital quantum number. $s$ is the scaled radius: $s=2r/a$ where $a=\hbar^2/me^2$ is the Bohr radius. $v=n-(l+1)$.

where $\alpha=2l+1$, $v=n-(l+1)$, $s=2r/a$, and $a=\hbar^2/me^2$ is the Bohr radius. In the last step we have used Theorem 7.27 of Section 7.8. (Be careful to distinguish between the two ways we have used the symbol $m$; it stands for the mass of the electron and for one of the indices in spherical harmonics.) In the last integral, go to the independent variable $x=s/n=(2/an)r$, so $r=(an/2)x$ and $s=nx$. In terms of $x$, that integral becomes

$$\frac{a^3 n^{2l+3}}{8}\int_0^\infty (xL_v^{(\alpha)}(x))(L_v^{(\alpha)}(x))x^\alpha e^{-x}\,dx \qquad (7.9.27)$$

This integral is the inner product of $xL_v^{(\alpha)}$ and $L_v^{(\alpha)}$ in the Hilbert space $L^2(0,\infty;x^\alpha e^{-x})$. Its value is given in Exercise 25 of Section 7.10. Putting that value into (7.9.26) and remembering that $\alpha=2l+1$ and $v=n-(l+1)$, we find that (7.9.27) equals

$$\frac{a^3 n^{2l+4}}{4}\frac{(n+l)!}{(n-(l+1))!} \qquad (7.9.28)$$

Combine (7.9.28) with (7.9.26) to get

$$\|\Psi_n^{l,m}\|^2 = \frac{\pi a^3 n^{2l+4}}{(2l+1)}\frac{(l+|m|)!}{(l-|m|)!}\frac{(n+l)!}{(n-(l+1))!} \qquad (7.9.29)$$

thus

$$c = \| \Psi_n^{l,m} \|^{-1} = \frac{1}{\sqrt{\pi} \, a^{3/2} n^{l+2}} \left[ \frac{(2l+1)(l-|m|)!(n-(l+1))!}{(l+|m|)!(n+l)!} \right]^{1/2} \qquad (7.9.30)$$

Finally, some comments on the multiplicity of the eigenvalues of Schrödinger's equation $H(\Psi) = E\Psi$. The eigenvalue $E_n = (1/n^2)E_1$, where $E_1 = -me^4/2\hbar^2$, has multiplicity $n^2$. In the language of quantum mechanics, the energy level $E_n$ has "degeneracy" $n^2$. Because all the $n^2$ independent eigenstates that belong to $E_n$ have the same energy, it is not possible to distinguish one of these states from another by energy measurements alone.

This high degree of degeneracy can be traced to the underlying geometric symmetry of the problem. We saw a similar thing happen in the case of the eigenvalues of the square (Section 7.5), where we found that the multiplicity of the square's eigenvalues, $\lambda_{mn} = (m^2 + n^2)\pi^2$, did become arbitrarily large. In the case of the square, we showed that when we perturbed the square ever so slightly by changing its sides by a *very* small amount but keeping its area the same, then each multiple eigenvalue, of multiplicity $n$ say, split into $n$ simple eigenvalues which were closely spaced. We might expect the same thing to happen to the energy levels of the hydrogen atom if we were somehow to introduce an asymmetry.

A weak magnetic field or a weak electric field applied to the atom does produce such an asymmetry. These fields create a small force on the electron that alters the central symmetry of the Coulomb force attracting the electron to the massive proton. Indeed, in the presence of such external electric and magnetic fields each degenerate energy level does split into closely spaced levels. This splitting can be observed by the spectroscopic techniques described in Section 5.1. The splitting of the spectroscopic lines in a magnetic field is called the Zeeman effect, that in an electric field the Stark effect.

Schrödinger, in the third of his series of papers "Quantization as a problem of eigenvalues," used perturbation methods to compute the quantitative results of the Stark effect, and his numbers agreed very well with the experiment. A fully satisfactory quantitative analysis of the Zeeman effect had to wait for a proper understanding of the fact that the electron itself had a magnetic moment, or "spin," a fact that was discovered at about the same time that Schrödinger was creating his theory.

### Exercises

The general Stokes theorem,

$$\int_{\partial D} \omega = \int_D d\omega,$$

(see equation (7.2.2) of Section 7.2) is valid under the following interpretation of its symbols: $\omega$ is a 2-form; $d\omega$ its exterior derivative, which is a 3-form; $D$ a 3-dimensional region; and $\partial D$ its boundary, which is a closed surface or a collection of closed surfaces.

The corresponding formula for integration by parts:

$$\int_D f \, d\omega = \int_{\partial D} f\omega - \int_D (df)\omega$$

(see equation (7.2.5) of Section 7.2) is valid with the same interpretation of its symbols. Here $f = f(x, y, z)$ is an ordinary function on $\mathbf{R}^3$.

1. (a) Show that the differential expression $l$ for the energy operator $H$ of the idealized hydrogen atom is

$$l(\Psi) \equiv \frac{-\hbar^2}{2m}\left(\frac{\partial^2 \Psi}{\partial x^2} + \frac{\partial^2 \Psi}{\partial y^2} + \frac{\partial^2 \Psi}{\partial z^2}\right) - \frac{e^2 \Psi}{\sqrt{x^2 + y^2 + z^2}}$$

in Cartesian coordinates.

(b) Let $\omega = -(\partial\Psi/\partial x)\, dy\, dz - (\partial\Psi/\partial y)\, dz\, dx - (\partial\Psi/\partial z)\, dx\, dy$ (a 2-form). Show that

$$l(\Psi)\, dx\, dy\, dz = \frac{\hbar^2}{2m} d\omega - \frac{e^2 \Psi}{\sqrt{x^2 + y^2 + z^2}}\, dx\, dy\, dz$$

2. (Continuation) (a) Choose $0 < \varepsilon < R < \infty$ and let $D = \{(x, y, z): \varepsilon \leqslant |x| \leqslant R,\ \varepsilon \leqslant |y| \leqslant R,\ \text{and } \varepsilon \leqslant |z| \leqslant R\}$. Describe and sketch the 3-dimensional region $D$. Also describe and sketch its boundary $\partial D$, which is a pair of closed surfaces.

(b) Show that, for any pair of functions $f, g$ that are compatible with the differential expression $l$ of Exercise 1, we have

$$\int_D l(f)\bar{g}\, dx\, dy\, dz = \frac{\hbar^2}{2m}\int_D \left(\frac{\partial f}{\partial x}\frac{\partial \bar{g}}{\partial x} + \frac{\partial f}{\partial y}\frac{\partial \bar{g}}{\partial y} + \frac{\partial f}{\partial z}\frac{\partial \bar{g}}{\partial z}\right) dx\, dy\, dz$$

$$- e^2 \int_D \frac{f\bar{g}}{\sqrt{x^2 + y^2 + z^2}}\, dx\, dy\, dz$$

$$- \frac{\hbar^2}{2m}\int_{\partial D} \bar{g}\left(\frac{\partial f}{\partial x}\, dy\, dz + \frac{\partial f}{\partial y}\, dz\, dx + \frac{\partial f}{\partial z}\, dx\, dy\right)$$

[Hint: Recall that compatibility means that both $f$ and $l(f)$ belong to $L^2(\mathbf{R}^3)$. This implies that both $f$ and its first partial derivatives are continuous at every point of $\mathbf{R}^3\backslash(0, 0, 0)$.]

3. (Continuation) Let $H_1$ denote the operator in $L^2(\mathbf{R}^3)$ which has the differential expression $l$ of Exercise 1 and has for domain the subspace of $L^2(\mathbf{R}^3)$ consisting of functions that are compatible with $l$, have

$f(x, y, z)/(x^2 + y^2 + z^2)^{1/4} \in L^2(\mathbf{R}^3)$, and for which the integral over $\partial D$ in Exercise 2b tends to zero as $\varepsilon \to 0$, $R \to \infty$. Prove that

$$\langle H_1(f), g \rangle = \frac{\hbar^2}{2m} \left[ \left\langle \frac{\partial f}{\partial x}, \frac{\partial g}{\partial x} \right\rangle + \left\langle \frac{\partial f}{\partial y}, \frac{\partial g}{\partial y} \right\rangle + \left\langle \frac{\partial f}{\partial z}, \frac{\partial g}{\partial z} \right\rangle \right]$$

$$- e^2 \left\langle \frac{f}{(x^2 + y^2 + z^2)^{1/4}}, \frac{g}{(x^x + y^2 + z^2)^{1/4}} \right\rangle$$

for any $f$, $g$ in the domain of $H_1$. Using this formula, prove that $H_1$ is Hermitian. [Note: The inner product here is that of $L^2(\mathbf{R}^3)$.]

4. Prove: If $\Psi(r, \theta, \phi) = R(r)F(\theta, \phi)$ belong to $L^2(\mathbf{R}^3)$ and if $\|\Psi\| \neq 0$, then $R(r)$ belongs to $L^2(0, \infty; r^2)$ and $F(\theta, \phi)$ belongs to $L^2(S)$ where $S$ is the unit sphere. Prove also the converse; if $R(r) \in L^2(0, \infty; r^2)$ and $F(\theta, \phi) \in L^2(S)$, then $R(r)F(\theta, \phi) \in L^2(\mathbf{R}^3)$.

5. Show that, under the substitution $R = h(s)y$, the eigenvalue problem

$$\frac{1}{r}\frac{d}{ds}\left( p \frac{dR}{ds} \right) + \frac{q}{r} R = \lambda R \qquad \text{in } L^2(0, \infty; r)$$

becomes the eigenvalue problem

$$\frac{1}{h^2 r}\frac{d}{ds}\left( h^2 p \frac{dy}{ds} \right) + \frac{1}{hr}\frac{d}{ds}\left( p \frac{dh}{ds} \right) y + \frac{q}{r} y = \lambda y \qquad \text{in } L^2(0, \infty; h^2 r)$$

6. (Continuation) Using Exercise 5 with $h(s) = s^k e^{-\beta s}$, derive equation (7.9.15) (with (7.9.14)) from equation (7.9.11) (with (7.9.12)).

7. Using the format of Table 7.3, construct your own table of the non-normalized eigenfunctions $\Psi_n^{l, m}$ for $n = 4$. (There are 16 entries.)


# 7.10  PROJECT ON LAGUERRE'S OPERATOR AND LAGUERRE POLYNOMIALS

Laguerre's operator is a linear second order differential operator in Hilbert space which has as its eigenfunctions a set of orthogonal polynomials. The theory of Laguerre's operator and the associated Laguerre polynomials parallels the theory of Legendre's operator and the Legendre polynomials (Sections 4.6 through 4.8) and that of Hermite's operator and the Hermite polynomials (Sections 4.9 through 4.11). Inasmuch as we have explained the latter two theories in fine detail, it would seem reasonable to ask you to develop the Laguerre theory by yourself, using the previous expositions as a guide. The following sequence of exercises leads you through that development. The material in Chapter 4 should help you solve the exercises. Do them in the order given. In addition to the exercises, I have included relevant definitions and the statements of some facts which you may take as known.

Fix a real number $\alpha$, $\alpha \geqslant 0$. In this section, we shall be concerned with the Hilbert space $L^2(0, \infty; x^\alpha e^{-x})$ consting of all (real- or) complex-valued functions $f(x)$ for which

$$\int_0^\infty |f(x)|^2 x^\alpha e^{-x} \, dx < \infty$$

The inner product in $L^2(0, \infty; x^\alpha e^{-x})$ is

$$\langle f, g \rangle = \int_0^\infty f(x)\overline{g(x)} x^\alpha e^{-x} \, dx$$

We shall use this inner product exclusively in this section. Bear this in mind, as we shall not keep repeating it.

**Exercise 1.** (a) Show that the Hilbert space $L^2(0, \infty; x^\alpha e^{-x})$ contains all powers $x^n$, $n = 0, 1, 2, \ldots$ (Compare Exercise 6 of Section 3.9.)

(b) Using your result in (a), without any further calculations, show that the Hilbert space $L^2(0, \infty; x^\alpha e^{-x})$ contains all polynomials.

(c) Does this Hilbert space contain $e^{\beta x}$ for any positive $\beta$? If so, which $\beta$?

**Definition 7.29.** Fix a real number $\alpha$, $\alpha \geqslant 0$. Consider the Hilbert space $L^2(0, \infty; x^\alpha e^{-x})$ with inner product

$$\langle f, g \rangle = \int_0^\infty f(x)\overline{g(x)} x^\alpha e^{-x} \, dx$$

*Laguerre's operator*, $A_\alpha$, has the differential expression

$$l_\alpha(y) \equiv \frac{-1}{x^\alpha e^{-x}} \frac{d}{dx}\left(x^{\alpha+1} e^{-x} \frac{dy}{dx}\right) = -x \frac{d^2 y}{dx^2} - (\alpha + 1 - x)\frac{dy}{dx}$$

and domain consisting of functions $y = y(x)$ that are compatible with $l_\alpha$ and satisfy the boundary condition that $\lim_{x \to 0} x^{\alpha/2} y(x)$ is finite.

**Exercise 2.** (a) Explain what it means to say that $y(x)$ is compatible with Laguerre's differential expression $l_\alpha$.

(b) Show that, for every $\alpha \geqslant 0$, the domain of Laguerre's operator $A_\alpha$ contains all polynomials.

**Theorem 7.30.** *For any pair of functions $f(x)$ and $g(x)$ in the domain of Laguerre's operator $A_\alpha$ we have*

$$\lim_{x \to 0, \infty} x^{\alpha+1} e^{-x} f'(x)\overline{g(x)} = 0$$

*and*

$$\sqrt{x}f'(x) \in L^2(0, \infty; x^\alpha e^{-x})$$

(This theorem is proved in the appendix to this chapter. You are not responsible for its proof, but you will need the result in the following exercises.)

**Exercise 3.** (a) Let $0 < r < s < \infty$. Show that, for all $f, g$ in the domain of Laguerre's operator $A_\alpha$, we have

$$\int_r^s l_\alpha(f)\bar{g}x^\alpha e^{-x}\,dx = r^{\alpha+1}e^{-r}f'(r)\overline{g(r)} - s^{\alpha+1}e^{-s}f'(s)\overline{g(s)}$$

$$+ \int_r^s f'(x)\overline{g'(x)}x^{\alpha+1}e^{-x}\,dx$$

(b) Using your result in (a), together with Theorem 7.30, prove that

$$\langle A_\alpha(f), g\rangle = \langle\sqrt{x}f'(x), \sqrt{x}g'(x)\rangle$$

for all $f, g$ in the domain of $A_\alpha$.

(c) Using your result in (b), prove that Laguerre's operator is Hermitian positive semidefinite.

**Exercise 4.** Prove: The eigenvalues of Laguerre's operator are all $\geq 0$, and eigenfunctions belonging to different eigenvalues are orthogonal.

**Exercise 5.** Let $f(x) = \int_1^x (e^t/t)\,dt$, $x > 0$. Check that $f(x)$ satisfies $l_0(f) = -e^x(xe^{-x}f')' = 0$, thus is an eigenfunction belonging to the eigenvalue $\lambda = 0$. Check that $g(x) = x - 1$ satisfies $l_0(g) = g$, thus is an eigenfunction belonging to the eigenvalue $\lambda = 1$. As $0 \neq 1$, these eigenfunctions belong to different eigenvalues, so they must be orthogonal; i.e.,

$$\langle f(x), g(x)\rangle = \int_0^\infty f(x)(x - 1)e^{-x}\,dx = 0$$

But $f(x)(x-1)$ is positive for all $x > 0$ except at $x = 1$, so this integral cannot be zero! What is wrong?

**Exercise 6.** Show that if *Laguerre's differential equation*

$$x\frac{d^2y}{dx^2} + (\alpha + 1 - x)\frac{dy}{dx} + vy = 0$$

has a power series solution $y = \sum_{k=0}^{\infty} a_k x^k$, then the coefficients $a_k$, $k \geqslant 1$, are uniquely determined by $a_0$:

$$a_k = \frac{v(v-1)(v-2)\cdots(v-k+1)}{(\alpha+k)(\alpha+k-1)\cdots(\alpha+1)} \frac{(-1)^k}{k!} a_0, \qquad k = 1, 2, 3, \ldots$$

Show that this series converges for all values of $x$.

**Exercise 7.** Show that if $v = n$, a nonnegative integer, then Laguerre's differential equation (see Exercise 6) has a polynomial solution of degree $n$. Show also conversely that if Laguerre's differential equation has a polynomial solution of degree $n$, $n$ a nonnegative integer, then $v = n$. Show that, with appropriate normalization, the degree $n$ polynomial solution to Laguerre's equation may be written

$$L_n^{(\alpha)}(x) = \sum_{k=0}^{n} \binom{n+\alpha}{n-k} \frac{(-x)^k}{k!} = \frac{(n+\alpha)(n+\alpha-1)\cdots(\alpha+1)}{n!} - \cdots + (-1)^n \frac{x^n}{n!}$$

These are the *Laguerre polynomials*. List the first five ($L_0^{(\alpha)}$ through $L_4^{(\alpha)}$) for general $\alpha$ and for $\alpha = 0$ (write $L_n$ for $L_n^{(0)}$). Here $\binom{\lambda}{k}$ is the generalized binomial coefficient

$$\binom{\lambda}{k} = \begin{cases} 1 & \text{when } k = 0 \\ \dfrac{\lambda(\lambda-1)(\lambda-2)\cdots(\lambda-k+1)}{k!} & \text{when } k \geqslant 1 \end{cases}$$

$k$ a nonnegative integer, $\lambda$ any real or complex number. The first five Laguerre polynomials are

$$L_0^{(\alpha)} = 1$$

$$L_1^{(\alpha)}(x) = (1+\alpha) - x$$

$$L_2^{(\alpha)}(x) = \frac{(2+\alpha)(1+\alpha)}{2!} - (2+\alpha)x + \frac{x^2}{2!}$$

$$L_3^{(\alpha)}(x) = \frac{(3+\alpha)(2+\alpha)(1+\alpha)}{3!} - \frac{(3+\alpha)(2+\alpha)}{2!}x + (3+\alpha)\frac{x^2}{2!} - \frac{x^3}{3!}$$

$$L_4^{(\alpha)}(x) = \frac{(4+\alpha)(3+\alpha)(2+\alpha)(1+\alpha)}{4!} - \frac{(4+\alpha)(3+\alpha)(2+\alpha)}{3!}x$$

$$+ \frac{(4+\alpha)(3+\alpha)}{2!}\frac{x^2}{2!} - (4+\alpha)\frac{x^3}{3!} + \frac{x^4}{4!}$$

For $\alpha = 0$:

$L_0(x) = 1$

$L_1(x) = 1 - x$

$L_2(x) = 1 - 2x + \dfrac{x^2}{2}$

$L_3(x) = 1 - 3x + \dfrac{3}{2}x^2 - \dfrac{x^3}{6}$

$L_4(x) = 1 - 4x + 3x^2 - \dfrac{2}{3}x^3 + \dfrac{x^4}{24}$

**Exercise 8.** Prove that Laguerre's operator $A_\alpha$ has the eigenvalues $\lambda = n$, $n = 0, 1, 2, \ldots$. Prove that the Laguerre polynomial $L_n^{(\alpha)}(x)$ is an eigenfunction belonging to the eigenvalue $\lambda = n$; i.e.,

$$A_\alpha(L_n^{(\alpha)}) = nL_n^{(\alpha)}$$

**Exercise 9.** Prove that

$$\int_0^\infty L_m^{(\alpha)}(x)L_n^{(\alpha)}(x)x^\alpha e^{-x}\,dx = 0 \qquad \text{when } m \neq n$$

**Exercise 10.** Let $f(x) \in L^2(0, \infty; x^\alpha e^{-x})$. For what values of the constants $c_0, c_1, \ldots, c_n$ is the integral

$$\int_0^\infty \left| f(x) - \sum_{k=0}^n c_k L_k^{(\alpha)}(x) \right|^2 x^\alpha e^{-x}\,dx$$

an absolute minimum? [Ans.: $c_k = \langle f, L_k^{(\alpha)} \rangle / \langle L_k^{(\alpha)}, L_k^{(\alpha)} \rangle$.]

**Exercise 11.** Prove that any polynomial of degree $n$ is a linear combination of the first $n+1$ Laguerre polynomials $L_0^{(\alpha)}, L_1^{(\alpha)}, \ldots, L_n^{(\alpha)}$.

**Exercise 12.** Prove: If $Q(x)$ is a polynomial of degree $m$, then $\langle Q, L_n^{(\alpha)} \rangle = 0$ for $n > m$.

**Exercise 13.** Prove that the Laguerre polynomials are uniquely determined up to scalar multiples by their orthogonality properties. That is, prove the following result: If $R_n$, $n = 0, 1, 2, \ldots$, is a sequence of polynomials such that (1) $R_n$ is of degree $n$ for each $n = 0, 1, 2, \ldots$, and (2) $\langle R_m, R_n \rangle = 0$ when $m \neq n$, then $R_n = \mu_n L_n^{(\alpha)}$, $n = 0, 1, 2, \ldots$, where the $\mu_n$ are nonzero scalars.

**Theorem 7.31.**  *Fix a real number* $\alpha$, $\alpha \geqslant 0$. *The Laguerre polynomials* $L_n^{(\alpha)}(x)$, $n=0, 1, 2, \ldots$, *form an orthogonal basis of* $L^2(0, \infty; x^\alpha e^{-x})$.

(This theorem is proved in J. R. Higgins's book, *Completeness and Basis Properties of Sets of Special Functions*, §2.6.4. You are not responsible for its proof, but you will need the result in the following exercises.)

**Exercise 14.**  Explain what it means to say that the Laguerre polynomials $L_n^{(\alpha)}(x)$, $n=0,1,2,\ldots$, form an orthogonal basis of the Hilbert space $L^2(0, \infty; x^\alpha e^{-x})$.

**Exercise 15.**  Prove: If the function $f(x)$ belongs to $L^2(0, \infty; x^\alpha e^{-x})$ and is orthogonal to every $L_n^{(\alpha)}(x)$, $n=0,1,2,\ldots$, then $f=0$.

**Exercise 16.**  Prove that Laguerre's operator $A_\alpha$ has no eigenvalues other than $\lambda = n$, $n=0,1,2,\ldots$, and that each of these eigenvalues is simple.

**Exercise 17.**  Suppose $\nu \neq n$ for any $n=0,1,2,\ldots$. Is it possible to find a solution $y(x)$ to Laguerre's differential equation $xy'' + (\alpha+1-x)y' + \nu y = 0$ that satisfies the following two conditions?

$$\int_0^\infty |y(x)|^2 x^\alpha e^{-x}\,dx < \infty \qquad \text{and} \qquad \lim_{x \to 0} x^{\alpha/2} y(x) \text{ is finite}$$

Explain why or why not.

**Exercise 18.**  Using the formula

$$\frac{d^l}{dx^l} x^{n+\alpha} = (n+\alpha)(n-1+\alpha) \cdots (n-l+1+\alpha)x^{n-l+\alpha}$$

and Leibniz's formula

$$\frac{d^n}{dx^n} fg = \sum_{k=0}^n \binom{n}{k} \frac{d^{n-k}f}{dx^{n-k}} \frac{d^k g}{dx^k}$$

prove *Rodrigues' formula for the Laguerre polynomials*:

$$L_n^{(\alpha)}(x) = \frac{x^{-\alpha}e^x}{n!} \frac{d^n}{dx^n}(x^{n+\alpha}e^{-x})$$

The gamma function, $\Gamma(x)$, is defined by the integral

$$\Gamma(x) = \int_0^\infty t^{x-1} e^{-t}\,dt$$

which is convergent for $x > 0$. In Exercise 1 of Section 6.10 the following facts were established: $\Gamma(1) = 1$, $\Gamma(x+1) = x\Gamma(x)$, and $\Gamma(n+1) = n!$ for every integer $n = 0, 1, 2, \ldots$. You may take these properties for granted here.

**Exercise 19.** Using the "functional equation" of the gamma function, $\Gamma(x+1) = x\Gamma(x)$ (*not* the integral), prove that

$$\Gamma(n + x + 1) = (n + x)(n - 1 + x) \cdots (1 + x)\Gamma(1 + x)$$

for any integer $n \geqslant 0$.

**Exercise 20.** Using Rodrigues' formula (Exercise 18) prove that

$$\langle x^n, L_n^{(\alpha)} \rangle = (-1)^n(n+\alpha)(n-1+\alpha) \cdots (1+\alpha)\Gamma(1+\alpha)$$

$$= (-1)^n n! \binom{n+\alpha}{n} \Gamma(1+\alpha)$$

Using this result, prove that

$$\langle L_n^{(\alpha)}, xL_{n-1}^{(\alpha)} \rangle = -n \binom{n+\alpha}{n} \Gamma(1 + \alpha)$$

and that

$$\langle L_n^{(\alpha)}, L_n^{(\alpha)} \rangle = \binom{n + \alpha}{n} \Gamma(1 + \alpha)$$

Note that this last fact implies that $\langle L_n, L_n \rangle = 1$. (We write $L_n$ for $L_n^{(0)}$.)

**Exercise 21.** Prove that $nL_n^{(\alpha)} + xL_{n-1}^{(\alpha)}$ is a polynomial of degree $\leqslant n-1$, so that

$$nL_n^{(\alpha)} + xL_{n-1}^{(\alpha)} = \sum_{k=0}^{n-1} c_k L_k^{(\alpha)}$$

(see Exercise 11). Prove that $c_k = 0$, $0 \leqslant k \leqslant n-3$. Determine $c_{n-2}$ and $c_{n-1}$ and thus prove the *recursion formula for Laguerre's polynomials*

$$nL_n^{(\alpha)} = (2n + \alpha - 1 - x)L_{n-1}^{(\alpha)} - (n - 1 + \alpha)L_{n-2}^{(\alpha)}, \qquad n \geqslant 2$$

Put $\alpha = 0$, $L_0 = 1$, $L_1 = 1 - x$ and use the recursion formula to compute $L_2$, $L_3$, and $L_4$ (write $L_n$ for $L_n^{(0)}$).

**Exercise 22.** Using the recursion relation for Laguerre polynomials (Exercise 21), show that the function

$$h(x, w) = \sum_{n=0}^{\infty} L_n^{(\alpha)}(x)w^n$$

regarded as a function of $w$ with $x$ held fixed, satisfies the differential equation

$$\frac{dh}{h} = \left(\frac{1+\alpha}{1-w} - \frac{x}{(1-w)^2}\right)dw$$

Solve this equation to show that $h = A/(1-w)^{1+\alpha}\exp(-x/(1-w))$, where $A$ is a function of $x$ alone. Determine $A$, and *deduce the generating function for the Laguerre polynomials*:

$$\frac{1}{(1-w)^{1+\alpha}}\exp\left(-\frac{xw}{1-w}\right) = \sum_{n=0}^{\infty} L_n^{(\alpha)}(x)w^n, \qquad |w| < 1$$

**Exercise 23.** From the generating function of the Laguerre polynomials (Exercise 22) deduce that

$$\int_0^{\infty} x^{\alpha}\exp\left(\frac{-x}{1-w}\right)L_n^{(\alpha)}(x)\,dx = \binom{n+\alpha}{n}\Gamma(1+\alpha)w^n(1-w)^{1+\alpha}$$

and that

$$\int_0^{\infty}\exp\left(-\frac{x}{1-w}\right)L_n(x)\,dx = w^n(1-w)$$

**Exercise 24.** From the generating function of the Laguerre polynomials (Exercise 22) deduce that

$$e^{x/3} = \frac{3}{2}\sum_{n=0}^{\infty}\frac{(-1)^n}{2^n}L_n(x)$$

On three-cycle semilog graph paper, compare $e^{x/3}$, the five-term partial sum of its Laguerre series ($L_0$ through $L_4$), and the five-term partial sum of its power series ($x^0$ through $x^4$) for $0 \le x \le 15$.

**Exercise 25.** Prove that

$$\langle xL_n^{(\alpha)}, L_n^{(\alpha)}\rangle = (2n + \alpha + 1)\binom{n+\alpha}{n}\Gamma(1+\alpha)$$

[Hint: Combine the recursion formula with the normalization from Exercise 20.]

## 7.11 LAPLACE'S EQUATION AND HARMONIC POLYNOMIALS

The kernel of a second order linear ordinary differential expression $l(y) \equiv a(x)y'' + b(x)y' + c(x)y$ is 2-dimensional. Put another way, the homogeneous equation $l(y) = 0$ always has two linearly independent solutions, say $f_1(x)$ and $f_2(x)$, and any solution of $l(y) = 0$ is a linear combination $y = c_1 f_1 + c_2 f_2$ for suitable constants $c_1$ and $c_2$. This fact, originally stated as Corollary 2.10 in Section 2.3, was crucial in our study of second order ordinary linear differential equations and in our theory of ordinary linear differential operators in Hilbert space. It is therefore worth noting that nothing like this holds for linear *partial* differential expressions. Their kernels are generally infinite-dimensional.

We shall establish this fact for the simplest second order linear partial differential expression, the Laplacian. Because in the present situation we shall be working only with the differential expression, not with any operator it might determine, we shall for the moment denote the Laplacian differential expression by $\Delta$, following common usage. When we revert to the study of the Laplacian operator we shall return to our policy of distinguishing the expression from the operator by using different symbols for each ($\delta$ and $\Delta$).

The kernel of the Laplacian differential expression consists of all solutions $u$ of the partial differential equation $\Delta u = 0$. This is called *Laplace's equation*. Functions $u$ that satisfy it are called *harmonic*. The proof that the solution space of Laplace's equation is infinite-dimensional, or, equivalently, that the space of harmonic functions is infinite-dimensional, is contained in Exercises 1 through 12. These exercises are designed to impress you with the enormous abundance and variety of harmonic functions, in both two variables and three variables. These exercises show explicitly the dramatic difference between linear second order ordinary (one variable) differential expressions, whose kernels are always two-dimensional, and linear second order partial (two or more variable) differential expressions, whose kernels are generally infinite-dimensional.

Let us return now to the study of the Laplacian *operator* $\Delta$, specifically the Laplacian operator on $L^2(S)$, where $S$ is the unit sphere. In Sections 7.7 and 7.8 we proved that $\Delta$ is a Hermitian operator on the Hilbert space $L^2(S)$; found its eigenvalues, $n(n+1)$, $n = 0, 1, 2, \ldots$; and showed that each eigenvalue had multiplicity $2n+1$. Hence, for each fixed nonnegative integer $n$, the space of solutions $f$ to the equation $\Delta(f) = n(n+1)f$ has dimension $2n+1$. This eigenspace, which is a subspace of the domain of $\Delta$ (which is in turn a subspace of the Hilbert space $L^2(S)$), is spanned by the $2n+1$ spherical harmonics (see (7.8.10) of Section 7.8)

$$\tfrac{1}{2}P_n(\cos\theta), \cos\phi P_n^1(\cos\theta), \cos 2\phi\, P_n^2(\cos\theta), \ldots, \cos n\phi\, P_n^n(\cos\theta)$$
$$\sin\phi\, P_n^1(\cos\theta), \quad \sin 2\phi\, P_n^2(\cos\theta), \ldots, \sin n\phi\, P_n^n(\cos\theta)$$

$$(7.11.1)$$

(In this section we shall use these real-valued spherical harmonics in preference to the complex-valued ones that were handy for the hydrogen atom (Section 7.9).)

**Definition 7.32.** Let $\Delta$ be the Laplacian operator on the Hilbert space $L^2(S)$, $S$ the unit sphere. We denote by $H(n)$, $n = 0, 1, 2, \ldots$, the eigenspace of $\Delta$ belonging to the eigenvalue $n(n + 1)$.

Any nonzero function in $H(n)$ is an eigenfunction of the Laplacian $\Delta$ belonging to the eigenvalue $n(n + 1)$. Put another way, if $f = f(\theta, \phi)$ is any nonzero linear combination of the functions in (7.11.1), then $f \in \text{domain}(\Delta)$ and $\Delta(f) = n(n + 1)f$.

We say that the subspace $H(n)$ is *invariant* for the operator $\Delta$ because every $H(n)$ is mapped onto itself by $\Delta$. In fact, every $f \in H(n)$ is mapped by $\Delta$ to the multiple $n(n + 1)$ of itself.

When $k \neq n$, the spaces $H(k)$ and $H(n)$ are mutually orthogonal in $L^2(S)$ because eigenspaces belonging to different eigenvalues of a Hermitian operator are orthogonal. To the orthogonal expansion of a general $f(\theta, \phi) \in L^2(S)$ in terms of spherical harmonics ((7.8.11) of Section 7.8),

$$f(\theta, \phi) = \tfrac{1}{2}a_{00} + \sum_{n=1}^{\infty} \left( \tfrac{1}{2}a_{0n}P_n(\cos\theta) + \sum_{m=1}^{n} (a_{mn}\cos m\phi + b_{mn}\sin m\phi)P_n^m(\cos\theta) \right)$$

$$(7.11.2)$$

we can give a new interpretation: each $f(\theta, \phi) \in L^2(S)$ admits an expansion

$$f = \sum_{n=0}^{\infty} f_n \qquad \text{where } f_n \in H(n), \ n = 0, 1, 2, \ldots \qquad (7.11.3)$$

i.e., we can express every square-integrable function on the unit sphere $S$ as an (infinite) orthogonal sum of functions $f_n$, one from each of the orthogonal subspaces $H(n)$. The $f_n$ are called the *projections* of $f$ on $H(n)$. They are given by

$$f_0 = \tfrac{1}{2}a_{00}$$

$$(7.11.4)$$

$$f_n = \tfrac{1}{2}a_{0n}P_n(\cos\theta) + \sum_{m=1}^{n} (a_{mn}\cos m\phi + b_{mn}\sin m\phi)P_n^m(\cos\theta), \qquad n \geq 1$$

with the coefficients given by

$$\left.\begin{array}{c} a_{mn} \\ b_{mn} \end{array}\right\} = \frac{2n+1}{2}\frac{(n-m)!}{(n+m)!} \int_{-\pi}^{\pi} d\phi \int_{0}^{\pi} d\theta \sin\theta f(\theta, \phi)\substack{\cos \\ \sin} m\phi \, P_n^m(\cos\theta), \quad 0 \leqslant m \leqslant n$$

$$(7.11.5)$$

(see (7.8.15) of Section 7.8).

We may use the orthogonal invariant subspaces $H(n)$ to gain a new perspective regarding the action of the Laplacian. Given a function $f$ in the domain of $\Delta$, we may apply $\Delta$ term by term to the expansion (7.11.3), thus

$$\Delta f = \Delta\left(\sum_{n=0}^{\infty} f_n\right) = \sum_{n=0}^{\infty} \Delta(f_n) = \sum_{n=0}^{\infty} n(n+1)f_n$$

Thus the subspaces $H(n)$ themselves come to the fore in this perspective. The $2n+1$ functions in (7.11.1) form a convenient orthogonal basis of $H(n)$. But, as *every* nonzero function in $H(n)$ is an eigenfunction belonging to the eigenvalue $n(n+1)$, other orthogonal bases of $H(n)$ could have served as well. We turn now to another way of generating the invariant subspaces $H(n)$.

By way of introduction, we list in Table 7.4 the first nine spherical harmonics in both spherical and Cartesian coordinates. (The evaluations of associated Legendre functions were taken from Exercise 6 of Section 7.8.) Take note in Table 7.4 of the particularly simple form of the spherical harmonics when expressed in Cartesian coordinates—they are just polynomials in the variables $x$, $y$, and $z$. It is our main purpose here to show that all the spherical harmonics can indeed be characterized as such special polynomials in the variables $x$, $y$, and $z$.

A polynomial in the variables $x$, $y$, and $z$ is a linear combination of "terms" like $x^2yz^3$—for example, the polynomial $6xy - 7xyz + 12x^2$. The *degree* of a term $x^a y^b z^c$ is $a+b+c$. (Here $a$, $b$, and $c$ are nonnegative integers.) Thus the terms $x^3$, $x^2y$, and $xyz$ each have degree 3, while $xy$, $y^2$, and $yz$ have degree 2.

**Table 7.4**  The First Nine Real Spherical Harmonics

$$Y_n^m(\theta, \phi) = \varepsilon_n \frac{\cos m\phi}{\sin m\phi} P_n^m(\cos\theta), \qquad \varepsilon_n = \begin{cases} \frac{1}{2} & n = 0 \\ 1 & n \geq 1 \end{cases}$$

Expressed in Both Spherical Coordinates $(\theta, \phi)$ and Cartesian Coordinates $(x, y, z)$, $x = \cos\phi \sin\theta$, $y = \sin\phi \sin\theta$, $z = \cos\theta^a$

| $n$ | $m$ | | Spherical | Cartesian |
|---|---|---|---|---|
| 0 | 0 | $Y_0^0$ | $\frac{1}{2}$ | $\frac{1}{2}$ |
| 1 | 0 | $Y_1^0$ | $\frac{1}{2}\cos\theta$ | $z/2$ |
| | 1 | $Y_1^1$ | $\begin{cases} \cos\phi \sin\theta \\ \sin\phi \sin\theta \end{cases}$ | $x$ |
| | | | | $y$ |
| 2 | 0 | $Y_2^0$ | $\frac{1}{4}(3\cos^2\theta - 1)$ | $\frac{1}{4}(2z^2 - x^2 - y^2)$ |
| | 1 | $Y_2^1$ | $\begin{cases} 3\cos\phi \cos\theta \sin\theta \\ 3\sin\phi \cos\theta \sin\theta \end{cases}$ | $3xz$ |
| | | | | $3yz$ |
| | 2 | $Y_2^2$ | $\begin{cases} 3\cos 2\phi \sin^2\theta \\ 3\sin 2\phi \sin^2\theta \end{cases}$ | $3(x^2 - y^2)$ |
| | | | | $6xy$ |

$^a$Note that $x^2 + y^2 + z^2 = 1$.

A polynomial is *homogeneous of degree n* if all its terms have degree $n$. For example, the polynomial $\frac{3}{2}x^3 - 2xyz + 4y^2z$ is homogeneous of degree 3, while $6xy - 7xyz + 12x^2$ is not homogeneous.

The set of all terms of a fixed degree $n$ consists of all $x^a y^b z^c$ where $a$, $b$, and $c$ are nonnegative integers such that $a + b + c = n$. There are $(n+1)(n+2)/2$ of these (Exercise 13), and they are linearly independent over the real or complex numbers (Exercise 14). Hence, as every homogeneous polynomial of degree $n$ is a linear combination of these terms, we see that *the set of homogeneous polynomials of degree n is a vector space of dimension $(n+1)(n+2)/2$.*

Look again at the spherical harmonics $Y_n^m$ listed in Table 7.4. Note that the polynomials in $x$, $y$, $z$ corresponding to harmonics having $n = 0$ are homogeneous of degree 0, those corresponding to $n = 1$ are homogeneous of degree 1, and those corresponding to $n = 2$ homogeneous of degree 2. This is the general situation.

**Theorem 7.33.** *Every homogeneous harmonic polynomial of degree n, when restricted to the unit sphere $S = \{(x, y, z); x^2 + y^2 + z^2 = 1\}$, is an eigenfunction of the spherical Laplacian belonging to the eigenvalue $n(n+1)$. Conversely, every nonzero function in $H(n)$ is the restriction to S of some homogeneous harmonic polynomial of degree n.*

*Proof.* Suppose that $p = p(x, y, z)$ is a homogeneous harmonic polynomial. Now, to say that $p$ is harmonic is to say that $\Delta_3(p) = 0$ where $\Delta_3$ is the three-dimensional Laplacian. And this must hold for any coordinate system for $\mathbf{R}^3$. Choose spherical coordinates $(r, \theta, \phi)$, and express $p(x, y, z)$ in spherical coordinates by substituting $x = r \cos \phi \sin \theta$, $y = r \sin \phi \sin \theta$, $z = r \cos \theta$. As $p$ is homogeneous of degree $n$, a term $r^n$ will be common to each term in $p$, so we will get $p = r^n q(\theta, \phi)$. The Laplacian $\Delta_3$, expressed in spherical coordinates, is (see (7.7.12) of Section 7.7)

$$\Delta_3(p) = \frac{-1}{r^2} \frac{\partial}{\partial r} \left( r^2 \frac{\partial p}{\partial r} \right) + \frac{1}{r^2} \Delta_s(p) \tag{7.11.6}$$

where $\Delta_s$ is the Laplacian on the unit sphere $S$,

$$\Delta_s(p) = \frac{-1}{\sin \theta} \frac{\partial}{\partial \theta} \left( \sin \theta \frac{\partial p}{\partial \theta} \right) - \frac{1}{\sin^2 \theta} \frac{\partial^2 p}{\partial \phi^2}$$

Now substitute $p = r^n q(\theta, \phi)$ into (7.11.6):

$$\Delta_3(p) = -r^{n-2} n(n+1)q + r^{n-2} \Delta_s(q)$$

Since $\Delta_p(p) = 0$, we get

$$r^{n-2}(\Delta_s(q) - n(n+1)q) = 0 \tag{7.11.7}$$

As (7.11.7) must hold for all $(r, \theta, \phi)$, we get $\Delta_s(q) = n(n+1)q$, which says that $q(\theta, \phi)$ is an eigenfunction of $\Delta_s$ belonging to the eigenvalue $n(n+1)$. Now observe that, when we put $r = 1$ in $p = r^n q(\theta, \phi)$, we get $p = q$. But $r = 1$ means $x^2 + y^2 + z^2 = 1$, which means in turn that we have restricted the polynomial $p(x, y, z)$ to the unit sphere. Thus we have proved the first part of Theorem 7.33: every homogeneous harmonic polynomial $p(x, y, z)$, when restricted to the unit sphere $x^2 + y^2 + z^2 = 1$, is an eigenfunction of the spherical Laplacian belonging to the eigenvalue $n(n+1)$.

To complete the proof, we need to show that every function in the eigenspace $H(n)$ arises this way, as the restriction to the unit sphere of some homogeneous harmonic polynomial of degree $n$. The bulk of the proof is set as a series of exercises. In Exercises 15, 16, and 17 you will show that each of the $2n+1$ spherical harmonics $Y_n^m$ is the restriction to the unit sphere of a homogeneous polynomial of degree $n$. Exercise 18 proves that if a homogeneous polynomial of degree $n$ becomes an eigenfunction of the Laplacian when it is restricted to the unit sphere, then the polynomial was in fact harmonic. These facts combined imply that every spherical harmonic $Y_n^m$ is the restriction to $S$ of a homogeneous *harmonic* polynomial of degree $n$. The last part of Theorem 7.33 is clearly a consequence of this.

### Exercises

The following exercises, numbers 2 through 8, provide examples of *harmonic functions in two variables*, i.e., functions $u = u(x, y)$ that satisfy Laplace's equation

$$-\Delta u = \frac{\partial^2 u}{\partial x^2} + \frac{\partial^2 u}{\partial y^2} = 0$$

These examples were all generated by one method: if $f(z)$ is an analytic function of one complex variable, then its real part $u(x, y) = \mathrm{Re}(f(z))$ and its imaginary part $v(x, y) = \mathrm{Im}(f(z))$ satisfy the Cauchy-Riemann equations and are therefore harmonic (Exercise 1). But you don't need to know any complex variables to do these exercises; either verify directly that the functions specified satisfy $\Delta u = \Delta v = 0$, or check that they satisfy the Cauchy-Riemann equations and then cite Exercise 1 (or do both).

1. A pair of functions, $u(x, y)$ and $v(x, y)$ satisfy the *Cauchy-Riemann equations* when

$$\frac{\partial u}{\partial x} = \frac{\partial v}{\partial y} \quad \text{and} \quad \frac{\partial u}{\partial y} = -\frac{\partial v}{\partial x}$$

Prove that if $u$ and $v$ satisfy the Cauchy-Riemann equations, then $u$ is harmonic and so is $v$.

2. Prove that $u(x, y) = e^x \cos y$ is harmonic and that $v(x, y) = e^x \sin y$ is harmonic.

3. Prove that $u(x, y) = x^2 - y^2$ is harmonic and that $v(x, y) = 2xy$ is harmonic.

4. Prove that $u(x, y) = x/(x^2 + y^2)$ is harmonic and that $v(x, y) = -y/(x^2 + y^2)$ is harmonic.

5. Prove that $u(x, y) = \sin x \cosh y$ is harmonic and that $v(x, y) = \cos x \sinh y$ is harmonic.

6. Let $n$ be a fixed nonnegative integer. Prove that the function $u(r, \theta) = r^n \cos n\theta$, which is expressed in polar coordinates, is harmonic. (Use the polar form of the Laplacian (Section 6.1)). Do the same for $v(r, \theta) = r^n \sin n\theta$.

7. Prove that if $u_1, u_2, \ldots, u_n$ are harmonic and $c_1, c_2, \ldots, c_n$ are any constants, then $u = c_1 u_1 + c_2 u_2 + \cdots c_n u_n$ is harmonic.

8. (a) Prove that if $u(x, y)$ is harmonic, then so are both $\partial u/\partial x$ and $\partial u/\partial y$. On what is this deduction based?

   (b) Prove that if $u(x, y)$ is harmonic, then so is

$$\frac{\partial^n u}{\partial x^a \, \partial y^b}, \qquad n = a + b$$

   for any $n = 0, 1, 2, \ldots$.

Exercises 9 through 12 deal with harmonic functions in three variables.

9. Show that the function $u(r, \theta, \phi) = r^n Z_n^m(\theta, \phi)$, which is expressed in spherical coordinates, is harmonic, i.e., satisfies Laplace's equation $\Delta u = 0$. [Hint: Use the spherical form of the Laplacian, (7.7.12) of Section 7.7. $Z_n^m(\theta, \phi)$ is a spherical harmonic; see Section 7.9.]

10. Prove that $u = 1/r$ is harmonic (spherical coordinates).

11. (a) Prove: If $u = u(x, y, z)$ is harmonic, then so are $\partial u/\partial x$, $\partial u/\partial y$, and $\partial u/\partial z$. On what is this deduction based?

    (b) Prove: If $u = u(x, y, z)$ is harmonic, so is

$$\frac{\partial^n u}{\partial x^a \, \partial y^b \, \partial z^c}, \qquad n = a + b + c$$

   for any $n = 0, 1, 2, \ldots$.

12. Combining Exercises 10 and 11, show that

$$\frac{\partial^n}{\partial x^a \, \partial y^b \, \partial z^c} \left( \frac{1}{r} \right), \qquad n = a + b + c$$

   is harmonic.

The following exercises deal with homogeneous harmonic polynomials.

13. (a) List all terms $x^a y^b z^c$, $a+b+c=n$, for $n=0$, $n=1$, $n=2$, and $n=3$.
    (b) Explaining carefully how you get your answer, determine how many different terms there are of a fixed degree $n$.

14. (Continuation) Show that the different terms of fixed degree $n$ are linearly independent.

15. (a) Show that the polynomial $3z^2 - 1$ is the restriction to the unit sphere of a homogeneous polynomial of degree 2.
    (b) Same for $5z^3 - 3z$ for a homogeneous polynomial of degree 3.
    (c) Same for $35z^4 - 30z^2 + 3$, degree 4.

16. (Continuation) Show that a polynomial $f(z) = a_0 + a_2 z^2 + a_4 z^4 + \cdots + a_{2n} z^{2n}$, $a_{2n} \neq 0$, that contains only even powers of $z$ is a restriction to the unit sphere $x^2 + y^2 + z^2 = 1$ of a homogeneous polynomial $p(x, y, z)$ of degree $2n$. State and prove a similar result for polynomials that contain only odd powers of $z$.

17. (Continuation) Show that the spherical harmonics $(1/2)P_n(z)$, $\cos m\phi \, P_n^m(z)$, and $\sin m\phi \, P_n^m(z)$, $1 \leqslant m \leqslant n$, are each the restriction to the unit sphere $x^2 + y^2 + z^2 = 1$ of a homogeneous polynomial of degree $n$. [Hint: Use Euler's formula $(e^{i\phi})^m = \cos m\phi + i \sin m\phi$ in the form

$$\left( \frac{x}{\sqrt{1-z^2}} + i \frac{y}{\sqrt{1-z^2}} \right)^m = \cos m\phi + i \sin m\phi$$

to express $\cos m\phi$ and $\sin m\phi$ in terms of $x$, $y$, and $z$.]

18. Let $p = p(x, y, z)$ be a homogeneous polynomial of degree $n$. Prove: If $p$ restricted to the unit sphere $S = \{(x, y, z): x^2 + y^2 + z^2 = 1\}$ satisfies $\Delta_s(p) = n(n+1)p$, then $p(x, y, z)$ is harmonic.

## APPENDIX 7.A    THE LEGENDRE, LAGUERRE, AND SCHRÖDINGER OPERATORS

### 7.A.1    Legendre's Operator $L_m$

Legendre's operator $L_m$ is based on the differential expression $l_m(g) \equiv -((1-z^2)g')' + m^2(1-z^2)^{-1}g$ and acts in the Hilbert space $L^2(-1, 1)$; see Definition 7.21. The case $m=0$ was given full and circumstantial treatment in Appendix 4.A. Because the analysis of the case $m \geqslant 1$ shares many of the same details with that of the case $m=0$, we can afford to be brief here. Appendix 4.A is available for expanded discussions as well as some informal comments concerning the underlying motivation that guides our approach.

The domain of $L_m$ consists of all functions $g(z)$ that are compatible with the differential expression $l_m(g) \equiv -((1-z^2)g')' + m^2(1-z^2)^{-1}g$ and are finite at $z = \pm 1$. The compatibility condition breaks down to these four assertions (compare Appendix 4.A): (1) $g \in L^2(-1, 1)$, (2) $g(z)$ is differentiable with finite derivative $g'(z)$ at each point $z$ of the open interval $-1 < z < 1$, (3) $(1-z^2)g'(z)$ is absolutely continuous in the open interval $-1 < z < 1$, and (4) $l_m(g) \in L^2(-1, 1)$. Because we have already done the case $m = 0$ in Appendix 4.A, we shall assume here $m \geq 1$. We shall also assume that we are dealing with real-valued functions to avoid inessential complications.

Let $g(z)$ be in the domain of $L_m$, and let $-1 < r < s < 1$. Integrating by parts, we get

$$\int_r^s l_m(g)g \, dz = (1 - r^2)g'(r)g(r) - (1 - s^2)g'(s)g(s)$$

$$+ \int_r^s (1 - z^2)(g')^2 \, dz + m^2 \int_r^s \frac{g^2}{1 - z^2} \, dz \qquad (7.A.1)$$

As either $s \to 1$ or $r \to -1$, the limit of the integral on the left side of (7.A.1) exists and is finite, by virtue of the fact that both $g$ and $l_m(g)$ belong to $L^2(-1, 1)$. We shall concentrate on the case $s \to 1$. The case $r \to -1$ is treated by exactly the same methods.

Because the left side of (7.A.1) has a finite limit as $s \to 1$, so must the right side. If either of the last two integrals on the right side of (7.A.1) were to diverge as $s \to 1$, it would have to go to $+\infty$ as the integrands are nonnegative. Hence, to maintain a finite limit on the right side of (7.A.1), we would then have $(1 - s^2)g'(s)g(s) \to +\infty$ as $s \to 1$. So there would be an $s_0$ such that for all $z$ satisfying $s_0 \leq z < 1$, $(1 - z^2)g'(z)g(z) > 1$. This would entail

$$\tfrac{1}{2}(g^2)' = g'g > \frac{1}{1 - z^2} = \tfrac{1}{2}\left(\frac{1}{1 - z} + \frac{1}{1 + z}\right) > \tfrac{1}{2}\frac{1}{1 - z}$$

thus $(g(z)^2)' > (1 - z)^{-1}$ for $s_0 \leq z < 1$. Integrating, we would have

$$g^2(s) - g^2(s_0) > \ln\left(\frac{1}{1 - s}\right) - \ln\left(\frac{1}{1 - s_0}\right)$$

which implies that $g^2(s) \to +\infty$ as $s \to 1$. But this contradicts the fact that $g(1) = \lim_{s \to 1} g(s)$ is finite. This contradiction resulted from the assumption that one of the integrals on the right side of (7.A.1) diverged as $s \to 1$. Hence both those integrals tend to a finite limit as $s \to 1$.

Using exactly the same technique, we can prove that both integrals on the right side of (7.A.1) tend to finite limits as $r \to -1$.

We conclude: for any function $g(z)$ in the domain of Legendre's operator $L_m$, both integrals

$$\int_{-1}^1 (1 - z^2)(g'(z))^2 \, dz \qquad \text{and} \qquad \int_{-1}^1 \frac{g(z)^2}{1 - z^2} \, dz$$

are finite. We restate this conclusion in another way.

**Lemma 7.34.** *For any function $g(z)$ in the domain of Legendre's operator $L_m$, $m \geqslant 1$, both $(1-z^2)^{1/2}g'(z)$ and $(1-z^2)^{-1/2}g(z)$ belong to $L^2(-1,1)$.*

Now let $g$ and $h$ represent two functions in the domain of $L_m$. Integrating by parts again, we get

$$\int_r^s l_m(g)h\, dz = (1-r^2)g'(r)h(r) - (1-s^2)g'(s)h(s)$$

$$+ \int_r^s (1-z^2)g'h'\, dz + m^2 \int_r^s \frac{gh}{1-z^2}\, dz \qquad (7.A.2)$$

Let $s \to 1$. Remember that the product of two square-integrable functions is integrable (see 4.57 of Appendix 4.A, as well as Section 3.9). Using this fact, together with Lemma 7.34, deduce that all three integrals in (7.A.2) converge to a finite limit as $s \to 1$. Hence

$$\lim_{s \to 1} (1-s^2)g'(s)h(s)$$

exists and is finite. Let $c$ denote this limit. We shall prove that $c=0$.

Suppose $c > 0$. Then there would exist an $s_0 < 1$ such that for $s_0 \leqslant z < 1$, we would have $(1-z^2)g'(z)h(z) > c/2$, thus $g'(z)h(z) > c/2(1-z^2) > c/4(1-z)$. Integrate this inequality:

$$\int_{s_0}^s g'(z)h(z)\, dz > \frac{c}{2}\left[\ln\left(\frac{1}{1-s}\right) - \ln\left(\frac{1}{1-s_0}\right)\right]$$

This implies that

$$\int_{s_0}^s g'(z)h(z)\, dz \to +\infty \qquad \text{as } s \to 1$$

But the integral

$$\int_{s_0}^s g'(z)h(z)\, dz = \int_{s_0}^s (\sqrt{1-z^2}\, g'(z))\left(\frac{h(z)}{\sqrt{1-z^2}}\right) dz$$

must have a finite limit as $s \to 1$ because its integrand is integrable over $[-1,1]$, being the product of two square-integrable functions (see Lemma 7.34). So the assumption $c > 0$ is untenable. Using exactly the same kind of argument, we rule out $c < 0$. Thus $c = 0$. We summarize our results:

**Theorem 7.35.** *For any pair of functions $g, h$ in the domain of Legendre's operator $L_m$, $m \geqslant 1$, we have*

$$\lim_{z \to \pm 1} (1-z^2)g'(z)\overline{h(z)} = 0 \qquad (7.A.3)$$

$$\sqrt{1-z^2}\,g'(z)\in L^2(-1,1) \tag{7.A.4}$$

$$\frac{g(z)}{\sqrt{1-z^2}}\in L^2(-1,1) \tag{7.A.5}$$

### 7.A.2 Laguerre's Operator $A_\alpha$

**Definition 7.36.** Fix a real number $\alpha$, $\alpha\geqslant0$. Laguerre's operator $A_\alpha$ has the differential expression

$$l_\alpha(y)\equiv\frac{-1}{x^\alpha e^{-x}}(x^{\alpha+1}e^{-x}y')' = -xy'' - (\alpha+1-x)y'$$

and domain consisting of those functions $y=y(x)$ in the Hilbert space $L^2(0,\infty;x^\alpha e^{-x})$ that are compatible with $l_\alpha$ and for which $\lim_{x\to0}x^{\alpha/2}y(x)$ is finite.

The objective of the following analysis corresponds to that of the preceding analysis of Legendre's operator: to derive the properties of the functions that lie in the domain of Laguerre's operator so that, using these properties, the student may then prove that Laguerre's operator is Hermitian (Exercise 3 of Section 7.10).

To say that a function $y(x)$ is compatible with the differential expression $l_\alpha$ is to say that the following four conditions are satisfied: (1) $y(x)\in L^2(0,\infty;x^\alpha e^{-x})$, (2) $y(x)$ is differentiable with finite derivative $y'(x)$ at each point $x$ of the open interval $0<x<\infty$, (3) $x^{\alpha+1}e^{-x}y'(x)$ is absolutely continuous in the open interval $0<x<\infty$, and (4) $l_\alpha(y)\in L^2(0,\infty;x^\alpha e^{-x})$.

In addition to that compatibility requirement for membership in the domain of $A_\alpha$, we also require that $y(x)$ satisfy the boundary condition that $\lim_{x\to0}x^{\alpha/2}y(x)$ be finite. If $\alpha=0$, this just means that $y(0)$ is finite. If $\alpha>0$, then $y(x)$ is allowed to have a singularity at $x=0$ as long as the product $x^{\alpha/2}y(x)$ tends to a finite limit as $x\to0$.

We shall prove that the functions that comprise the domain of $A_\alpha$, which are exactly those functions that satisfy both the compatibility requirement and the boundary condition at $x=0$, must be rather well behaved in certain aspects. Here is the precise statement:

**Theorem 7.37.** *Fix a real number $\alpha$, $\alpha\geqslant0$. For any pair of functions $f(x)$ and $g(x)$ in the domain of Laguerre's operator $A_\alpha$, we have*

$$\lim_{x\to0,\infty}x^{\alpha+1}e^{-x}f'(x)\overline{g(x)} = 0 \tag{7.A.6}$$

$$\sqrt{x}\,f'(x)\in L^2(0,\infty;x^\alpha e^{-x}) \tag{7.A.7}$$

The proof of this theorem follows exactly the same pattern as that for Legendre's operator (Appendix 4.A and this appendix), Hermite's operator (Appendix 4.A), and Bessel's operator (Appendix 6.A). Therefore I shall be *very* brief here and really just outline the proof.

To avoid inessential complications, assume all functions real valued.

Suppose now that $f(x)$ belongs to the domain of $A_\alpha$. Then both $f$ and $l_\alpha(f)$ will be absolutely continuous on every finite closed interval $[r, s]$, where $0 < r < s < \infty$. So we may integrate by parts:

$$\int_r^s l_\alpha(f)f x^\alpha e^{-x}\, dx = f(r)f'(r)r^{\alpha+1}e^{-r} - f(s)f'(s)s^{\alpha+1}e^{-s}$$

$$+ \int_r^s (f'(x))^2 x^{\alpha+1}e^{-x}\, dx \tag{7.A.8}$$

Consider the integral on the left side of (7.A.8). Because both $l_\alpha(f)$ and $f$ are square integrable against the weight function $x^\alpha e^{-x}$, their product is integrable against the same weight function. Hence the integral on the left of (7.A.8) will converge as $r \to 0$ and $s \to \infty$. This fact provides the key to this stage of the proof.

The integral on the right side of (7.A.8) has a positive integrand. So, as $r \to 0$, it will either converge to a finite limit or diverge to $+\infty$. If the latter were to occur, we would have to have $f(r)f'(r)r^{\alpha+1}e^{-r} \to -\infty$. But one can show readily that this latter fact is incompatible with the boundary condition that $r^{\alpha/2}f(r)$ has a finite limit when $r \to 0$. Hence the last integral in (7.A.8) tends to a finite limit as $r \to 0$. A somewhat more involved argument proves that the same integral converges to a finite limit as $s \to \infty$. Therefore the integral

$$\int_0^\infty (f'(x))^2 x^{\alpha+1}e^{-x}\, dx = \int_0^\infty (\sqrt{x}\, f'(x))^2 x^\alpha e^{-x}\, dx$$

is finite, which is (7.A.7) of Theorem 7.37.

Now let $f(x)$ and $g(x)$ represent two functions in the domain of $A_\alpha$. Integrate by parts again:

$$\int_r^s l_\alpha(f)g x^\alpha e^{-x}\, dx = r^{\alpha+1}e^{-r}f'(r)g(r) - s^{\alpha+1}e^{-s}f'(s)g(s)$$

$$+ \int_r^s f'(x)g'(x)x^{\alpha+1}e^{-x}\, dx \tag{7.A.9}$$

Once again, the integral on the left side of (7.A.9) converges as $r \to 0$ and $s \to \infty$, because the product of two square-integrable functions is integrable. Also, by what has just been proved, the integral on the right side of (7.A.8) converges. Hence the first and second integrated terms in (7.A.9) tend to finite limits as

$r \to 0$ and $s \to \infty$, respectively. The final step is to show that both these limits must be zero. That is (7.A.6) of Theorem 7.37.

## 7.A.3 Schrödinger's Operator

**Definition 7.38.** Let $l$ denote an integer $\geq 0$. Schrödinger's operator $M_l$ has the differential expression

$$m(f) = -\frac{1}{s^2}\frac{d}{ds}\left(s^2\frac{df}{ds}\right) + \left(\frac{l(l+1)}{s^2} - \frac{1}{s}\right)f$$

and domain consisting of those functions $f(s)$ in the Hilbert space $L^2(0, \infty; s^2)$ that are compatible with $m$ and for which $f(0)$ is finite.

To say that a function $f(s)$ is compatible with the differential expression $m$ is to say that the following four conditions are satisfied:

1. $\int_0^\infty |f(s)|^2 s^2\,ds < \infty$.
2. $f(s)$ is differentiable with finite derivative $f'(s)$ at each point $s$ of the open interval $0 < s < \infty$.
3. $s^2 f(s)'$ is absolutely continuous in the open interval $0 < s < \infty$.
4. $\int_0^\infty |m(f)|^2 s^2\,ds < \infty$.

To avoid inessential complications, assume all functions real valued.

**Theorem 7.39.** *Fix an integer $l$, $l \geq 0$. For every pair of functions $f(s)$ and $g(s)$ in the domain of Schrödinger's operator $M_l$, we have*

$$f'(s) \in L^2(0, \infty; s^2) \tag{7.A.10}$$

$$\lim_{s \to 0, \infty} s^2 f'(s)\overline{g(s)} = 0 \tag{7.A.11}$$

The proof begins with the usual valid integration by parts. Choose $0 < a < b < \infty$. Then for any $f(s)$ in the domain of $M_l$,

$$\int_a^b m(f)fs^2\,ds = a^2 f'(a)f(a) - b^2 f'(b)f(b) + \int_a^b (f'(s))^2 s^2\,ds$$

$$+ \int_a^b (f(s))^2(l(l+1) - s)\,ds \tag{7.A.12}$$

The integral on the left is finite and will remain finite as $a \to 0$ and $b \to \infty$ because its integrand is the product of two square-integrable functions. Hence the right side of (7.A.12) must also remain finite as $a \to 0$ and $b \to \infty$.

Consider first the case $b \to \infty$. The integral furthest right in (7.A.12) converges as $b \to \infty$ because the absolute value of its integrand is less than $(f(s)s)^2$ for sufficiently large $s$, and $f(s)$ belongs to $L^2(0, \infty; s^2)$. If the other integral on the right of (7.A.12) were to diverge, it would have to diverge to $+\infty$ because its integrand is nonnegative. Hence, to balance, we must have $b^2 f'(b)f(b) \to +\infty$ as $b \to \infty$. So there would exist a $b_0$ such that $b^2 f'(b)f(b) = (b^2/2)(f^2(b))' > 1$ for $b \geq b_0$. Thus $(f^2(b))' > 2/b^2$ for $b > b_0$. Thus

$$\int_{b_0}^{s} (f^2(b))' \, db > 2 \int_{b_0}^{s} b^{-2} \, db$$

or

$$f^2(s) - f^2(b_0) > 2\left(\frac{1}{b_0} - \frac{1}{s}\right) > \frac{1}{b_0} \qquad \text{for } s > 2b_0$$

which leads to

$$s^2 f^2(s) > s^2 \left(f^2(b_0) + \frac{1}{b_0}\right) \qquad \text{for } s > 2b_0$$

which contradicts the integrability of $s^2 f^2(s)$. So both integrals on the right of (7.A.12) converge as $b \to \infty$.

Now consider the case $a \to 0$. As we are assuming that $f(0)$ is finite, the last integral on the right of (7.A.12) will certainly converge as $a \to 0$. If the other integral were to diverge, we must have $a^2 f'(a)f(a) \to -\infty$ as $a \to 0$ to compensate. But this would contradict the finiteness of $f(0)$. Hence

$$\int_{0}^{\infty} (f'(s))^2 s^2 \, ds < \infty$$

which is (7.A.10) of Theorem 7.39. The proof of (7.A.11) follows the same pattern as the other proofs we have done. With Theorem 7.39 in hand, it is easy to prove that Schrödinger's operator is Hermitian.

# 8

# The Fourier Transform

## 8.1 COMPLEX METHODS IN FOURIER SERIES; THE FOURIER TRANSFORM

In Section 3.4 we discussed Fourier series over intervals $-A \leqslant x \leqslant A$. The Fourier series of a function $f(x)$ defined on $-A \leqslant x \leqslant A$ is

$$f(x) = \frac{a_0}{2} + \sum_{n=1}^{\infty} a_n \cos \frac{n\pi x}{A} + b_n \sin \frac{n\pi x}{A} \tag{8.1.1}$$

The functions $1/2$, $\cos(\pi x/A)$, $\sin(\pi x/A)$, $\cos(2\pi x/A)$, $\sin(2\pi x/A)$,... constitute an orthogonal basis of $L^2(C)$, where $C$ is the circle of radius $A/\pi$ formed by removing the interval $-A \leqslant x \leqslant A$ from the real line, bending it round, and joining (identifying) the point $x = -A$ to the point $x = A$. The representation (8.1.1) is valid in the mean-square sense for all $f(x)$ in $L^2(C)$.

As the point $x$ moves round and round the circle $C$, the function values $f(x)$ trace and retrace the same graph. We say that $f(x)$ is periodic, of period $2A$ ($2A$ is the circumference of the circle $C$). We get an equivalent picture if we imagine the circle $C$ rolled back out onto the interval $-A \leqslant x \leqslant A$ and the graph of $f(x)$ then transplanted to all translated intervals $A \leqslant x \leqslant 3A$, $3A \leqslant x \leqslant 5A$, $-3A \leqslant x \leqslant -A$, etc. Then $f(x)$ is defined on the whole real line $\mathbf{R}$ and satisfies the periodicity condition $f(x + 2kA) = f(x)$ for all $k = 0$, $\pm 1$, $\pm 2$,... and for all real $x$. Then, for such a periodic function $f(x)$, the Fourier series representation (8.1.1) has meaning for all real $x$ and converges to $f(x)$ in the mean-square metric of $L^2(-A, A)$, or of $L^2(a, b)$ for any interval $a \leqslant x \leqslant b$ of length $2A$.

If we imagine $A$ as very large so that the points $x = A$ and $x = -A$ have receded over the horizon, as it were, and if we further imagine ourselves as standing at the origin, then the periodic repetitions of the function $f(x)$ would be beyond our fancied perspective, and we might hope that the representation (8.1.1) could be used to represent general nonperiodic $f(x)$ in $L^2(-\infty, \infty)$. Such hope would be vain, however, because as long as $A$ is finite, no matter how large, the function $f(x)$ remains periodic and repeats itself infinitely often as $x \to \pm \infty$. Faced with this picture, students frequently ask if it is possible, by somehow letting $A \to \infty$, to secure a representation analogous to (8.1.1) for nonperiodic general $f(x)$. The answer is yes. The Fourier integral representation can be realized as a kind of limit of (8.1.1). It is the purpose of this section to explain how we can look at the Fourier transform as a kind of limiting case of Fourier series.

We begin by converting the Fourier series (8.1.1) into an equivalent complex form. Recall Euler's formula $e^{iz} = \cos z + i \sin z$ connecting the exponential and trigonometric functions, valid for all complex numbers $z = x + iy$, $x$, $y$ real. If we substitute $-z$ for $z$ in Euler's formula and remember that $\cos(-z) = \cos z$, $\sin(-z) = -\sin z$ (from their power series expansions) we get $e^{-iz} = \cos z - i \sin z$. Then add and subtract these equations to express the sine and cosine in terms of the exponential: $\sin z = (e^{iz} - e^{-iz})/2i$, $\cos z = (e^{iz} + e^{-iz})/2$. We have used these formulas several times in earlier chapters. Now substitute $\cos(n\pi x/A) = (\exp(in\pi x/A) + \exp(-in\pi x/A))/2$, $\sin(n\pi x/A) = (\exp(in\pi x/A) - \exp(-in\pi x/A))/2i$ into (8.1.1), and do a little rearranging to get

$$f(x) = \sum_{n=-\infty}^{n=\infty} c_n e^{in\pi x/A} \tag{8.1.2}$$

where the $c_n$ are complex constants (see Exercise 1). A similar series was discussed in Exercise 10 of Section 3.6 and Exercise 12 of Section 3.9.

The functions $\exp(in\pi x/A)$, $n = 0, \pm 1, \pm 2, \ldots$, form an orthogonal basis of the complex Hilbert space $L^2(-A, A)$. In the inner product of $L^2(-A, A)$,

$$\langle f, g \rangle = \int_{-A}^{A} f(x)\overline{g(x)}\, dx$$

we have $\|\exp(in\pi x/A)\|^2 = 2A$, so the general Fourier coefficients $c_n$ in (8.1.2) are given by

$$c_n = \frac{1}{2A} \langle f, \exp(in\pi x/A) \rangle = \frac{1}{2A} \int_{-A}^{A} f(x) e^{-in\pi x/A}\, dx$$

We want to set things up so we can perceive a pattern as $A \to \infty$. Define

$$\hat{f}\left(\frac{n\pi}{A}\right) = \frac{2A}{\sqrt{2\pi}}c_n = \frac{1}{\sqrt{2\pi}}\int_{-A}^{A} f(x)e^{-in\pi x/A}\,dx \tag{8.1.3}$$

and combine (8.1.2) and (8.1.3):

$$f(x) = \frac{1}{\sqrt{2\pi}}\sum_{n=-\infty}^{\infty} \hat{f}\left(\frac{n\pi}{A}\right)e^{i(n\pi/A)x}\left(\frac{\pi}{A}\right) \tag{8.1.4}$$

$$\hat{f}\left(\frac{n\pi}{A}\right) = \frac{1}{\sqrt{2\pi}}\int_{-A}^{A} f(x)e^{-i(n\pi/A)x}\,dx \tag{8.1.5}$$

Equation (8.1.4) is our complex Fourier series for $f(x)$ in $L^2(-A, A)$, and (8.1.5) is the formula for the scaled Fourier coefficients of $f(x)$. These are the equations that suggest how to get sensible formulas when we pass to the limit $A \to \infty$.

Suppose now that $A$ is very very large and fixed. Then, as $n$ varies in integral steps, $n\pi/A$ varies in very very small steps so that we may look on it as a continuous variable, say $u = n\pi/A$. Then, noting that $\pi/A = n\pi/A - (n-1)\pi/A$, we recognize (8.1.4) as an approximating Riemann sum for the integral

$$\frac{1}{\sqrt{2\pi}}\int_{-\infty}^{\infty} \hat{f}(u)e^{iux}\,du$$

Equation (8.1.5) is

$$\hat{f}(u) = \frac{1}{\sqrt{2\pi}}\int_{-A}^{A} f(x)e^{-iux}\,dx$$

These arguments suggest that, as $A \to \infty$, (8.1.4) and (8.1.5) become

$$f(x) = \frac{1}{\sqrt{2\pi}}\int_{-\infty}^{\infty} \hat{f}(u)e^{iux}\,du \tag{8.1.6}$$

$$\hat{f}(u) = \frac{1}{\sqrt{2\pi}}\int_{-\infty}^{\infty} f(x)e^{-iux}\,dx \tag{8.1.7}$$

Equation (8.1.6) is the limiting case of the Fourier series and equation (8.1.7) the limiting case of the formula for the Fourier coefficients. Equation (8.1.7) defines the *Fourier transform*, $\hat{f}(u)$, of the function $f(x)$.

A closer examination of the Fourier series given by equations (8.1.4) and (8.1.5) will help us gain a better understanding of the Fourier transform, which is given by equations (8.1.6) and (8.1.7). So look back again at equation (8.1.5); it tells you how to calculate the coefficients, $\hat{f}(n\pi/A)$, $n = 0, \pm 1$,

$\pm 2, \ldots$, from the given function $f(x) \in L^2(-A, A)$. So we start with a function $f(x)$ in the Hilbert space $L^2(-A, A)$, and we end up with a *sequence* $\hat{f}(n\pi/A)$, $n = 0, \pm 1, \pm 2, \ldots$. Think back. Every time we expanded a function $f(x)$ in a Hilbert space $L^2(a, b; r(x))$ with respect to an orthogonal basis of that space, say $\phi_1, \phi_2, \ldots$, we did the same thing: we converted a function $f(x) \in L^2(a, b; r(x))$ into the *sequence* of its generalized Fourier coefficients $c_n$ given by

$$c_n = \frac{\langle f, \phi_n \rangle}{\langle \phi_n, \phi_n \rangle}$$

All information about the function $f(x)$ is encoded in the sequence $c_n$, $n = 1, 2, \ldots$. Knowing the sequence $c_n$, we retrieve the function $f(x)$ from its general Fourier expansion $f(x) = \Sigma c_n \phi_n(x)$. (All this assumes that an orthogonal basis has been chosen and fixed.)

Apply these comments to our Fourier series (8.1.4) and (8.1.5). Here we have fixed an orthogonal basis of $L^2(-A, A)$, the functions $\exp(in\pi x/A)$, $n = 0$, $\pm 1, \pm 2, \ldots$. And we can interpret Fourier series as arranging a correspondence between each function $f(x)$ in $L^2(-A, A)$ and its sequence of Fourier coefficients $\hat{f}(n\pi/A)$, $n = 0, \pm 1, \pm 2, \ldots$. These sequences constitute a Hilbert space—see Exercise 11 of Section 3.9. Moreover, if we take the inner product of each side of (8.1.4) with itself we get

$$\int_{-A}^{A} |f(x)|^2 \, dx = \sum_{n=-\infty}^{\infty} \left| \hat{f}\left(\frac{n\pi}{A}\right) \right|^2 \frac{\pi}{A} \tag{8.1.8}$$

which relates the norm of $f(x)$ in $L^2(-A, A)$ to the norm of its sequence of Fourier coefficients. On the right side of (8.1.8) we have an approximating Riemann sum for the integral.

$$\int_{-\infty}^{\infty} |\hat{f}(u)|^2 \, du$$

So, as $A \to \infty$, it is plausible to take

$$\int_{-\infty}^{\infty} |f(x)|^2 \, dx = \int_{-\infty}^{\infty} |\hat{f}(u)|^2 \, du \tag{8.1.9}$$

as the limiting case of (8.1.8).

Now in our passage to the limit $A \to \infty$, $L^2(-A, A)$ becomes $L^2(-\infty, \infty)$. So this would be the appropriate Hilbert space for our functions $f(x)$. Equation (8.1.9) then says that the Fourier transform $\hat{f}(u)$ also belongs to $L^2(-\infty, \infty)$ and, in fact, has the same norm as $f(x)$. So, unlike the Fourier series, which converted a function to a sequence, the Fourier transform appears to map the Hilbert space $L^2(-\infty, \infty)$ onto itself. If we define a linear operator $U$ on $L^2(-\infty, \infty)$ by

$$U(f)(u) = \hat{f}(u) = \frac{1}{\sqrt{2\pi}} \int_{-\infty}^{\infty} f(x)e^{-iux} dx$$

then equation (8.1.9) says that

$$\langle f, f \rangle = \langle Uf, Uf \rangle$$

for all $f$ in $L^2(-\infty, \infty)$. From this equation it follows automatically that

$$\langle f, g \rangle = \langle Uf, Ug \rangle$$

for all $f, g$ in $L^2(-\infty, \infty)$ (Exercise 4). Thus, if our reasoning is correct, the Fourier transform not only is a linear operator mapping the Hilbert space $L^2(-\infty, \infty)$ onto itself but also preserves the inner product!

Let us summarize our conclusions:

---

### The Fourier Transform

The Fourier transform $U(f) = \hat{f}$ of a function $f(x)$ in $L^2(-\infty, \infty)$ is defined by

$$\hat{f}(u) = \frac{1}{\sqrt{2\pi}} \int_{-\infty}^{\infty} f(x)e^{-iux} dx \qquad (8.1.10)$$

We have $\hat{f} \in L^2(-\infty, \infty)$ and, in fact,

$$\langle \hat{f}, \hat{g} \rangle = \langle f, g \rangle \qquad (8.1.11)$$

for all $f, g$ in $L^2(-\infty, \infty)$. The function $f$ is recovered from its Fourier transform $\hat{f}$ by

$$f(x) = \frac{1}{\sqrt{2\pi}} \int_{-\infty}^{\infty} \hat{f}(u)e^{iux} du \qquad (8.1.12)$$

---

Equation (8.1.10) is the analogue of the expression for the Fourier coefficients. Equation (8.1.12) is the analogue of the expansion of $f(x)$ in its Fourier series.

The conclusions we have summarized in the box are all in fact true, once we put the proper interpretation on the infinite integrals. We shall do this in the following section, where we shall also test out our equations on some explicit simple functions $f(x)$.

## Exercises

1. Show that, with the substitutions

$$\cos\frac{n\pi x}{A} = \frac{1}{2}(e^{in\pi x/A} + e^{-in\pi x/A}) \qquad \sin\frac{n\pi x}{A} = \frac{1}{2i}(e^{in\pi x/A} - e^{-in\pi x/A})$$

the Fourier series (8.1.1) goes into the complex form

$$f(x) = \sum_{n=-\infty}^{\infty} c_n e^{in\pi x/A}$$

where

$$c_n = \begin{cases} \frac{1}{2}(a_{-n} + ib_{-n}), & n < 0 \\ \frac{1}{2}a_0, & n = 0 \\ \frac{1}{2}(a_n - ib_n), & n > 0 \end{cases}$$

2. Show that for the inner product in $L^2(-A, A)$ given by

$$\langle f, g \rangle = \int_{-A}^{A} f(x)\overline{g(x)}\, dx$$

we have

$$\langle e^{im\pi x/A}, e^{in\pi x/A} \rangle = \begin{cases} 0, & m \neq n \\ 2A, & m = n \end{cases}$$

(Integrate the complex exponential directly; do not convert to sines and cosines.)

3. Let $a$ be a real or complex number, not an integer. Show that the complex Fourier series of $e^{iax}$ in $L^2(-\pi, \pi)$ is

$$e^{iax} = \sum_{n=-\infty}^{\infty} \frac{\sin(a-n)\pi}{(a-n)\pi} e^{inx}$$

What happens as $a \to$ integer?

4. (a) Prove the "polarization identity" for a complex inner product space $V$:

$$4\langle f, g \rangle = \|f + g\|^2 - \|f - g\|^2 + i\|f + ig\|^2 - i\|f - ig\|^2$$

(b) Using the polarization identity, prove that if $U$ is a linear transformation of $V$ onto itself such that $\|Uf\|^2 = \|f\|^2$ for all $f$ in $V$, then $\langle Uf, Ug \rangle = \langle f, g \rangle$ for all $f, g$ in $V$.

## 8.2  PLANCHEREL'S THEOREM; EXAMPLES OF FOURIER TRANSFORMS

Consider the characteristic function $\chi_{[a,b]}(x)$ of a closed interval $a \leqslant x \leqslant b$:

$$\chi_{[a,b]}(x) = \begin{cases} 1, & a \leqslant x \leqslant b \\ 0, & \text{otherwise} \end{cases}$$

According to equation (8.1.10) of the previous section, the Fourier transform of $\chi_{[a,b]}(x)$ is given by

$$\hat{\chi}_{[a,b]}(u) = \frac{1}{\sqrt{2\pi}} \int_{-\infty}^{\infty} \chi_{[a,b]}(x) e^{-iux} \, dx = \frac{1}{\sqrt{2\pi}} \int_{a}^{b} e^{-iux} \, dx$$

$$\hat{\chi}_{[a,b]}(u) = \frac{1}{\sqrt{2\pi}} \frac{e^{-ibu} - e^{-iau}}{-iu} \tag{8.2.1}$$

In particular, if $b > 0$ and $a = -b$ so that our interval is $[-b, b]$, then we have

$$\hat{\chi}_{[-b,b]}(u) = \left(\frac{2}{\pi}\right)^{1/2} \frac{\sin bu}{u}$$

According to equation (8.1.12) we should be able to recover the function $\chi_{[a,b]}(x)$ from (8.2.1) by applying Fourier inversion:

$$f(x) = \frac{1}{\sqrt{2\pi}} \int_{-\infty}^{\infty} \hat{f}(u) e^{iux} \, du$$

Let's test that:

$$\frac{1}{\sqrt{2\pi}} \int_{-\infty}^{\infty} \hat{\chi}_{[a,b]}(u) e^{iux} \, du = \frac{1}{2\pi} \int_{-\infty}^{\infty} \frac{e^{-ibu} - e^{-iau}}{-iu} e^{iux} \, du$$

$$= \frac{1}{2\pi} \int_{-\infty}^{\infty} \frac{e^{i(x-b)u} - e^{i(x-a)u}}{-iu} \, du$$

$$= \frac{-1}{2\pi i} \int_{-\infty}^{\infty} \frac{\cos(x-b)u + i\sin(x-b)u - (\cos(x-a)u + i\sin(x-a)u)}{u} \, du$$

$$= \frac{-1}{2\pi i} \int_{-\infty}^{\infty} \frac{\cos(x-b)u - \cos(x-a)u}{u} \, du - \frac{1}{2\pi} \int_{-\infty}^{\infty} \frac{\sin(x-b)u - \sin(x-a)u}{u} \, du$$

$$\tag{8.2.2}$$

At this point we should focus some attention on the more subtle mathematical aspects of the Fourier transform. The Fourier transform and its inverse are expressed in terms of infinite integrals. We have dealt with infinite integrals before—in studying the Hilbert space $L^2(-\infty, \infty)$, for example. The integrals that are involved in defining membership in the $L^2$ spaces are what are called "absolutely convergent" integrals. We dealt with the concept of absolute convergence in Section 3.9: it means that the integral of the *absolute value* of the function converges. But the integrals involved in the Fourier transform—including those in equation (8.2.2) in particular—are not absolutely convergent, in general. So these integrals have to be handled somewhat differently.

Here is the correct method for evaluating the integrals that occur in the evaluation of the Fourier transform. We'll state the facts as a formal theorem, but without proof.

**Theorem 8.1, Plancherel's Theorem.** *Let there be given a function $f(x)$ in $L^2(-\infty, \infty)$. Then, for each fixed $A > 0$, the function of the variable u defined by*

$$\hat{f}_A(u) = \frac{1}{\sqrt{2\pi}} \int_{-A}^{A} f(x)e^{-iux}\, dx$$

*belongs to the Hilbert space $L^2(-\infty, \infty)$, and the mean-square limit*

$$\hat{f}(u) = \frac{1}{\sqrt{2\pi}} \lim_{A \to \infty} \int_{-A}^{A} f(x)e^{-iux}\, dx = \lim_{A \to \infty} \hat{f}_A(u) \tag{8.2.3}$$

*exists and defines a function $\hat{f}(u)$ in $L^2(-\infty, \infty)$. We have also*

$$f(x) = \frac{1}{\sqrt{2\pi}} \lim_{A \to \infty} \int_{-A}^{A} \hat{f}(u)e^{iux}\, du \tag{8.2.4}$$

*Moreover, for any $f, g$ in $L^2(-\infty, \infty)$, we have*

$$\langle \hat{f}, \hat{g} \rangle = \langle f, g \rangle \tag{8.2.5}$$

*where $\langle f, g \rangle = \int_{-\infty}^{\infty} f(x)\overline{g(x)}\, dx$ is the inner product in $L^2(-\infty, \infty)$.*

Two comments about Plancherel's theorem:

1.  The assertion that $\hat{f}(u)$ is the mean-square limit of the $\hat{f}_A(u)$ refers to the limit in the intrinsic metric of the Hilbert space $L^2(-\infty, \infty)$:

$$\|\hat{f}(u) - \hat{f}_A(u)\| = \left[ \int_{-\infty}^{\infty} |\hat{f}(u) - \hat{f}_A(u)|^2\, du \right]^{1/2} \to 0 \qquad \text{as } A \to \infty$$

We are well acquainted with this mean-square convergence since we have used it exclusively in all our orthogonal function expansions. It represents nothing new.

2. More novel, and of more immediate practical impact, is the rule that tells us to evaluate all Fourier integrals as the limit of *symmetric* partial integrals of the form

$$\int_{-A}^{A}$$

rather than those of the form

$$\int_{N}^{M}$$

involving two independent passages to the limit, $M \to \infty$ and $N \to -\infty$. The latter type are appropriate to absolutely convergent integrals. The former type are to be used in evaluating Fourier integrals—that is one of the main rules to learn from Plancherel's theorem.

Let's return to our example and apply our new rule to the evaluation of the two integrals in (8.2.2). As for the first integral, our rule about symmetric partial integrals tells us to first evaluate

$$\int_{-A}^{A} \frac{\cos(x - b)u - \cos(x - a)u}{u} \, du$$

for $A > 0$. But the integrand is an odd function of $u$, so its integral over the symmetric interval $-A \leqslant u \leqslant A$ is zero for every $A$ (Section 3.2, Exercises 1 through 4, and Exercise 1 of this section). Thus the limit as $A \to \infty$ is zero, so the first integral in (8.2.2) drops out. So we are left with

$$\frac{1}{\sqrt{2\pi}} \int_{-\infty}^{\infty} \hat{\chi}_{[a,b]}(u)e^{iux} \, du$$

$$= \frac{1}{2\pi} \left\{ \int_{-\infty}^{\infty} \frac{\sin(x - a)u}{u} \, du - \int_{-\infty}^{\infty} \frac{\sin(x - b)u}{u} \, du \right\} \tag{8.2.6}$$

and thus with the problem of evaluating the integral

$$\int_{-\infty}^{\infty} \frac{\sin \alpha u}{u} \, du = \lim_{A \to \infty} \int_{-A}^{A} \frac{\sin \alpha u}{u} \, du \tag{8.2.7}$$

If $\alpha = 0$ this integral is obviously 0. If $\alpha \neq 0$, then the substitution $t = \alpha u$ leads to

$$\int_{-\infty}^{\infty} \frac{\sin \alpha u}{u} \, du = \text{sign}(\alpha) \int_{-\infty}^{\infty} \frac{\sin t}{t} \, dt \tag{8.2.8}$$

So we are now faced with the task of evaluating the integral on the far right of (8.2.8). At this point we are entitled to surmise that the Fourier transform of even the simplest functions leads to some very interesting definite integrals. In Appendix 8.A we shall prove that

$$\int_{-\infty}^{\infty} \frac{\sin t}{t}\, dt = \pi \qquad (8.2.9)$$

Combining (8.2.6), (8.2.8), and (8.2.9), we get

$$\frac{1}{\sqrt{2\pi}} \int_{-\infty}^{\infty} \hat{\chi}_{[a,b]}(u)e^{iux}\, du = \begin{cases} 0 & \text{when } x < a \\ \frac{1}{2} & x = a \\ 1 & a < x < b \\ \frac{1}{2} & x = b \\ 0 & x > b \end{cases}$$

which is our original function $\chi_{[a,b]}(x)$, except that the function values at the two points of discontinuity have been replaced by the average of the right- and left-hand limits.

Thus equation (8.2.4) of Plancherel's theorem checks out, at least for the function $f(x) = \chi_{[a,b]}(x)$. This is reassuring, but unnecessary. Because Theorem 8.1 is exactly that, a mathematical theorem, true always and without exception. So we can rely on equations (8.2.3), (8.2.4), and (8.2.5); they are invariably true. This is important because these equations usually are very difficult to check directly in specific instances. Hence they offer us new information not available by other means. You will be using equations (8.2.3), (8.2.4), and (8.2.5) in the exercises to evaluate some very interesting definite integrals.

However, when you use equations (8.2.3) and (8.2.4), both in the exercises and in practice, write them as infinite integrals, without the "lim," just as we did in Section 8.1:

$$\hat{f}(u) = \frac{1}{\sqrt{2\pi}} \int_{-\infty}^{\infty} f(x)e^{-iux}\, dx \qquad (8.2.3')$$

and

$$f(x) = \frac{1}{\sqrt{2\pi}} \int_{-\infty}^{\infty} \hat{f}(u)e^{iux}\, du \qquad (8.2.4')$$

As long as we understand how these integrals are to be interpreted, we simplify the formulas and aid in their application by keeping the limits involved tacit.

Equations (8.2.3') and (8.2.4') are formally the same except for the sign of the exponent. An elementary analysis based on this symmetry shows that: *if the Fourier transform of $f(x)$ is $\hat{f}(u)$, then the Fourier transform of $\hat{f}(x)$ is*

$f(-u)$. (Exercise 12). Thus Fourier transforms occur in simply related pairs (see Exercises 8 through 11).

## Exercises

Do not use a table of integrals in any of these exercises.

1. Consider the function

$$f(u) = \frac{\cos(x - b)u - \cos(x - a)u}{u}$$

   that appears in equation (8.2.2). Show that $f(0)=0$ and that $f(u)$ is odd: $f(-u)=-f(u)$.

2. (a) By applying equation (8.2.5) to

   $$f(x)=g(x)=\chi_{[a,b]}(x), \qquad \hat{f}(u)=\hat{g}(u)=\hat{\chi}_{[a,b]}(u)$$
   derive the formula

   $$\int_{-\infty}^{\infty} \frac{1 - \cos xu}{u^2}\, du = \pi x, \qquad x > 0$$

   From this, deduce

   $$\int_{-\infty}^{\infty} \frac{1 - \cos xu}{u^2}\, du = \pi|x|, \qquad x \text{ real}$$

   (b) By differentiating the formula in (a) with respect to $x$, derive

   $$\int_{-\infty}^{\infty} \frac{\sin xu}{u}\, du = \begin{cases} \pi, & x > 0 \\ -\pi, & x < 0 \end{cases}$$

3. (a) Show that

   $$\int_{-\infty}^{\infty} \frac{\sin bu \cos xu}{u}\, du = \begin{cases} \pi, & |x| < b \\ 0, & |x| > b \end{cases}$$

   where $b$ is a positive constant. (Use equation (8.2.4).)

   (b) Show also that

   $$\int_{-\infty}^{\infty} \frac{\sin^2 bu}{u^2}\, du = \pi|b|$$

   for $b$ real.

4. (a) Let $f(x)=0$ for $x<0$, $f(x)=e^{-ax}$ for $x>0$, where $a$ is a positive constant. Show that

   $$\hat{f}(u) = \frac{1}{\sqrt{2\pi}(a + iu)}$$

(b) Deduce that

$$\int_{-\infty}^{\infty} \frac{e^{iux}}{a+iu}\, du = \begin{cases} 0, & x < 0 \\ 2\pi e^{-ax}, & x > 0 \end{cases}$$

(c) Deduce also that

$$\int_{-\infty}^{\infty} \frac{du}{a^2 + u^2} = \frac{\pi}{a}, \qquad a > 0$$

5. Show that

$$\frac{1}{\pi}\int_{-\infty}^{\infty} \frac{\sin bu}{(a+iu)u}\, du = \frac{1}{\pi}\int_{-\infty}^{\infty} \frac{\sin bu}{(a-iu)u}\, du = \frac{1-e^{-ab}}{a}$$

where $a$ and $b$ are positive. From this equation, derive

$$\frac{1}{\pi}\int_{-\infty}^{\infty} \frac{\sin bu}{u(a^2 + u^2)}\, du = \frac{1-e^{-ab}}{a^2}, \qquad a > 0, b > 0$$

6. (a) Let $f(x) = e^{-a|x|}$ where $a$ is a positive constant. Show that

$$\hat{f}(u) = \left(\frac{2}{\pi}\right)^{1/2} \frac{a}{a^2 + u^2}$$

(b) Using (a), show that

$$\frac{\pi}{a} e^{-a|x|} = \int_{-\infty}^{\infty} \frac{\cos ux}{a^2 + u^2}\, du$$

Put $x = 0$ to deduce again 4c.

(c) Show that

$$\int_{-\infty}^{\infty} \frac{du}{(a^2 + u^2)^2} = \frac{\pi}{2a^3}$$

7. In this exercise, you will check Fourier inversion as in the example in the text. Throughout, $a$ denotes a positive constant.

(a) Show that the Fourier transform of the triangular pulse

$$f(x) = \begin{cases} 1 - |x|/a, & |x| \leqslant a \\ 0, & |x| > a \end{cases}$$

is

$$\hat{f}(u) = \left(\frac{2}{\pi}\right)^{1/2} \frac{1-\cos au}{au^2} \left( = \left(\frac{2}{\pi}\right)^{1/2} \frac{2\sin^2(au/2)}{au^2} \right)$$

(b) Check Fourier inversion for this function in the following steps

$$\frac{1}{\sqrt{2\pi}} \int_{-\infty}^{\infty} \hat{f}(u) e^{iux} \, du$$

$$= \frac{2}{\pi a} \int_{0}^{\infty} \frac{(1 - \cos au) \cos ux}{u^2} \, du \quad \text{(even-odd considerations)}$$

$$= \frac{-2}{\pi a} \int_{0}^{\infty} (\cos ux - \tfrac{1}{2}\cos(x+a)u - \tfrac{1}{2}\cos(x-a)u) \, d\left(\frac{1}{u}\right) \quad \text{(trig id)}$$

$$= \frac{2}{\pi a} \int_{0}^{\infty} \frac{-x \sin ux + \tfrac{1}{2}(x+a)\sin(x+a)u + \tfrac{1}{2}(x-a)\sin(x-a)u}{u} \, du \quad \text{(parts)}$$

Then follow the method of the text example.

8. What is the Fourier transform of $g(x) = (\sin bx)/x$, $b > 0$?
9. What is the Fourier transform of $g(x) = 1/(a+ix)$, $a > 0$? (Refer to Exercise 4.)
10. What is the Fourier transform of $g(x) = a/(a^2 + x^2)$, $a > 0$? (Refer to Exercise 6.)
11. What is the Fourier transform of $g(x) = (1 - \cos ax)/ax^2$, $a > 0$? (Refer to Exercise 7.)
12. (Principle behind Exercises 8 through 11) Show that if the Fourier transform of $f(x)$ is $g(u)$, then the Fourier transform of $g(x)$ is $f(-u)$.
13. Suppose that $f(x)$ belongs to $L^2(-\infty, \infty)$. Show that

$$g(u) = \int_{-\infty}^{\infty} f(x) e^{-2\pi iux} \, dx$$

also belongs to $L^2(-\infty, \infty)$ and that

$$f(x) = \int_{-\infty}^{\infty} g(u) e^{2\pi iux} \, du$$

(This definition of the Fourier transform exchanges the $2\pi$ factor in front of the integral for a $2\pi$ factor in the exponent.)

## 8.3 FOURIER SINE AND COSINE TRANSFORMS

In Section 3.3 we studied the odd and even extensions of a function $f(x)$ defined on $0 \leqslant x \leqslant \pi$ and the Fourier sine and cosine series expansions that correspond thereto. These concepts find their analogues in the theory of the Fourier transform.

Suppose we are given a real-valued function $f(x)$ defined on $0 \leqslant x < \infty$, and in $L^2(0, \infty)$, which we wish to represent by a Fourier transform. If we extend the definition of $f(x)$ to $-\infty < x < \infty$ so that the extended function belongs to $L^2(-\infty, \infty)$, then we may deal with the problem in the framework of our theory already developed in Section 8.2. There are two natural ways to extend $f$, as an odd function and as an even function.

*Odd extension*: in $-\infty < x \leqslant 0$ define $f(x) = -f(-x)$.
*Even extension*: in $-\infty < x \leqslant 0$ define $f(x) = f(-x)$.

In both cases, the extended function belongs to $L^2(-\infty, \infty)$ (Exercise 1). The odd extension will be an odd function on $-\infty < x < \infty$; thus its Fourier transform,

$$\hat{f}(u) = \frac{1}{\sqrt{2\pi}} \int_{-\infty}^{\infty} f(x) e^{-iux} dx$$

$$= \frac{1}{\sqrt{2\pi}} \left[ \int_{-\infty}^{\infty} f(x) \cos ux \, dx - i \int_{-\infty}^{\infty} f(x) \sin ux \, dx \right] \tag{8.3.1}$$

reduces to

$$\hat{f}(u) = \frac{-i}{\sqrt{2\pi}} \int_{-\infty}^{\infty} f(x) \sin ux \, dx = \frac{1}{i} \left( \frac{2}{\pi} \right)^{1/2} \int_{0}^{\infty} f(x) \sin ux \, dx \tag{8.3.2}$$

because the integrands on the right side of (8.3.1) are odd and even, respectively. We speak of the expression on the far right of (8.3.2) (delete the $1/i$) as the *Fourier sine transform* of $f(x)$ on $0 \leqslant x < \infty$, which equals $i$ times the Fourier transform of its odd extension. Likewise, the even extension of $f(x)$ will be an even function on $-\infty < x < \infty$; thus its Fourier transform, given by (8.3.1), reduces to

$$\hat{f}(u) = \frac{1}{\sqrt{2\pi}} \int_{-\infty}^{\infty} f(x) \cos ux \, dx = \left( \frac{2}{\pi} \right)^{1/2} \int_{0}^{\infty} f(x) \cos ux \, dx \tag{8.3.3}$$

because the integrands on the right side of (8.3.1) are, in this case, even and odd, respectively. We speak of the expression on the far right of (8.3.3) as the *Fourier cosine transform* of $f(x)$ on $0 \leqslant x < \infty$, which equals the Fourier transform of its even extension.

In the case of Fourier series, we could expand a function defined on $0 \leqslant x \leqslant \pi$ in either a sine or cosine series depending on whether we used its odd or even extension to $[-\pi, \pi]$. In the case of the Fourier transform, we can "expand" a function $f(x)$ defined on $0 \leqslant x < \infty$ as either a sine or cosine integral depending on whether we choose its odd or even extension. For example, suppose $f(x) = e^{-ax}$, $x \geqslant 0$, where $a$ is a positive constant. Its even extension is $e^{-a|x|}$, and the Fourier transform of $e^{-a|x|}$ is the Fourier cosine transform of $e^{-ax}$. Thus

$$e^{-ax} = \frac{2a}{\pi} \int_0^\infty \frac{\cos ux}{a^2 + u^2}\, du, \qquad x \geq 0,\, a > 0$$

This computation was done in Exercise 6 of Section 8.2.

On the other hand, the Fourier transform of the odd extension of $f(x) = e^{-ax}$ is $-i$ times the Fourier sine transform of $e^{-ax}$. Thus

$$e^{-ax} = \frac{2}{\pi} \int_0^\infty \frac{u \sin ux}{a^2 + u^2}\, du, \qquad x > 0,\, a > 0$$

This computation is assigned as Exercise 2.

### Exercises

1. Show that if $f(x)$ is defined on $0 \leq x < \infty$ and if we extend $f$ as an odd function to $-\infty < x < \infty$, then

$$\int_{-\infty}^\infty |f(x)|^2\, dx = 2 \int_0^\infty |f(x)|^2\, dx$$

   Conclude that if $f(x)$ belongs to $L^2(0, \infty)$, then its odd extension belongs to $L^2(-\infty, \infty)$. Formulate and prove a corresponding assertion for the even extension of a function $f(x)$ in $L^2(0, \infty)$.

2. Show that the Fourier sine transform of $e^{-ax}$ is $(2/\pi)^{1/2} u/(a^2 + u^2)$. From this, deduce

$$e^{-ax} = \frac{2}{\pi} \int_0^\infty \frac{u \sin ux}{a^2 + u^2}\, du, \qquad x > 0,\, a > 0$$

3. For real-valued $f(x)$ in $L^2(0, \infty)$, denote its Fourier sine and cosine transforms by $s(f)$ and $c(f)$, respectively:

$$s(f)(u) = \left(\frac{2}{\pi}\right)^{1/2} \int_0^\infty f(x) \sin ux\, dx,$$

$$c(f)(u) = \left(\frac{2}{\pi}\right)^{1/2} \int_0^\infty f(x) \cos ux\, dx$$

   Show that $s(f)$ is odd and $c(f)$ is even. Show that

$$f(x) = \left(\frac{2}{\pi}\right)^{1/2} \int_0^\infty s(f)(u) \sin ux\, du = \left(\frac{2}{\pi}\right)^{1/2} \int_0^\infty c(f)(u) \cos ux\, du$$

4. Let $0 \leq a < b < \infty$, and let $\chi_{[a,b]}(x)$ be the characteristic function of the interval $[a, b]$ (see Section 8.2).

(a) Sketch the odd and even extensions of $\chi_{[a,b]}(x)$.

(b) Compute the Fourier sine transform of $\chi_{[a,b]}(x)$. From your result, deduce that

$$\int_0^\infty \frac{(\cos ua - \cos ub) \sin ux}{u} \, du$$

$$= \begin{cases} \pi/2, & a < x < b \\ 0, & 0 \leqslant x < a \text{ or } x > b \end{cases}$$

(c) Compute the Fourier cosine transform of $\chi_{[a,b]}(x)$. From your result, deduce that

$$\int_0^\infty \frac{(\sin ub - \sin ua) \cos ux}{u} \, du = \begin{cases} \pi/2, & a < x < b \\ 0, & 0 \leqslant x < a \text{ or } x > b \end{cases}$$

5. Let $a > 0$, and define

$$f(x) = \begin{cases} 1 - x/a, & 0 \leqslant x \leqslant a \\ 0, & x > a \end{cases}$$

(a) Sketch the odd and even extensions of $f$.

(b) Compute the Fourier sine transform of $f$. From your result, deduce that

$$\int_0^\infty \frac{ua - \sin ua}{au^2} \sin ux \, du = \begin{cases} (\pi/2)(1 - x/a), & 0 \leqslant x \leqslant a \\ 0, & x > a \end{cases}$$

(c) Compute the Fourier cosine transform of $f$. Compare with Exercise 7 of Section 8.2.

## 8.4  THE FOURIER TRANSFORM IS A UNITARY OPERATOR ON $L^2(-\infty, \infty)$

The Fourier transform $\hat{f}(u)$ of an $f(x)$ in $L^2(-\infty, \infty)$ also belongs to $L^2(-\infty, \infty)$ according to Plancherel's theorem (Theorem 8.1). The Fourier transform of the sum of two functions is the sum of the Fourier transforms:

$$\frac{1}{\sqrt{2\pi}} \int_{-\infty}^\infty (f(x) + g(x)) e^{-iux} \, dx$$

$$= \frac{1}{\sqrt{2\pi}} \int_{-\infty}^\infty f(x) e^{-iux} \, dx + \frac{1}{\sqrt{2\pi}} \int_{-\infty}^\infty g(x) e^{-iux} \, dx$$

And the Fourier transform of $cf(x)$, where $c$ is a real or complex constant, is $c\hat{f}(u)$:

$$\frac{1}{\sqrt{2\pi}} \int_{-\infty}^{\infty} cf(x)e^{-iux}\,dx = c\frac{1}{\sqrt{2\pi}} \int_{-\infty}^{\infty} f(x)e^{-iux}\,dx$$

If we denote by $U$ the operator that takes $f(x)$ to its Fourier transform $\hat{f}(u)$, a notation that we have already anticipated in Section 8.1, then we can summarize the facts above by writing $U(f+g)=U(f)+U(g)$ and $U(cf)=cU(f)$ for all $f$, $g$ in $L^2(-\infty, \infty)$ and all real or complex constants $c$. Thus $U$ is a linear operator (in the sense of Definition 4.9) on the complex Hilbert space $L^2(-\infty, \infty)$. We shall henceforth make the term "Fourier transform" do double duty, using it to label both the mapping $U$ and the image function $\hat{f}(u)$.

Our previous study of Hilbert space operators focused on Hermitian differential operators. Differential operators are unbounded (Definition 4.10) and cannot be everywhere defined. Right away we see a dramatic difference between a differential operator and the Fourier transform, because the Fourier transform is everywhere defined on $L^2(-\infty, \infty)$ (according to Plancherel's theorem). And, not only is the Fourier transform bounded, but it has many other remarkable properties:

**Theorem 8.2.** *The Fourier transform $U$, defined by*

$$U(f)(u) = \frac{1}{\sqrt{2\pi}} \int_{-\infty}^{\infty} f(x)e^{-iux}\,dx \tag{8.4.1}$$

*is an everywhere defined bounded linear operator that maps the Hilbert space $L^2(-\infty, \infty)$ onto itself. The operator $U$ is one-to-one, and its inverse is given by*

$$U^{-1}(f)(x) = \frac{1}{\sqrt{2\pi}} \int_{-\infty}^{\infty} f(u)e^{iux}\,du \tag{8.4.2}$$

*We also have the identities*

$$\langle Uf, Ug \rangle = \langle f, g \rangle \qquad \text{for all } f, g \text{ in } L^2(-\infty, \infty) \tag{8.4.3}$$

*and*

$$U^4 = I \tag{8.4.4}$$

*where $I$ is the identity operator on $L^2(-\infty, \infty)$.*

This theorem is, for the most part, a restatement of Plancherel's Theorem

8.1. The bulk of the proof is left for you to do in Exercise 1. When you do that exercise, and indeed in all manipulations with the Fourier transform, base your arguments on Plancherel's theorem. Understand the assertions of that theorem, accept them, and use them. The theorem of Plancherel provides a beautifully succinct account of the main features of the Fourier transform. If you understand that theorem, you will have a good working knowledge of the Fourier transform, and you will be able to use it in applications. The little book *Fourier Transforms* by R. R. Goldberg (Camb. Univ. Press, 1962) contains one of the more accessible proofs of Plancherel's theorem, but we won't get into the proof in this book.

The only part of Theorem 8.2 not covered in Exercise 1 is the formula $U^4 = I$, and we shall prove that now. Remember that $I$ is the operator on $L^2(-\infty, \infty)$ defined by $I(f) = f$ for every $f$ in $L^2(-\infty, \infty)$, so we have to prove that $U^4 f = f$ for every $f$ in $L^2(-\infty, \infty)$. In our old notation $U(f) = \hat{f}$, so $U^2(f) = U(U(f)) = U(\hat{f})$. If we display the independent variable and use equation (8.2.3') of Section 8.2, we get

$$U(\hat{f})(x) = \frac{1}{\sqrt{2\pi}} \int_{-\infty}^{\infty} \hat{f}(t) e^{-itx}\, dt$$

But according to equation (8.2.4')

$$\frac{1}{\sqrt{2\pi}} \int_{-\infty}^{\infty} \hat{f}(t) e^{-itx}\, dt = \frac{1}{\sqrt{2\pi}} \int_{-\infty}^{\infty} \hat{f}(t) e^{it(-x)}\, dt = f(-x)$$

Putting these results together, we get

$$U^2(f)(x) = f(-x) \tag{8.4.5}$$

valid for every $f$ in $L^2(-\infty, \infty)$, an interesting equation in its own right. Equation (8.4.5) says that if we apply the Fourier transform twice to a function $f(x)$ in $L^2(-\infty, \infty)$, the net result is simply $f(-x)$, whose graph is easily described: interchange the graph of $f(x)$ for positive $x$ with that for negative $x$. If we define a function $f^-$ by $f^-(x) = f(-x)$, then equation (8.4.5) says simply $U^2(f) = f^-$. Then $U^2(f^-) = f$, because $U^2(f^-)(x) = f^-(-x) = f(-(-x)) = f(x)$. Finally, $U^4(f) = U^2(U^2(f)) = U^2(f^-) = f$, which proves equation (8.4.4) of Theorem 8.2.

It is a fact that, given an everywhere defined bounded linear operator $T$ on a Hilbert space $H$, there is another bounded everywhere defined linear operator $T^*$, called the *adjoint* of $T$, which satisfies $\langle Tf, g \rangle = \langle f, T^*g \rangle$ for all $f, g$ in $H$. $T^*$ is uniquely determined by that condition. We have already alluded to the adjoint in Section 4.4. We take up the question: *What is the adjoint $U^*$ of the Fourier transform $U$?*

To answer this question, we need to determine the form of the operator $U^*$ that validates the equation $\langle Uf, g \rangle = \langle f, U^*g \rangle$ for all $f, g$ in $L^2(-\infty, \infty)$. For any $g \in L^2(-\infty, \infty)$ there is a unique $h \in L^2(-\infty, \infty)$ with $g = U(h)$. This follows from the "one-to-one" and "onto" statements in Theorem 8.2. Then $U^{-1}(g) = h$, as also follows from Theorem 8.2. Assemble these relations and use (8.4.3):

$$\langle Uf, g \rangle = \langle Uf, Uh \rangle = \langle f, h \rangle = \langle f, U^{-1}g \rangle$$

Thus $U^* = U^{-1}$. We have answered our question: *the adjoint of the Fourier transform is its inverse.*

A selfadjoint operator $T$ is characterized by the equation $T^* = T$; it is its own adjoint. The selfadjoint operators are a subclass of the Hermitian operators we have dealt with exclusively in earlier chapters. The Fourier transform $U$ has a character quite different from that of a selfadjoint operator because it obeys the equation $U^* = U^{-1}$ rather than $T^* = T$.

Operators that satisfy $U^* = U^{-1}$ or, equivalently, $UU^* = U^*U = I$ are called *unitary*. With this terminology we can summarize our result: *the Fourier transform is a unitary operator on* $L^2(-\infty, \infty)$–which gives meaning to the title of this section. A unitary operator can be interpreted as a rotation (Exercise 7).

The adjoint $U^*$ of $U$ has an interesting alternative formula that involves complex conjugation:

$$U^*(f) = \overline{U(\bar{f})} \tag{8.4.6}$$

(Exercise 8). From this, we derive $U(f) = \overline{U^*(\bar{f})}$. Both formulas are valid for all $f$ in $L^2(-\infty, \infty)$.

The equation $\langle Uf, g \rangle = \langle f, U^*g \rangle$ has practical applications to the evaluation of definite integrals. In integral form, this equation says

$$\int_{-\infty}^{\infty} Uf\bar{g}\, dx = \int_{-\infty}^{\infty} f\overline{U^*g}\, dx$$

Replacing $g$ by $\bar{g}$ and using (8.4.6) and its consequences, we get

$$\int_{-\infty}^{\infty} \hat{f}(x)g(x)\, dx = \int_{-\infty}^{\infty} f(x)\hat{g}(x)\, dx \tag{8.4.7}$$

In the applications of mathematics to physics and engineering, one frequently needs to evaluate complicated definite integrals. If you can recognize your integrand as the product of two $L^2$ functions, one of which is a known Fourier transform, then equation (8.4.7) will transform your integral dramatically and often put it in a form that can be evaluated. For example, consider the problem of evaluating the integral

$$\int_{-\infty}^{\infty} \frac{\sin bx}{x} e^{-a|x|} dx, \qquad a > 0, b > 0$$

We recognize the integrand as a product $\hat{f}(x)g(x)$ where $f(x) = (\pi/2)^{1/2} \times \chi_{[-b,b]}(x)$ (Section 8.2) and $g(x) = e^{-a|x|}$. Hence, using (8.4.7) and Exercise 6a of Section 8.2,

$$\int_{-\infty}^{\infty} \frac{\sin bx}{x} e^{-a|x|} dx = \int_{-\infty}^{\infty} \left[ \left(\frac{\pi}{2}\right)^{1/2} \chi_{[-b,b]}(x) \right] \left[ \left(\frac{2}{\pi}\right)^{1/2} \frac{a}{a^2 + x^2} \right] dx$$

$$= 2a \int_0^b \frac{dx}{a^2 + x^2} = 2 \arctan \frac{b}{a}$$

Similar evaluations are included as Exercises 9 through 12.

We close this section with some other helpful rules obeyed by the Fourier transform operator $U$. If $f(x)$ belongs to $L^2(-\infty, \infty)$, then $f(ax)$, $f(-x)$, and $f(x+c)$ also belong to $L^2(-\infty, \infty)$, where $a > 0$ and $c$ is real. The Fourier transforms of these functions are related to that of $f(x)$ as follows:

$$U(f(ax)) = \frac{1}{a} U(f)\left(\frac{u}{a}\right), \qquad a > 0 \tag{8.4.8}$$

$$U(f(-x)) = U(f)(-u) \tag{8.4.9}$$

$$U(f(x + c)) = e^{iuc} U(f)(u), \qquad c \text{ real} \tag{8.4.10}$$

In our alternative notation, we can write these formulas $\widehat{f(ax)} = (1/a)\hat{f}(u/a)$, $\widehat{f(-x)} = \hat{f}(-u)$, and $\widehat{f(x+c)} = e^{iuc}\hat{f}(u)$, respectively. Proofs and applications of these formulas are given as Exercises 13, 14, and 15.

Given functions $f(x)$, $g(x)$ in $L^2(-\infty, \infty)$, we define their *convolution* $f * g(x)$ by

$$f * g(x) = \frac{1}{\sqrt{2\pi}} \int_{-\infty}^{\infty} f(t)g(x - t) dt \tag{8.4.11}$$

For example, if $a > 0$, $b > 0$,

$$\chi_{[-b,b]}(x) * \frac{a}{a^2 + x^2} = \frac{1}{\sqrt{2\pi}} \int_{-\infty}^{\infty} \chi_{[-b,b]}(t) \frac{a}{a^2 + (x - t)^2} dt$$

$$= \frac{a}{\sqrt{2\pi}} \int_{-b}^{b} \frac{dt}{a^2 + (x - t)^2} = \frac{a}{\sqrt{2\pi}} \int_{x-b}^{x+b} \frac{dt}{a^2 + t^2}$$

$$= \frac{1}{\sqrt{2\pi}} \left[ \arctan \frac{x + b}{a} - \arctan \frac{x - b}{a} \right]$$

As the product of two $L^2$ functions is absolutely integrable, the integral in
(8.4.11) is absolutely convergent. But the function it represents need not be $L^2$.
So, if we wish to stay within our $L^2$ framework, we must add the assumption
that $f * g(x)$ belongs to $L^2(-\infty, \infty)$.

**Theorem 8.3.**  *If the functions $f(x)$, $g(x)$ and their convolution $f * g(x)$ all belong
to $L^2(-\infty, \infty)$, then*

$$U(f * g) = U(f)U(g) \tag{8.4.12}$$

*In other words, the Fourier transform of the convolution is the product of the
Fourier transforms.*

You will do the proof as Exercise 18. The proof relies mainly on an
interchange of the order of integration, which you may do without
justification.

Equation (8.4.12) has the equivalent form $U^{-1}(\hat{f}\hat{g}) = f * g$. In other words,
*the inverse Fourier transform of the product $\hat{f}\hat{g}$ is the convolution of $f$ and $g$.* For
example, what is the inverse Fourier transform of $(\sin bu)/u(a^2 + u^2)$, $a > 0$,
$b > 0$? Write this function as a product

$$\left( \frac{\sin bu}{u} \right) \left( \frac{1}{a^2 + u^2} \right)$$

then recognize that

$$\hat{f}(u) = \frac{\sin bu}{u} \qquad \text{where } f(x) = \left( \frac{\pi}{2} \right)^{1/2} \chi_{[-b, b]}(x)$$

and

$$\hat{g}(u) = \frac{1}{a^2 + u^2} \qquad \text{where } g(x) = \left( \frac{\pi}{2} \right)^{1/2} \frac{1}{a} e^{-a|x|}$$

the latter from Exercise 6a of Section 8.2. Thus

$$U^{-1}\left[ \frac{\sin bu}{u(a^2 + u^2)} \right](x) = f * g(x) = \frac{1}{\sqrt{2\pi}} \left( \frac{\pi}{2a} \right) \int_{-\infty}^{\infty} \chi_{[-b, b]}(t) e^{-a|x - t|} dt$$

$$= \frac{1}{2a} \left( \frac{\pi}{2} \right)^{1/2} \int_{x-b}^{x+b} e^{-a|u|} du$$

$$= \frac{1}{a^2} \left( \frac{\pi}{2} \right)^{1/2} \times \begin{cases} 1 - e^{-ab} \cosh ax, & |x| \leqslant b \\ e^{-a|x|} \sinh ab, & |x| \geqslant b \end{cases}$$

Remembering equation (8.4.2) and using the fact that $(\sin bu)/u(a^2+u^2)$ is an even function of $u$, we can rewrite the result just obtained as

$$\int_0^\infty \frac{\sin bu \cos xu}{u(a^2+u^2)}\, du = \frac{\pi}{2a^2} \times \begin{cases} 1-e^{-ab}\cosh ax, & |x| \leq b \\ e^{-a|x|}\sinh ab, & |x| \geq b \end{cases}$$

## Exercises

In all of these exercises $U$ stands for the Fourier transform and all function symbols $f, g, \ldots$ stand for functions in $L^2(-\infty, \infty)$.

1. (Proof of Theorem 8.2)
   (a) Equation (8.4.3) of Theorem 8.2, namely $\langle Uf, Ug \rangle = \langle f, g \rangle$, is just a restatement of an equation from Plancherel's Theorem 8.1. Which equation?
   (b) Using the equation $\langle Uf, Ug \rangle = \langle f, g \rangle$, prove that $U$ is bounded (Definition 4.10). In fact, show that the inequality involved is actually an equality, and find the constant $M$. What is the domain of $U$?
   (c) Using your result from (b), prove that $\|Uf\| = 0$ implies $\|f\| = 0$. In the implication "$\|f\| = 0 \Rightarrow f = 0$," what meaning do we attach to the equality $f = 0$? What is $\ker(U)$? Prove that $U$ is one-to-one. (Refer to Exercise 8 of Section 4.2.)
   (d) Using Plancherel's Theorem 8.1, show that the operator $U^{-1}$ as given in equation (8.4.2) of Theorem 8.2 satisfies $U^{-1}(Uf) = f$ for all $f$ in $L^2(-\infty, \infty)$. Show also that $U^{-1}f = \hat{f}(-u)$ and, using this, show that $U(U^{-1}f) = f$ for all $f$ in $L^2(-\infty, \infty)$.

2. (a) Show that
   $$U^2\left(\frac{\sin x}{x}\right) = \frac{\sin x}{x} \quad \text{and that} \quad U^2\left(\frac{1}{1+x^2}\right) = \frac{1}{1+x^2}$$
   (b) Show that $U^2(f) = f$ if $f$ is even and $U^2f = -f$ if $f$ is odd.

3. Show that $U^5 = U$ and that $U^3 = U^{-1}$.

4. (a) From memory, write down the definition of eigenvalue and eigenfunction for the operator $U$. Check with Definition 4.4.
   (b) Show that any eigenvalue $\lambda$ of $U$ must satisfy the equation $\lambda^4 = 1$. Thus there are only four possible eigenvalues of the Fourier transform. Find them.

5. Consider the finite-dimensional Hilbert space $\mathbb{C}^n$ consisting of all $n$-tuples with real or complex entries, and put the usual inner product on $\mathbb{C}^n$: if $a = (\alpha_1, \alpha_2, \ldots, \alpha_n)$ and $b = (\beta_1, \beta_2, \ldots, \beta_n)$, then
   $$\langle a, b \rangle = \alpha_1 \bar{\beta}_1 + \alpha_2 \bar{\beta}_2 + \cdots + \alpha_n \bar{\beta}_n$$
   Suppose $M = (\mu_{ij})$ is the matrix of a linear operator $T$ with respect to the

standard basis $e_1=(1,0,0,\ldots,0)$, $e_2=(0,1,0,\ldots,0),\ldots$, $e_n=(0,0,0,\ldots,1)$. Show that, if $N=(\eta_{ij})$ is the matrix of $T^*$, then $\eta_{ij}=\bar{\mu}_{ji}$. Thus $N$ is the conjugate transpose of $M$.

6. (Continuation)  Show that the operator $T$ is unitary if, and only if, the $n$ column vectors of its matrix are orthonormal and the $n$ row vectors of its matrix are also orthonormal. (Remarkably enough, in this finite-dimensional case, either one of these conditions implies the other.) Show that, if $T$ is unitary, then $|\det(T)|=1$.

7. (Continuation)  Specialize to $\mathbf{R}^2$. Show that if $T$ is unitary and $\det(T)=1$, then the matrix of $T$ has the form

$$\begin{bmatrix} \cos\theta & \sin\theta \\ -\sin\theta & \cos\theta \end{bmatrix}$$

Thus $T$ is a rotation through angle $\theta$. (A unitary operator on a real space is frequently called "orthogonal.")

8. Show that $U^*(f)=\overline{U(\bar{f})}$. From this, deduce that $U(f)=\overline{U^*(\bar{f})}$.

9. Evaluate

$$\int_0^\infty \left(\frac{1-\cos bx}{x^2}\right) e^{-ax}\,dx, \qquad a>0, b>0$$

in two different ways.

10. Evaluate

$$\int_{-\infty}^\infty \frac{\sin bx}{x(a+ix)}\,dx, \qquad a>0, b>0$$

and check your answer against that of Exercise 5 of Section 8.2.

11. Let $a>0$, $b>0$. Show that

$$\int_{-\infty}^\infty \frac{(\sin bx)(1-\cos ax)}{x^3}\,dx$$

$$= \begin{cases} \pi a^2/2, & \text{if } a\leqslant b \\ (\pi ab/2)(2-b/a), & \text{if } a\geqslant b \end{cases}$$

12. Show that

$$\int_0^b \frac{1-\cos ax}{ax^2}\,dx = \int_0^a \left(1-\frac{x}{a}\right)\frac{\sin bx}{x}\,dx, \qquad a>0, b>0$$

13. Derive equations (8.4.8), (8.4.9), and (8.4.10).

14. Using equation (8.4.8) and the fact that $U(\chi_{[-1,1]}(x)) = (2/\pi)^{1/2}(\sin u)/u$, find again $U(\chi_{[-b,b]}(x))$, $b > 0$.

15. Find the real numbers $c$ and $d$ such that $\chi_{[a,b]}(x) = \chi_{[-d,d]}(x+c)$. Then, using equation (8.4.10) and $U(\chi_{[d,d]}(x)) = (2/\pi)^{1/2}(\sin du)/u$, compute again $U(\chi_{[a,b]}(x))$. Reconcile your answer with that in Section 8.2.

16. (a) Prove that $f * g = g * f$.
    (b) Prove that if $f$ and $g$ are even so is $f * g$.

17. (a) On graph paper, draw a careful graph of $(\pi/2)^{1/2}\chi_{[-1,1]} * \chi_{[-2,2]}(x)$.
    (b) Given $0 < a < b$, evaluate $(\pi/2)^{1/2}(1/a)\chi_{[-a,a]} * \chi_{[-b,b]}(x)$.

18. Prove that $U(f * g) = \hat{f}\hat{g}$.

19. Find the function $h(x)$ that has the Fourier transform

$$\hat{h}(u) = \frac{\sin bu}{u(a + iu)}, \qquad a > 0, b > 0$$

## 8.5 THE FOURIER TRANSFORM CONVERTS DIFFERENTIATION INTO MULTIPLICATION BY THE INDEPENDENT VARIABLE

We shall continue to use the symbol $U$ to stand for the Fourier transform. Keep in mind that the Fourier transform is a unitary operator on the Hilbert space $L^2(-\infty, \infty)$: an everywhere defined linear operator that satisfies $\langle U(f), U(g) \rangle = \langle f, g \rangle$ for all $f, g$ in $L^2(-\infty, \infty)$. In this section we shall prove and exploit one of its more remarkable and useful properties: *the Fourier transform converts the operation of differentiation into multiplication by the independent variable.*

We begin by describing exactly what "differentiation" means in the Hilbert space context. The following definition uses once again the prescription stated in Section 4.1 and applied successively in the definitions of the Legendre, Hermite, Bessel, Laplacian, and Laguerre operators, namely the prescription differential operator = differential expression + domain.

The crucial distinction between a differential *expression* and the Hilbert space differential *operator* induced by that expression has been a dominant theme throughout this book. Differentiation, by itself, is an extraordinarily powerful process applicable in a broad range of situations, as we know from calculus. When using differentiation outside of the Hilbert space context, use it as you always have. But, within the Hilbert space context, to define the Hilbert space operator that goes with differentiation we must specify the domain of the operator—the subspace of functions on which differentiation acts.

**Definition 8.4.** The differentiation operator $D$ on $L^2(-\infty, \infty)$ has the

*Differential expression* $l(f) \equiv (1/i)/df/dx$ and
*Domain* consisting of all $f$ in $L^2(-\infty, \infty)$ that are compatible with $l$

The domain of the differentiation operator $D$ of Definition 8.4 is as large as possible. It consists of all functions $f$ in $L^2(-\infty, \infty)$ that are compatible with $l(f) \equiv (1/i)f'$, i.e., all $f \in L^2(-\infty, \infty)$ such that also $l(f) \in L^2(-\infty, \infty)$. There are no explicit boundary conditions, but, as we shall see in a moment, the compatibility requirement forces $f(x) \to 0$ as $x \to \pm\infty$. So any function $f(x)$ in the domain of $D$ does automatically satisfy those boundary conditions.

There is also a necessary condition inherent in the requirement that $l(f) \in L^2(-\infty, \infty)$, namely that $f(x)$ be continuous on $-\infty < x < \infty$. If $f(x)$ shows a discontinuity, then $l(f) \equiv (1/i)f'(x)$ would involve a delta function, thus not belong to $L^2(-\infty, \infty)$ (refer back to Section 4.1). So, for $f(x)$ to belong to the domain of $D$, it must at least be true that $f(x)$ is continuous everywhere. Also, of course, the derivative $f'(x)$ must exist at most points. But we don't require differentiability at every point—that would be too restrictive. A function $f(x)$ can fail to be differentiable at finitely many points and still belong to the domain of $D$ (Exercises 1 and 2). (The accurate technical term here is that $f(x)$ should be absolutely continuous, but we are steering clear of those technical mathematical details.)

**Lemma 8.5.** *If $f(x)$ belongs to the domain of the differentiation operator $D$ of Definition 8.4, then $f(x) \to 0$ as $x \to \pm\infty$.*

*Proof.* As both $f$ and $f'$ are square integrable over $-\infty < x < \infty$, the products $f\bar{f}'$ and $\bar{f}f'$ are both absolutely integrable (refer to Section 3.9), thus also the sum $f\bar{f}' + \bar{f}f'$. Now $|f|^2 = f\bar{f}$ so $(|f|^2)' = f\bar{f}' + f'\bar{f}$, thus $(|f|^2)'$ is also absolutely integrable. Hence

$$\lim_{x \to \infty} \int_0^x (|f|^2)' \, dt = \lim_{x \to \infty} |f(x)|^2 - |f(0)|^2$$

exists and is finite. We have $|f(x)|^2 \geq 0$, so to say that $\lim_{x \to \infty} |f(x)|^2$ exists and is finite means that, as $x$ gets very, very large, the graph of $|f(x)|^2$ levels out to a constant nonnegative value. But the integral

$$\int_{-\infty}^{\infty} |f(x)|^2 \, dx$$

which measures the area under the graph of $|f(x)|^2$, is finite because $f(x) \in L^2(-\infty, \infty)$. Hence we must have $|f(x)| \to 0$ as $x \to \infty$, because if $|f(x)|$ approached a positive constant as $x \to \infty$, the area under the graph of $|f(x)|^2$

would be infinite. A similar argument shows that $f(x)\to 0$ as $x\to-\infty$. That completes the proof of Lemma 8.5.

**Theorem 8.6.** *The differentiation operator D of Definition 8.4 is Hermitian. It is not positive semidefinite. It has no eigenvalues.*

The proof, which uses Lemma 8.5, is assigned as Exercises 5, 6, and 7.

Consider now the operator $D^2$, which signifies the differentiation operator $D$ applied twice: $D^2(f)=D(D(f))$. Its domain, by definition, consists of all functions $f$ in the domain of $D$ such that $D(f)$ also belongs to the domain of $D$. Thus the domain of $D^2$ will be smaller than the domain of $D$ (Exercise 8). On its domain, the action of $D^2$ is given by $l^2$,

$$l^2(f) = l(l(f)) = \frac{1}{i}(l(f))' = \frac{1}{i}\left(\frac{1}{i}f'\right)' = -f''$$

which is the differential expression for the one-dimensional Laplacian (Definition 4.13). Moreover, to say that $Df=(1/i)f'$ belongs to the domain of $D$ is to say that $-f''\in L^2(-\infty,\infty)$, i.e., that $f$ is compatible with $l^2(f)=-f''$. We shall write $\Delta$ for $D^2$ and shall refer to this operator as the *one-dimensional Laplacian in* $L^2(-\infty,\infty)$, thus broadening the applicability of the symbol $\Delta$ and the term Laplacian as these were defined in Section 4.5.

**Theorem 8.7.** *The one-dimensional Laplacian $\Delta=D^2$ in $L^2(-\infty,\infty)$, which has the*

*Differential expression $l^2(f)\equiv-f''$ and*
*Domain($\Delta$)= all functions in $L^2(-\infty,\infty)$ compatible with $l^2(f)\equiv-f''$*

*is Hermitian and positive definite. It has no eigenvalues.*

Proof in Exercises 9 and 10. To do these exercises you will need this fact:

**Lemma 8.8.** *If $f(x)$ belongs to the domain of the one-dimensional Laplacian $\Delta$ in $L^2(-\infty,\infty)$, then $f'(x)\in L^2(-\infty,\infty)$ and both $f(x)\to 0$ and $f'(x)\to 0$ as $x\to\pm\infty$.*

*Proof.* If $f(x)$ belongs to the domain of $\Delta=D^2$, then $f$ belongs to the domain of $D$ and $Df=(1/i)f'$ also belongs to the domain of $D$. Because $f$ belongs to the domain of $D$, $f'=il(f)$ belongs to $L^2(-\infty,\infty)$ (compatibility) and $f(x)\to 0$ as $x\to\pm\infty$ (Lemma 8.5). Because $Df=(1/i)f'$ belongs to the domain of $D$, $f'(x)\to 0$ as $x\to\pm\infty$ by Lemma 8.5 again. That completes the proof. An alternative proof is given in 4.63 in Appendix 4.A.

Note also that a necessary condition for a function $f(x)$ to belong to the domain of the Laplacian $\Delta$ is that $f'(x)$ be continuous for every $x$ (Sections 4.1 and Appendix 4.A).

**Theorem 8.9.** *Let $\phi(x)$ be a real-valued function defined on $-\infty < x < \infty$. The operator $M_\phi$ on $L^2(-\infty, \infty)$ defined by*

*Law of action $M_\phi(f) = \phi(x)f(x)$*
*Domain$(M_\phi)$ = all functions $f(x)$ in $L^2(-\infty, \infty)$ such that also $\phi(x)f(x) \in L^2(-\infty, \infty)$*

*is Hermitian.*

The proof is assigned as Exercise 13. We shall refer to the operator $M_\phi$ as *multiplication by $\phi$*. We shall be mainly concerned with $M_x$, often referred to as *multiplication by the independent variable*. It is an unbounded operator (Exercise 12).

We come now to the remarkable connection, alluded to in the first paragraph, between the Fourier transform $U$, the differentiation operator $D$, and multiplication $M_x$ by the independent variable $x$.

**Theorem 8.10.** *If the function $f$ belongs to the domain of $D$, then $U(f)$ belongs to the domain of $M_x$ and $UD(f) = M_x U(f)$. Briefly*

$$\boxed{UD = M_x U} \tag{8.5.1}$$

*Proof.* If $f$ belongs to the domain of $D$, then $D(f) = (1/i)f'$ belongs to $L^2(-\infty, \infty)$ so that we may apply the Fourier transform to it,

$$UD(f)(x) = U\left(\frac{f'}{i}\right)(x) = \frac{1}{i\sqrt{2\pi}} \lim_{A \to \infty} \int_{-A}^{A} f'(t)e^{-ixt}\, dt$$

$$UD(f)(x) = \frac{1}{i\sqrt{2\pi}} \lim_{A \to \infty} \left[ (e^{-ixA}f(A) - e^{ixA}f(-A)) - \int_{-A}^{A} (-ix)f(t)e^{-ixt}\, dt \right] \tag{8.5.2}$$

Consider the integrated term in (8.5.2) (which resulted from an integration by parts). Because $|e^{\pm ixA}| = 1$ we have $|e^{-ixA}f(A) - e^{ixA}(f(-A))| \leqslant |f(A)| + |f(-A)|$. Since $f(\pm A) \to 0$ as $A \to \infty$ (by Lemma 8.5), the entire term vanishes in the limit. Thus, in the limit, (8.5.2) becomes

$$UD(f)(x) = \frac{x}{\sqrt{2\pi}} \int_{-\infty}^{\infty} f(t)e^{-ixt}\, dt = xU(f)(x) \tag{8.5.3}$$

As $UD(f)(x)$ belongs to $L^2(-\infty, \infty)$, so does the right side of (8.5.3). Hence $U(f) \in \text{domain}(M_x)$ and $UD(f)(x) = M_x U(f)(x)$. That completes the proof.

We can cast equation (8.5.1) in an alternative form:

$$\frac{1}{\sqrt{2\pi}} \int_{-\infty}^{\infty} \left(\frac{1}{i} f'(t)\right) e^{-ixt} dt = \frac{x}{\sqrt{2\pi}} \int_{-\infty}^{\infty} f(t) e^{-ixt} dt \qquad (8.5.4)$$

valid for all $f$ in the domain of the differentiation operator $D$—which is to say, for all functions $f$ in $L^2(-\infty, \infty)$ whose derivative $f'$ also belongs to $L^2(-\infty, \infty)$.

Here are two immediate applications of Theorem 8.10.

**Application 1.**  Suppose $f$ belongs to the domain of the differentiation operator $D$, and suppose we know its Fourier transform $\hat{f}(u)$. What is the Fourier transform of $D(f)$? It is simply $u\hat{f}(u)$ because $UD(f) = M_u U(f) = M_u(\hat{f}) = u\hat{f}(u)$. For example, the triangular pulse, $f(x) = 1 - |x|/a$ when $|x| \leqslant a$, $f(x) = 0$ when $|x| \geqslant a$, lies in the domain of $D$ (Exercise 1). Its Fourier transform is $\hat{f}(u) = (2/\pi)^{1/2}(1 - \cos au)/au^2$ (Exercise 7 of Section 8.2). Then, to compute the Fourier transform of $D(f)$, which is an awkward double-step function $(D(f)(x) = -i, \ -a < x < 0; \ D(f)(x) = +i, \ 0 < x < a; \ D(f)(x) = 0$ elsewhere), we need do no computation at all. The Fourier transform of $D(f)$ is $u\hat{f}(u) = (2/\pi)^{1/2}(1 - \cos au)/au$. Exercises 15 through 17 provide other instances of Application 1.

**Application 2.**  Suppose we are given a function $g$ in $L^2(-\infty, \infty)$ and we wish to solve the operator equation $D(f) = g$. That is, we wish to find a function $f$ in the domain of the differentiation operator $D$ such that $D(f) = g$. According to Exercise 4b, this is not always possible. (If it is possible, there is only one solution $f$. See Exercises 4a and 18.) Our equation (8.5.1), namely $UD = M_x U$, leads to an interesting necessary condition on the function $g$ in order that $D(f) = g$ be solvable: if $D(f) = g$ has a solution, then $\hat{g}(u)/u \in L^2(-\infty, \infty)$. To prove this, apply the Fourier transform $U$ to the equation $D(f) = g$: $UD(f) = U(g) = \hat{g}$. Then use equation (8.5.1): $UD(f) = M_u U(f) = uU(f)$. (We have labeled our transform variable $u$ instead of $x$.) Combine these equations to get $U(f) = \hat{g}(u)/u$. Because $U(f) \in L^2(-\infty, \infty)$, it follows that $\hat{g}(u)/u \in L^2(-\infty, \infty)$, which proves the assertion. In contrapositive form, our necessary condition becomes: if $g \in L^2(-\infty, \infty)$ and $\hat{g}(u)/u \notin L^2(-\infty, \infty)$, then $D(f) = g$ has no solution.

Those are two direct applications of our formula $UD = M_x U$. Using a few simple manipulations, we can derive other formulas that have powerful and surprising implications. We shall take those up in the next section.

## Exercises

1. Let $a > 0$.
   (a) Sketch the graph of the triangular pulse

   $$f(x) = \begin{cases} 1 - |x|/a, & |x| \leqslant a \\ 0, & |x| \geqslant a \end{cases}$$

   (b) Explain why $f(x)$ is continuous at every $x$, $-\infty < x < \infty$.
   (c) Does $f(x)$ belong to $L^2(-\infty, \infty)$? Prove your answer.
   (d) Using your graph in (a), deduce that $f(x)$ fails to be differentiable at exactly three points, and find those points. Find $f'(x)$ where it exists.
   (e) Prove that $f(x)$ is in the domain of the differentiation operator $D$ of Definition 8.4. [Hint: Changing the values of a function $g(x)$ at finitely many points cannot alter the value of the integral $\int_{-\infty}^{\infty} |g(x)|^2 \, dx$.]

2. (a) Show that $e^{-|x|}$ belongs to the domain of the operator $D$. Is $e^{-|x|}$ differentiable everywhere?
   (b) Show that if $p(x)$ is any polynomial, then $p(x)e^{-x^2/2}$ belongs to the domain of $D$.

3. Let $f(x) = \sin(x^2)/x$, $x \neq 0$; $f(0) = 0$.
   (a) Show that $f(x)$ is continuous at $x=0$; i.e., show that $\lim_{x \to 0} f(x) = f(0)$. Conclude that $f(x)$ is continuous everywhere.
   (b) By showing that $\lim_{h \to 0} (f(0+h) - f(0))/h$ exists, prove that $f(x)$ is differentiable at $x=0$. Find $f'(0)$.
   (c) Show that for $x \neq 0$, $f'(x) = 2\cos(x^2) - (\sin(x^2))/x^2$. Using this fact and (b), conclude that $f(x)$ is differentiable everywhere.
   (d) Show that $f(x)$ belongs to $L^2(-\infty, \infty)$. [Hint: Write $\int_{-\infty}^{\infty} |f(x)|^2 \, dx = 2\int_0^1 |f(x)|^2 \, dx + 2\int_1^{\infty} |f(x)|^2 \, dx$, and show that both integrals on the right are finite.)
   (e) Does $f(x)$ belong to the domain of the differentiation operator $D$ of Definition 8.4? Prove your answer. (Most of the credit for this exercise rides on this last proof.)

4. (a) Show that $\ker(D) = 0$. Prove that $D$ is one-to-one.
   (b) Show that $x/(1+x^2)$ belongs to $L^2(-\infty, \infty)$. Does the equation $Df = x/(1+x^2)$ have a solution $f$? Prove your answer. Is $D$ an "onto" map?

5. Suppose that $f$ and $g$ belong to the domain of the differentiation operator $D$ of Definition 8.4. Let $-\infty < r < 0 < s < \infty$. Show that

   $$\int_r^s D(f)\bar{g} \, dx = \frac{1}{i}(f(s)\overline{g(s)} - f(r)\overline{g(r)}) + \int_r^s f\overline{D(g)} \, dx$$

   Using this equation, together with Lemma 8.5, prove that $D$ is Hermitian;

i.e., prove that $\langle D(f), g \rangle = \langle f, D(g) \rangle$ for every $f, g$ in the domain of $D$, where $\langle \cdot, \cdot \rangle$ is the inner product in $L^2(-\infty, \infty)$.

6. (a) (Continuation) Prove that $\langle D(f), f \rangle = 0$ for every *real-valued f* in the domain of the differentiation operator $D$.

   (b) Let $a$ be a positive constant, let

   $$g(x) = \begin{cases} a - |x| & \text{when } |x| \leqslant a \\ 0 & \text{when } |x| \geqslant a \end{cases}$$

   and define $h(x) = g(x) + ig(x - a)$. (Note that, by Exercise 1, $g(x)$ belongs to the domain of $D$; hence $h(x) = g(x) + ig(x - a)$ also belongs.) By considering $\langle D(h), h \rangle$ and $\langle D(\bar{h}), \bar{h} \rangle$, prove that $\langle D(f), f \rangle$ takes on *all* positive and negative values as $f$ runs through the domain of $D$. Is the differentiation operator $D$ positive semidefinite?

7. (Continuation) (a) Show that if $D$ has an eigenvalue, then that eigenvalue must be real.

   (b) Prove that the differentiation operator $D$ has no eigenvalues; i.e., prove that if $D(f) = \lambda f$ for a function $f$ in the domain of $D$, then $f = 0$.

8. Consider the one-dimensional Laplacian $\Delta$ on $L^2(-\infty, \infty)$ ($\Delta$ is defined in Theorem 8.7). Show that the triangular pulse of Exercise 1 does not belong to the domain of $\Delta$.

9. Using Theorems 8.6 and 8.7, show that for all $f, g$ in the domain of the Laplacian $\Delta$, we have $\langle \Delta(f), g \rangle = \langle D(f), D(g) \rangle$. Using this equation, prove the following part of Theorem 8.7: $\Delta$ is Hermitian and positive definite.

10. (Continuation) (a) Show that if $\Delta$ has an eigenvalue, then that eigenvalue must be real and positive.

   (b) Prove that the one-dimensional Laplacian $\Delta$ of Theorem 8.7 has no eigenvalues; i.e., prove that if $\Delta(f) = \lambda f$ for a function $f$ in the domain of $\Delta$, then $f = 0$.

11. Show that if $\phi(x)$ is a bounded function, i.e., $|\phi(x)| \leqslant M$, $-\infty < x < \infty$, then $M_\phi$ is a bounded operator (Definition 4.10).

12. (a) For $a > 0$, let $f_a(x) = \chi_{[-a,a]}(x)$ (refer to Section 8.2). Compute $\|M_x(f_a)\| / \|f_a\|$. Is $M_x$ a bounded operator? Explain.

   (b) Give an example of a function in $L^2(-\infty, \infty)$ but not in the domain of $M_x$.

13. (a) Prove that, if $\phi$ is real valued, then $M_\phi$ is Hermitian.

   (b) Show that $M_x$ is not positive semidefinite.

14. Prove that $M_x$ has no eigenvalues.

15. What is the Fourier transform of $D(\sin bx/x) = (bx \cos bx - \sin bx)/ix^2$, $b > 0$? [Ans.: $(\pi/2)^{1/2} u \chi_{[-b,b]}(u)$.]

16. What is the Fourier transform of $-1/(a + ix)^2$, $a > 0$?

17. What is the Fourier transform of $2iax/(a^2+x^2)^2$, $a>0$?

18. Using Exercise 4a, prove that if the operator equation $D(f)=g$ has a solution, then that solution is unique. That is, prove that if $D(f_1)=g$ and $D(f_2)=g$ then $f_1=f_2$.

19. Does $D(f) = \chi_{[-b,b]}(x)$ have a solution? [Hint: $\hat{\chi}_{[-b,b]}(u) = (2/\pi)^{1/2}(\sin bu)/u$.]

# 8.6  THE EIGENVALUES AND EIGENFUNCTIONS OF THE FOURIER TRANSFORM

In Theorem 8.10 of the preceding section we proved the formula $UD=M_xU$, where $D$ is the differentiation operator, $U$ is the Fourier transform, and $M_x$ is multiplication by the independent variable $x$. This simple-looking symbolic formula, translated into words, says this: the Fourier transform converts differentiation into multiplication by the independent variable. In the last section we studied a couple of direct consequences of this fact. Additional surprising and powerful consequences can be derived from $UD=M_xU$ by our operator methods.

**Theorem 8.11.** *The following four operator equations are true:*

$$U^2M_{x^2} = M_{x^2}U^2 \tag{8.6.1}$$

$$U\Delta = M_{x^2}U \tag{8.6.2}$$

$$\Delta U = UM_{x^2} \tag{8.6.3}$$

$$U(\Delta + M_{x^2}) = (\Delta + M_{x^2})U \tag{8.6.4}$$

*where $U$ is the Fourier transform, $\Delta=D^2$ the one-dimensional Laplacian on $L^2(-\infty,\infty)$ (Theorem 8.7), and $M_{x^2}$ multiplication by $x^2$ (Theorem 8.9).*

*Proof.* In Section 8.4, equation (8.4.5) and the discussion following it, we showed that the operator $U^2$, which means applying the Fourier transform twice in succession, has the net effect of simply replacing the independent variable $x$ by its negative, $-x$: $U^2(f)(x)=f(-x)$. Thus

$$U^2M_{x^2}(f)(x) = U^2(x^2f(x)) = (-x)^2f(-x) = x^2f(-x)$$

and

$$M_{x^2}U^2(f)(x) = M_{x^2}(f(-x)) = x^2f(-x)$$

Hence $U^2M_{x^2}=M_{x^2}U^2$, which proves (8.6.1). We say that $U^2$ *commutes with* $M_{x^2}$.

To prove (8.6.2), multiply our basic equation $UD = M_xU$ on the right by $D$. You will get $U\Delta = M_xUD = M_x(UD)$. (Remember that $D^2 = \Delta$—see Theorem 8.7.) Substitute $UD = M_xU$ to get $U\Delta = M_x(M_xU) = M_xM_xU = M_{x^2}U$.

To prove (8.6.3), multiply (8.6.2) on the right by $U$ and use (8.6.1): $U\Delta U = M_{x^2}U^2 = U^2M_{x^2}$. Then multiply on the left by $U^{-1}$ to get $\Delta U = UM_{x^2}$, which is (8.6.3). Equation (8.6.4) follows from (8.6.2) and (8.6.3).

While the "proof" of Theorem 8.11 that we have just given is slick and neat, we ought to note that it really doesn't pay close enough attention to the domains of the various unbounded operators that appear there. This is particularly true of the operator we have written $\Delta + M_{x^2}$ because we have not even defined the sum of two unbounded operators. But the procedure we have consistently used throughout the book can be effectively applied here to give precise meaning to $\Delta + M_{x^2}$: just note that the corresponding differential expression is $l(y) \equiv -y'' + x^2y$, hence we define $\Delta + M_{x^2}$ as the operator with that differential expression and domain consisting of all functions in $L^2(-\infty, \infty)$ that are compatible with $l$. If we were to be absolutely mathematically rigorous, we would have to prove (8.6.4) of Theorem 8.11 using that definition of $\Delta + M_{x^2}$. But we shall be satisfied with the formal proof as given, because our focus in this book is on the principal ideas and applications rather than the construction of faultlessly logical proofs.

**Theorem 8.12.** *Consider the Hilbert space $L^2(-\infty, \infty)$ and the operator $\Delta + M_{x^2}$ defined as follows: it has the differential expression $l(y) \equiv -y'' + x^2y$ and domain consisting of all functions in $L^2(-\infty,\infty)$ that are compatible with $l$. The operator $\Delta + M_{x^2}$, so defined, is Hermitian and has the simple eigenvalues $2n + 1$, $n = 0, 1, 2, \ldots,$ with corresponding eigenfunctions $\phi_n(x) = 2^{n/2}H_n(\sqrt{2}x)e^{-x^2/2}$ where $H_n$ is the nth Hermite polynomial. The $\phi_n(x)$, $n = 0, 1, 2, \ldots,$ form an orthogonal basis of $L^2(-\infty, \infty)$.*

*Proof.* We have seen the operator $\Delta + M_{x^2}$ before. It is essentially the Schrödinger operator for the quantum linear oscillator. In Section 5.5 we analyzed this operator by making a substitution that transferred the Hilbert space $L^2(-\infty, \infty)$ to the Hilbert space $L^2(-\infty, \infty; e^{-x^2/2})$ and simultaneously transferred Schrödinger's operator to Hermite's operator. We shall do essentially the same thing here, except a little more carefully. This proof follows the pattern of Exercises 6 through 9 of Section 4.5.

Consider the Hilbert space $L^2(-\infty, \infty)$ with inner product

$$\langle f, g \rangle = \int_{-\infty}^{\infty} f(t)\overline{g(t)} \, dt$$

and the Hilbert space $L^2(-\infty, \infty; e^{-x^2/2})$ with inner product

$$[f, g] = \int_{-\infty}^{\infty} f(x)\overline{g(x)}e^{-x^2/2}\,dx$$

Given a function $y(t)$ in $L^2(-\infty, \infty)$ we define another function $z(x)$ by

$$z(x) = 2^{-1/4}y\left(\frac{x}{\sqrt{2}}\right)e^{x^2/4} \tag{8.6.5}$$

*Now observe this: $y$ belongs to $L^2(-\infty, \infty)$ if, and only if, $z$ belongs to $L^2(-\infty, \infty; e^{-x^2/2})$.* Because

$$\int_{-\infty}^{\infty} |z(x)|^2 e^{-x^2/2}\,dx = \int_{-\infty}^{\infty} \left|2^{-1/4}y\left(\frac{x}{\sqrt{2}}\right)e^{x^2/4}\right|^2 e^{-x^2/2}\,dx$$

$$= 2^{-1/2}\int_{-\infty}^{\infty} \left|y\left(\frac{x}{\sqrt{2}}\right)\right|^2 dx$$

$$\int_{-\infty}^{\infty} |z(x)|^2 e^{-x^2/2}\,dx = \int_{-\infty}^{\infty} |y(t)|^2\,dt \qquad \left(t = \frac{x}{\sqrt{2}}\right) \tag{8.6.6}$$

Notice that the substitution (8.6.5) combines a change of the independent variable, $t = x/\sqrt{2}$, with a change of the dependent variable. If $y$ goes to $z$ according to (8.6.5), then $z$ goes back to $y$ by

$$y(t) = 2^{1/4}z(\sqrt{2}\,t)e^{-t^2/2} \tag{8.6.7}$$

Next, we note the following fact: *if both $f$ and $g$ belong to $L^2(-\infty, \infty)$, and if $u(x) = 2^{-1/4}f(x/\sqrt{2})e^{x^2/4}$ and $v(x) = 2^{-1/4}g(x/\sqrt{2})e^{x^2/4}$ then $[u, v] = \langle f, g \rangle$.* This is because

$$[u, v] = \int_{-\infty}^{\infty} u(x)\overline{v(x)}e^{-x^2/2}\,dx = 2^{-1/2}\int_{-\infty}^{\infty} f\left(\frac{x}{\sqrt{2}}\right)\overline{g\left(\frac{x}{\sqrt{2}}\right)}\,dx$$

$$= \int_{-\infty}^{\infty} f(t)\overline{g(t)}\,dt = \langle f, g \rangle$$

In other words, the substitution (8.6.5) preserves the inner product.

What happens to the differential expression $l(y) \equiv -d^2y/dt^2 + t^2y$ when we make the substitution (8.6.5)? In this calculation we have to pay careful attention to the change of independent variable, $t = x/\sqrt{2}$, or $x = \sqrt{2}\,t$. Keeping in mind this change of independent variable, we can write the inverse substitution (8.6.7) as

$$y(t) = 2^{1/4}z(x)e^{-x^2/4} \qquad (x = \sqrt{2}\,t) \tag{8.6.8}$$

Using the chain rule, $d/dt = (d/dx)(dx/dt) = \sqrt{2}d/dx$, we get

$$\frac{d^2y}{dt^2} = 2^{5/4}\left(\frac{d^2z}{dx^2} - x\frac{dz}{dx} - \frac{z}{2} - \frac{x^2}{4}z\right)e^{-x^2/4}$$

which leads to

$$-\frac{d^2y}{dt^2} + t^2y = \left(-\frac{d^2z}{dx^2} + x\frac{dz}{dx} + \frac{z}{2}\right)2^{5/4}e^{-x^2/4} \tag{8.6.9}$$

The differential expression $h(z) \equiv -d^2z/dx^2 + x\,dz/dx$ in $L^2(-\infty, \infty; e^{-x^2/2})$ belongs to Hermite's operator (Section 4.9). The differential expression $l(y) \equiv -d^2y/dt^2 + t^2y$ on the left of (8.6.9) belongs to $\Delta + M_{x^2}$ (Theorem 8.12). Summarizing: if $z(x) = 2^{-1/4}y(x/\sqrt{2})e^{x^2/4}$ then

$$l(y) = \left(h(z) + \frac{z}{2}\right)2^{5/4}e^{-x^2/4} \tag{8.6.10}$$

Integrating equation (8.6.10), we get

$$\int_{-\infty}^{\infty}|l(y)|^2\,dt = 4\int_{-\infty}^{\infty}\left|h(z) + \frac{z}{2}\right|^2 e^{-x^2/2}\,dx \qquad \left(t = \frac{x}{\sqrt{2}}\right) \tag{8.6.11}$$

Equation (8.6.11) tells us that $l(y)$ belongs to $L^2(-\infty, \infty)$ if, and only if, $h(z) + z/2$ belongs to $L^2(-\infty, \infty; e^{-x^2/2})$. From this it follows easily that $y(t)$ *belongs to the domain of* $\Delta + M_{x^2}$ *if, and only if,* $z(x) = 2^{-1/4}y(x/\sqrt{2})e^{x^2/4}$ *belongs to the domain of Hermite's operator.*

Next we shall prove that $\Delta + M_{x^2}$ is Hermitian. For convenience, let us write $L$ for the operator $\Delta + M_{x^2}$. We need to show $\langle L(y_1), y_2\rangle = \langle y_1, L(y_2)\rangle$ for all $y_1, y_2$ in the domain of $L$. Use the symbol $H$ for Hermite's operator (we used $L$ in Section 4.9). Equation (8.6.10) says that if $z$ corresponds to $y$ under (8.6.5), then $L(y) = (H(z) + z/2)2^{5/4}e^{-x^2/4}$. So, if $z_1$ corresponds to $y_1$ and $z_2$ to $y_2$, then, using (8.6.7) and the equation $x = \sqrt{2}t$, we get

$$\langle L(y_1), y_2\rangle = \int_{-\infty}^{\infty} L(y_1)\bar{y}_2\,dt$$

$$= \int_{-\infty}^{\infty}\left(\left(H(z_1) + \frac{z_1}{2}\right)2^{5/4}e^{-x^2/4}\right)(2^{1/4}\overline{z_2(x)}e^{-x^2/4})\,dt$$

$$= 2\int_{-\infty}^{\infty}\left(H(z_1) + \frac{z_1}{2}\right)\bar{z}_2 e^{-x^2/2}\,dx$$

$$\langle L(y_1), y_2\rangle = 2[H(z_1), z_2] + [z_1, z_2] \tag{8.6.12}$$

Because Hermite's operator $H$ is Hermitian in $L^2(-\infty, \infty; e^{-x^2/2})$ (Theorem

4.40 of Chapter 4), equation (8.6.12) shows that the operator $L$ is Hermitian in $L^2(-\infty, \infty)$. Thus we have proved the first assertion of Theorem 8.12.

As for the eigenvalues and eigenfunctions of $L$, the equation

$$L(y) = \lambda y, \qquad y \in \text{domain}(L) \tag{8.6.13}$$

is equivalent under the substitution (8.6.5) to

$$H(z) = \left(\frac{\lambda - 1}{2}\right) z, \qquad z \in \text{domain}(H) \tag{8.6.14}$$

a fact you are asked to prove as Exercise 3. The eigenvalues of Hermite's operator $H$ are $0, 1, 2, \ldots$, and these eigenvalues are simple (Theorem 4.46 of Section 4.9). Hence the eigenvalues of $L = \Delta + M_{t^2}$ are given by $(\lambda - 1)/2 = n$, or $\lambda = 2n + 1$, $n = 0, 1, 2, \ldots$. The corresponding eigenfunctions for $H$ are $z_n(x) = H_n(x)$, the Hermite polynomials (Theorems 4.42 and 4.46 of Section 4.9). Hence the eigenfunctions of $L$ are $y_n(t) = 2^{1/4} H_n(\sqrt{2}t) e^{-t^2/2}$, $n = 0, 1, 2, \ldots$ (use (8.6.7)). The fact that the $y_n$ form an orthogonal basis of $L^2(-\infty, \infty)$ follows from the known fact that the Hermite polynomials $H_n$ are an orthogonal basis of $L^2(-\infty, \infty; e^{-x^2/2})$ (Theorem 4.44 of Section 4.9).

The Hermite polynomials $H_n(x)$ are polynomials of degree $n$ with integer coefficients containing only even powers of $x$ when $n$ is even, only odd powers when $n$ is odd (Section 4.11). Thus, when $n$ is odd, $H_n(\sqrt{2}t)$ will contain awkward half-integer powers of 2. The polynomials $2^{n/2} H_n(\sqrt{2}t)$ will have integer coefficients for every $n$, so it is customary to use these polynomials in place of the $H_n(\sqrt{2}t)$.

Now replace the factor $2^{1/4}$ in $y_n$ by $2^{n/2}$ (because scalar factors don't matter in eigenfunctions), write the letter $x$ in place of the letter $t$ and $\phi$ in place of $y$, and recap: The eigenvalues of the Hermitian operator $\Delta + M_{x^2}$ are $2n + 1$, $n = 0, 1, 2, \ldots$; each of these eigenvalues is simple; the eigenfunction $\phi_n(x) = 2^{n/2} H_n(\sqrt{2}x) e^{-x^2/2}$ belongs to the eigenvalue $2n + 1$; and the $\phi_n(x)$, $n = 0, 1, 2, \ldots$, are an orthogonal basis of $L^2(-\infty, \infty)$. That completes the proof of Theorem 8.12.

We shall use Theorem 8.12 to ascertain the eigenvalues and eigenfunctions of the Fourier transform. Already, from Exercise 4 of Section 8.4, we know that there are only four possible eigenvalues for the Fourier transform, namely $\pm 1$ and $\pm i$. (Knowing $U^4 = I$ (Theorem 8.2), we deduce from $U(\phi) = \lambda \phi$ that $\lambda^4 = 1$; thus $\lambda = \pm 1$ or $\pm i$.) Of the four numbers $\pm 1$, $\pm i$, which are actually eigenvalues of the Fourier transform?

**Theorem    8.13.**    $U(\phi_n) = (-i)^n \phi_n$, $\quad n = 0, 1, 2, \ldots, \quad$ *where* $\quad \phi_n(x) = 2^{n/2} H_n(\sqrt{2}x) e^{-x^2/2}$, $H_n$ *being the nth Hermite polynomial. Thus each of the four* *numbers* $\pm 1$, $\pm i$ *is an eigenvalue of the Fourier transform. Each has infinite multiplicity.*

The equation $U(\phi_n) = (-i)^n \phi_n$, written in equivalent integral form, becomes

$$\frac{1}{\sqrt{2\pi}} \int_{-\infty}^{\infty} H_n(\sqrt{2}x)e^{-x^2/2}e^{-iux}\,dx$$

$$= (-i)^n H_n(\sqrt{2}u)e^{-u^2/2}, \qquad n = 0, 1, 2, \ldots \tag{8.6.15}$$

(I have canceled a factor $2^{n/2}$ from both sides of equation (8.6.15). You may or may not include it.) When $n = 0$, equation (8.6.15) becomes

$$\frac{1}{\sqrt{2\pi}} \int_{-\infty}^{\infty} e^{-x^2/2}e^{-iux}\,dx = e^{-u^2/2} \tag{8.6.16}$$

which, put into words, says that the Gaussian $e^{-x^2/2}$ is its own Fourier transform!

*Proof of Theorem* 8.13. Start with equation (8.6.4) of Theorem 8.11: $U(\Delta + M_{x^2}) = (\Delta + M_{x^2})U$. For convenience, write $L = \Delta + M_{x^2}$, a notation already introduced in the proof of Theorem 8.12. Then our equation becomes $UL = LU$. We know from Theorem 8.12 that $L(\phi_n) = (2n+1)\phi_n$, $n = 0, 1, 2, \ldots$. Apply the Fourier transform $U$ to both sides of this relation to get $UL(\phi_n) = (2n+1)U(\phi_n)$. But $UL = LU$, so

$$LU(\phi_n) = (2n + 1)U(\phi_n) \tag{8.6.17}$$

Equation (8.6.17) says that $U(\phi_n)$ is also an eigenfunction of $L$ belonging to the eigenvalue $2n+1$. But $L$ has only simple eigenvalues, so $U(\phi_n)$ must be a scalar multiple of $\phi_n$:

$$U(\phi_n) = \mu_n \phi_n, \qquad n = 0, 1, 2, \ldots \tag{8.6.18}$$

We shall prove $\mu_n = (-i)^n$ by induction. For $n = 0$, equation (8.6.18) becomes

$$\frac{1}{\sqrt{2\pi}} \int_{-\infty}^{\infty} e^{-x^2/2}e^{-iux}\,dx = \mu_0 e^{-u^2/2} \tag{8.6.19}$$

Put $u = 0$ in (8.6.19) to get $\mu_0 = 1 = (-i)^0$. That starts the induction. We then have to verify the induction hypothesis: if $\mu_n = (-i)^n$, then $\mu_{n+1} = (-i)^{n+1}$. We shall need the following lemma.

**Lemma 8.14.** *The functions* $\phi_n(x) = 2^{n/2}H_n(\sqrt{2}x)e^{-x^2/2}$ *satisfy the following equations*:

$$\phi_n' = n\phi_{n-1} - \tfrac{1}{2}\phi_{n+1} \tag{8.6.20}$$

$$x\phi_n = \tfrac{1}{2}\phi_{n+1} + n\phi_{n-1} \tag{8.6.21}$$

The proof is left for you to do as Exercise 4.

The functions $\phi_n$ evidently belong to the domain of the differentiation operator $D$ (Definition 8.4), so we may apply our equation $UD = M_x U$ (Theorem 8.10): $UD(\phi_n) = M_x U(\phi_n) = x\mu_n\phi_n$. Then use $D(\phi_n) = (1/i)\phi_n'$ together with (8.6.20) and (8.6.21) to get

$$\frac{1}{i} U(n\phi_{n-1} - \tfrac{1}{2}\phi_{n+1}) = \mu_n(\tfrac{1}{2}\phi_{n+1} + n\phi_{n-1})$$

which leads to

$$\frac{1}{i}(n\mu_{n-1}\phi_{n-1} - \tfrac{1}{2}\mu_{n+1}\phi_{n+1}) = \tfrac{1}{2}\mu_n\phi_{n+1} + n\mu_n\phi_{n-1} \tag{8.6.22}$$

The $\phi_n$'s are orthogonal, therefore linearly independent, so coefficients of like terms in (8.6.22) must be equal:

$$\left(\frac{1}{i}\right)n\mu_{n-1} = n\mu_n, \qquad -\frac{1}{2i}\mu_{n+1} = \frac{1}{2}\mu_n$$

The second of these equations says $\mu_{n+1} = (-i)\mu_n$. Hence, if $\mu_n = (-i)^n$, then $\mu_{n+1} = (-i)(-i)^n = (-i)^{n+1}$. Our induction hypothesis and Theorem 8.13 are proved.

Theorem 8.13 gives us an entirely new way to compute the Fourier transform of a function $f \in L^2(-\infty, \infty)$. First, expand $f$ in its "Fourier" series with respect to the orthogonal basis $\phi_n$:

$$f = \sum_{n=0}^{\infty} c_n\phi_n, \qquad c_n = \frac{\langle f, \phi_n\rangle}{\langle \phi_n, \phi_n\rangle} = \frac{1}{2^n\sqrt{\pi}\,n!}\int_{-\infty}^{\infty} f(x)\phi_n(x)\,dx \tag{8.6.23}$$

where we have used the formula

$$\langle \phi_n, \phi_n\rangle = 2^n\sqrt{\pi}\,n!$$

from Exercise 7. Then, to get the Fourier transform of $f$, simply apply $U$ term by term to (8.6.23):

$$U(f) = \sum_{n=0}^{\infty} (-i)^n c_n\phi_n \tag{8.6.24}$$

This works for every $f$ in $L^2(-\infty, \infty)$ because the Fourier transform $U$ is a bounded operator whose domain is the entire Hilbert space $L^2(-\infty, \infty)$.

## Exercises

1. Derive the formulas $UDU^{-1} = M_x$ and $U\Delta U^{-1} = M_{x^2}$.

2. Derive equation (8.6.7); i.e., show that if $z(x)=2^{-1/4}y(x/\sqrt{2})e^{x^2/4}$, then $y(t)=2^{1/4}z(\sqrt{2}t)e^{-t^2/2}$.

3. Using the notation of the proof of Theorem 8.12, prove the equivalence of equations (8.6.13) and (8.6.14); i.e., show that $L(y)=\lambda y$ if and only if $H(z)=(\lambda-1)z/2$ where $L=\Delta+M_{x^2}$, $H$ is Hermite's operator, and $y$ and $z$ correspond via (8.6.5) and (8.6.7).

4. Let $\phi_n(x)=2^{n/2}H_n(\sqrt{2}x)e^{-x^2/2}$, where $H_n$ is the $n$th Hermite polynomial. Using the relations $dH_n(t)/dt=nH_{n-1}(t)$ and $tH_n(t)=H_{n+1}(t)+nH_{n-1}(t)$ (both from Exercise 6 of Section 4.11), prove the relations (8.6.20) and (8.6.21):

$$\frac{d}{dx}\phi_n(x) = n\phi_{n-1}(x) - \tfrac{1}{2}\phi_{n+1}(x)$$

$$x\phi_n(x) = \tfrac{1}{2}\phi_{n+1}(x) + n\phi_{n-1}(x)$$

5. Let $a>0$. Show that the Fourier transform of $e^{-ax^2}$ is $e^{-u^2/4a}/\sqrt{2a}$.

6. (a) Derive the formula

$$\frac{1}{\sqrt{2\pi}}\int_{-\infty}^{\infty} H_n(x)e^{-x^2/2}e^{-iux}\,dx = (-i)^n u^n e^{-u^2/2}$$

where $H_n(x)$ is the $n$th Hermite polynomial. [Hint: Use the last equation of Section 4.11.]

   (b) Using your result from part (a), derive the formula

$$\frac{1}{\sqrt{2\pi}}\int_{-\infty}^{\infty} x^n e^{-x^2/2}e^{-iux}\,dx = (-i)^n H_n(u)e^{-u^2/2}$$

   [Hint: If $\hat{f}(u)$ is the Fourier transform of $f(x)$, then $f(-u)$ is the Fourier transform of $\hat{f}(x)$—see Section 8.2.]

7. Prove that $\langle\phi_n, \phi_n\rangle=2^n\sqrt{\pi}\,n!$.

# APPENDIX 8.A   $\int_{-\infty}^{\infty}((\sin x)/x)\,dx=\pi$

We shall prove the equation of the title by complex methods. You are therefore not responsible for this proof, only for the result.

The function $e^{iz}/z$ is analytic everywhere except at $z=0$. It is therefore analytic inside and on the closed curve $C$ pictured in Figure 8.1. Hence, by Cauchy's theorem

$$\int_C \frac{e^{iz}}{z}\,dz = 0 \qquad\qquad (8.A.1)$$

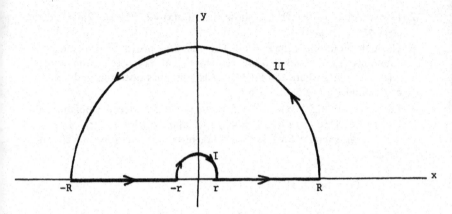

**Figure 8.1**   The closed integration contour for $e^{iz}/z$.

Write the integral over $C$ as the sum of four integrals:

$$\int_C \frac{e^{iz}}{z}\,dz = \int_{-R}^{-r} \frac{e^{ix}}{x}\,dx + \int_I \frac{e^{iz}}{z}\,dz + \int_r^R \frac{e^{ix}}{x}\,dx + \int_{II} \frac{e^{iz}}{z}\,dz \qquad (8.A.2)$$

Using $e^{ix} = \cos x + i \sin x$ and remembering that $\cos x/x$ is odd and $\sin x/x$ is even, we get

$$\int_{-R}^{-r} \frac{e^{ix}}{x}\,dx + \int_r^R \frac{e^{ix}}{x}\,dx = 2i \int_r^R \frac{\sin x}{x}\,dx \qquad (8.A.3)$$

Now combine (8.A.1), (8.A.2), and (8.A.3).

$$2i \int_r^R \frac{\sin x}{x}\,dx = -\int_I \frac{e^{iz}}{z}\,dz - \int_{II} \frac{e^{ix}}{z}\,dz \qquad (8.A.4)$$

where $I$ and $II$ are semicircles of radii $r$ and $R$, respectively, as shown in Figure 8.1. For the integral over $I$, write $z = re^{i\theta}$, $dz = ire^{i\theta}\,d\theta$, and note also that $|z| = r$, so $e^{iz} \to 1$ as $r \to 0$. So

$$\int_I \frac{e^{iz}}{z}\,dz = i \int_{\theta=\pi}^{\theta=0} e^{iz}\,d\theta \to -i \int_0^\pi d\theta = -\pi i \qquad \text{as } r \to 0$$

An easy estimate shows that the integral over $II$ tends to 0 as $R \to \infty$. So, letting $r \to 0$ and $R \to \infty$ in (8.A.4) and canceling an $i$, we get

$$2 \int_0^\infty \frac{\sin x}{x}\,dx = \pi$$

which is equivalent to the equation of the title.

# Index of Symbols

# Index